AF718456

Me	medial amygdaloid nucleus
ME	median eminence
me5	mesencephalic trigeminal tract
Med	medial cerebellar nucleus
MeRF	mesencephalic reticular formation (deep mesencephalic nucleus)
mfb	medial forebrain bundle
MG	medial geniculate nucleus
MHb	medial habenular nucleus
Mi	mitral cell layer of the olfactory bulb
ml	medial lemniscus
ML	medial mammillary nucleus, lateral part
mlf	medial longitudinal fasciculus
MM	medial mammillary nucleus, medial part
MnPO	median preoptic nucleus
MnR	median raphe nucleus
Mo5	motor trigeminal nucleus
mp	mammillary peduncle
MP	medial mammillary nucleus, posterior part
MPT	medial pretectal nucleus
Ms	medial septal nucleus
MSO	medial superior olive
mt	mammillothalamic tract
MVe	medial vestibular nucleus
MVPO	medioventral periolivary nucleus
ON	olfactory nerve layer
Op	optic nerve layer of the superior colliculus
opt	optic tract
OPT	olivary pretectal nucleus
OT	nucleus of the optic tract
ox	optic chiasm
PaV	paraventricular hypothalamic nucleus, ventral part
PB	parabrachial nuclei
PBL	lateral parabrachial nucleus
PBM	medial parabrachial nucleus
pc	posterior commissure
PC	paracentral thalamic nucleus
PcRt	parvocellular reticular nucleus
Pe	periventricular hypothalamic nucleus
PF	parafascicular thalamic nucleus
PH	posterior hypothalamic area
Pir	piriform cortex
PMD	premammilary nucleus, dorsal part
PMn	paramedian reticular nucleus
PMV	premammillary nucleus, ventral part
pn	pontine nuclei
PnC	pontine reticular nucleus, caudal part
PnO	pontine reticular nucleus, oral part
PO	posterior thalamic nuclear group
PP	peripeduncular nucleus
PPT	posterior pretectal nucleus
Pr5	principal sensory trigeminal nucleus
PrH	prepositus hypoglossal nucleus
PT	paratenial thalamic nucleus
PV	paraventricular thalamic nucleus
py	pyramidal tract
R	red nucleus
Re	reuniens thalamic nucleus
Rh	rhomboid thalamic nucleus
RL	retrolemniscal nucleus
RMg	raphe magnus nucleus
ROb	raphe obscurus nucleus
RPa	raphe pallidus nucleus
RPO	rostral periolivary region
rs	rubrospinal tract
Rt	reticular thalamic nucleus
RtTg	reticulotegmental nucleus of the pons
s5	sensory root of the trigeminal nerve
SCh	suprachiasmatic nucleus
scp	superior cerebellar nucleus
SFi	septofimbrial nucleus
SHi	septohippocampal nucleus
SI	substantia innominate
sm	stria medullaris of the thalamus
SNC	substantia nigra, compact part
SNR	substantia nigra, reticular part
SO	supraoptic nucleus
sol	solitary tract
Sol	nucleus of the solitary tract
sp5	spinal trigeminal tract
Sp5	spinal trigeminal nucleus
SPO	superior paraolivary nucleus
SpVe	spinal vestibular nucleus
st	stria terminalis
SuG	superficial gray layer of the superior colliculus
SuM	supramammillary nucleus
SuVe	superior vestibular nucleus
Tu	olfactory tubercle
Tz	nucleus of the trapezoid body
tz	trapezoid body
unc	uncinate fasciculus
VC	ventral cochlear nucleus
VDB	nucleus of the vertical limb of the diagonal band
VH	ventral horn of the spinal cord
VL	ventrolateral thalamic nucleus
VLG	ventral lateral geniculate nucleus
VLL	ventral nucleus of the lateral lemniscus
VM	ventromedial thalamic nucleus
VMH	ventromedial hypothalamic nucleus
VN	vomeronasal nerve layer
VP	ventral pallidum
VPL	ventral posterolateral thalamic nucleus
VPM	ventral posteromedial thalamic nucleus
VTA	ventral tegmental area (Tsai)
VTg	ventral tegmental area (Gudden)
ZI	zona incerta

Atlas of Neuroactive Substances and Their Receptors in the Rat

Atlas of Neuroactive Substances and Their Receptors in the Rat

Edited by
MASAYA TOHYAMA and KOICHI TAKATSUJI

Translated from the Japanese by
SACHI SRI KANTHA

This publication was supported by a generous donation from
the Daido Life Foundation
and the Sinya Osaka University Medical School
International Foundation

Oxford New York Tokyo
OXFORD UNIVERSITY PRESS
1998

Oxford University Press, Great Clarendon Street, Oxford OX2 6DP
Oxford New York Athens Auckland Bangkok Bogota Bombay Buenos Aires Calcutta Cape Town Dar es Salaam Delhi Florence Hong Kong Istanbul Karachi Kuala Lumpur Madras Madrid Melbourne Mexico City Nairobi Paris Singapore Taipei Tokyo Toronto
and associated companies in Berlin Ibadan

Oxford is a trade mark of Oxford University Press

Published in the United States by Oxford University Press Inc., New York

First published by Oxford University Press 1998

A catalogue record for this book is available from the British Library

Library of Congress Cataloging in Publication Data

Atlas of neuroactive substances and their receptors in the rat / edited by Masaya Tohyama and Koichi Takatsuji; translated from the Japanese by Sachi Sri Kantha.
Includes index.
1. Neurotransmitters—Atlases. 2. Neurotransmitter receptors—Atlases. 3. Rats—Nervous system—Atlases. I. Tohyama, M. (Masaya) II. Takatsuji, Koichi. III. Kantha, Sachi Sri.
QP364.7.A85 1998 599'.0188—dc20 96–23128

ISBN 0 19 852372 6

Typeset by Footnote Graphics, Warminster, Wilts
Printed in Hong Kong

Preface

In the twentieth century, our knowledge of neuroscience has advanced almost beyond imagination, and many scientists believe that it will remain the most important area of research in the life sciences in the twenty-first century. The number of scientists who enter the discipline to elucidate the function of the brain continues to increase tremendously. However, their task is made harder by the structural and functional complexity of the central nervous system. Several approaches have been devised in order to achieve a breakthrough in the unexplored domains of neuroscience. One powerful approach in elucidating brain function is the analysis of the protein and gene expression mechanisms in each neuronal system. For this purpose, it is fundamental to understand the distribution of the various proteins and gene expressions which are related to neural transmission, neurogenesis and neural regeneration.

The introduction of advanced biochemical and molecular biological methods to neuroscience has resulted in remarkable progress in neural morphology. Immunohistochemistry is the most useful tool for the identification of protein localization. *In situ* hybridization histochemistry is powerful technique for the identification of the cells which produce each protein. Heterogeneity of cellular composition, the most remarkable feature of the nervous system, made it inevitable that morphological techniques would be used in neuroscience. Based on the results obtained from such techniques, we can comprehend the detailed localization of the protein and protein-producing cells in the complex neuronal system.

In this book, we have focused on substances relating to neurotransmission and neuronal growth, and have described their localization, gene expression, and receptor localization in the rat brain. We will be pleased if this book becomes a useful aid for researchers studying the diverse aspects of the mammalian nervous system.

July 1997

M. T.
K. T.

Contents

List of contributors ix
List of abbreviations xi
Key to figures xii

Part I The neurotransmission mechanism—present and future

1 Introduction 2
2 Neurotransmitters 2
Classification of neurotransmitters/neuromodulators 2
Distribution of neurotransmitters/neuromodulators in the brain 3
Chemical neuro-networks and functions of neurotransmitters/neuromodulators 10
Trophic (growth factor) functions of neurotransmitters/neuromodulators 13
3 Receptors 13
Structure and mechanisms of action 13
Heterogeneity of receptors 15
4 Transporters 16
5 Feedback of information from postsynaptic to presynaptic region 16
6 Acceptors other than neuronal receptors 17
7 Non-synaptic transmission 17

Part II Neuroactive systems

1 The catecholaminergic system 20
Dopamine 20
Dopamine receptors 24
Noradrenergic–adrenergic systems 32
Receptors for noradrenaline and adrenaline 36
2 The serotonin system 42
Serotonin 42
Serotonin receptors 46
3 The histaminergic system 54
Histamine 54
Histamine receptors 58
4 The acetylcholinergic system 66
Acetylcholine 66
Acetylcholine receptors 70
5 The γ-aminobutyric acid system 98
γ-Aminobutyric acid 98
γ-Aminobutyric acid receptors 102
6 The glycinergic system 118
Glycine 118
Glycine receptor 122
7 The glutamatergic system 130
Glutamic acid 130
Glutamic acid receptor 134
8 The aspartate system 144
Aspartate 144
Asparaginergic innervation 144
9 The taurine system 148
Taurine 148
Localization 148
10 The adenosine system 154
Adenosine 154
Adenosine receptor 158
11 The nerve growth factor system 164
Nerve growth factor 164
Nerve growth factor receptor 164
12 The fibroblast growth factor system 168
Fibroblast growth factor 168
Fibroblast growth factor receptor 172
13 The pleiotrophin system 176
Pleiotrophin 176
14 The tachykinin system 182
Tachykinin 182
Tachykinin receptor 186
15 The calcitonin gene related peptide system 192
Calcitonin gene related peptide 192
Calcitonin gene related peptide receptor 196
16 The cholecystokinin system 200
Cholecystokinin 200
Cholecystokinin receptor 204
17 The neurotensin system 208
Neurotensin 208
Neurotensin receptor 212
18 The angiotensin II system 218
Angiotensin II 218
Angiotensin II receptor 222
19 The corticotropin releasing factor system 226
Corticotropin releasing factor 226
Corticotropin releasing factor receptor 230

20 The bombesin-like peptide system 234
Bombesin-like peptides 234
Bombesin-like peptide receptors 238
21 The opioid peptide system 242
Opioid peptides 242
Opioid receptors 248
22 The galanin system 254
Galanin 254
Galanin receptor 255
23 The somatostatin system 260
Somatostatin 260
Somatostatin receptor 264
24 The vasopressin–oxytocin system 268
Vasopressin and oxytocin 268
Vasopressin and oxytocin receptors 274
25 The luteinizing hormone releasing hormone system 280
Luteinizing hormone releasing hormone (gonadotrophin releasing hormone) 280
Luteinizing hormone releasing hormone receptor 281
26 The thyrotropin releasing hormone system 286
Thyrotropin releasing hormone 286
Thyrotropin releasing hormone receptor 290
27 The neuropeptide Y system 294
Neuropeptide Y 294
Neuropeptide Y receptor 298
28 The natriuretic peptide system 302
Natriuretic peptide 302
Natriuretic peptide receptor 306
29 The vasoactive intestinal peptide system 310
Vasoactive intestinal peptide 310
Vasoactive internal peptide receptor 311
30 The arginosuccinate system 318
Arginosuccinate 318

Part III Cerebellum

1 Cerebellofugal neurotransmitters and neuromodulators 323
2 Neurons and neurotransmitters/neuromodulators in the cerebellar circuit 323
3 Output system from cerebellum and neurotransmitters/neuromodulators 324
4 Bargmann glia 324

Index

335

Contributors

Toshiyuki Araki Department of Anatomy and Neuroscience, Osaka University Medical School, Osaka, Japan
(Chapter 5)

Masahiro Fujita Department of Anatomy and Neuroscience, Osaka University Medical School, Osaka, Japan
(Chapters 1A and B)

Tatsuo Furuyama Department of Molecular Genetic Research, National Institute for Longevity Sciences, Aichi, Japan
(Chapters 15, 16)

Kunio Hiwada Department of Internal Medicine, Ehime University School of Medicine, Ehime, Japan
(Chapter 18)

Shinobu Inagaki Laboratory of Neurobiology, School of Allied Health Sciences, Osaka University Faculty of Medicine, Osaka University Medical School, Osaka, Japan
(Chapters 15, 16)

Hidemasa Katoh Department of Anatomy, Asahikawa Medical College, Asahikawa, Japan
(Chapter 13)

Yoshinori Kawai Department of Anatomy and Neurobiology, Wakayama Medical College, Wakayama, Japan
(Chapter 22)

Hiroshi Kiyama Department of Anatomy, Asahikawa Medical College, Asahikawa, Japan
(Chapters 1A and B, 2, 14, 17)

Yasuko Li Department of Internal Medicine, Ehime University School of Medicine, Ehime, Japan
(Chapter 18)

Hiroshi Maeno Division of Degenerative Neurological Diseases, National Institute of Neuroscience, Tokyo, Japan
(Chapters 2, 14)

Seiji Matsuda Department of Anatomy and Neuroscience, Ehime University School of Medicine, Ehime, Japan
(Chapter 20)

Akifumi Matsuyama Department of Anatomy and Neuroscience, Osaka University Medical School, Osaka, Japan
(Chapters 11, 12)

Shiro Nakagawa Department of Anatomy, Faculty of Medicine, Kagoshima University, Kagoshima, Japan
(Chapter 30)

Koichi Noguchi Department of Anatomy and Neuroscience, Hyogo College of Medicine, Hyogo, Japan
(Chapter 21)

Hiroki Otsuka Department of Anatomy and Neuroscience, Ehime University School of Medicine, Ehime, Japan
(Chapter 19)

Masahiro Sakanaka Department of Anatomy and Neuroscience, Ehime University School of Medicine, Ehime, Japan
(Chapters 18, 19)

Masayuki Sasaki Department of Pharmacology I, Tohoku University School of Medicine, Miyagi, Japan
(Chapter 3)

Kohji Satoh Department of Neuroanatomy, Biomedical Research Center, Osaka University Medical School, Osaka, Japan
(Chapters 6, 7)

Emiko Senba Department of Anatomy and Neurobiology, Wakayama Medical College, Wakayama, Japan
(Chapters 10, 21)

Shoichi Shimada Department of Anatomy, Nagoya City University Medical School, Nagoya, Japan
(Chapters 1C and D)

Sadao Shiosaka Department of Structural Cellular Biology, Nara Institute of Science and Technology, Nara, Japan
(Chapters 12, 23, 25)

Koichi Takatsuji Osaka Prefectural College of Nursing, Osaka, Japan
(Chapters 1C and D, 29)

Masaya Tohyama Department of Anatomy and Neuroscience, Osaka University Medical School, Osaka, Japan
(Introduction, Chapters 4, 8, 9, and Cerebrum)

Akio Wanaka Department of Cell Science, Institute of Biomedical Sciences, Fukushima Medical College, Fukushima, Japan
(Chapters 1C and D, 11, 12, 13)

Takehiko Watanabe Department of Pharmacology I, Tohoku University School of Medicine, Miyagi, Japan
(Chapter 3)

Mariko Yamano Department of Anatomy, Osaka Prefectural College of Health Sciences, Osaka, Japan
(Chapters 26, 27, 28)

Kazuhiko Yanai Department of Pharmacology I, Tohoku University School of Medicine, Miyagi, Japan
(Chapter 3)

Shigetaka Yoshida Department of Structural Cellular Biology, Nara Institute of Science and Technology, Nara, Japan
(Chapter 24)

Abbreviations

AC	adenylate cyclase
ACh	acetylcholine
AD	adrenaline
ADAase	adenosine deaminase
Ang II	angiotensin II
AMPA	α-amino-3-hydroxy-5-methyl-4-isoazole propionate
ANP	atrial natriuretic peptide
BDNF	brain derived neutrogtropic factor
BN	bombesin
BNP	brain natriuretic peptide
cAMP	cyclic AMP
CCK	cholecystokinin
CGRP	calcitonin gene related peptide
CNP	C-type natriuretic peptide
CRF	corticotropin-releasing factor
DA	dopamine
DA-R	dopamine receptor
DBH	dopamine-β-hydroxylase
DG	diacylglycerol
DOPA	dihydroxyphenylalanine
DYN	dynorphin
EKS	ethylketocyclazocine
Enk	enkephalin
FGF	fibroblast growth factor
FGF-R	fibroblast growth factor receptor
FSH	follicle stimulating hormone
GABA	γ-aminobutyric acid
GAD	glutamic acid decarboxylase
GLAST	glutamate-aspartate transporter
GRH	growth hormone releasing hormone
GRIF	growth hormone release inhibitory factor
GRP	gastrin-releasing peptide
His	histamine
IL-1	interleukin-1
IP_3	inositol triphosphate
LH	luteinizing hormone
LHRH	luteinizing hormone releasing hormone
NA	noradrenaline
NKA	neurokinin A
NKB	neurokinin B
NMU	neuromedin U
NMDA	*N*-methyl-D-aspartate
NOS	nitric oxide synthetase
NPK	neuropeptide K
NPY	neuropeptide Y
NT-3	neurotropin-3
Oxy	oxytocin
PGE_2	prostaglandin E_2
PHI	peptide histidine isoleucine
PIP_2	phosphatidylinositol diphosphate
PKC	protein kinase C
PLA_2	phopholipase A_2
PNMT	phenylethanolamine-*N*-methyl transferase
POMC	proopiomelanocortin
PP	pancreatic peptide
PPEA	preproenkephalin-A
PPD	preprodynorphin
PPT	preoprotachykinin
PTN	pleiotrophin
PYY	peptide YY
SNA	substantia nigra pars compacta
SP	substance P
Tau	taurine
TH	tyrosine hydroxylase
TNF	tumour necrosis factor
TRH	thyrotropin releasing hormone
TSH	thyroid stimulatinghormone
Vas	vasopressin
VIP	vasoactive intestinal peptide
VTA	ventral tegmental area

Key to figures

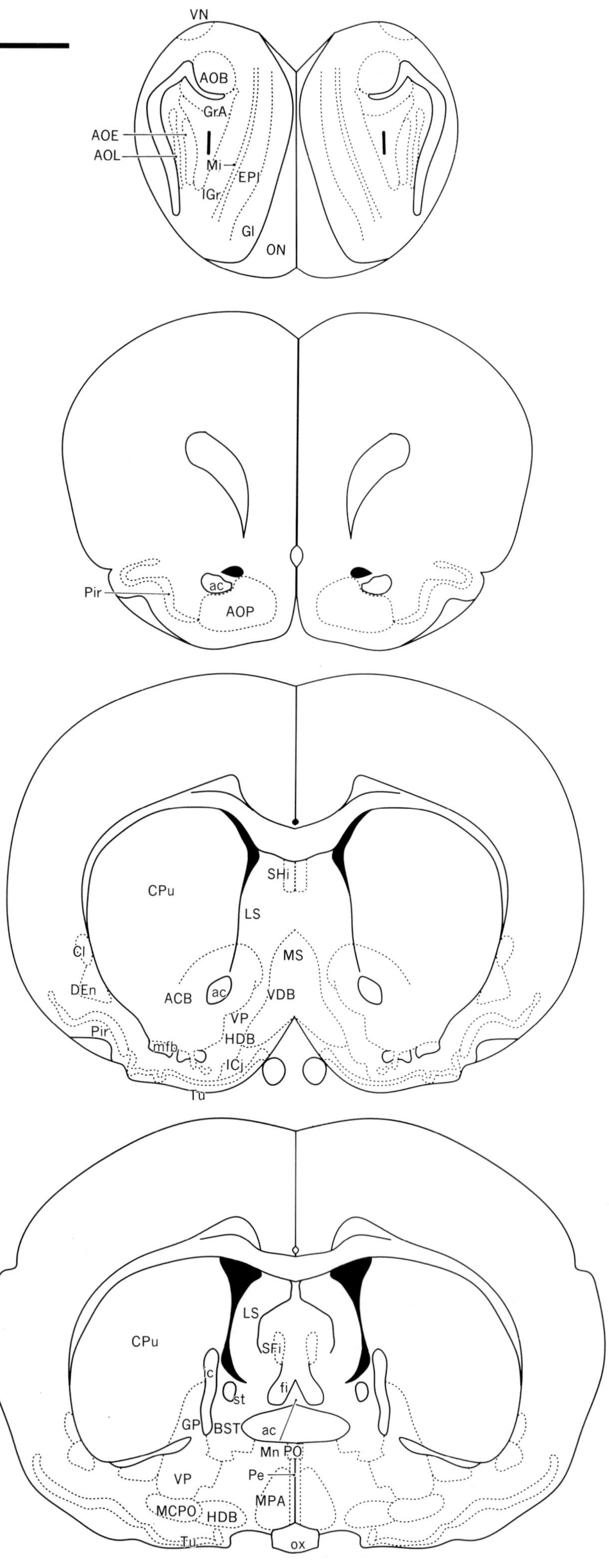

The structure of the atlas in this book is based on that in *The rat brain in stereotaxic coordinates* by G. Paxinos and C. Watson (Academic Press). Each panel depicts brain sections containing representative nuclei and pannels are arranged in a retrocaudal order (olfactory bulb, telencephalon, diencephalon, mesencephalon, pons, medulla oblongata, spinal cord).

The positive cells for neurotransmitters and their receptors indicated in this atlas were based on the data obtained by immunocytochemistry and *in situ* hybridization (ISH). The positive cells labelled by immunocytochemistry and *in situ* hybridization are indicated by red circles, those positive only for ISH are displayed as filled red squares, while those positive for immunocytochemistry are shown as red stars. Positive fibres are indicated by fine dots.

In the panels showing the data of receptor binding studies, the level of binding density is classified into three classes (high, medium, and low), which are coloured red, gray, and blue, respectively.

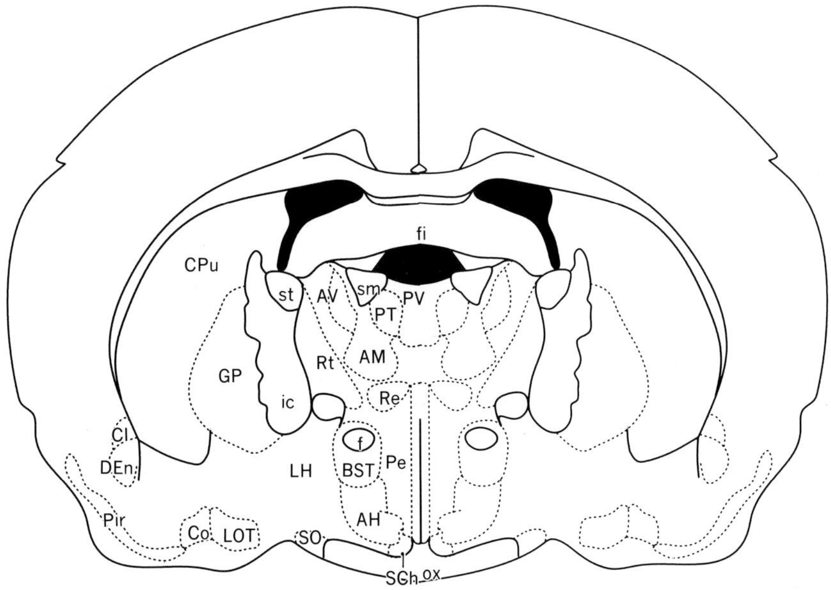
fi
CPu
st
AV
sm
PV
PT
AM
Rt
GP
ic
Re
Cl
f
DEn
LH
BST
Pe
Pir
AH
Co
LOT
SO
SCh
ox

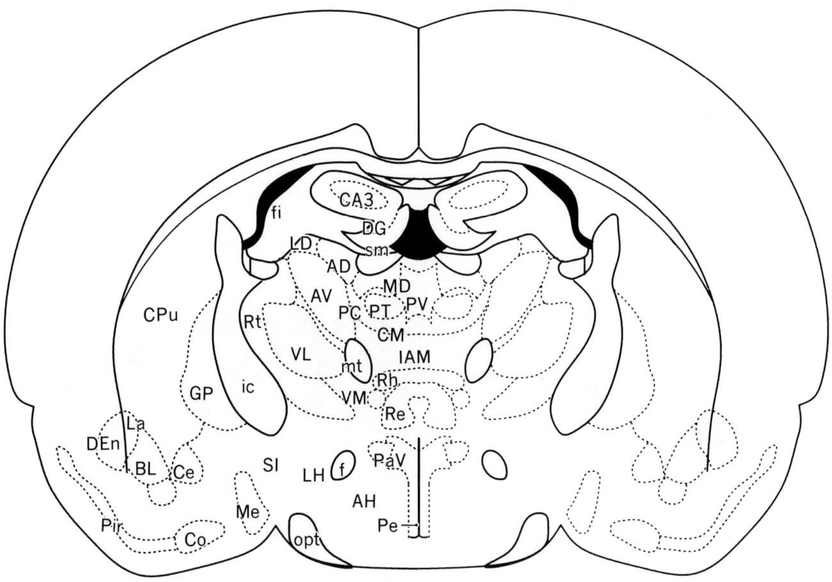
CA3
fi
DG
LD
sm
AD
MD
AV
PV
CPu
PC
PT
Rt
CM
VL
mt
IAM
GP
ic
Rh
VM
Re
La
DEn
SI
PaV
BL
Ce
LH
f
AH
Me
Pir
Pe
Co
opt

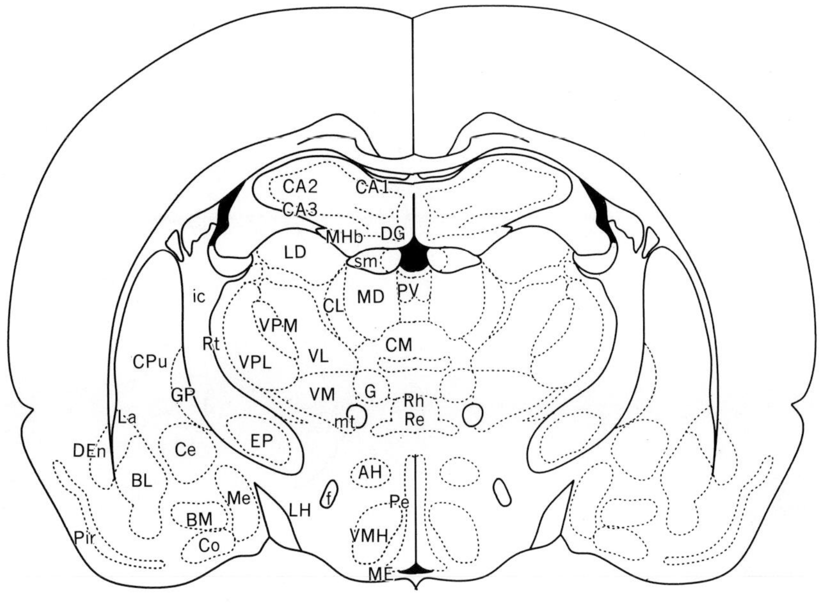
CA2
CA1
CA3
MHb
DG
LD
sm
PV
ic
MD
CL
VPM
Rt
CM
CPu
VPL
VL
GP
VM
G
Rh
La
mt
Re
DEn
Ce
EP
BL
AH
Me
f
Pe
LH
BM
VMH
Pir
Co
ME

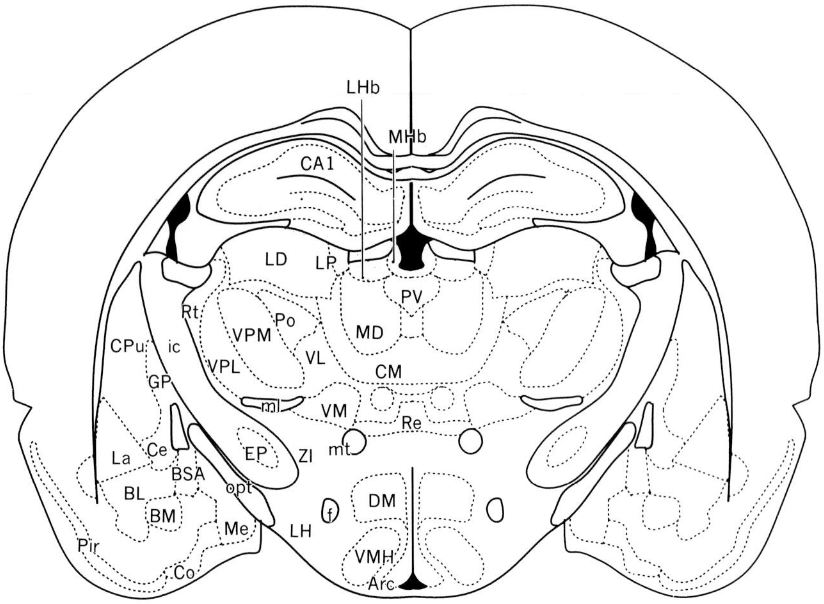
LHb
MHb
CA1
LD
LP
PV
Rt
Po
VPM
MD
CPu
ic
VPL
VL
CM
GP
ml
VM
Re
La
Ce
EP
ZI
mt
BSA
BL
opt
DM
BM
f
Me
LH
Pir
VMH
Co
Arc

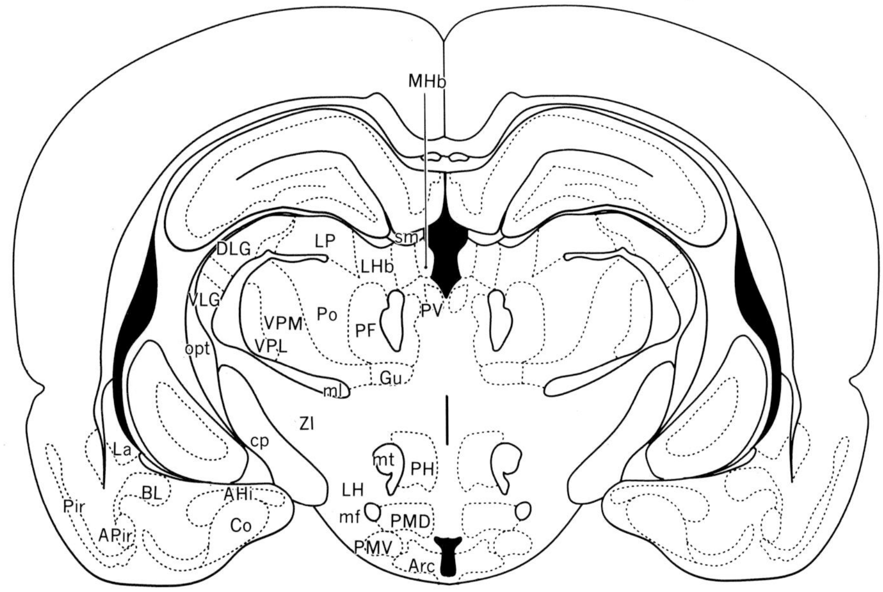
MHb
sm
DLG
LP
LHb
VLG
PV
VPM
Po
PF
VPL
opt
Gu
ml
ZI
cp
La
mt
PH
LH
AHi
BL
Pir
mf
PMD
Co
APir
PMV
Arc

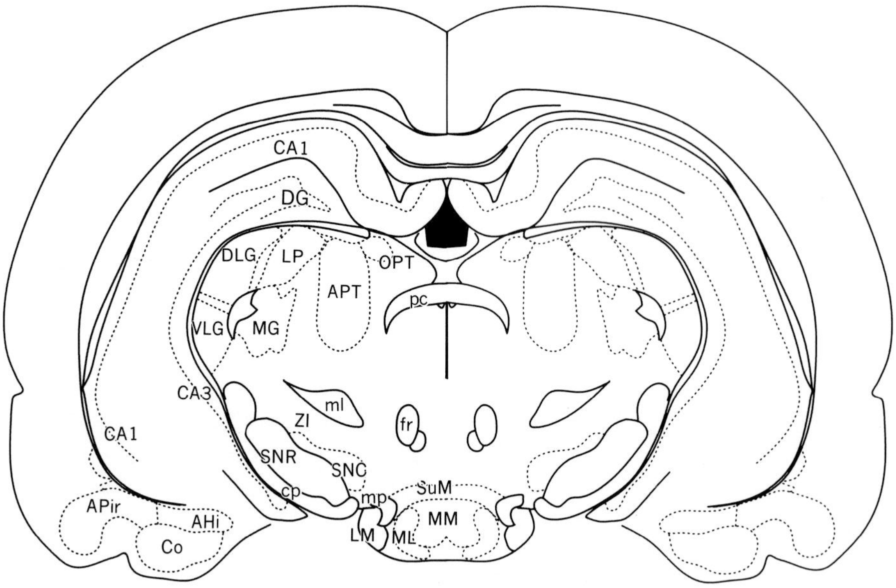
CA1
DG
DLG
LP
OPT
APT
pc
VLG
MG
CA3
ml
ZI
fr
CA1
SNR
SNC
cp
SuM
APir
mp
MM
AHi
LM
ML
Co

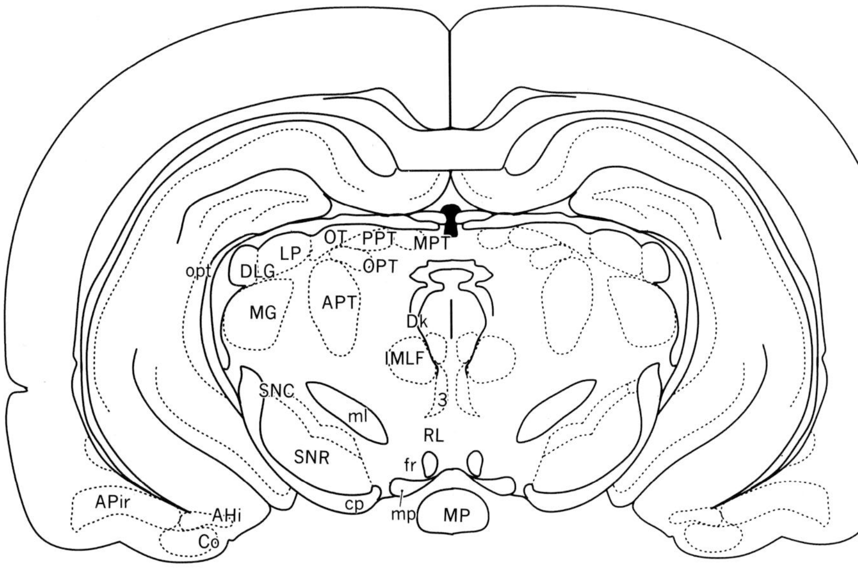
OT
PPT
MPT
LP
OPT
opt
DLG
MG
APT
Dk
IMLF
SNC
3
ml
RL
SNR
fr
APir
AHi
cp
mp
MP
Co

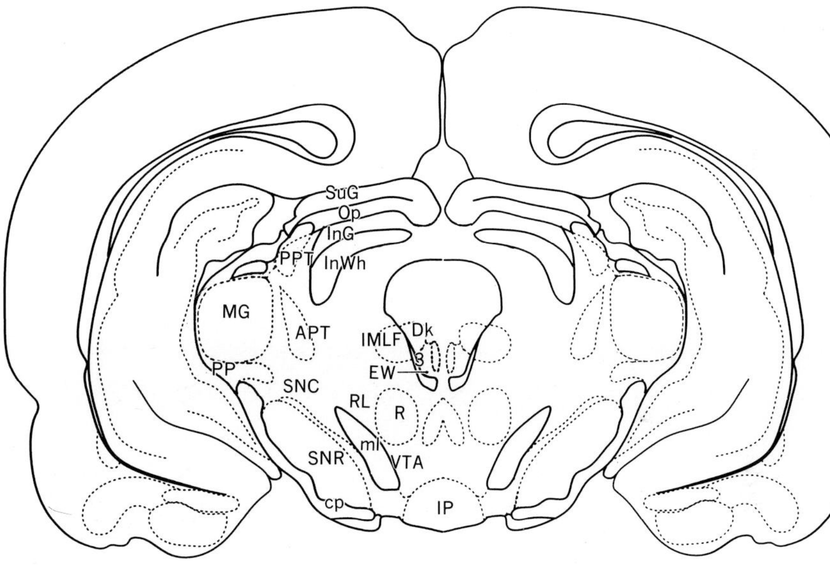
SuG
Op
InG
PPT
InWh
MG
APT
Dk
IMLF
3
EW
PP
SNC
RL
R
ml
SNR
VTA
cp
IP

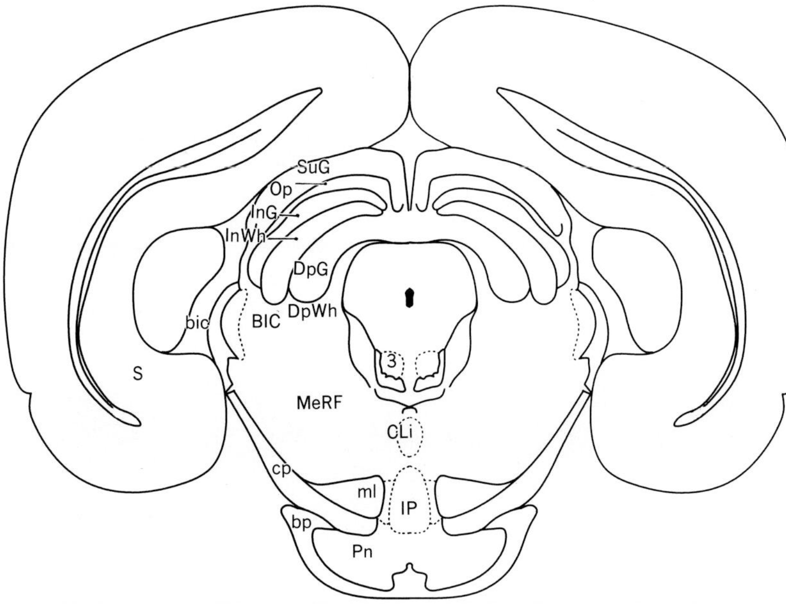
SuG
Op
InG
InWh
DpG
bic
BIC
DpWh
3
S
MeRF
CLi
cp
ml
IP
bp
Pn

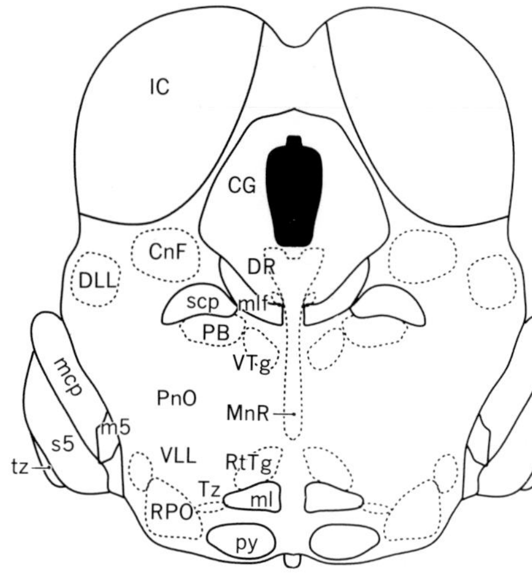
IC
CG
CnF
DR
DLL
scp
mlf
PB
VTg
mcp
PnO
MnR
s5
m5
tz
VLL
RtTg
Tz
ml
RPO
py

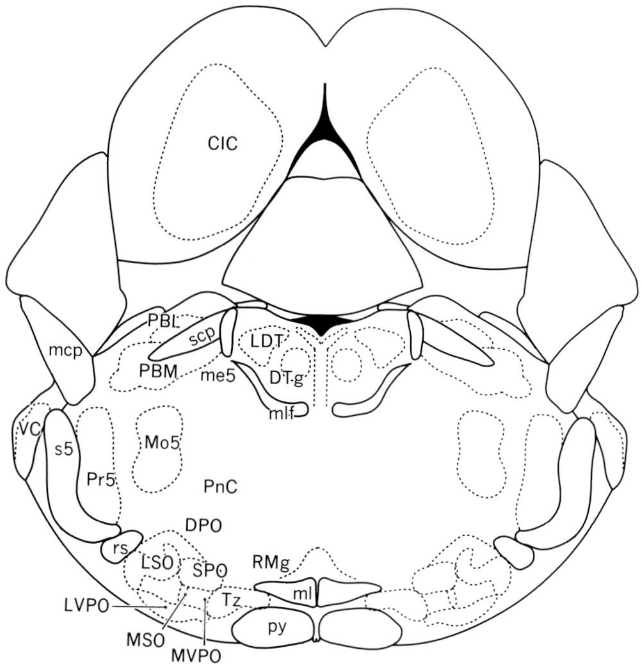
CIC
PBL
scp
LDT
mcp
PBM
me5
DTg
mlf
VC
s5
Mo5
Pr5
PnC
DPO
rs
LSO
SPO
RMg
ml
Tz
LVPO
py
MSO
MVPO

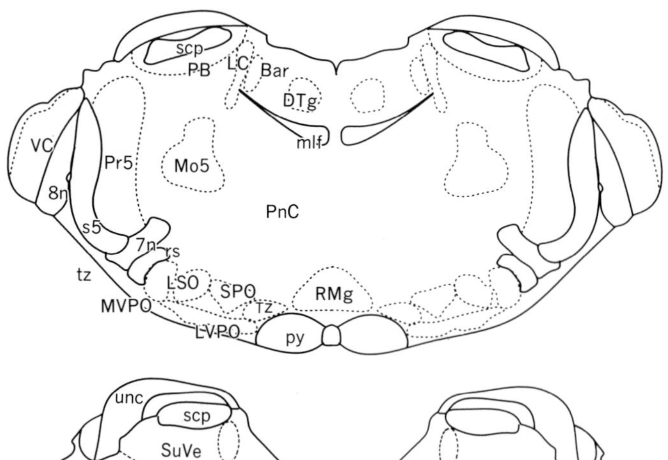
scp
PB
LC
Bar
DTg
mlf
VC
Pr5
Mo5
8n
PnC
s5
7n
rs
tz
LSO
SPO
RMg
MVPO
Tz
LVPO
py

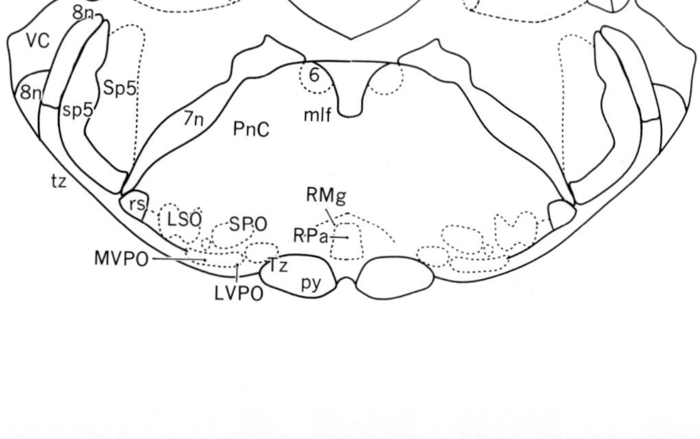
unc
scp
SuVe
8n
VC
6
8n
Sp5
sp5
7n
PnC
mlf
tz
RMg
rs
LSO
SPO
RPa
MVPO
Tz
py
LVPO

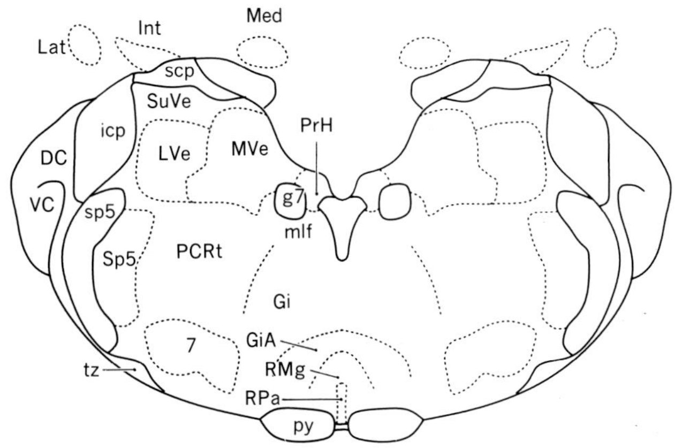
Lat
Int
Med
scp
SuVe
icp
DC
VC
LVe
MVe
PrH
g7
mlf
sp5
Sp5
PCRt
Gi
7
GiA
RMg
RPa
tz
py

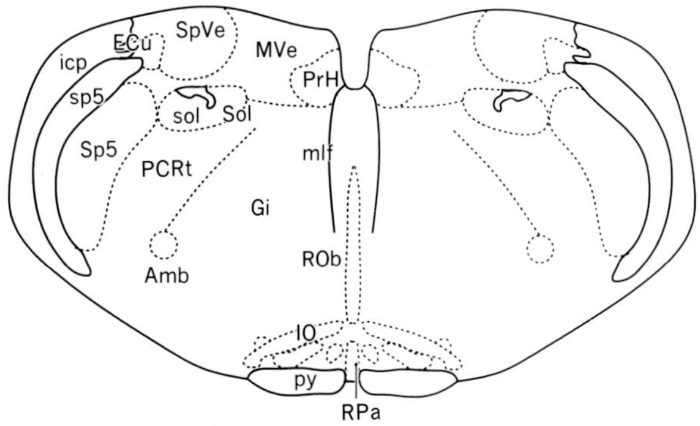
ECu
SpVe
MVe
icp
PrH
sp5
sol
Sol
Sp5
PCRt
mlf
Gi
Amb
ROb
IO
py
RPa

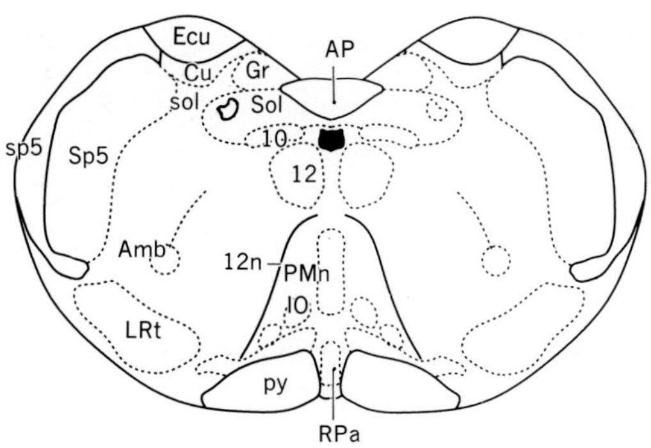
Ecu
AP
Cu
Gr
sol
Sol
10
sp5
Sp5
12
Amb
12n
PMn
IO
LRt
py
RPa

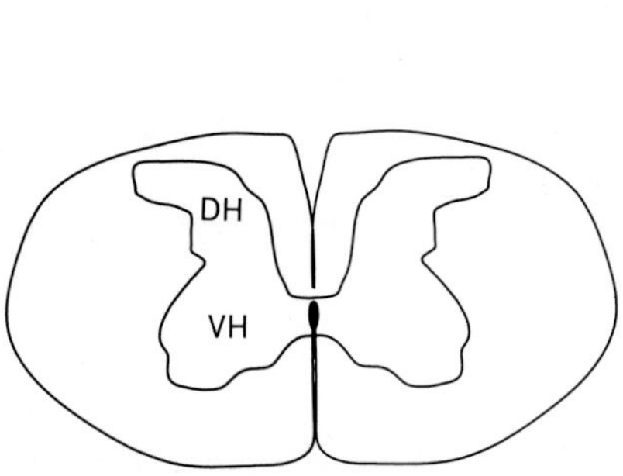
DH
VH

The neurotransmission mechanism—present and future

1 Introduction

The brain is composed of two types of cells: *neurons* and *glial cells*. Neurons have two kinds of processes, called *axons* and *dendrites*. Thus, the brain has been understood as integrated networks of tens of billions of neurons connected with each other by these processes. In principle, dendrites receive information from other neurons, while axons transmit information to other neurons. This mechanism is referred to as *neurotransmission*, with the link junction called a *synapse*. Two types of synapses are now known: chemical and electrical. Neurotransmitters (and neuromodulators) transmit information via chemical synapses. They are produced within the cell-body of neurons when information is inputted and transported along the axon to its end. There, they are released on stimulation, binding to their specific receptors in the cell membrane of the target neurons, and exert their relevant effects. The existence of receptors is therefore a prerequisite for the action of any neurotransmitter.

Thus, it follows that the most important basis for understanding the mechanisms of neurotransmission is to identify: (i) where the neurotransmitters and neuromodulators are produced (i.e. their distribution and that of their mRNA), (ii) where these endogenous substances are released (i.e. the distribution of their terminal sites), and (iii) localization of their receptors and the production sites of the receptors (i.e. distribution of the target neurons of neurotransmitters within the brain) (Fig. 1). This book is intended to integrate our current knowledge of the features listed above. One of the most interesting findings of recent years is the mechanism of re-uptake or inactivation of the released neurotransmitters and neuromodulators. The agents responsible for these actions are proteins which exist in the presynpatic membrane or glial cells. These are distinguished from receptors in the three-dimensional structure. Recent advances on these groups, neurotransmitter transporters, are also included in this book.

The conventional approach to elucidating the mechanism of neuronal transmission has placed much emphasis on information flow between the synapses from their 'pre' to 'post' sites. Recently, the opposite direction (from 'post' to 'pre') has also been considered. An important neurotransmitter which acts on this contrary flow is nitric oxide, which has received much attention recently. Chapter 30 decribes the knowledge we have gained about nitric oxide. It has also been recently established that some of the neurotransmitters and neuromodulators act as 'trophic' (growth) factors. These aspects are also described in this book, and one chapter is devoted to the localization of representative (and also recently identified) growth factors and their receptors.

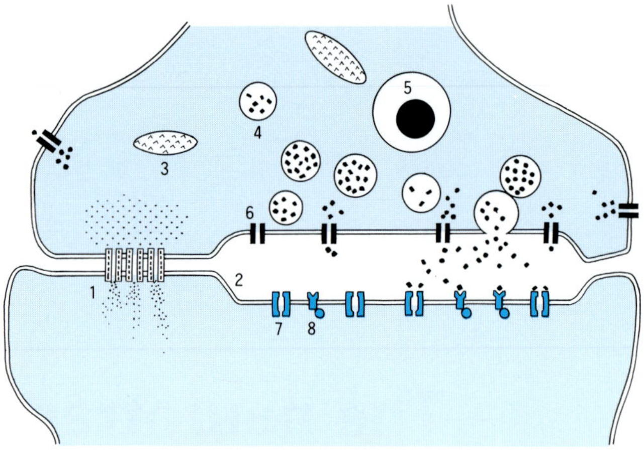

Fig. 1 Synaptic transmission. Two types of synaptic transmission are distinguished: electrical (1) and chemical (2). At electrical synapses, transmission occurs through gap junctions. Three important substances are known at chemical synapses. They are neurotransmitters or neuromodulators, receptors, and transporters. At presynaptic sites, various types of synaptic vesicles exist (3,4,5). Transporters are in general expressed at the presynaptic membrane and re-uptake neurotransmitters/neuromodulators which are released into the synaptic cleft. At the postsynaptic membrane, neuroreceptors are expressed. They are divided into ionotoropic receptors, G-protein coupled receptors, and receptors with tyrosine kinase activity.

Table 1. Neurotransmitters/neuromodulators

Amines	**Peptides**
Dopa	Somatostatin
Dopamine	GRH
Noradrenaline	VIP/PHI
Serotonin (5-HT)	Tachykinin (SP, NKA, NKB)
Histamine	Enkephalin
	POMC-derived peptides
Actylcholine	Dynorphin
Amino acids	Neuropeptide Y
Glutamate	CRF
GABA	Neurotensin
Glycine	CGRP
Aspartate	Cholecystokinin
Taurine	LHRH
	Glucagon
Adenosine	Vasopressin
NO.CO	Oxycytocin
	Angiotensin
	Bombesin-like peptides

2 Neurotransmitters

Classification of neurotransmitters/ neuromodulators

Around 1970, the neurotransmission system was thought to consist mainly of two subsystems. One was the acetylcholinergic (cholinergic) system, and the other was the monaminergic system. γ-Aminobutyric acid (GABA) and glycine were regarded as minor neurotransmitters. However, many new substances were identified as neurotransmitters and neuromodulators during the 1970s and 1980s and, so far, more than a hundred have been discovered and identified (some are listed in Table 1). This list may expand substantially in the future. The monamines, acetylcholine (Ach) and aminoacids are conventionally called 'classical' neurotransmitters.

Arguments often arise on the question whether a particular substance deserves the designation of neurotransmitter or neuromodulator. The classical paradigm of a neurotransmitter was Ach, which is released from the terminal of motor neurons. However, it is too inflexible to maintain a position that any substance that does not fulfil the listed criteria is not a neurotransmitter. Instead, one has to look for the real functions of these compounds in the information processing between neurons. Here we adopt a pragmatic view that any endogenous compound which plays a major role in neural transmission is a neurotransmitter; neuromodulators are substances which modulate the functions of neurotransmitters. Readers should take note of the possibility that the same substance which acts as a neurotransmitter in some types of synapse may act as a neuromodulator in other synapses. We should focus our attention on real biological processes, rather than academic definitions.

Distribution of neurotransmitters/ neuromodulators in the brain

More than 100 neurotransmitters/neuromodulators have currently been identified. The initial step towards understanding the physiological functions of these compounds is to identify their localization within the brain, and their sites of production and release. In fact, the architecture of the brain is so complex that cells located only a few microns apart often have widely differing characteristics. Also, recent investigations have made it clear that the functionally uniform groups of cells called *nuclei* are actually composed of several subgroups of cells with distinct functions.

Immunocytochemistry is the most powerful method of localizing the neurotransmitters/neuromodulators in the brain. Recently, however, the *in situ* hybridization method, by which the mRNAs of neurotransmitters/receptors are localized within the cells, has become well enough developed that it has enabled us to visualize the tiny amounts of mRNA present in neurons. This method is utilized to investigate the possibility that a given cell is capable of producing a given substance. These morphological methods have revealed many new concepts concerning the dynamics of neurotransmitters and neuromodulators. Some of these concepts are described below.

Coexistence of multiple neurotransmitters/ neuromodulators in the same cells

One well-established neurological concept in the 1970s and early 1980s was the 'one neuron—one neurotransmitter' hypothesis. However, immunocytochemical investigations have firmly refuted this hypothesis, and established evidence for the contrary viewpoint. When initial experimental evidence was reported for the presence of several neurotransmitters/neuromodulators, researchers believed this to be an exceptional situation. However, recent research has confirmed that this situation, rather than being an exception, is the normal rule. Table 2 presents representative examples.

It is an interesting question whether there are some principles or patterns in such a coexistence of more than one compound. Figure 2 shows a pattern of coexistence in the peripheral nervous system. It is striking to find that in this system, the 'classical' neurotransmitters usually coexist with peptides. The possible meaning of such a coexistence in now being intensively investigated.

In motor neurons, Ach and calcitonin gene-related peptides (CGRP) coexist. CGRP potentiates the contraction of striated muscle induced by the Ach. Though this action is mediated by the increase of cAMP concentration within the muscle, CGRP does not exert its effects through nicotinic (n) Ach receptors or β-receptors. Thus, CGRP indirectly potentiates, or is complementary to, the action of Ach in inducing the contraction of striated muscle.

The postganglionic sympathetic neurons, which are one type of noradrenergic (NA) neuron, contain neuropeptide Y (NPY) in addition to NA. Both can induce contraction of blood vessels; contraction of NA is quicker and transient, while that of NPY is slower and long-lasting (Fig. 3). Thus, the effects of these two compounds are synergistic. On the other hand, in the seminal vesicles, NPY inhibits the effect of NA in inducing their contraction. This

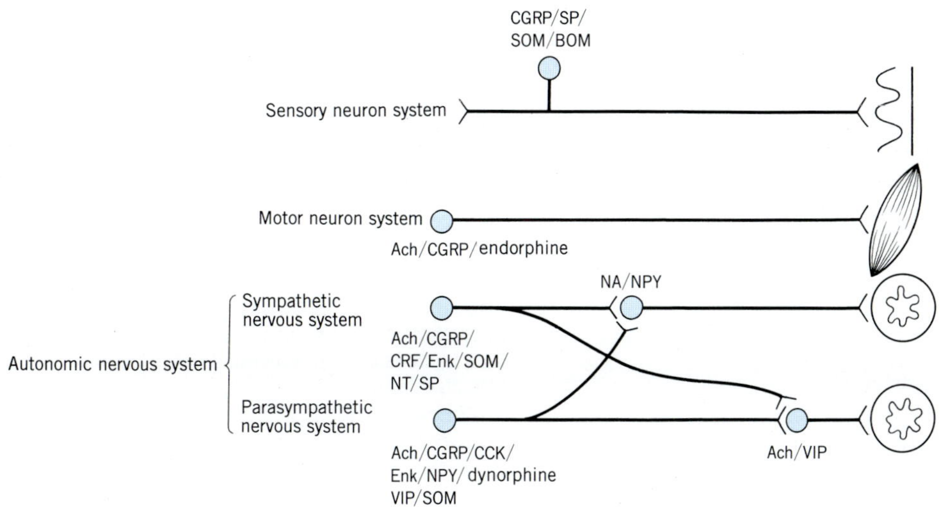

Fig. 2 Coexistence of neurotransmitters or neuromodulators in single cells in the peripheral nervous system.

Table 2. Examples of neurons containing two or more neurotransmitters/neuromodulators. See abbreviations, p. xi.

	Anatomical location of neurons
Peptides originating from the same precursor	
POMC-derived peptides (ACTH, Endorphin, MSH, etc.)	Arcuate nucleus
SP– –NKA– –NKB	Thyroid gland cell
CGRP– –Calcitonin	Thyroid gland cell
NT– –Neuromedin N	Thyroid gland cell
VIP– –PHI	Thyroid gland cell
Coexistence of classical neurotransmitters or of neurotransmitters and peptides	
Dopamine– –CCK	A9,A10
Dopamine– –NT	A10
Dopamine– –GRF	A12 (arcuate nucleus)
Dopamine– –GABA	A12 (arcuate nucleus)
Dopamine– –ANP	SIF cells
Dopamine– –5-HT– –EnK	SIF cells
Noradrenaline– –NPY	Postganglionic neurons of the autonomic nervous system (A1,A2<A4,A6)
Noradrenaline– –Enk	A6
Noradrenaline– –Vasopressin	A6
Noradrenaline– –Galanin	A2,A4,A6
Adrenaline– –NPY	C1,C2 (magnocellular part)
Adrenaline– –SP	C2 (parvocellular part)
Adrenaline– –NT/CCK	C2 (parvocellular part)
Serotonin (5-HT)– –SP	Caudal raphe nuclei
Serotonin (5-HT)– –TRH	Caudal raphe nuclei
Serotonin (5-HT)– –SP, TRH	Caudal raphe nuclei
Serotonin (5-HT)– –CCK	Caudal raphe nuclei
Serotonin (5-HT)– –Enk	Caudal raphe nuclei
Serotonin (5-HT)– –GABA	Caudal raphe nuclei
Histamine– –GABA– –Gal	Posteriohypothalamic area
Acetylcholine– –CGRP– –Endorphin	Motor neurons
Acetylcholine– –SP	Laterodorsal tegmental nucleus
Acetylcholine– –CGRP	Parabigeminal nucleus
Acetylcholine– –VIP	Cerebral cortex, Parasympathetic preganglionic neurons
Acetylcholine– –Galanin	Basal nucleus of Mynert
Acetylcholine– –Enk– –CGR– –Dynorphine	Olivocochlear bundle
Acetylcholine– –Enk– –SOM– –NT– –SP– –CRF	Sympathetic preganglionic neurons
Acetylcholine– –Enk–VIP–Dynorphine	Parasympathetic preganglionic neurons
Acetylcholine– –SOM– –NT– –CCK– –CGRP	Parasympathetic preganglionic neurons
GABA– –CGRP	Purkinje cells
GABA– –SOM	Cerebral cortex, Hippocampus
GABA– –CCK	Cerebral cortex
GABA– –NPY	Cerebral cortex
GABA– –Gal	Hypothalamus
GABA– –Enk	Retina, Striatum, Ventral nucleus of the lateral geniculate nucleus
GABA– –Glycine	Terminals within the cochlear nucleus
Glycine– –NT	Retina
Coexistence of peptides	
CGRP– –SP	Dorsal root ganglion
CGRP– –SOM	Dorsal root ganglion
CGRP– –BOM	Dorsal root ganglion
CGRP– –SP– –SOM	Dorsal root ganglion
CGRP– –SP– –BOM	Dorsal root ganglion
CGRP– –SOM– –BOM	Dorsal root ganglion
CGRP– –SOM– –BOM– –SP	Dorsal root ganglion
CGRP– –NT– –SP	Parabrachial nucleus
CCK	Parabrachial nucleus
SP– –Enk	Hypothalamus, Posterior horn
SP– –NT	Hypothalamus, Posterior horn
SP– –Dynorphine	Bed nucleus of stria terminalis, Amygdala
SP– –Enk	Olfactory bulb
SP– –CCK	Midbrain central grey matter
Vasopressin– –Dynorphine– –Angiotensin II– –Enk– –CRF	Paraventricular nucleus
Oxytocin– –CCK– –Dynorphine– –CRF– –Glucagon– –Prolactin	Paraventricular nucleus
NPY– –SOM	Cerebral cortex
SP– –NPY	Retina
SP– –Glucagon	Retina
Somatostatin– –Enk	Retina

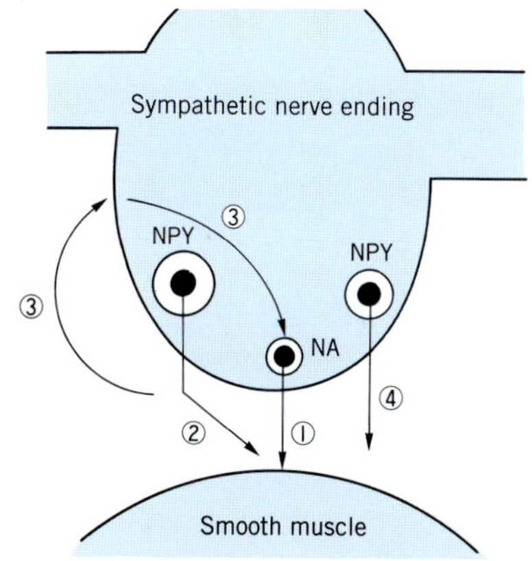

Fig. 3 Role of NA and NPY in the sympathetic nervous system.
(1) NA contracts smooth muscle;
(2) NPY increases the effect of NA;
(3) NPY inhibits the release of NA presynaptically;
(4) NPY itself contracts smooth muscle.
(modified from Fig. 8 on p. 249 of *Progress in Brain Res.* **68** (1986) by Lundberg and Hökfelt)

phenomenon is found to be caused by the inhibition of NPY, acting on the 'pre' terminal from where NA is released from the same terminal. Interestingly, the release of NPY is increased by the inhibitors for α receptors. This indicates that the release of NPY is regulated presynaptically by NA.

The postganglionic neurons of the parasympathetic system, which was established as cholinergic, also contain vasoactive intestinal peptide (VIP). VIP requires stronger stimulation than Ach to be released.

The dorsal root ganglia contains various peptides, of which the best known are substance P (SP) and CGRP. SP plays an important role in the transmission of noxious stimuli from the primary sensory neurons to the spinal cord. CGRP increases the release of SP. Recent findings indicate that SP itself is a neuromodulator of the transmission of noxious stimuli, of which the neurotransmitter is glutamate.

In view of these patterns of coexistence and the physiological actions of individual agents, a picture arises that the neurotransmitters in the peripheral neural system are the 'classical' neurotransmitters, whereas the coexisting peptides act as neuromodulators. This principle, however, is probably not valid for the central nervous system. One simple reason is that there are many neurons within the brain which do not contain any classical neurotransmitters. Although a number of researchers have examined the possibility of functional coordination between the classical neurotransmitters and peptides, not much is known about the mechanism of actions of coexisting peptides in one single cell.

Dissociation between the quantities of intracellular mRNA and protein

In immunocytochemistry, localization of protein is visualized in microscopic sections. In contrast, what is visualized in *in situ* hybridization is localization of mRNA in these cells. While immunocytochemistry visualizes cell bodies, terminals and occasionally the passing fibres, *in situ* hybridization marks only the cell bodies. Thus, if there are mRNAs within the cells, there would be corresponding translated proteins. Therefore, it would be expected that the results obtained by these two methods should essentially correspond.

However, as seen in many illustrations presented in this atlas, we quite often encounter cases where the results obtained from these two methods show discrepancies. It is often impossible to detect the protein corresponding to mRNA which has been confirmed to exist. One explanation for this phenomenon is that the proteins translated are transported continuously to the terminal, thereby maintaining their concentration within the cell wall below the limit of detection for immunocytochemistry. However, if this explanation is correct, we should expect to be able to visualize the proteins by this method after the ligation of axons, which effectively blocks the axonal transport. This has not proved to be the case. Explanations should therefore be sought in such possibilities as immediate degradation following protein synthesis, or inhibition of translation by some mechanism. This remains one of the great puzzles that await a solution in the future.

Chemical neuro-networks

Immunocytochemistry is one of the most powerful tools for localizing the neurotransmitters and neuromodulators. The aim of this book is to compile all the available information concerning the distribution of neurons and fibres containing these endogenous substances. One fact that should be emphasized is that it is not usually the case that a given endogenous substance is strictly localized in one specific area of the brain.

The brain can be considered as a large mass of integrated networks. Thus, the real function of one endogenous substance will not be understood unless its neural circuits are known from a viewpoint of neurotransmitters or neuromodulators. A combination of immunocytochemistry and other methods such as tissue destruction experiments and the retrograde tracer method will elucidate the complex projection pathways.

The presence of the synaptic contact between neurons and inputs to them which contain different or similar neurotransmitters or neuromodulators are revealed by the mirror method and multiple staining method at the electron microscopic level. Target cells are identified by the combination of the anterograde tracing method and immunocytochemistry or by the mirror method. Details of these techniques are described in other texts. To identify target neurons, intense investigations are currently being carried out by visualizing receptor proteins and their precursor mRNAs. The chemical neuronetworks in several regions of the brain are depicted in Figs. 4–13.

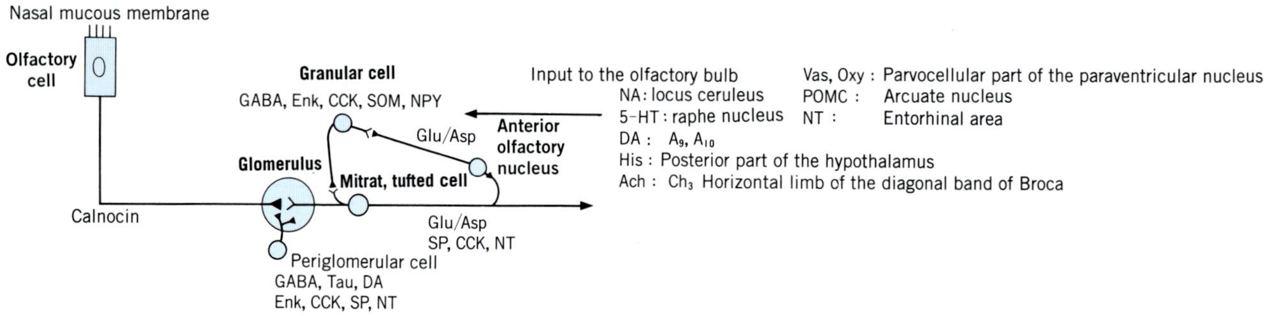

Fig. 4 Chemical neuroanatomy in the olfactory bulb.

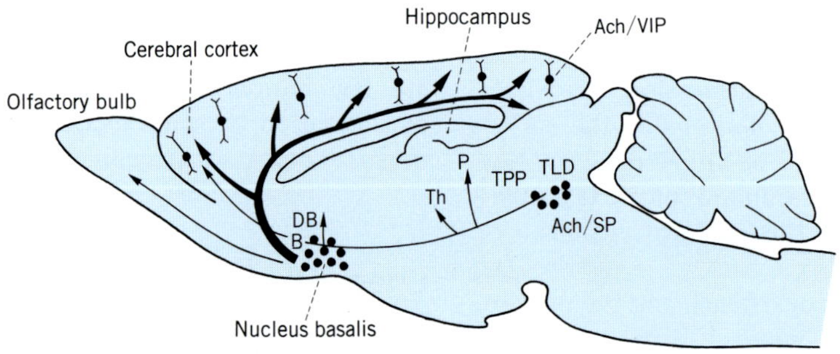

Fig. 5 Acetylcholinergic innervation in the cerebral cortex. Abbrevations: B, nucleus basalis; DB, diagonal band of Broca; P, pretectal area; Th, thalamus; TLD, nucleus laterodorsalis tegmenti; TPP, nucleus pedunculopontinus.

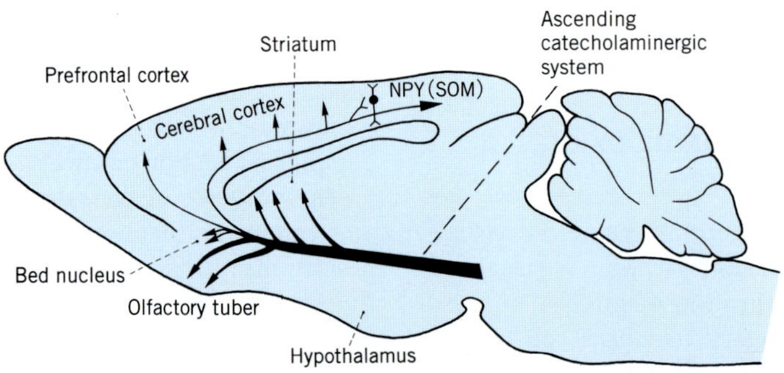

Fig. 6 Interaction between ascending catecholaminergic (CA) system and cortical neuropeptide Y-containing neuron system.

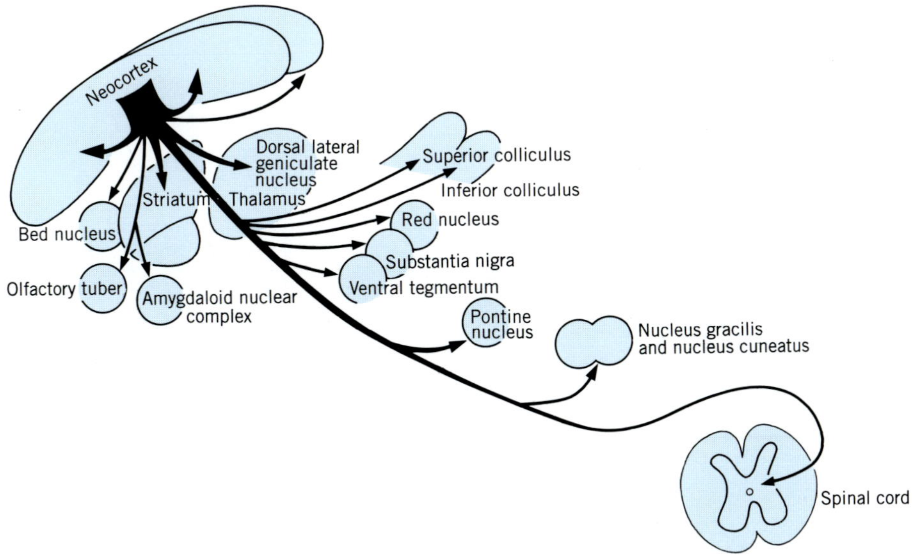

Fig. 7 Descending glutamatergic system from cerebral cortex.

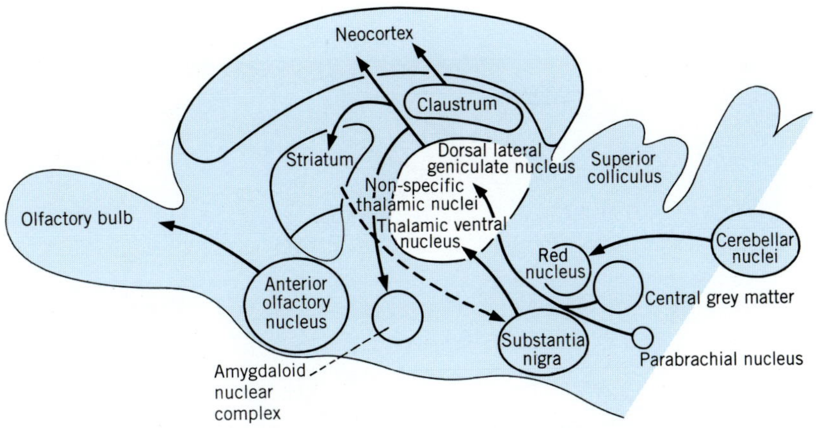

Fig. 8 Glutamatergic projection to the cerebral cortex.

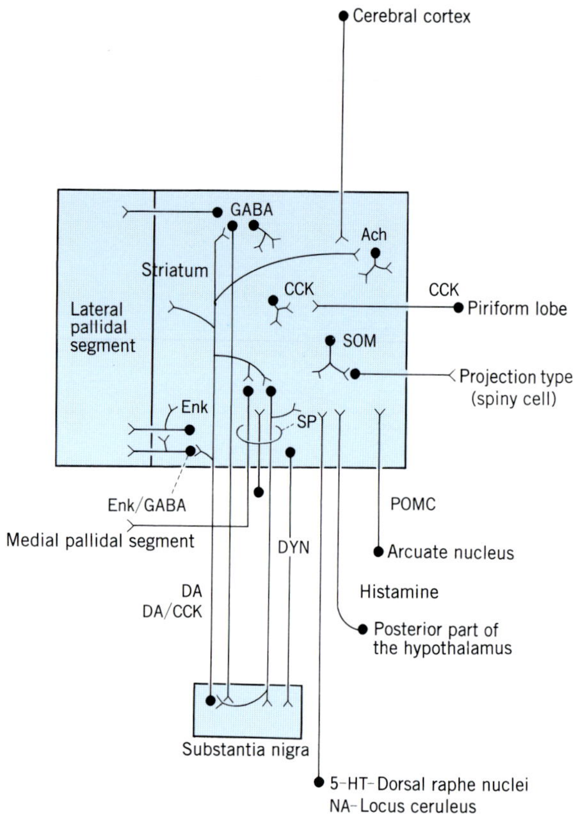

Fig. 9 Chemical neuroanatomy in the striatum and globus pallidus.

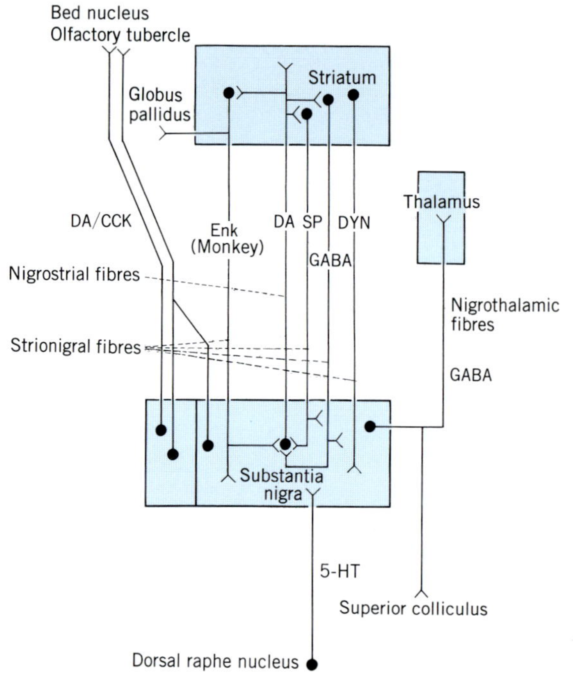

Fig. 10 Chemical neuroanatomy of the substantia nigra.

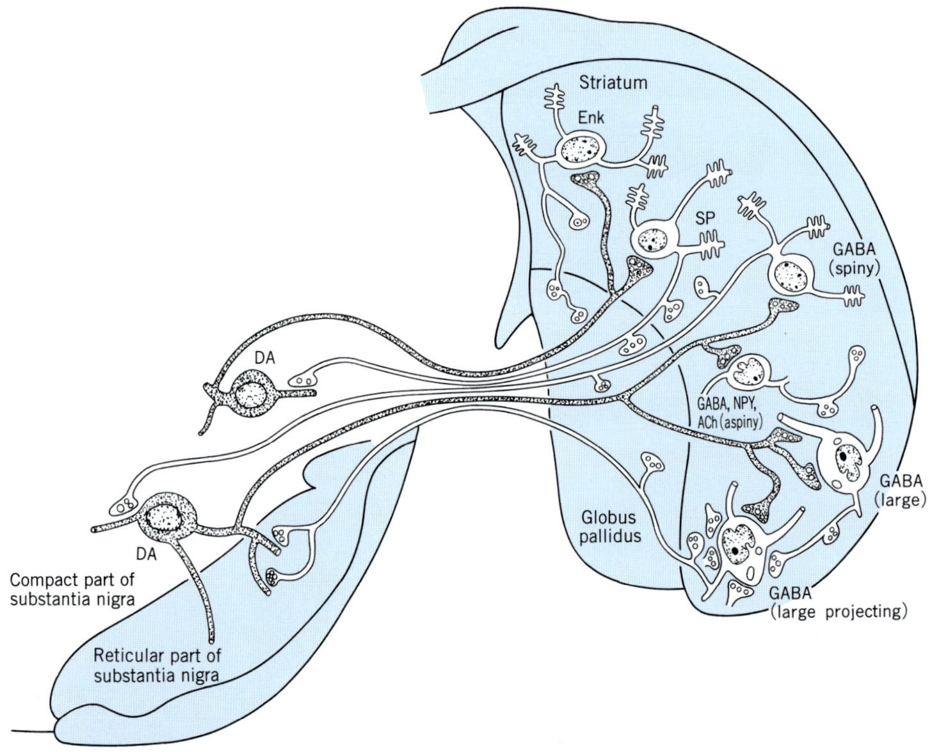

Fig. 11 Interaction between strionigral dopaminergic (DA) system and striatal neurons.

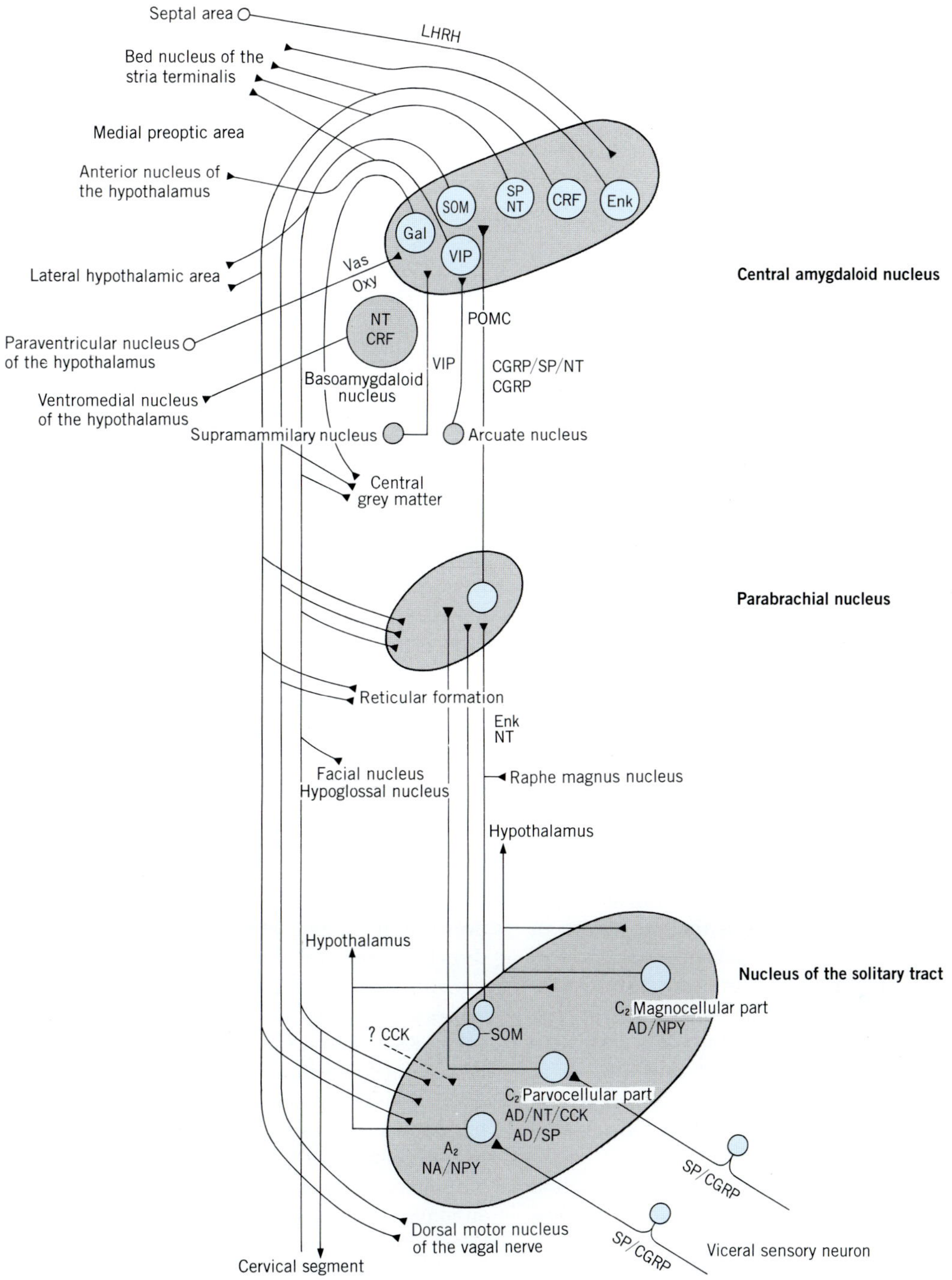

Fig. 12 Chemical neuroanatomy of the amygdaloid complex, parabrachial nucleus, and solitary tract nucleus.

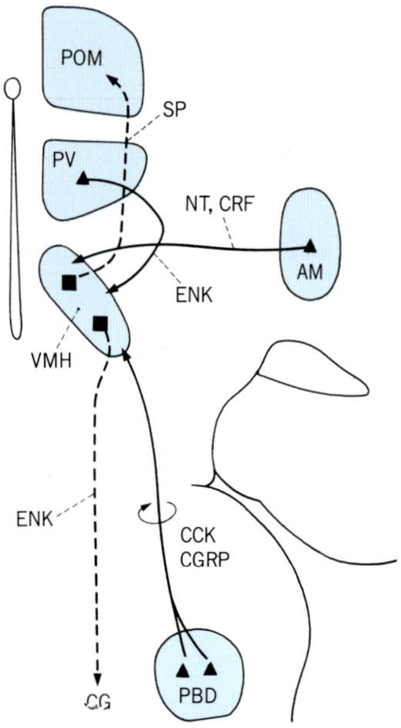

Fig. 13 Chemical neuroanatomy of the ventromedial hypothalamic nucleus.

Chemical neuro-networks and functions of neurotransmitters/neuromodulators

As stated above, neurotransmitters and neuromodulators are distributed widely in the brain. It is often asked what function a particular substance possesses. To answer such a question, an essential prerequisite is the knowledge of the site of location of that particular substance. It is not an exaggeration to note that if some neurotransmitter is distributed widely in the brain, one can assume that it possesses a variety of functions. Therefore, the first step toward understanding the function of a substance is to build a map of the chemical neuro-network involving the substance in question. Its exact role can be elucidated only in the context of such a network.

The most powerful tool in this direction is the method of *in situ* hybridization. An example is introduced below for this analysis, with the transmission pathway of noxious stimuli as a model.

Mechanisms of transmission of noxious stimuli to the central nervous system and its neurotransmitters/neuromodulators

There are various kinds of peptidergic cells in the dorsal horn and dorsal root ganglia, suggesting that they relate to transmission of noxious stimuli. However, the exact involvement of these peptides in the transmission of the noxious stimuli is not clear. In experimental animals which received a subcutaneous injection of formalin to the hind limbs, the SP precursor mRNA increased in small neurons in the dorsal root ganglion, and enkephalin (Enk) precursor mRNA was increased in local circuit neurons in the posterior horn (Fig. 14). The data indicate that in these animals, production of SP is increased in the posterior root ganglion and that of Enk is increased in the local circuit neurons of the posterior root. The increased SP is supposed to be transported in the central branch of neurons in dorsal root ganglia and then released at the posterior horn. In fact, the immunoreactivity of SP fibres there is enhanced by the noxious stimuli. These data support the hypothesis that SP in neurons of the dorsal root ganglion plays an important role in transmitting noxious stimuli. However, the results concerning Enk do not support the previous hypothesis concerning the mechanism of the action of Enk in the spinal cord. It is well established that Enk has an analgesic effect, and also that it inhibits the release of SP from primary sensory afferent fibres. The neuron in the spinothalamic tract branches off before it reaches the thalamus, to caudal raphe nuclei, reticular formation, and also the central grey matter in the midbrain. The neuron there sends descending fibres to the spinal cord, inhibiting the transmission of noxious stimuli from the primary sensory neurons to those at the spinothalamic neurons (which comprise a part of the descending inhibitory pathway). It has been assumed that the analgesic action of Enk is mediated by the descending inhibitory pathway, because many Enk-containing neurons exist at the original nuclei of this pathway (Fig. 15). However, in experimental animals the production of Enk is increased in the local circuit neurons. Therefore, the analgesic action of Enk is due, at least in part, to the function of Enk-containing local circuit neurons (Fig. 15). There are also reports showing that SP fibres synapse with Enk cells at the posterior horn. The overall picture that emerges for the mechanism of actions of Enk at the posterior horn of the spinal cord is shown diagrammatically in Fig. 16.

On the other hand, there are some suggestions that glutamate is the main transmitter of the small neurons of the dorsal root ganglia, which convey noxious stimuli, but the direct evidence for this is lacking. For example, since glutamate is found in both small and large neurons of the sensory ganglion, it is probable that glutamate is not solely related to nociception but is involved in the fundamental mechanism of sensory transmission. A study using *in situ* hybridization histochemistry revealed marked heterogeneity in subunit composition of AMPA receptors in the spinal cord, showing the postsynaptic effect of glutamate is characterized by the composition of the subunits. Thus, at present, glutamate is shown to relate to sensory transmission as a neurotransmitter.

Neurotransmitters/neuromodulators in the peripheral branches of sensory neurons

The sensory neurons receive noxious stimuli at their peripheral branch and convey the stimulation to the posterior horn through peripheral and central branches. Therefore, the neurotransmitters/neuromodulators produced in them should be transported only toward their central branches. However, in reality they are transported toward the peripheral branches too. In fact, a larger proportion is transported to peripheral than to central branches (Fig. 16). The representatives of these neurotransmitters/neuromodulators are SP and CGRP. In fact, nerve endings for SP and CGRP which are found abundantly in the end-organs disappear after removal of sensory neurons. In the case in which retrograde tracers are injected in the end-organs, SP and CGRP neurons at the dorsal root ganglia are labelled by them. However, it has been shown that sensory neurons distributed in skin are

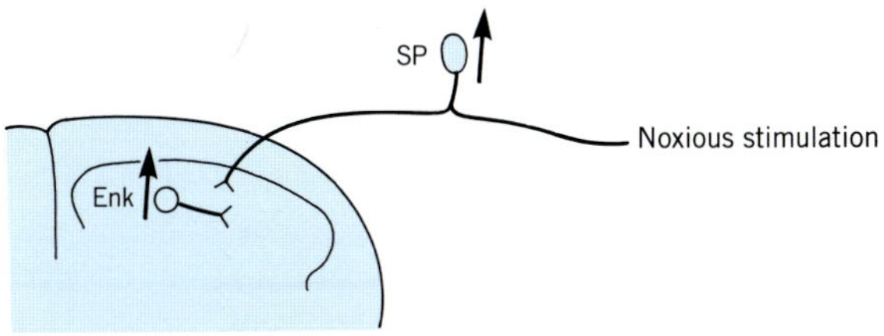

Fig. 14 Noxious stimuli increase production of SP and Enk in the dorsal root ganglion neurons and local circuit neurons of the posterior horn, respectively.

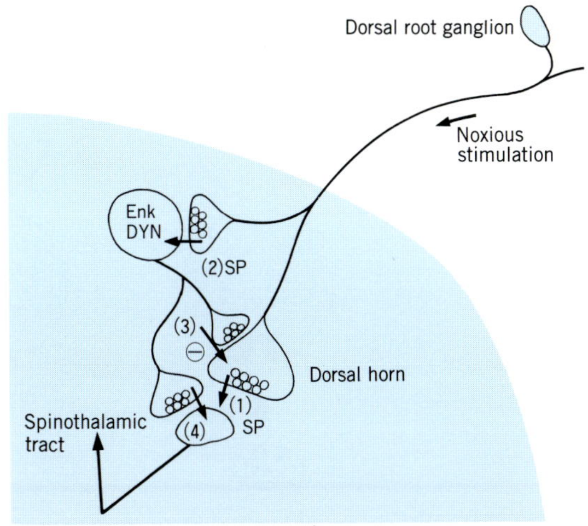

Fig. 16 How enkephalin (Enk)-containing local circuit neurons in the dorsal horn are involved in the analgesic effect of Enk. (1) Noxious stimuli cause the release of substance P (SP) from the primary afferents which activates the spinothalamic neurons. (2) SP released from the axon collaterals of the primary sensory afferents activates the Enk-containing local circuit neurons of the dorsal horn. (3) Enk released from the local circuit neurons inhibits the release of SP from the primary sensory afferents axo-axonically or (4) released Enk inhibits the activity of the spinothalmic neurons.

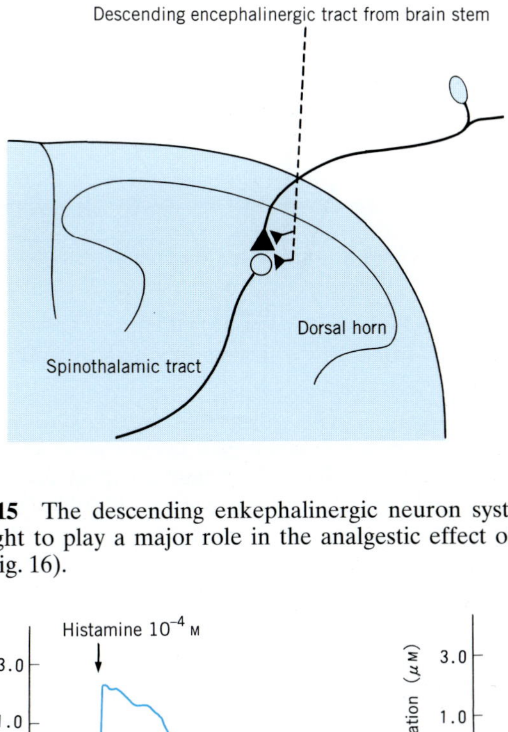

Fig. 15 The descending enkephalinergic neuron system is thought to play a major role in the analgestic effect of Enk (cf. Fig. 16).

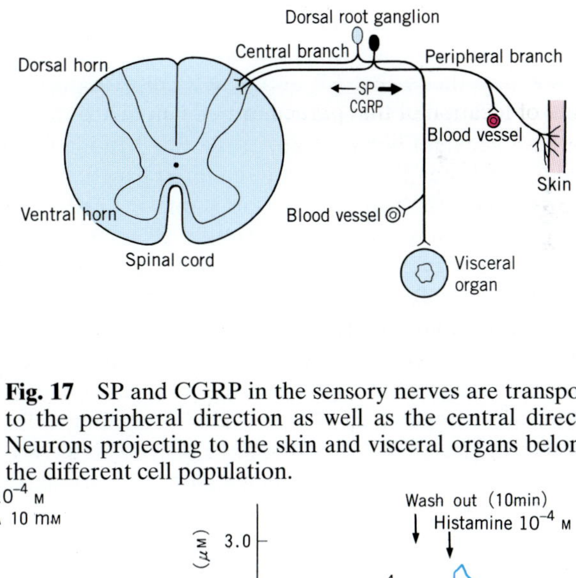

Fig. 17 SP and CGRP in the sensory nerves are transported to the peripheral direction as well as the central direction. Neurons projecting to the skin and visceral organs belong to the different cell population.

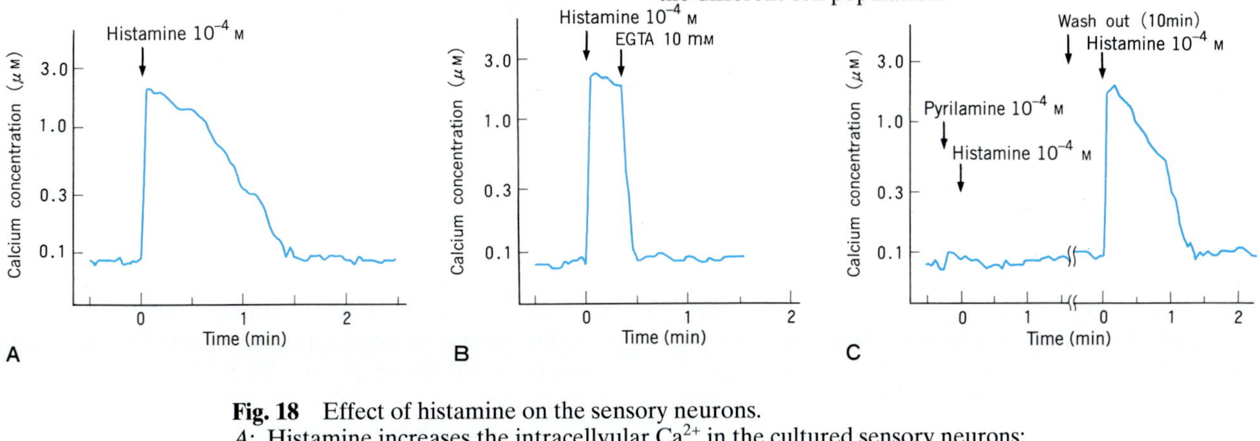

Fig. 18 Effect of histamine on the sensory neurons.
A: Histamine increases the intracellyular Ca^{2+} in the cultured sensory neurons;
B: Histamine-induced Ca^{2+} influx abolished by addition of EGTA;
C: Pretreatment with H_1-receptor antagonist pyrilamine inhibited the response, while the neurons showed increases in intracellular Ca^{2+} with histamine, after washing out pyrilamine.

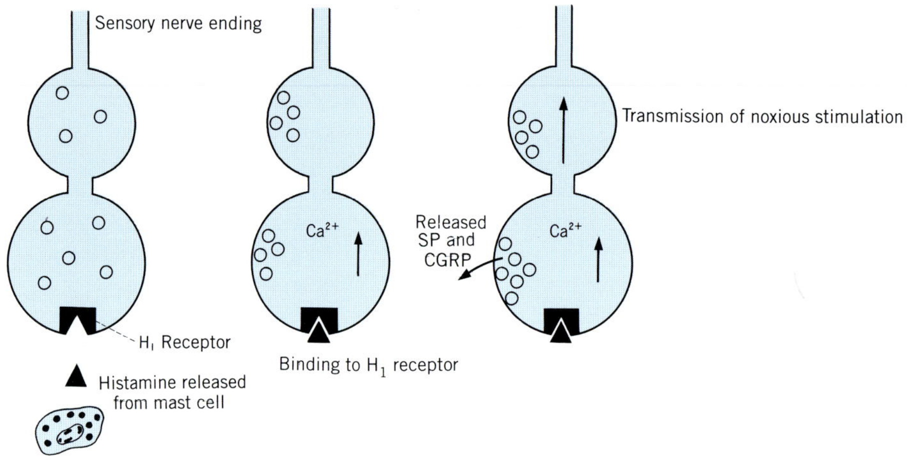

Fig. 19 Possible action of histamine on the release of SP and CGRP from the peripheral endings of the sensory nerves.

different from those of the internal organs (Fig. 17). This has been proved by the experiment in which when retrograde tracer X was injected in the skin while Y was injected into the internal organ, almost no neuron was doubly labelled by both X and Y. These results also support the view that arthralgia cannot be ascribed to the side branches of sensory neurons, but to the internal organs. What then are the exact roles of SP and CGRP in the peripheral branches of sensory neurons?

Pharmacologically, both SP and CGRP have vasodilatory actions. Also, they change the permeability of capillary vessels. Histamine released from mast cells induces further histamine release from mast cells. These facts support the hypothesis that SP and CGRP play important roles in the formation of oedema caused by pain (or itch)-provoking agents. Evidence supporting this hypothesis is as follows: if histamine is added to cultured ganglion cells at the posterior root, calcium ions flow into the cells, and increase their intracellular concentration (Fig. 18, A and B); the effect is inhibited by histamine H1 receptor (Fig. 18 C); and the cells where intracellular concentration of calcium ions is raised by histamine are stained for SP and CGRP, with very few exceptions. These results indicate that histamine released from the mast cells acts through histamin H1 receptors on SP- or CGRP-containing cells, releasing SP and CGRP from them. In fact, when a trace amount of histamine is applied on the nasal mucous membrane, the number and staining intensity of SP and CGRP fibres are increased, while immunoreactivity of both agents is decreased at the somatic level. However, the level of mRNA for SP and CGRP precursor remains unchanged. Taken together, one can conclude that histamine activates the axonal transport of SP and CGRP, by increasing the amount of neurotransmitters transported from the cell body to the nerve ending. Figure 19 summarizes the view that histamine directly binds to the nerve ending of sensory neurons, resulting in the release of SP and CGRP. In comparison, it is known that a stronger stimulus agent such as toluene increases the production of SP and CGRP in the cell body (increase of their precursor mRNA).

The next question, naturally, is what are the exact roles of the released SP and CGRP? There have been several speculations on this aspect (Fig. 20).

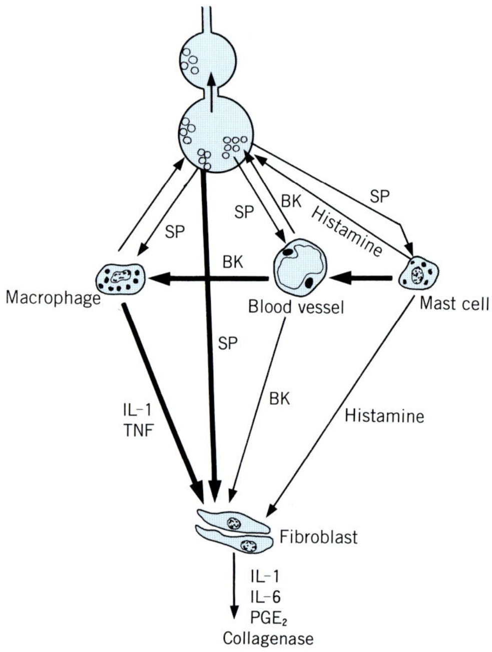

Fig. 20 SP released from the sensory ending may affect macrophages, blood vessels, mast cells, and fibroblasts.

- They act on mast cells, releasing histamine. The released histamine is believed to act on three different paths:
  (i) it acts on SP-terminal, releasing SP again
  (ii) it acts on blood vessels, increasing its permeability and releasing bradykinin
  (iii) it acts on fibroblasts to release IL-1, collagenase and PGE_2.
- SP acts on macrophages, releasing IL-1 and TNF, which in turn act on fibroblasts to release IL-1, collagenase and PGE_2.
- SP acts on blood vessels to release bradykinin (similar to the release of histamine). Bradykinin is released by the action of histamine too. Its release can lead to the following:
  (i) it acts on SP-terminals and induces the release of SP
  (ii) it acts on macrophages to release cytokines IL-1 and TNF, which in turn act on fibroblasts
  (iii) it can also act directly on fibroblasts.

Thus, there is neuroimmunological coordination at the terminals of sensory neurons.

Trophic (growth factor) functions of neurotransmitters/neuromodulators

When the ontogeny of various neuropeptides was examined, researchers found many areas where these substances are produced transiently just before and around the perinatal period. These include somatostatin in the cerebellum and cochlear nucleus, neurotensin in the cerebral cortex and mammillary bodies, cholecystokinin in the thalamus and the dorsal tegmental nucleus of Gudden. As neuro-networks are incompletely organized at the perinatal stage, the fact stated above suggested that these peptides are mostly involved with trophic (or growth) functions rather than in neurotransmission. These ontogenic investigations led to other studies which also indicated that these peptides play a role in neurotrophism as well as neurotransmission; for example, involvement of CGRP in regeneration of neurons, and that of SP in the locus ceruleus neurons. Apart from peptides, even GABA, Ach and others have also been reported to possess neurotrophic (growth-promoting) functions.

3 Receptors

There are three types of receptors for physiological agonists in the nervous system; the ion-channel type, the G-protein coupled type and the tyrosine kinase type. The neurotransmitters/neuromodulators belong to the former two types.

Traditionally, localization of receptors was studied by autoradiography, in which isotope-labelled ligands were used to identify the distribution of their binding sites. However, this method has been criticized for its low sensitivity and specificity, and the results obtained by this method have not been accepted as conclusive. Recent advances in biochemical and molecular biological techniques have made it possible to prepare highly purified receptor proteins and to elucidate their primary structures. Thus, immunohistochemical methods using antisera to receptor proteins are currently employed by researchers to examine the localization of receptors in the brain. It has also become possible to use cDNA, synthesized probes and RNA probes for *in situ* hybridization to localize the neurons where receptor proteins are produced. In this book, data obtained by autoradiography, immunohistochemistry and *in situ* hybridization on the localization of receptor proteins are presented.

Structure and mechanisms of action

Ion-channel linked receptors

The ion-channel linked receptors are oligomers composed of several mutually similar subunits and are integral proteins in cell membranes. All the receptors have four transmembrane perforation sites (Fig. 21*A*). When neurotransmitters/neuromodulators bind to one of the subunits, the protein configuration of the subunit and/or other subunits changes to make the channel open. If it is a $GABA_A$ receptor or glycine receptor, Cl^- flows in and the cell becomes hyperpolarized (inhibition). However, if it is a nACh receptor or AMPA-type receptor of

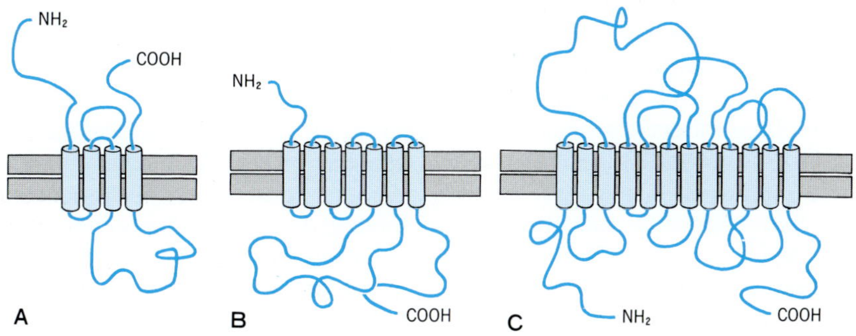

Fig. 21 Structure of the ionotropic receptors (*A*), G-protein coupled receptors (*B*), and transporters (*C*).

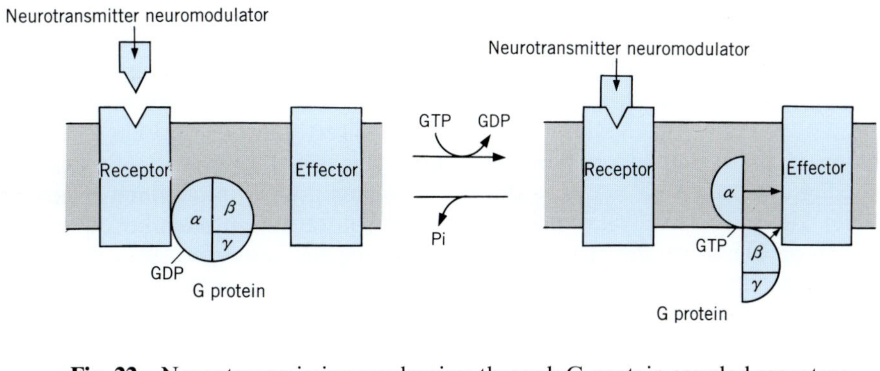

Fig. 22 Neurotransmission mechanism through G-protein coupled receptors.

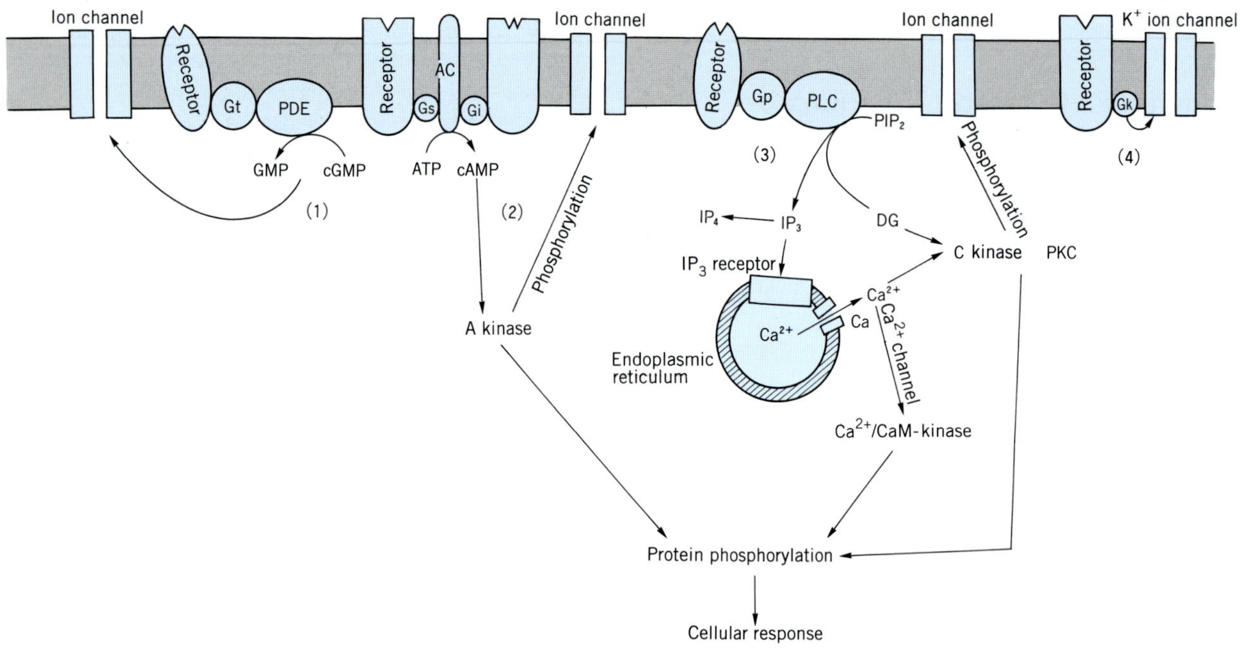

Fig. 23 Intracellular responses.

glutamate, Na^+ or K^+ flows in to make the cell depolarized (excitation).

G-protein coupled receptors

The receptors coupled to G proteins protrude their N-terminal end beyond the cell, and possess seven trans-membrane perforation sites (Fig. 21*B*). They form a complex with G proteins when neurons are at rest (Fig. 22). G proteins are composed of three subunits α, β and γ, with the α subunit having binding sites to GTP or GDP and hydrolytic activity of GTP (Fig. 22). GDP binds to the α subunit at rest. When neurotransmitters/neuromodulators bind to their receptors, GDP is converted to GTP in the presence of Mg^{2+}. The receptors dissociate from the G protein coupled with this conversion at the same time and dissociate into α-GTP complex and β–γ (Fig. 22). The dissociated subunits trigger the subsequent metabolic changes within the cell. The α-GTP complex, being GTPase itself, is reconverted back to α-GDP in the presence of Mg^{2+} and resumes its initial form bound to the receptor, combining with the β–γ subunits.

Thus, the subunits of the activated products of G proteins, namely α-GTP and β–γ, are molecules responsible for transducing the information carried by the ligand into the intracellular events. The events elicited are protean (Fig. 23). One of the major pathways is the system in which the intracellular concentration of cAMP is controlled through inhibition (Gi) or stimulation (Gs) of adenylate cyclase (AC) (Fig. 23(2)). cAMP activates the cAMP-dependent protein kinase (A kinase), which in turn phosphorylates various protein substrates. This event is regarded as essential for the cells to respond to various stimuli. In some tissues like atrium, A kinase directly phosphorylates the subunit of channel protein (in this case, Ca^{2+} channel), making the ion channel open (Fig. 23(2)). In olfactory cells, increased cAMP acts directly on the Na^+ channel to make it open. The activation of G protein also results in activation of phospholipase C, which hydrolyses phosphatidylinositol diphosphate (PIP_2) into diacylglycerol (DG) and inositol triphosphate (IP_3).

IP_3 binds to its receptor on the membrane of intracellular storage organelles of Ca^{2+} such as endoplastic reticulum (Fig. 23(2)). The stored Ca^{2+} ions then flow out of the organelles into cytoplasm through Ca^{2+} channels. On the other hand, DG activates Ca^{2+} phospholipid-dependent protein kinase C (PKC) in the presence of Ca^{2+}. It has been indicated that PKC, phosphorylating a variety of substrate proteins, can also phosphorylate the K^+ ion channel protein and regulate its opening (Fig. 23(3)). Another consequence of the activation of G protein is the increased activity of phospholipase A_2 (PLA_2), which results in the release of arachidonic acid. This event is the initial step of biosynthesis of various prostaglandins, the physiological effects of which in the brain are yet to be determined. In the retina, light activates cGMP phosphodiesterase via Gt (transalucin), decreasing the concentration of cGMP. The permeability of Na^+ ions at the sodium channel is decreased as a result, thus the retina cell become hyperpolarized (Fig. 23(1)). In the olfactory cells, activation of G proteins leads to activation of guanylate cyclase as is the case for AC, opening the cation channel. In cardiac muscles, activated G protein is known to act directly on K^+ and Ca^{2+} channels to make them open (Fig. 23(4)).

In summary, the intracellular events elicited by first messengers through G-protein linked receptors are as follows. Following the activation of G proteins, intracellular concentration of various second messengers (cAMP, cGMP, DG) or Ca^{2+} is increased. The second messengers activate various protein kinases, e.g. kinase A, cGMP-dependent protein kinase (G kinase), Ca^{2+}-calmodulin dependent protein kinase II and kinase C. The activation of these kinases leads to phosphorylation of various substrate proteins within the cells, which results in various cellular responses. For instance, release of neurotransmitters, which is inhibited by synapsin I binding to synaptic vesicles, is induced by phosphorylation of synapsin I by Ca^{2+}-calmodulin dependent protein kinase II. As opposed to this pathway, G proteins can act directly or indirectly on ion channels to regulate their opening. The indirect action is classified as the effect of the second messengers and that of the protein kinases activated by them, which phosphorylate ion channel proteins. In addition, the importance of the activation of other kinases such as MAP kinase has recently been recognized.

Heterogeneity of receptors

Heterogeneity of ion channel type receptors

Recent research has elucidated the primary structures of the subunits of several ion-channel receptors: $GABA_A$ receptor, glycine receptor, nACh receptor, and glutamate ion-channel type receptor. The characteristic feature of these receptors is that they more or less possess a subunit structure, which in turn has several variants. The most typical is $GABA_A$ receptor. It has been suggested that every group of neurons expressing $GABA_A$ receptor has its characteristic composition of $GABA_A$ receptor subunit variants. Furthermore, even if the composition of variants is similar, their quantitative expression differs in each group of neurons. Thus it is claimed that there are hundreds of subtypes of $GABA_A$ receptor and of the Ach receptor as well. The glycine receptor also possesses similar features, although its variation is not as pronounced as that of $GABA_A$ or Ach receptors (Fig. 24). Why should there be such variety in the receptors for one specific neurotransmitter? We have no answer available now, though research on the significance of variance of one particular subunit will lead to some progress on this question. It is an important problem to resolve if all the subunits expressed in a given neuron are assembled to construct a receptor in the same neuron.

Heterogeneity of G-protein coupled receptors

The amino acid sequences for the mACh receptor, receptors of various monoamines, tachykinins, somatostatin, neurotensin, NPY, TRH, angiotensin, VIP and metabotropic glutamate receptor have already been determined. Both the mACh receptors and various monoamine receptors have various subtypes; in many cases, multiple subunits are expressed in the same neuron, but the combination of their expression is not regular. Why the multiple subtypes are expressed in one neuron is not clear. A crucial key to this problem will be identified if immunocytochemical examination using antisera against each subunit resolved the question whether multiple subunits are expressed on the same postsynaptic membrane or at different sites. This situation is quite similar to that already discussed for ion-channel type receptors.

Cloning of peptide receptors is progressing rapidly. It has become clear that compared to the mACh receptors and the receptors of various monoamines, peptide receptors are less abundant in subtypes. For instance, tachykinins of mammals are clarified as SP, neurokinin A (NKA), and neurokinin B (NKB) and their corresponding receptors are NK-1, NK-2, and NK-3. Each receptor, in addition to binding to binding its corresponding peptide, also binds to other tachykinins. However, no subtypes have been reported for NK-1, NK-2, and NK-3 themselves. The recently cloned somatostatin receptor has two subtypes. But overall, it seems that peptide receptors have many fewer subtypes than mACh and monamine receptors. One can even consider this characteristic as typical of transmission of information by peptide neurotransmitters. This fact is quite intriguing for future analysis related to the anatomical and functional differences between peptide and Ach and monoaminergic neurotransmission of information.

Significance of subunits and subtypes

Many experiments have been carried out for ion-channel type receptors in which they were translated *in vitro*, expressed *in vitro*, and expressed in the oocyte of *Xenopus laevis*, the African clawed toad, and their functions were then examined. For example, the results obtained for $GABA_A$ receptor revealed that its γ subunit is a prerequisite for responses for benzodiazepine; the γ_2 subunit, when combined with the $\alpha6$ and $\beta2$ subunits can exert minimal function as a receptor; each α subunit responds to GABA. But one has to keep in mind that the real $GABA_A$ receptors are integrated into neural networks and therefore they are subjected to various modifications. In this situation

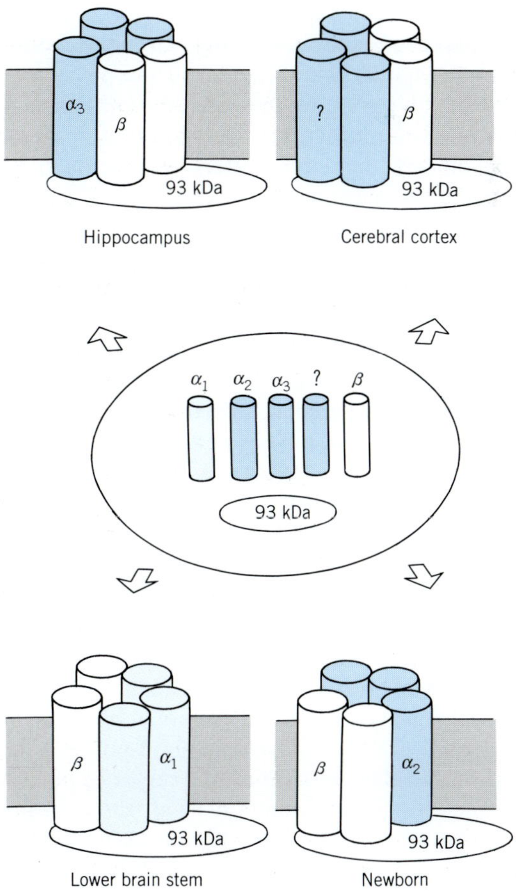

Fig. 24 Multiplicity of the subunit composition of the glycine receptor, which belongs to the class of ionotropic receptors.

we are faced with the familiar predicament of extrapolating the data obtained in cultured cells to the *in vivo* situation, though naturally the data indicate what kind of 'potential' functions the subunits in question possess.

It had been described earlier that several neurotransmitters/neuromodulators have trophic functions. It now appears that specific receptive molecules must present for these trophic functions to be manifested. For example, the α_2 subunit of glycine receptor, α_3, β_2, β_3, and γ_2 subunits of GAB_A receptors and δ and ϵ subunits of ACh receptors are such molecules. In contrast to this pattern, for neurotensin, while the same receptor acts for transmission of information in adult animals, it acts to mediate trophic function in the brain of immature forms. Also, it has been shown that γ subunits tend to exist more in GABA neurons, which suggest that the subunit is a constituent of an autoreceptor.

To resolve the function of various subunits/subtypes, it is mandatory in future to develop specific ligands to each subunit/subtype and examine their function *in vivo* in combination or as a whole.

4 Transporters

Transporters related to neural transmission can be divided into at least two types; transporters expressed at the synaptic cleft and those expressed at the vesicular membrane (vesicular transporters). The latter transporter takes up neurotransmitters or neuromodulators from axoplasm into synaptic vesicles. It has been thought the former type is expressed at presynaptic membrane and is responsible for re-uptake of neurotransmitters or neuromodulators into the presynaptic neural element. Thus, physiologically, their function is to remove the neurotransmitters or neuromodulators from the synaptic cleft so that transmission is finished quickly. The transporters to be cloned had 12 transmembrane perforation sites and are different from ion-channel type receptors (which have four transmembrane perforation sites) and G-protein coupled receptors (which have seven transmembrane sites) (Fig. 21).

However, recent progress in the study of the transporters requires changes to the concept mentioned above. First, it has been shown that multiple transporters are involved with a single neurotransmitter or neuromodulators. For example, these types of transporters have been cloned: GAT 1–3 for GABA; EAAC1, GLAST, GLT-1, and EAAT 4 for glutamate; and glygine transporter I and II for glycine. Second, findings showing that some of the transporters express at postsynaptic membrane or glial cells have been reported. For example, GAT-1 is expressed not only in the GABAergic neurons but also in non-GABAergic neurons. The latter findings show that GAT-1 is also expressed on the postsynaptic membrane at the GABAergic synapse. Furthermore, it has been shown that GLAST, GLT-1, and glycine transporter I are expressed on the glial cells. Third, the presence of 12 transmembrane sites is not common to all the transporters. Glutamate transporters have 6–10 transmembrane sites.

Recently, a peptide transporter which is specific in the brain and retina has been cloned. This transporter is proton dependent and re-uptakes di- or tripeptide. Although brain peptides have been thought to be degraded by peptidase, the presence of brain specific transporter suggests that peptidergic transmission is also influenced by the transporter.

It has been generally agreed that transporter expressed on the presynaptic site makes the synaptic transmission to be completed by uptaking the neurotransmitter or neuromodulator from the synaptic cleft into presynaptic sites. Some physiological responses to pharmacological agents can be accounted for by the actions of transporters expressed on the presynaptic site. Cocaine binds to the dopamine transporter and inhibits the re-uptake of dopamine to the presynaptic sites. This would result in accumulation of dopamine at the synaptic cleft, which naturally explains the habit of cocaine intoxication. Dopamine transporter also takes up pharmacological agents which have similar structure to dopamine, such as MPP^+ and 6-hydroxydopamine. Their uptake to dopaminergic neurons results in the cell death of dopamine neurons because they are neurotoxic for dopaminergic neurons. Desipramine, an antidepressant drug, is known to bind to noradrenaline (NA) transporter, inhibiting its re-uptake to exert its effects. The possibility that desipramine inhibits the serotonin transporter has been also suggested. These examples illustrate the importance of the functions of transporters expressed on the presynaptic sites.

Some of the functions of the transporters expressed in glial cells have been explored. It is likely that transporters expressed in the glial cells prevent the diffusion of neurotransmitters or neuromodulators out of the synaptic cleft by uptake of the neurotransmitters or neuromodulators.

Furthermore, GLAST has been shown to protect against cell death in the nervous system after ischemia. GLAST is expressed in the glial cells; Bergman glia in the cerebellum and astrocytes and Muller cells in the retina. In the retina, ischemia resulted in the up-regulation of the GLAST but had no effect on the production of GAT-1, glycine transporter I, and taurine transporter. This change was also confirmed at the protein level. Ischemia resulted in the degeneration of the neurons which cause the release of the glutamate (a neurotoxin) from the cell bodies. Released glutamate acts on surrounding neurons as a neurotoxin to cause severe damage. Up-regulation of the GLAST expressed on the glial cells resulted in decreasing the glutamate in the damaged tissue, which may save neurons from cell death.

5 Feedback of information from postsynaptic to presynaptic region

Neural information is usually transmitted from the presynaptic region to the postsynaptic region. Therefore the mainstream of research on the mechanism of neurotransmission has focused on the production, transport, and release of neurotransmitters/neuromodulators at the presynaptic region and on the receptors, G proteins, and intracellular events at the postsynaptic region. Only in the past few years have the regulatory mechanisms existing from the postsynaptic region to presynaptic regions by autoreceptors and transporters become areas of intense study. One of the most remarkable findings along this line of research is concerned with the question of whether information is really transmitted from the postsynaptic region to the presynaptic region. Nitric oxide is the agent which has attracted much attention in this regard (Fig. 25). Nitric oxide is synthesized from L-arginine by nitric oxide synthase. When the intracellular concentration of Ca^{2+} is increased due to transmission of neural information, nitric oxide synthetase is activated, which in turn increases the production of nitric oxide. The nitric oxide produced is released from the postsynaptic region and then acts there (or possibly on glia), activating guanylate cyclase. Production of cGMP from GTP is thus increased, which as a second messenger would induce various cellular responses. Though details are yet to be clarified, nitric oxide may certainly play an important role in the feedback of information from the postsynaptic to the presynaptic region.

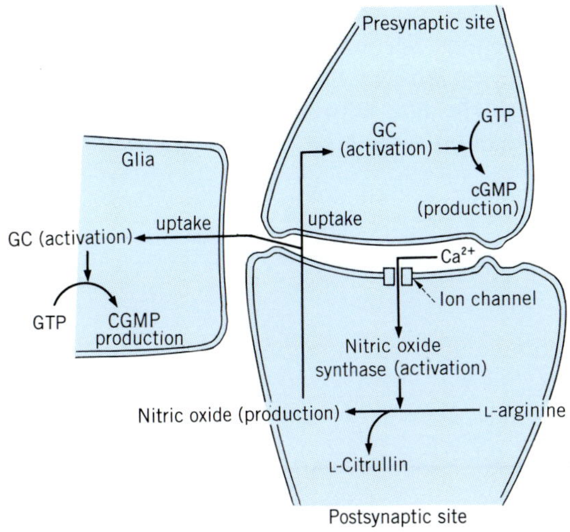

Fig. 25 Possible mechanism of nitric oxide in neural tissue.

6 Acceptors other than neuronal receptors

Neurotransmitters/neuromodulators are usually released from the presynaptic region and bind to receptors on the corresponding postsynaptic neurons. There is no doubt that the majority of the binding sites for them are located on neurons. However, recent evidence suggests that neurotransmitters/neuromodulators can act as acceptors other than receptors on the postsynaptic membranes. For instance, the enzyme lysing glycine is located exclusively in astrocytes. This fact suggests that glycine released into synaptic junctions as a neurotransmitter is removed to be metabolized in the adjacent astrocytes. From another viewpoint, there is the possibility that astrocytes are regulating the concentration of glycine at the synaptic junction or metabolizing it to prevent it from acting on neural elements to which glycine is not targeted (Fig. 26). It has long been recognized that astrocytes are capable of taking up radiolabelled neurotransmitters/neuromodulators *in vivo*. Thus it is reasonable to assume that one of the roles of astrocytes is to remove excess neurotransmitters/neuromodulators in the synaptic junction. In fact, GLAST (a kind of glutamate transporter) and glycine transporters are expressed on the glial cells. Receptors for neurotransmitters/neuromodulators are well known to be expressed in cultured glia. However, there are few reports showing the existence of receptors on glial cells *in vivo*. Thus it is premature to conclude that neurotransmitters/neuromodulators act on glia through receptors, and execute their function similar to how they function with neurons under physiological conditions. A recent report indicates that Glu 4 subunit of the AMPA-type glutamate receptor is expressed in Bergmann glia in the cerebellum *in vivo*. It would be of great interest to see if metabolic changes are induced in the glia in response to glutamate.

As an another example, let us look at the basic fibroblast growth factor (FGF) which belongs to the FGF family and is produced in neurons. The receptors corresponding to FGF belong to a family comprising three subsets, namely, flg, bek, FGFRT-3, and FGFR-4. It has been established that flg mRNA is normally expressed in neurons, while bek mRNA is also expressed in glia. This fact indicates that FGF can act on both neurons and glia, exerting a variety of effects through diversified receptors.

In addition, findings showing the importance of glial cells have been accumulated. For example, several genes which are not expressed in the normal brain are induced in glial cells during ischemia or brain injury. The *sgk* gene, a member of the serine/threomine protein kinase gene family, expressed in oligodendrocytes after brain injury, and serine protease inhibitor III, expressed in astrocytes after ischaemia, are examples. Furthermore, it has been shown that radicals stimulate IL-6 release from astrocyte, RA 301, and RA 410 synthesis in astrocytes. Thus, glial cells are not only supporting cells, but are also involved in the repair process in the damaged brain.

7 Non-synaptic transmission

An intriguing possibility has recently been raised concerning non-synaptic transmission of information by neurotransmitters/neuromodulators. This idea postulates that the neurotransmitters/neuromodulators leak out of the synaptic junction and influence the neurons surrounding it. Though it is certainly an attractive hypothesis to account for their broad range of functions, some questions need to be resolved before it is established as a fact. First,

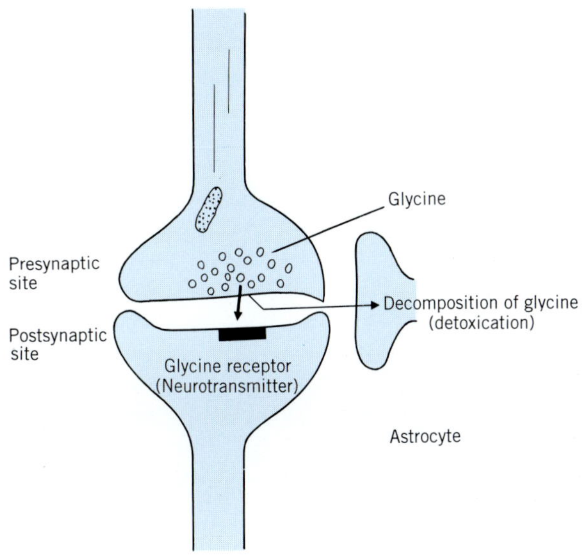

Fig. 26 Possible pathways of the inactivation of glycine released into the synaptic cleft.

it is not clear if neurotransmitters can really leak from the synaptic junctions. Secondly, even if such leakage occurs, it has to be established that the neurotransmitters can flow through the fine network of astrocytes. Thirdly, it should be demonstrated how the neurotransmitters could be directed to target neurons. Finally, how do they escape from being metabolized while they diffuse?

Note: We have decided not to include references for this chapter since it covers such a wide range.

Amines

1 The catecholaminergic system

Dopamine

Biosynthesis

Catecholamine is synthesized from tyrosine (Fig. 1.1). Tyrosine is hydroxylated by tyrosine hydroxylase (TH) to L-dihydroxyphenylalanine (L-DOPA), and subsequently converted to dopamine (DA), noradrenaline (NA), and adrenaline (AD). TH is the rate-limiting enzyme in this pathway and is therefore essential in cells where DA, NA, and AD are synthesized from tyrosine. Those cells which are stained immunocytochemically for TH are thus identified as those producing either DA, NA, or AD.

The synthesis of DA is regulated by controlling the activity of TH, the rate-limiting enzyme, by the following mechanisms.

1. The activity of TH is inhibited by the synthesized DA.
2. TH has two kinetic forms: a high affinity form and a low affinity form (for tetrahydrobiopterine). The conversion between the two forms is believed to be induced by phosphorylation of TH.
3. When the presynaptic dopamine receptor (autoreceptor) is stimulated by DA in the synaptic cleft, the activity of TH is inhibited. This stimulation inhibits the release of DA from the DA terminal at the same time.
4. The intensity of input stimulation to DA cells is proportional to the activity of TH.

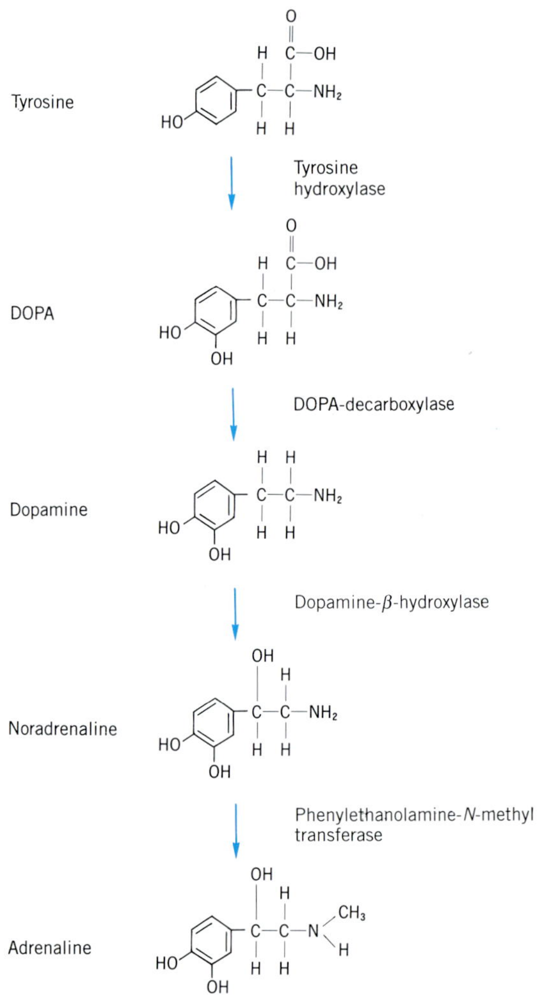

Fig. 1.1 Synthesis of catecholamines.

Distribution and projection pathway

TH-containing cells are classified according to the distribution of catecholamine synthetic enzymes and histofluorescence. DA cells are categorized as A8–A17 (Table 1.1). NA cells are categorized as A1–A7, and AD cells as C1–C3[2]. The A8–A17 regions are as follows (cf. the atlas):

A8	retrorubral nucleus
A9	substantia nigra
A10	ventral tegmental area
A11	dorsal and posterior hypothalamus, zona incerta, mesencephalic periaqueductal grey matter
A12	arcuate nucleus
A13	dorsomedial hypothalamus (zona incerta)
A14	periventricular hypothalamus
A15	ventral surface of the hypothalamus, ventral part of the bed nucleus of the stria terminalis
A16	olfactory bulb (periglomerular cell)
A17	retina (interplexiform cell)

A12 is further classified into A12d (dorsal area) and A12v (ventral area), where the former is considered to be made up of DA cells and the latter of DOPA-producing cells and not DA cells.[10]. Many researchers have reported that TH-positive cells are found in cerebral cortex and hippocampus in addition to those classified here, and yet to be established[4,5].

The following projection pathways for DA cells are known (see also Table 1.1 and Fig. 1.2)[1,2]: the mesostriatal, mesolimbocortrical, diencephalospinal, periventricular, incertohypothalamic, tuberohypophyseal, periglomerular, and retinal DA systems.

These pathways differ in their pharmacological properties[3]. For instance, the mesocortical system lacks autoreceptors and reacts easily in response to stimuli; the metabolic turnover rate of DA is fast; it responds only weakly to DA agonists and antagonists. These characteristics are quite different from other DA systems[3].

The DA system is the one whose links to pathological conditions has been most clearly demonstrated. Strong lines of evidence indicate that its abnormality is the main cause of Parkinson's disease and schizophrenia. Parkinson's disease is believed to be caused mainly by the destruction and loss of DA cells in the nigrostriatal system. This destruction of DA cells is also seen in the

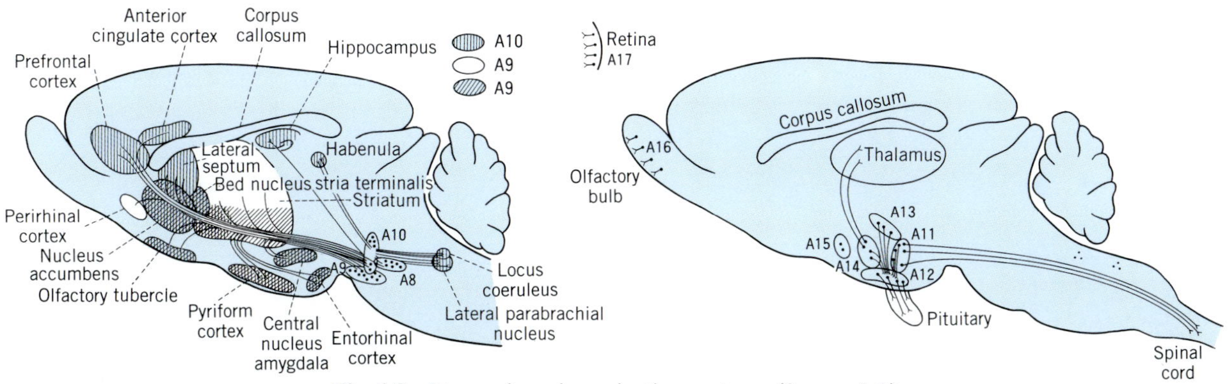

Fig. 1.2 Dopaminergic projection systems (from ref. 3).

Table 1.1. Major dopaminergic projection systems (from ref. 1)

System	Cells of origin	Projections
Mesostriatal system	Substantia nigra (A9), ventral tegmental area (A10), retrorubral nucleus (A8)	Striatal areas
Mesolimbocortical system	Ventral tegmental area, substantia nigra, retrorubral nucleus	Limbic and cortical areas
Diencephalospinal system	Dorsal and posterior hypothalamus, zona incerta, caudal thalamus (A11)	Spinal cord
Periventricular system	Mesencephalic periaqueductal gray, periventricular gray of caudal thalamus (A11)	Periqueductal gray, medial thalamus and hypothalamus
Incertohypothalamic system	Zona incerta, periventricular hypothalamus (A11, A13, A14)	Zona incerta, anterior, medial preoptic and periventricular hypothalamus, septum
Tuberohypophyseal system	Arcuate and periventricular hypothalamic nuclei (A12, A14)	Median eminence, pars nervosa and pars intermedia of the pituitary
Periglomerular dopamina neurons	Olfactory bulb (A16)	Dendritic processes into olfactory glomeruli
Retinal dopamine system	Mainly in the inner nuclear layer of the retina	Local dendritic projections

mesolimbic, mesocortical, and hypothalamic systems[7]. The link between the DA system and schizophrenia has not been as distinctly shown as in Parkinson's disease, but it has been demonstrated that several abnormalities, such as an increase in the number of D2 receptors, are associated with this disease[8].

Recent progress in positron emission tomography (PET) has made it possible to detect abnormalities in this system in patients suffering from these neuropathological conditions *in vivo*. For example, PET imaging has clearly shown the decrease of DA cells in Parkinson's disease by use of $[^{18}F]$-L-DOPA[9], a structural analogue of DOPA, and $[^{11}C]$nomifensine[6], a ligand for DA transporters existing in abundance at the terminal of DA neurons.

References

1. Björklund, A. and Lindvall, O. (1984). In *Handbook of chemical neuroanatomy*, Vol. 2: *Classical transmitters in the CNS*, Part I (ed. A. Björklund and T. Hökfelt) pp. 55–122. Elsevier, Amsterdam.
2. Hökfelt, T., Mårtensson, R., Björklund, A. *et al.* (1984). In *Handbook of chemical neuroanatomy*, Vol. 2: *Classical transmitters in the CNS*, Part I (ed. A. Björklund and T. Hökfelt) pp. 277–379. Elsevier, Amsterdam.
3. Cooper, J.R., Bloom, F.E., and Roth, R.H. (1991). In *The biochemical basis of neuropharmacology* (ed. J.R. Cooper *et al.*) pp. 285–337. Oxford University Press, New York.
4. Kiyama, H., Emson, P.C., and Ruth, J. (1990). *Eur. J. Neurosci.*, **2**, 512–24.
5. Kosaka, T., Hama, K., and Nagatsu, I. (1987). *Exp. Brain. Res.*, **68**, 393–405.
6. Kuhar, M.J., Sanchez-Roa, P.M., Wong, D.F. *et al.* (1990). *Eur. Neurol.*, **30**, 15–20.
7. Langston, J.W. (1989). *J. Neurol. Neurosurg. Psychiat., Suppl.*, 13–17.
8. Mackay, A.V.P., Iversen, L.L., Rossor, M. *et al.* (1982). *Arch. Gen. Psychiatr.*, **39**, 991–7.
9. Martin, W.R.W. (1990). In *Quantitative imaging: neuroreceptors, neurotransmitters, and enzymes* (ed. H.N. Wagner Jr and J.J. Frost) pp. 167–78. Raven Press, New York.
10. Meister, B., Hökfelt, T., Steinbush, H.W. *et al.* (1988). *J. Chem. Neuroanat.*, **1**, 59–64.

Tyrosine hydroxylase (immunocytochemistry)

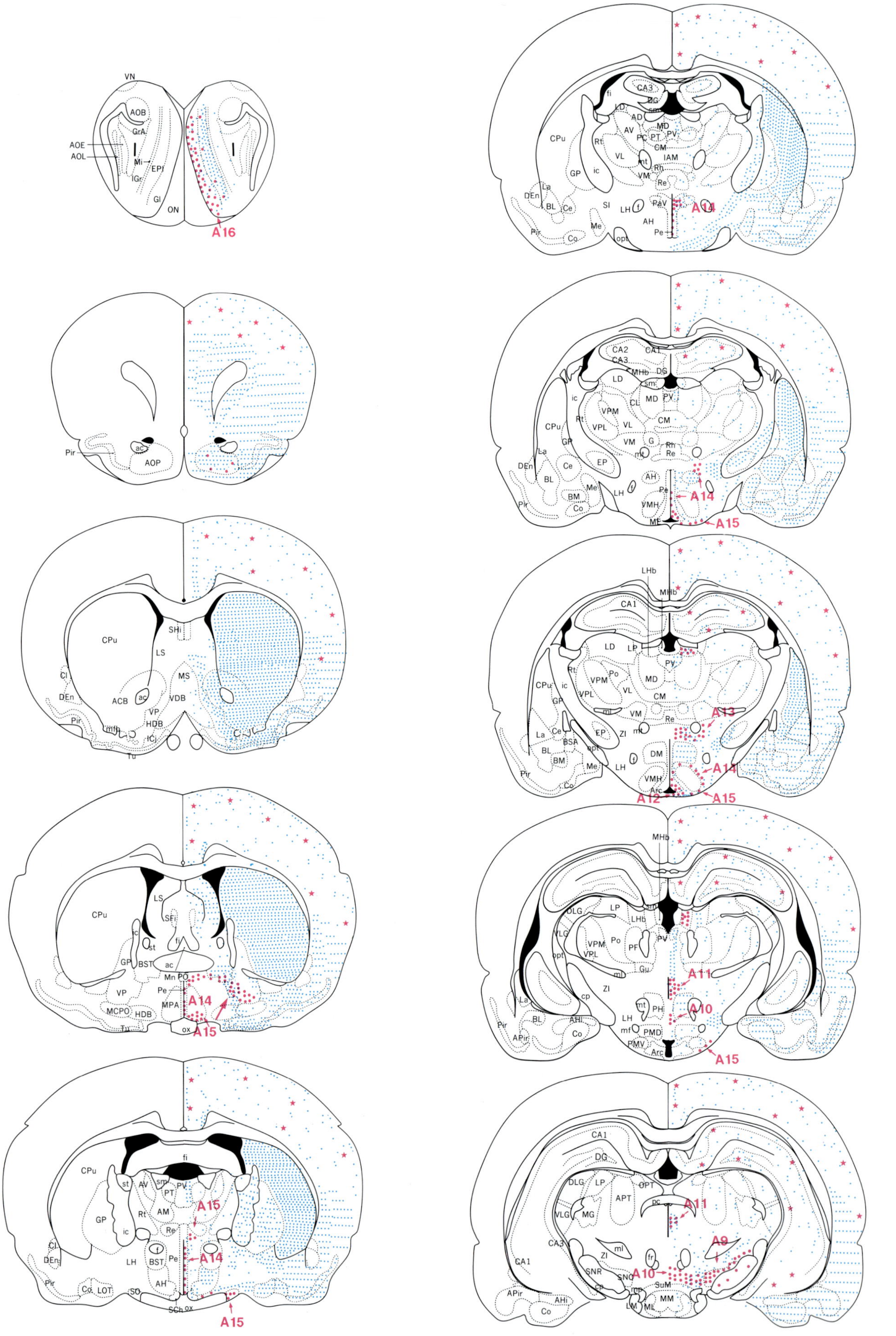

Tyrosine hydroxylase (immunocytochemistry)

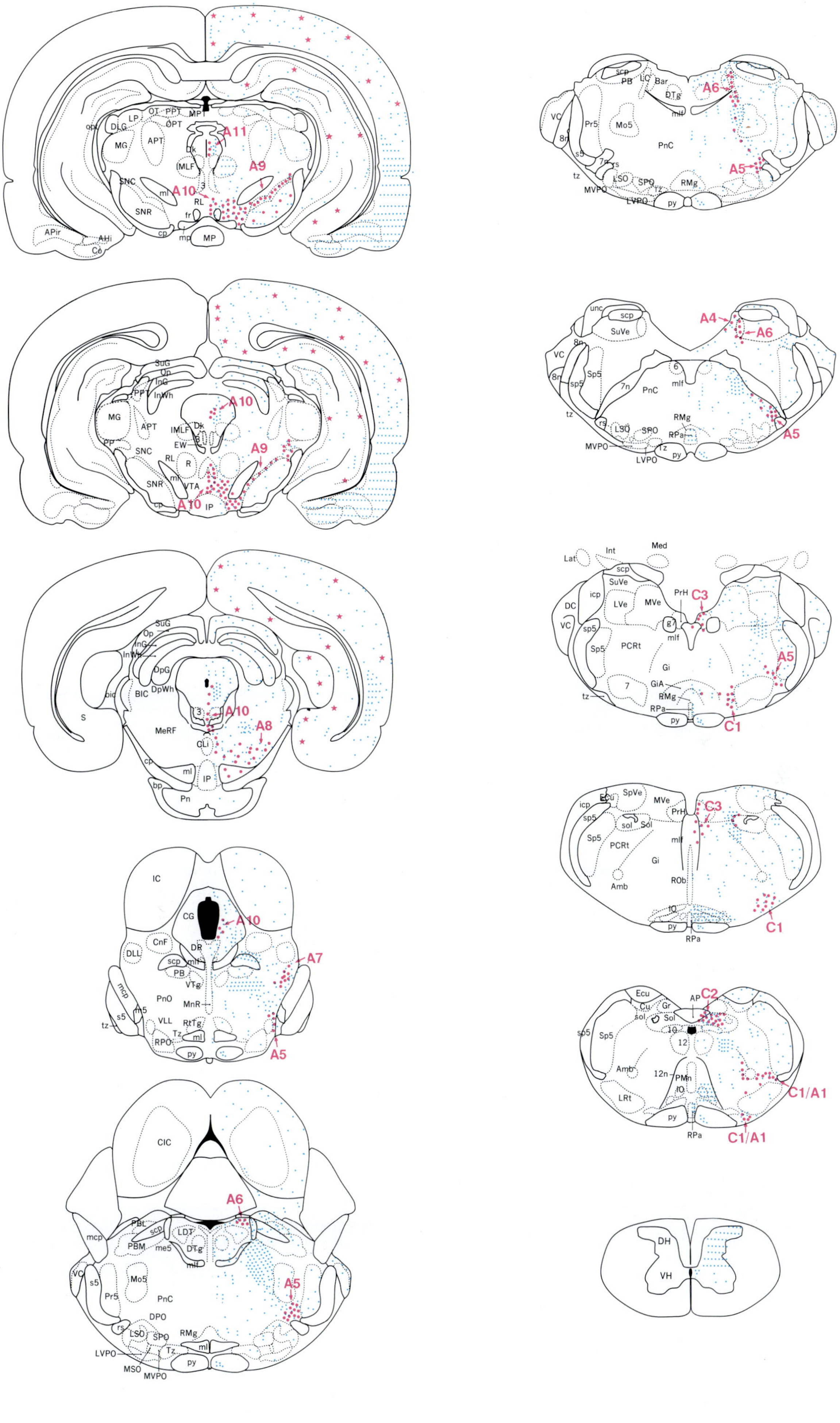

Dopamine receptors

Subtypes

Pharmacological (classical) classification

Dopamine receptors (DA-R) have been classified into those which are excitatorily coupled with adenylate cyclase (D_1 receptors) and those which are inhibitorily coupled with it (D_2 receptors)[6,14] (Fig. 1.3). Representative ligands in both groups are shown in Table 1.2. Radiolabelled ligands have been employed to determine the distribution of DA-Rs in the brain. The details presented in this atlas are based on the results obtained using [^{3}H]SKF83566[2], [^{125}I]SCH23982[8], and [^{3}H]SCH23390[23] for D_1 receptors, and those using [^{3}H]sulpiride[24] and [^{3}H]205-502[25] for D_2 receptors.

There are signal transmission pathways other than that mediated by adenylate cyclase. Both D_1 and D_2 receptors take part in the metabolism of inositol phosphate, and the latter plays a role in regulating Ca^{2+} and K^+ ion channels.

Molecular biological classification

To date, subtypes of DA-R as shown in Table 1.3 have been cloned[4,9,11,16,18,20,22,26]. They all have seven transmembrane portions, which is a characteristic structure of G-protein coupled receptors (Fig. 1.4). The subtypes D_{1A}, D_{1B} and D_5 in Table 1.3 correspond to the D_1 subtype, while the $D_{2A\text{-}L}$ (long), $D_{2A\text{-}S}$ (short), D_3 and D_4 correspond to the D_2 subtype in the pharmacological classification.

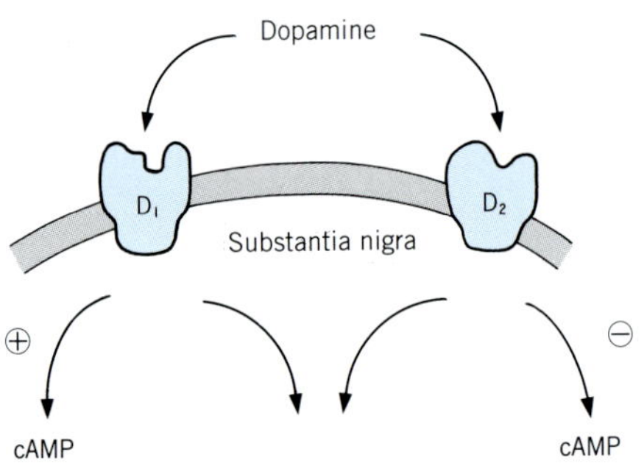

Fig. 1.3 Classification of dopamine receptors.

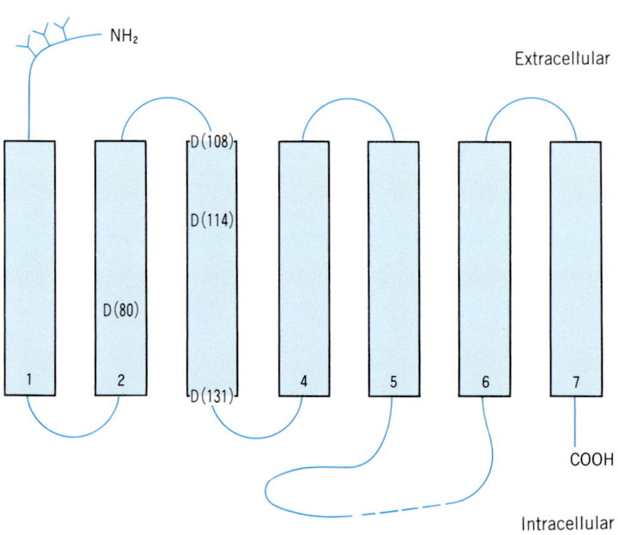

Fig. 1.4 Models for the structure of the D_2 dopamine receptor (ref. 17). The figure shows the approximate position of the conserved aspartic acid (D) residues which may be important for ligand binding.

The D_1 receptor was cloned by several groups[18,26] in 1990. Subsequently, a new DA-R with similar pharmacological property to D_1 receptors was cloned and named as D_{1B}[20]. Thus the previous clone D_1 was renamed D_{1A}. They have different distributions, as shown in Table 1.3. The detailed distribution of D_{1B} has not been worked out yet, so in this atlas only details of the distribution of D_{1A} are presented. The D_{2A} (long) and D_{2B} (short) are produced from the same gene by alternative splicing[11]. D_{2A} (long) differs from D_{2B} (short) only in possessing an insert sequence with 87 base pairs at the third intracellular loop, thus making it possible to have different coupling with the G proteins[21]. D_4 has high affinity for clozapine[22], while D_5 has higher affinity to dopamine as compared to D_{1A}[19]. Though D_{1B} exhibits 83 per cent homology in its sequence to D_5, it has much lower affinity to dopamine and its distribution in the brain is also different from that of the latter[20] (Table 1.3).

There have been detailed reports concerning the distribution of D_{1A} and D_2 receptors[10,13,25]. Information for distribution of D_{1A} in this atlas is based on Fremeau *et al.*[10] and Weiner *et al.*[25], and D_2 information is obtained from the reports of Mansour *et al.*[13] and Weiner *et al.*[25]. Whereas Walmsley *et al.*[24] employed an oliogonucleotide probe for the common sequence in D_{2A} (long) and D_{2B} (short), Mansour *et al.*[13] utilized a cRNA probe for D_{2B} (short). The data from the latter report have been incorporated frequently in this book. There is also a report on the localization of mRNA for D_3 receptor. These show that the receptors are localized more restrictively (mainly in the ventral striate body) than those of D_1 and D_2[1].

Autoreceptors

The autoreceptors exist in the cell body, dendrites, and terminals of DA cells, inhibiting the synthesis and release of DA and also excitation of DA cells. The D_2 receptors are regarded as functioning as presynaptic receptors as well as postsynaptic autoreceptors, for the following reasons[13]. First, the D_2 receptor mRNA is detected in substantia nigra and ventral tegmentum where DA cells are localized. Secondly, agonists to D_2 receptor augment their function as autoreceptors. It has also been shown that some D_3 receptors exist as autoreceptors[16]. In contrast, D_1 receptors are not thought to function as autoreceptors[3].

Table 1.2. Ligands for dopamine receptors

Receptors	D_1	D_2
Selective agonists	SKF 38393 Fenoldopam	N-0434 Bromocriptine
Selective antagonists	SCH-23390 SCH39166	1BZM YM-O9151-2 Domperidone
Radiolabelled ligands	[^{3}H]SCH 23390 [^{125}I]SCH23982 [^{3}H]SKF83566	[^{3}H]YMO91512 [^{3}H]sulpiride [^{3}H]N-0437

Table 1.3. Properties of cloned dopamine receptor subtypes

	D_1	D_5	D_{2S}/D_{2L}	D_3	D_4
Alternative nomenclature	D_{1A}	D_{1B}	D_{2A-S}/D_{2A-L}	D_{2B}	D_{2C}
Amino acids					
Human	446	477	414/443	400	387
Rat	446	475	415/444	446	–
Introns in gene	No	No	Yes	Yes	Yes
Human chromosome	5	4	11	3	11
Effector pathways	↑ cAMP	↑ cAMP	↓ cAMP ↑ K^+ channel ↓ Ca^{2+} channel		↓ cAMP
mRNA distribution in brain	Caudate-putamen Nucleus accumbens Olfactory tubercle	Hippocampus Hypothalamus	Caudate-putamen Nucleus accumbens Olfactory tubercle	Olfactory tubercle Hypothalamus Nucleus accumbens	Frontal cortex Medulla Midbrain

Relationship to diseases

It has been reported that in brains of patients with schizophrenia examined post-mortem, the number of D_2 receptors is increased, which is suggestive of a causal relationship[12]. In living patients, the state of D_2 receptors has been examined with a variety of ligands using PET, but as of now definite conclusions are unavailable[23].

Dopamine transporter

Dopamine transporter was cloned in 1991 by Shimada *et al.*[15]. This transporter is expressed specifically in the mesostriatal DA system and the mesolimbic system DA system and not in others. It terminates neurotransmission by the reuptake of DA into the presynaptic terminals in the striatum and nucleus accumbens. It has 12 transmembrane sites. It is also known that cocaine inhibits dopamine transport.

References

1. Bouthenet, M.-L., Souil, E., Martres, M.-P. *et al.* (1991). *Brain Res.* **564**, 203–19.
2. Boyson, S.J., McGonigle, P., and Molinoff, P.B. (1986). *J. Neurosci.*, **6**, 3177–88.
3. Brown, F., Campbell, W., Mitchell, P., and Randall, K. (1985). *Br. J. Pharmacol.*, **84**, 853–60.
4. Bunzow, J.R., Van Tol, H.H.M., Grandy, D.K. *et al.* (1988) *Nature*, **336**, 783–7.
5. Charuchinda, C., Supavilai, P., Karobath, M., and Palacios, J.M. (1987). *J. Neurosci.*, **7**, 1352–60.
6. Cooper, D.M., Bier-Laning, C.M., Halford, M.K. *et al.* (1986). *Mol. Pharmacol.*, **29**, 113–19.
7. Cooper, J.R., Bloom, F.E., and Roth, R.H. (1991). In *The biochemical basis of neuropharmacology* (ed. J.R. Cooper, *et al.*) pp. 285–337. Oxford University Press, New York.
8. Dawson, T.M., Barone, P., Sidhu, A. *et al.* (1988). *Neuroscience*, **26**, 83–100.
9. Dearry, A., Gingrich, J.A., Falardeau, P. *et al.* (1990). *Nature*, **347**, 72–5.
10. Fremeau, R.T., Jr., Duncan, G.E., Fornaretto, M.-G. *et al.* (1991). *Proc. Natl. Acad. Sci. USA* **88**, 3772–6.
11. Giros, B., Sokoloff, P., Martres, M.-P. *et al.* (1989). *Nature*, **342**, 923–6.
12. Mackay, A.V.P., Iversen, L.L., Rossor, M. *et al.* (1982) *Arch. Gen. Psychiatr.* **39**, 991–7.
13. Mansour, A., Meador-Woodruf, J.H., Bunzow, J.R. *et al.* (1990) *J. Neurosci*, **10**, 2587–2560.
14. Onali, P., Olianas, M.C. and Gessa, G.L. (1985). *Mol. Pharmacol.*, **28**, 138–45.
15. Shimada, S., Kitayama, S., Lin, C.-L. *et al.* (1991). *Science*, **254**, 576–8.
16. Sokoloff, P., Giros, B., Martres, M.-P. *et al.* (1998). *Nature*, **347**, 146–51.
17. Strange, P.G. (1990). *Trends Neurosci.* **13**, 373–8.
18. Sunahara, R.K., Niznik, H.B., Weiner, D.M. *et al.* (1990). *Nature*, **347**, 80–3.
19. Sunahara, R.K., Guan, H.C., O'Bowd, B.F. (1991). *Nature* **350**, 614–19.
20. Tieberi,M., Jarvie, K.R., Silvia, C. *et al.* (1991). *Proc. Natl. Acad. Sci. USA*, **88**, 7491–5.
21. Toso, R.D., Sommer, B., Ewert, M. *et al.* (1989). *J. Neurosci.*, **13**, 4025–34.
22. Van Tol, H.H.M., Bunzow, J.R., Guan, H.-C. (1991) *Nature* **350**, 610–14.
23. Wagner, H.N., Jr. and Wong, D.F. (1990) In *Quantitative imaging: Neuroreceptors, neurotransmitters, and enzymes* (ed. H.N. Wagner, Jr and J.J. Frost) pp. 97–108. Raven Press, New York.
24. Wamsley, J.K., Gehlert, D.R., Filloux, F.M., and Dawson, T.M. (1989). *J. Chem. Neuroanat.*, **2**, 119–37.
25. Weiner, D.V., Levey, A.I., Sunahara, R.K. *et al.* (1991). *Proc. Natl. Acad. Sci. USA*, **88**, 1859–63.
26. Zhou, Q.-Y., Grandy, D.K., Thambi, L. *et al.* (1990). *Nature*, **347**, 76–9.

Dopamine receptor (binding sites)

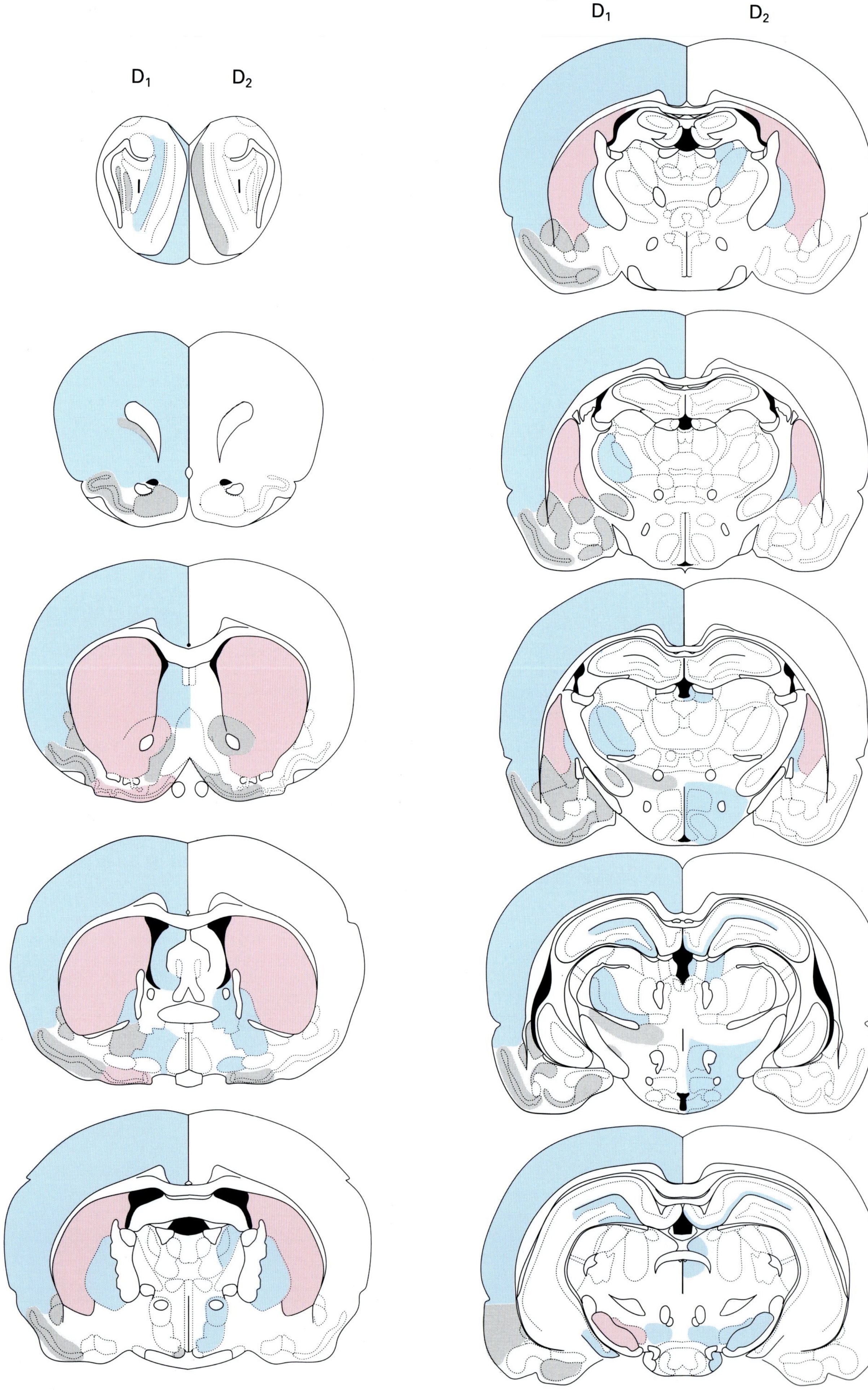

Dopamine receptor (ISH)

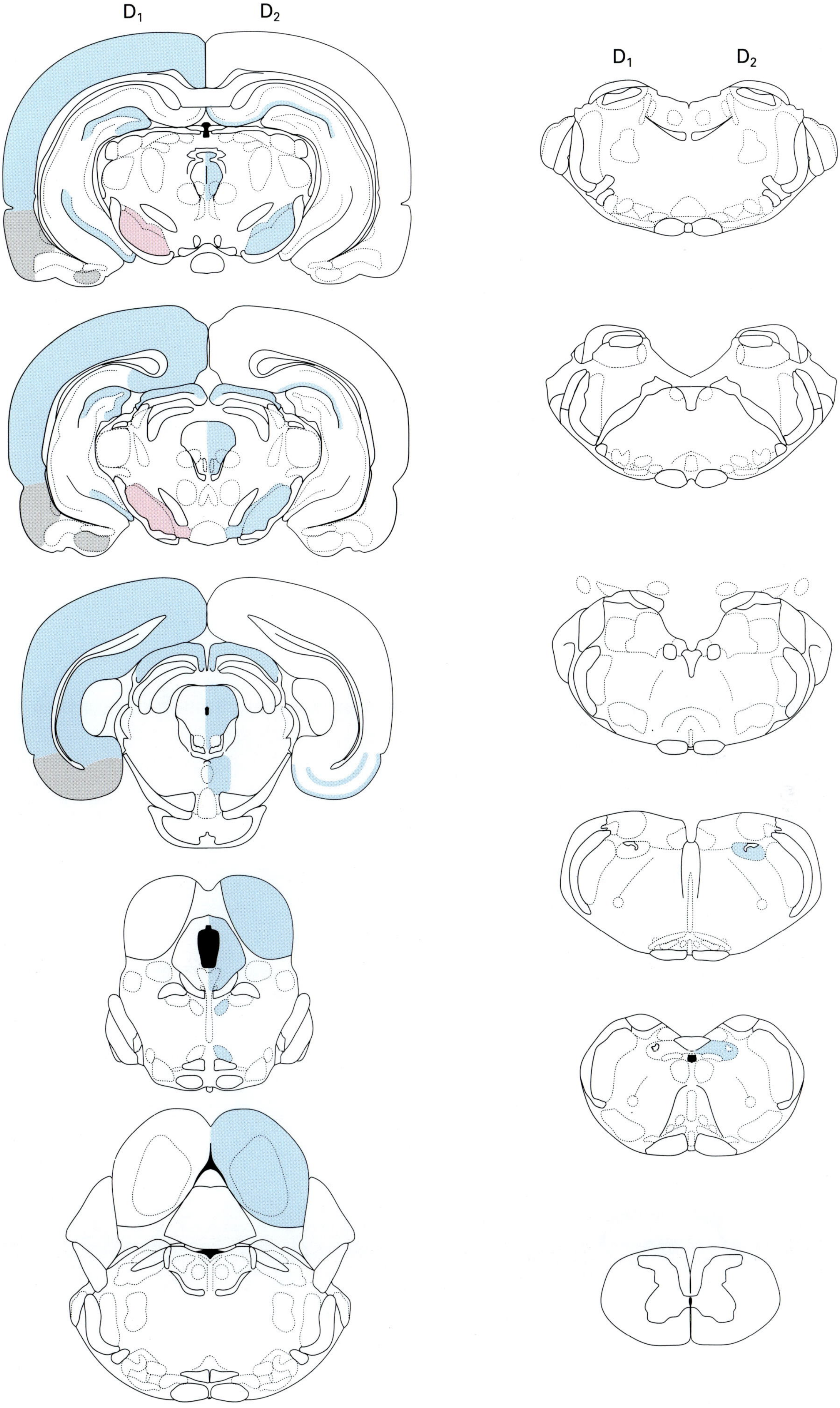

Dopamine receptor (ISH)

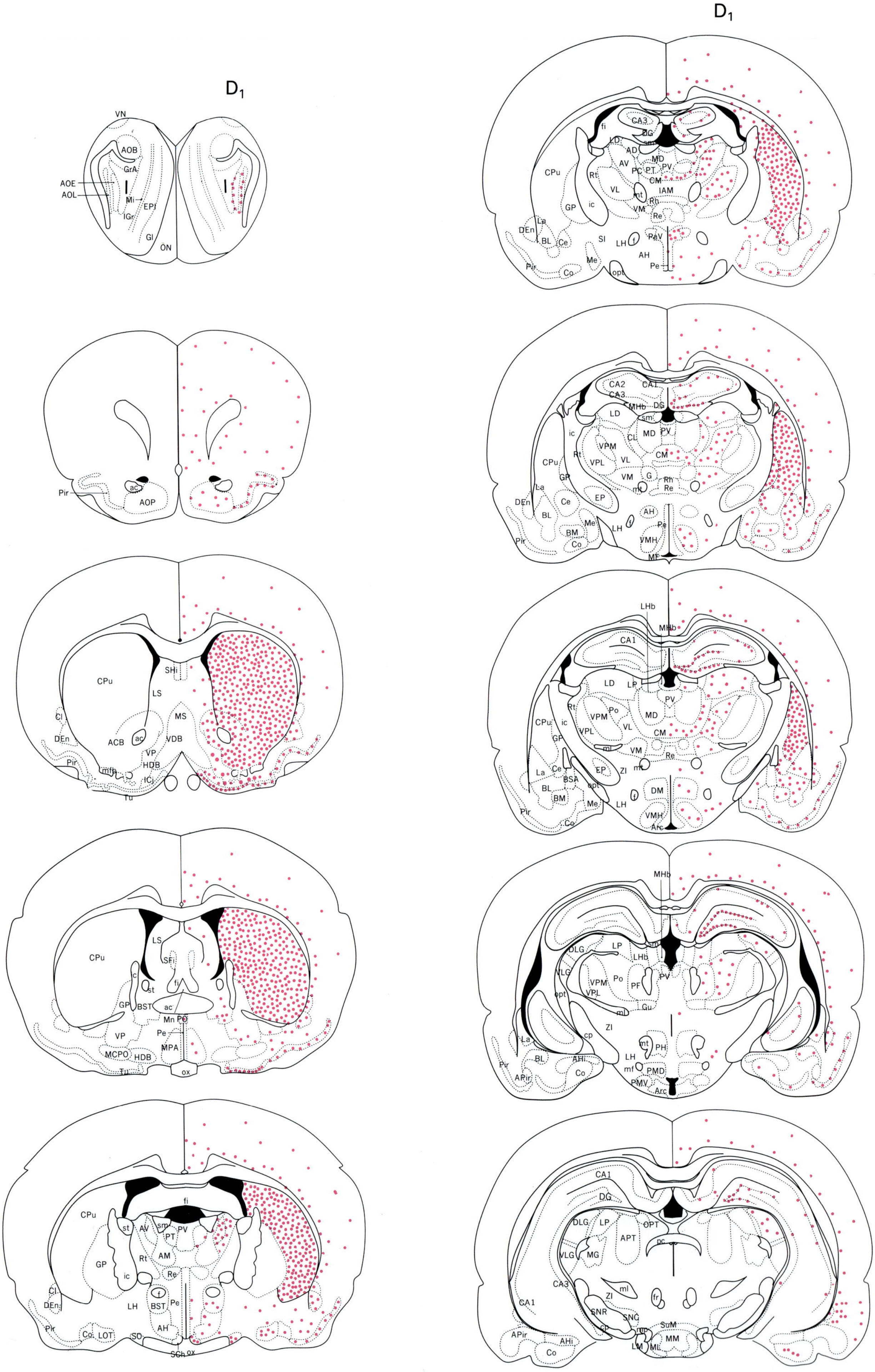

Dopamine receptor (binding sites)

D$_1$

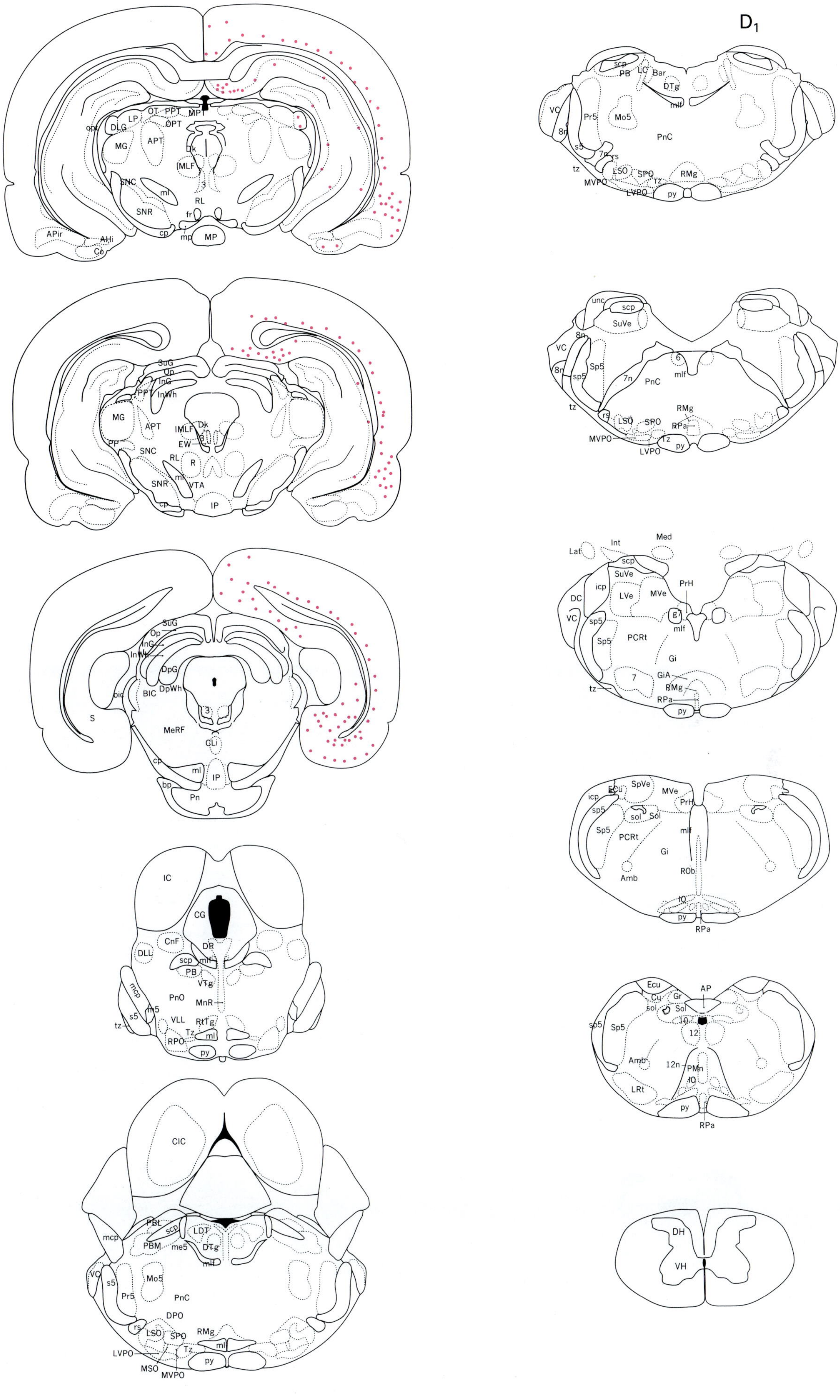

Dopamine receptor (ISH)

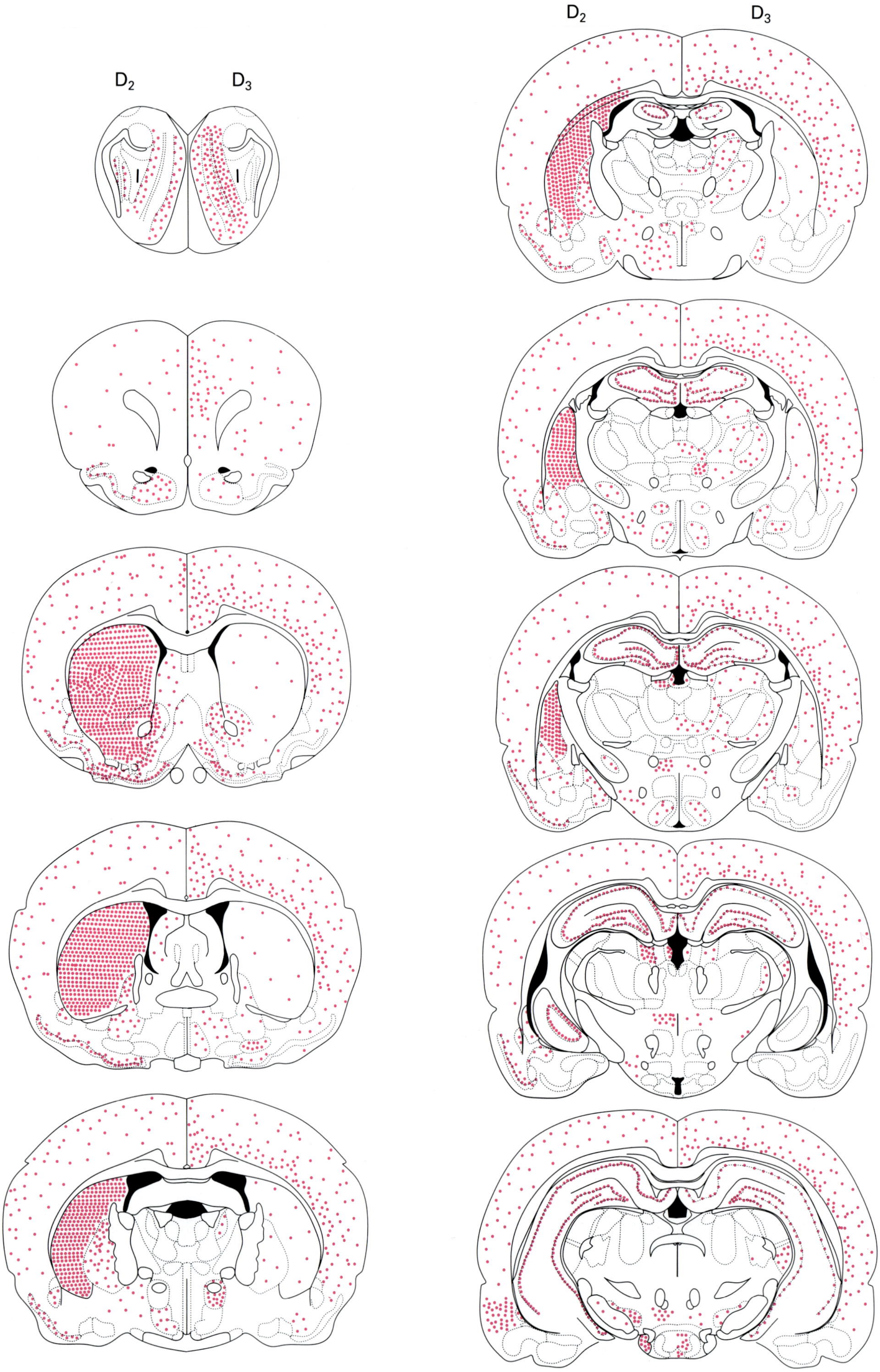

Dopamine receptor (ISH)

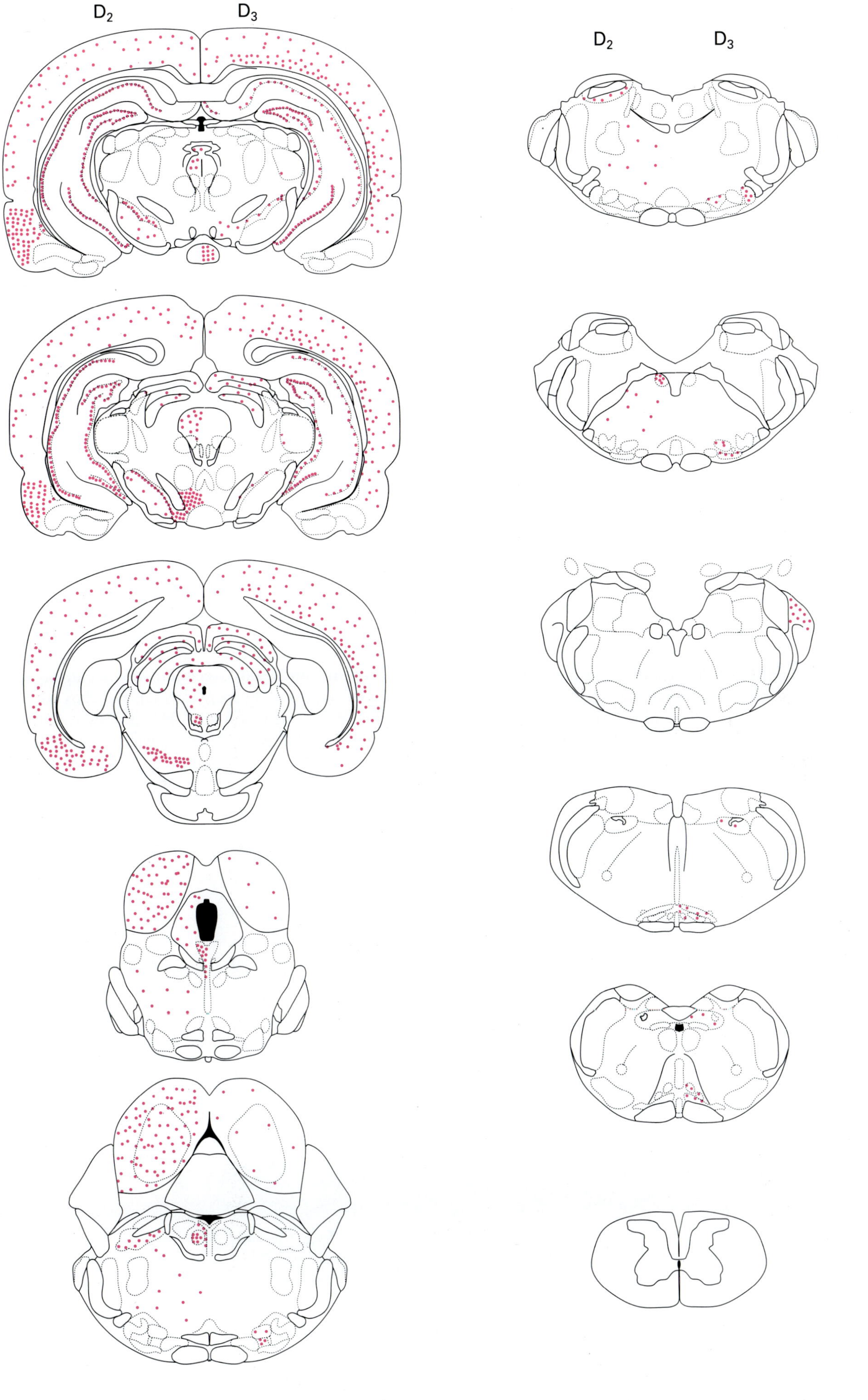

Noradrenergic–adrenergic systems

Biosynthesis[1]

Noradrenaline and adrenaline, together with dopamine, belong to the group of catecholamines, which possess two hydroxyl groups and one amine group bound to a benzene ring, as shown in Fig. 1.1. They are biosynthesized from tyrosine by the actions of several enzymes in the central nervous system, sympathetic nervous system and adrenal medulla. Tyrosine is first converted to DOPA by TH, which is the rate-limiting enzyme in this pathway. DOPA is then converted into dopamine by aromatic amino acid decarboxylase. In the cells containing noradrenaline, dopamine is converted to noradrenaline by dopamine β-hydroxylase (DBH), while in those containing adrenaline, the latter is *N*-methylated to adrenaline by phenylethanolamine-*N*-methyl transferase (PNMT).

The enzyme aromatic amino acid decarboxylase can act on several substrates other than DOPA or 5-HT. These include histidine, tyrosine, tryptophan, and phenylalanine. Its distribution is also not restricted to catecholamine-containing cells[1]. However, TH, DBH, and PNMT are restrictedly distributed in catecholamine-producing cells, thus making them usable as markers for dopamine, noradrenaline, and adrenaline.

Distribution and projection pathway of noradrenaline

Noradrenergic cells are classified as A1–A7. The A1 noradrenergic cell group is localized in the most posterior part of the noradrenaline-containing cells in the brain. At the level of the area postrema, there is the A1 noradrenergic cell group extending from the C1 adrenergic cell group.

The A2 noradrenergic cell group is distributed throughout the dorsal vagal complex. They also exist at the area postrema. Some of the cells in this group do not contain aromatic amino acid dehydroxylase and DBH, though they do contain TH, suggesting the existence of cells possessing dopamine or DOPA as neurotransmitters.

The A3 noradrenergic cell group is scattered in the medullary reticular formation and extends to the A5 and A7 noradrenergic cell groups.

The A4 noradrenergic cell group forms a dorsolateral continuation of the A6 group and surrounds the fourth ventricle dorsolaterally.

The A5 noradrenergic cell group is distributed in the ventrolateral part of the pons, with its most posterior part extending to the C1 adrenergic cell group.

The A6 noradrenergic cell group is mainly composed of the cells in the locus ceruleus. The term locus ceruleus was given to that region in human and other primate brains because of the pigments produced there. This nucleus has the richest density of adrenaline-producing cells in the brain; it is estimated that there are 1500 cells in rat and 12 000 in human brain hemispheres.

The A7 noradrenergic cell groups are the most anteriorly located among this type of cells. They are distributed at the lateral part of the pontine reticular formation, being very close to the lateral lemniscus.

The main noradrenergic projection pathways in the central nervous system are the dorsal and ventral bundle and the periventricular bundle (Fig. 1.5)[2].

The dorsal bundle is composed of the ascending fibres from the locus ceruleus. It begins from the nucleus, passing through the ventrolateral part of the central grey matter of the midbrain, ascending in the medial forebrain bundle, distributing many lateral branches to the amygdaloid complex, septal area, hippocampus, thalamus, and cerebral cortex. Also, there are two types of fibres ascend-

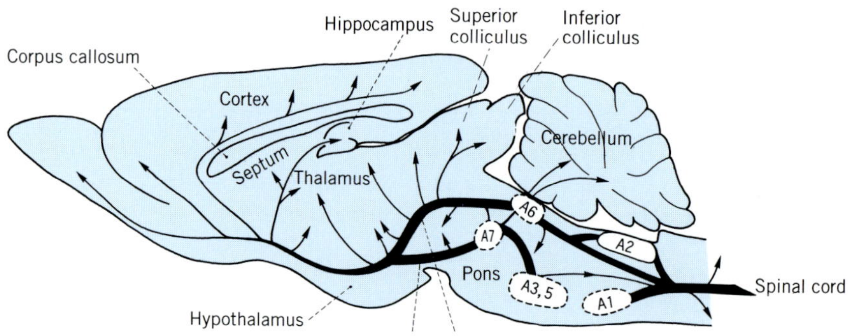

Fig. 1.5 Central neuronal pathways containing noradrenaline.

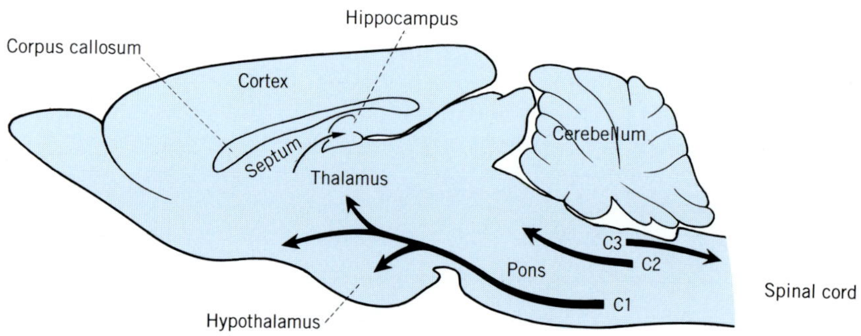

Fig. 1.6 Central pathways containing adrenaline.

ing in the medial forebrain bundle; one which terminates at the hypothalamus and olfactory bulb, and the other which runs through the septal area and terminates in the cerebral cortex. The ventral bundle is mainly composed of ascending fibres originating from the cell groups other than locus ceruleus. It ascends alongside the central tegmental tract, reaching into the medial forebrain bundle. It then terminates in the hypothalamus, amygdaloid complex, pyriform cortex, basal nucleus, and diagonal band. The periventricular bundle is composed of ascending fibres from the locus ceruleus and A2 to periventricular grey matter. Besides this ascending projection, there are descending bundles and descending fibres of A1, A2, A3, A5, A7 (and ventral A6 to the spinal cord) and the projection pathway to the cerebellum (the fibres originating from the A4, A6 noradrenergic cell groups to the cerebellar cortex, passing the uncinate fasciculus).

Distribution of adrenergic cells and their projection pathway

The adrenergic cells are classified as C1–C3. The C1 adrenergic cell group is located at the lateral part of the tegmentum, having a border with the A5 noradrenergic cell groups at the anterior part, and with the A1 noradrenergic cell groups at the posterior part. There are also adrenergic cells scattered between the C1 and C2 cell groups. The C2 adrenergic cell group is distributed in the medial aspects of solitary nucleus at the rostral part, and then in the medial border area of the dorsal vagal motor nucleus and also in the periventricular areas at the caudal part. The parvocellular adrenaline cell group, about 10 γm in diameter, appears in the dorsal area of the solitary nucleus. The C3 adrenergic cells are distributed in the dorsal midline area of the medial longitudinal fasciculus, reaching at its most dorsal part to the area surrounding the fourth ventricle.

The adrenergic cells have both ascending and descending projection pathways (Fig. 1.6), innervating the hypothalamus via the central tegmental tract, the periaqueductal central grey matter, the locus ceruleus, and the spinal cord.

Coexistence of noradrenaline/adrenaline and peptides within single cells

Catecholamines coexist with peptides within the same cells in both the peripheral and central nervous system, which enables transmission of complex information between neurons. Noradrenaline coexists with neuropeptide Y (NPY) in the sympathetic ganglion, A1–A3, and A6. Also in A6, noradrenaline coexists with enkephalin in the same cells[2]. Adrenaline coexists with NPY and neurotrensin in the same cells in C1–C3, and with cholecystokinin in C2[2,3].

References

1. Cooper, J.R., Bloom, F.E., and Robert, H.R. (1992). *The biochemical basis of neuropharmacology*. Oxford University Press.
2. Hökfelt, T., Johansson, O., and Robert, H.R. (1984). In *Handbook of chemical neuroanatomy,* Vol. 2 (ed. A. Björklund and T. Hökfelt) p. 157. Elsevier, Amsterdam.
3. Kawai, Y., Takagi, H., and Tohyama, M. (1988). *Neuroscience*, **24**, 227–36.
4. Moore, R.T., and Card, J.P. (1984). In *Handbook of chemical neuroanatomy*, Vol. 2 (ed. A. Björklund and T. Hökfelt) pp. 123–56. Elsevier, Amsterdam.

Noradrenaline–adrenaline (immunocytochemistry)

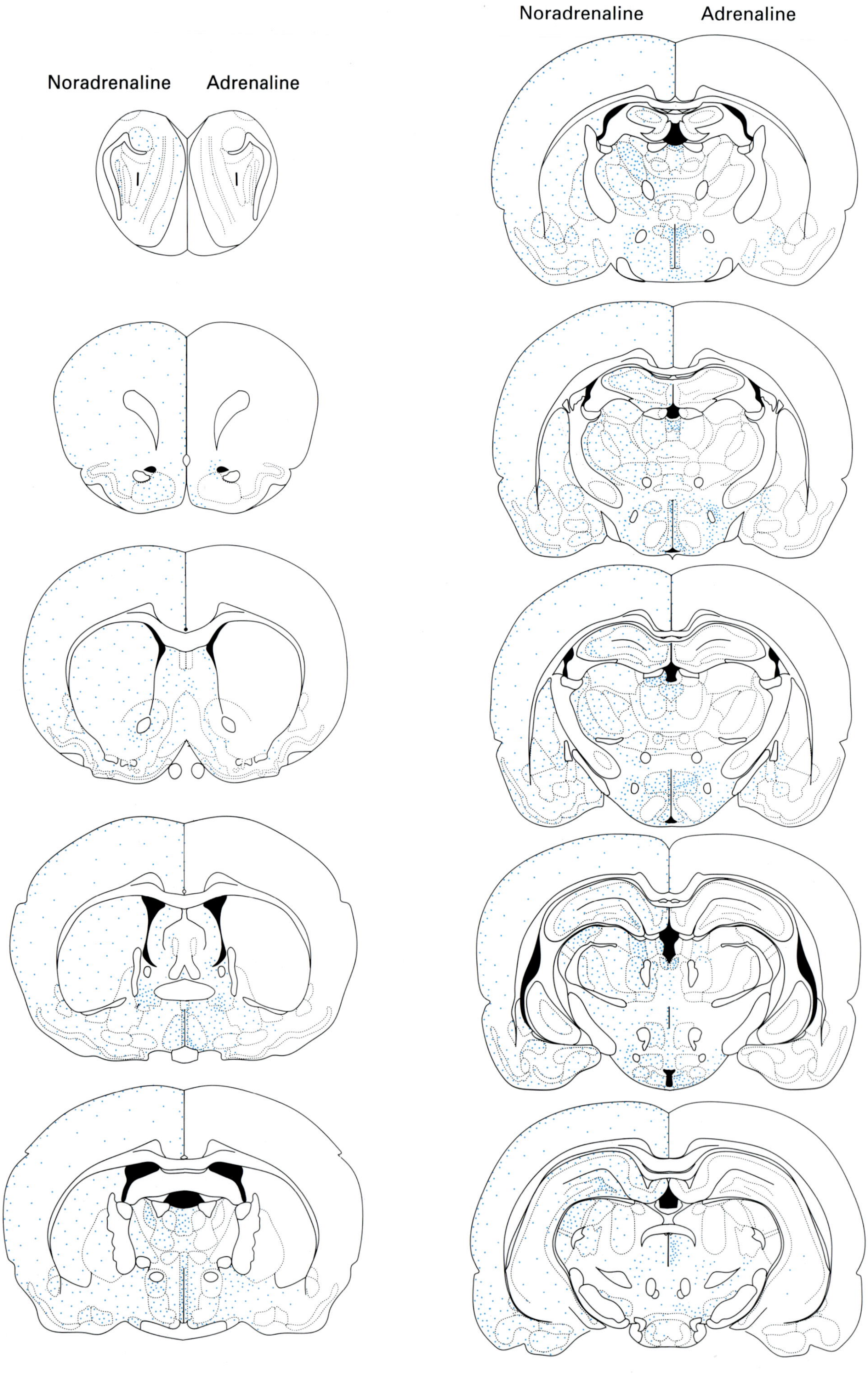

Noradrenaline–adrenaline (immunocytochemistry)

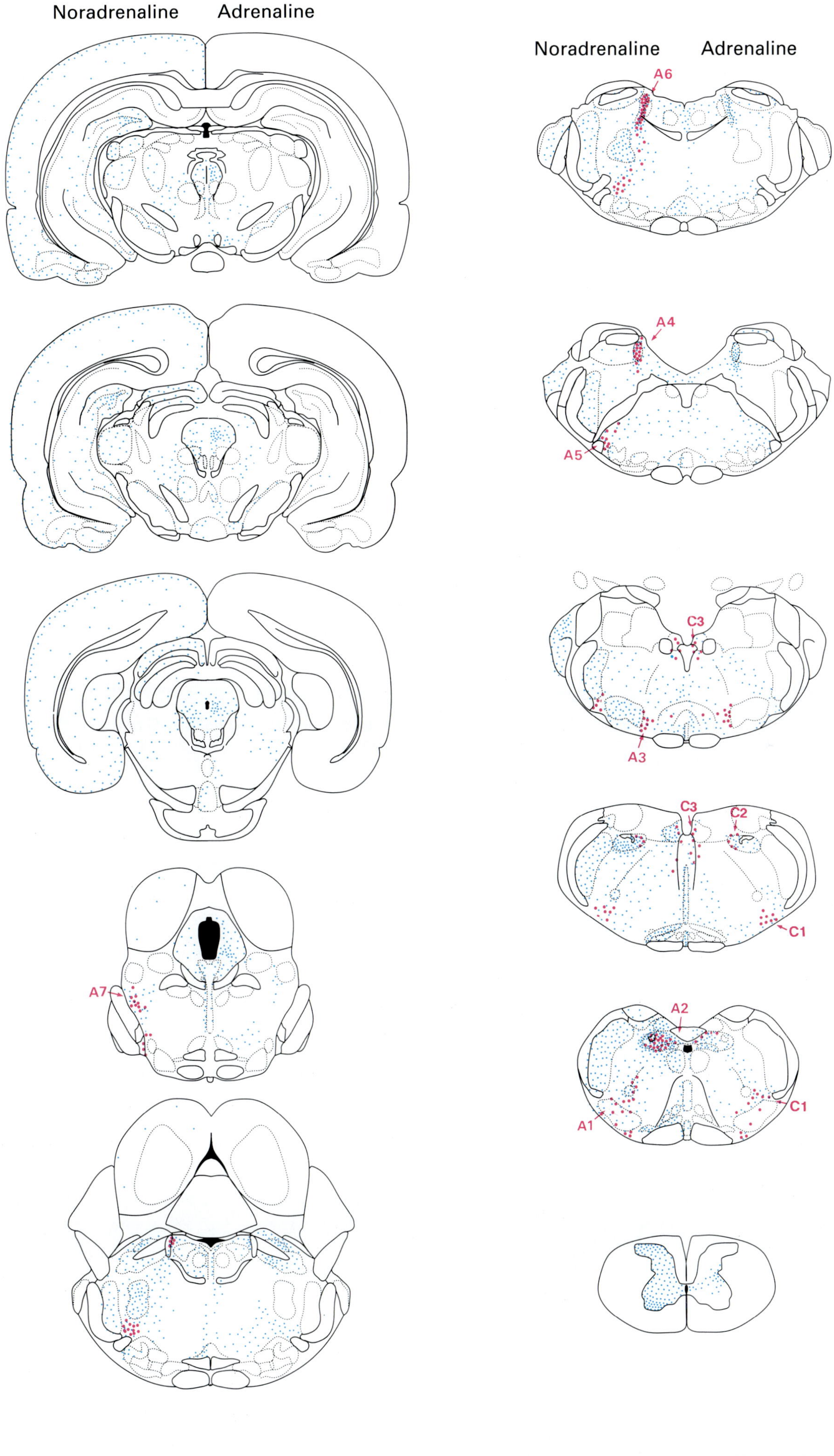

Receptors for noradrenaline and adrenaline

Subtypes

Ahlquist[1] first classified the receptors for catecholamine into α and β types, based upon their inhibitory and excitatory pharmacological effects on smooth muscles. The receptors for adrenaline were subsequently subclassified into α_1, α_2 and β_1, β_2 types, following the development of various synthetic ligands and examination of their effects[2]. As gene families for these receptors have been cloned in recent years, the classification of receptors has become much complicated. The recent classification shown in Table 1.4 lists subtypes of α_{1A}, α_{1B}, α_{1D}, α_{2A}, α_{2B}, α_{2C}, β_1, β_2 and β_3[7]. All cloned adrenergic receptors are G-protein coupled receptors and have seven putative transmembrane domains and have *N*-glycosylation sites at their N-terminal end. The α_1-adrenergic receptors activate phospholipase C. The α_2-adrenergic receptors decrease adenyl cyclase activity. The β-adrenergic receptors activate adenyl cyclase[7]. Some of these receptors are able to modify the ion channels via the G protein.

Based on the results of structure–function analysis on point mutation and deletion experiments of amino acid residues, the ligand binding sites in the β-adrenergic

Table 1.4a. α_1-Adrenoceptors (from refs 7,8,9)

Nomenclature	α_{1A}	α_{1B}	α_{1D}
Selective antagonists	WB4101	—	WB4101
Effector Syatem	IP_3/DG	IP_3/DG	IP_3/DG
Distribution	cerebellum cerebral cortex kidney submandibular gland	cerebral cortex thalamus pineal body liver spleen	cerebral cortex hippocampus aorta

Table 1.4B. α_2-Adrenoceptors (from refs 7,8)

Nomenclature	α_{2A}	α_{2B}	α_{2C}
Selective agonists	Oxymetazoline (partial agonist)	—	—
Selective antagonists	—	Prazosin ARC239	Prazosin ARC239
Effector Syatem	cAMP ↓ K^+ channel ↑ Ca^{2+} channel ↓	cAMP ↓ Ca^{2+} channel ↓ (These channels are regulated by G protein)	cAMP ↓
Distribution	cerebral cortex hippocampus Amygdala locus ceruleus spinal cord	thalamus cerebellum liver kidney	amyglada hippocampus olfactory tubercle

Table 1.4C. β-Adrenoceptors (from refs 7,8)

Nomenclature	β_1	β_2	β_3
Potency order	NA ⩾ adrenaline	Adrenaline > NA	NA > adrenaline
Selective agonists	Xamoterol	Procterol	BRL3733
Selective antagonists	CGP 20712 Betaxolol	ICI 118551	—
Effector Syatem	cAMP ↑	cAMP ↑	cAMP ↑
Distribution	cerebral cortex hippocampus pineal body heart	cerebral cortex lung uterus prostate	adipose tissue

receptors have been identified as Asp[113] at the third, Ser[204] and Ser[207] at the fifth and Phe[290] at the sixth transmembrane domain (Fig. 1.7)[4,6]. The receptor has a site necessary for coupling with the G protein at the third intracellular loop[4,6].

Near the C-terminal end of the receptor is located a site which is phosphorylated by both β-adrenoceptor kinase, and protein kinase A. Phosphorylation of this site results in desensitization[3].

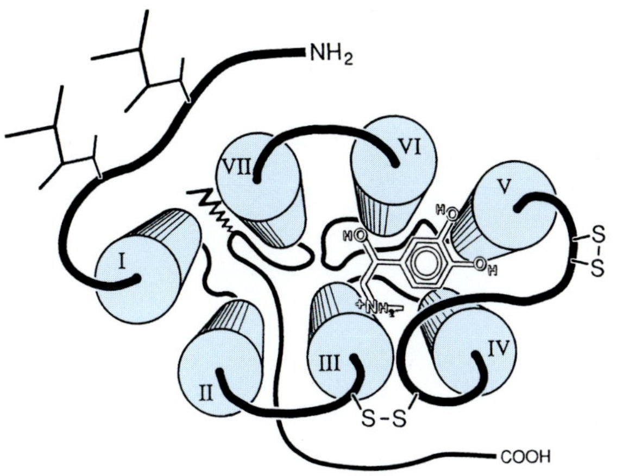

Fig. 1.7 A proposed model of the β_2-adrenergic receptor and the binding sites for epinephrine (from ref. 4).

Transporters for noradrenaline

Transporters play significant roles in synaptic transmission. They are membrane proteins which re-uptake transmitters back to the neuron, thus terminating the neurotransmission. Each transmitter has its corresponding transporter as well as receptors. The cDNA for the noradrenaline transporter was recently cloned and was shown to have 12 putative transmembrane domains (Fig. 1.8)[5]. Despramine, a tricyclic antidepressant, directly binds to the protein, inhibiting re-uptake of noradrenaline into the neurons.

References

1. Ahlquist, R.P. (1948). *Am. J. Physiol.*, **153**, 580–600.
2. Cooper, J.R., Bloom, F.E., and Roth, R.E. (1991). *The biochemical basis of neuropharmacology*. Oxford University Press.
3. Hausdorff, W.P., Caron, M.G., and Lefkowitz, R.J. (1990). *FASEB J.*, **4**, 2881–9.
4. Ostrowski, J., Kjelsberg, M.A., Caron, M.G., and Lefkowitz, R.J. (1992). *Annu. Rev. Pharmacol Toxicol.*, **32**, 167–83.
5. Pacholczyk, T., Blakely, R., and Amara, S.G. (1991). *Nature*, **350**, 350–1.
6. Strader, C.D., Sigal, I.S., and Dixon, R.A. (1989). *FASEB J.*, **3**, 1825–32.
7. Watson, S. and Abbott, A. (1991). *Trends Pharmacol. Sci.*, **12**, 4–6.
8. Bylund, D.B., Eikenberg, D.C., Hieble, J.P., Langer, S.Z., Lefkowitz, R.J., Minneman, K.P., Molinoff, P.B., Ruffolo, R.R. Jr., and Trendelenburg, U. (1994). *Pharmacol. Rev.*, **46**, 121–36.
9. Hieble, J.P., Bylund, D.B., Clarke, D.E., Eikenburg, D.C., Langer, S.Z., Lefkowitz, R.J., Minneman, K.P., and Ruffolo, R.R. Jr. (1995). *Pharmacol. Rev.*, **47**, 267–70.

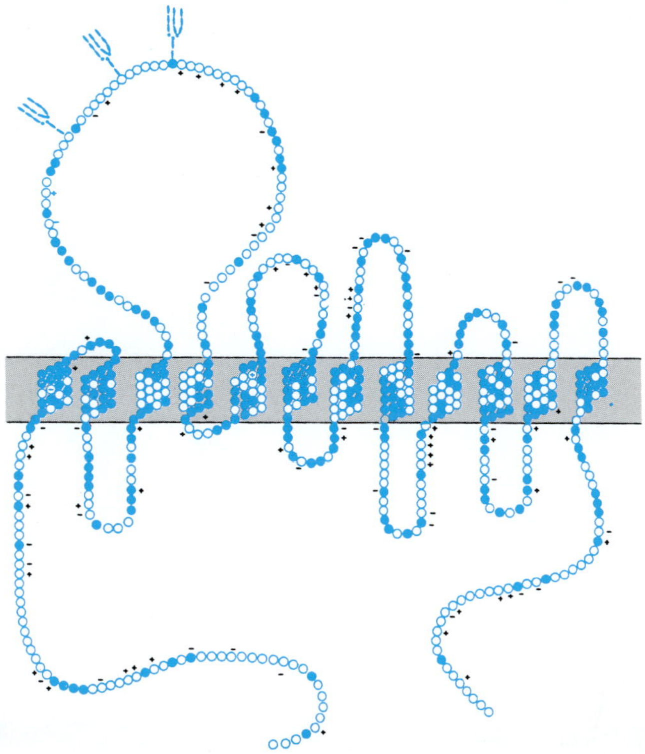

Fig. 1.8 The norepinephrine transporter: twelve putative transmembrane domains and a large extracellular loop with three glycosylation sites. Darker circles represent residues conserved in the GABA transporter (from ref. 5, Copyright 1991 Macmillan Magazines Ltd.).

β-Adrenoceptors (binding sites)

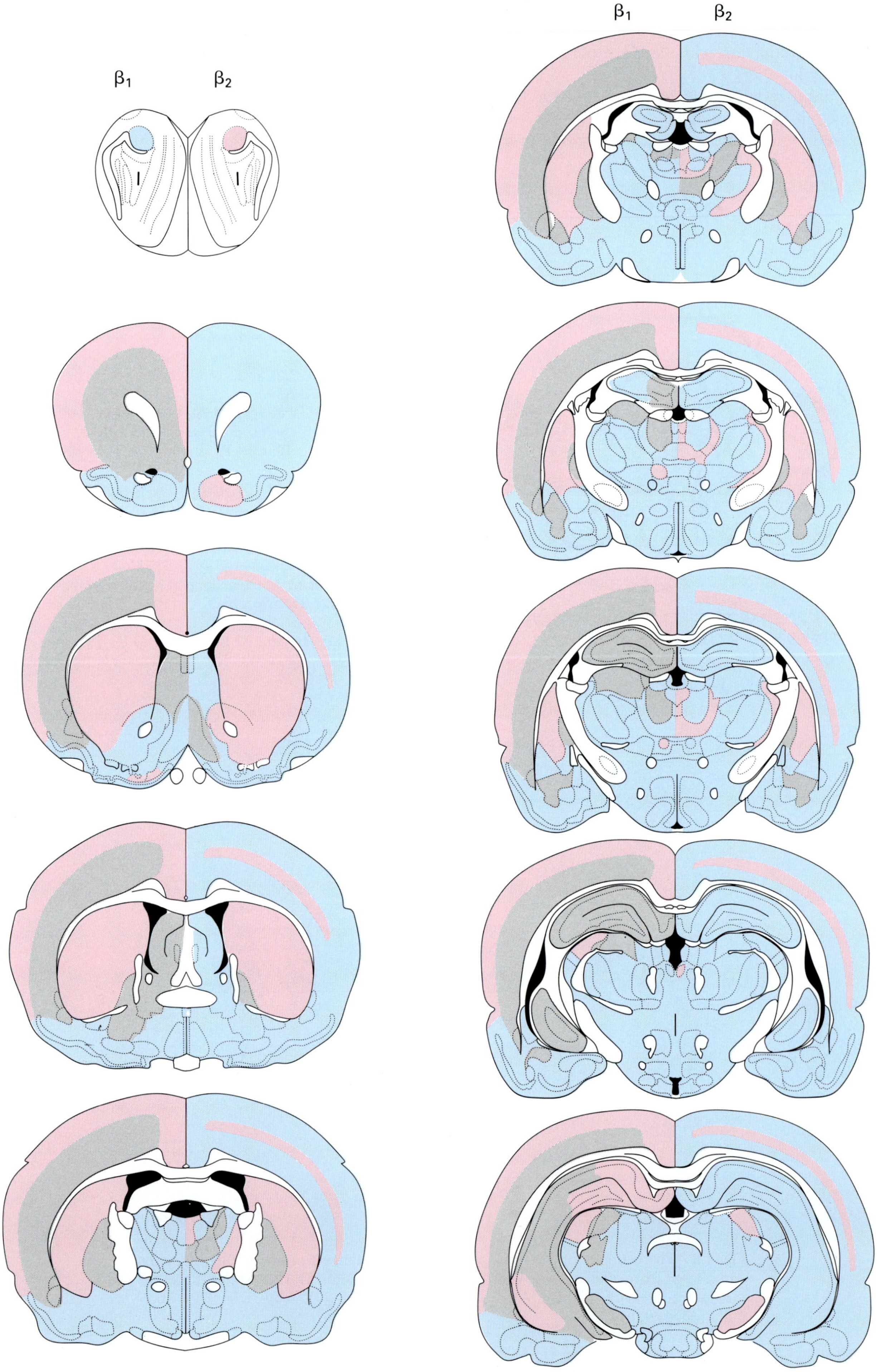

β-Adrenoceptors (binding sites)

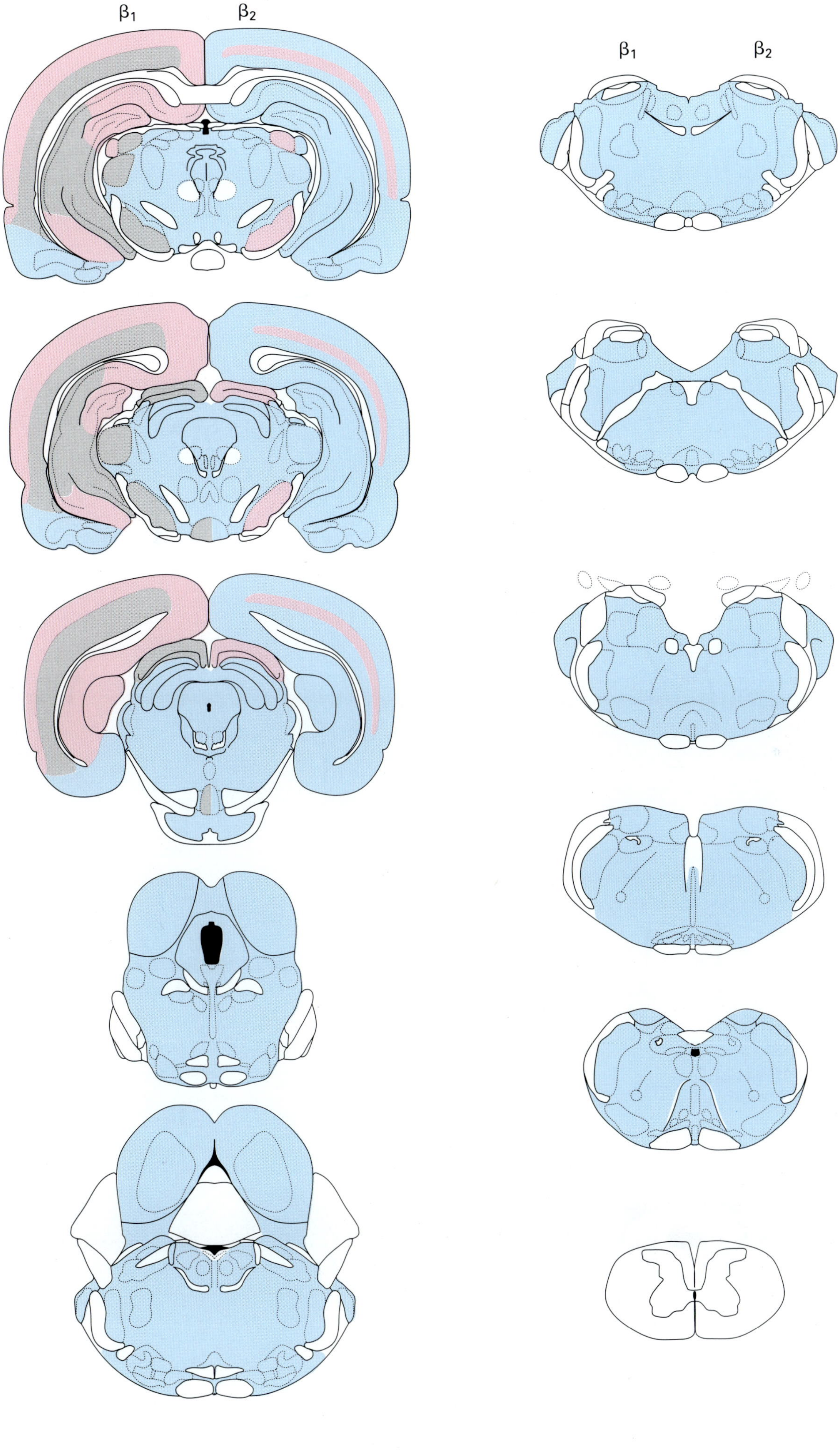

β-Adrenoceptor (immunocytochemistry)

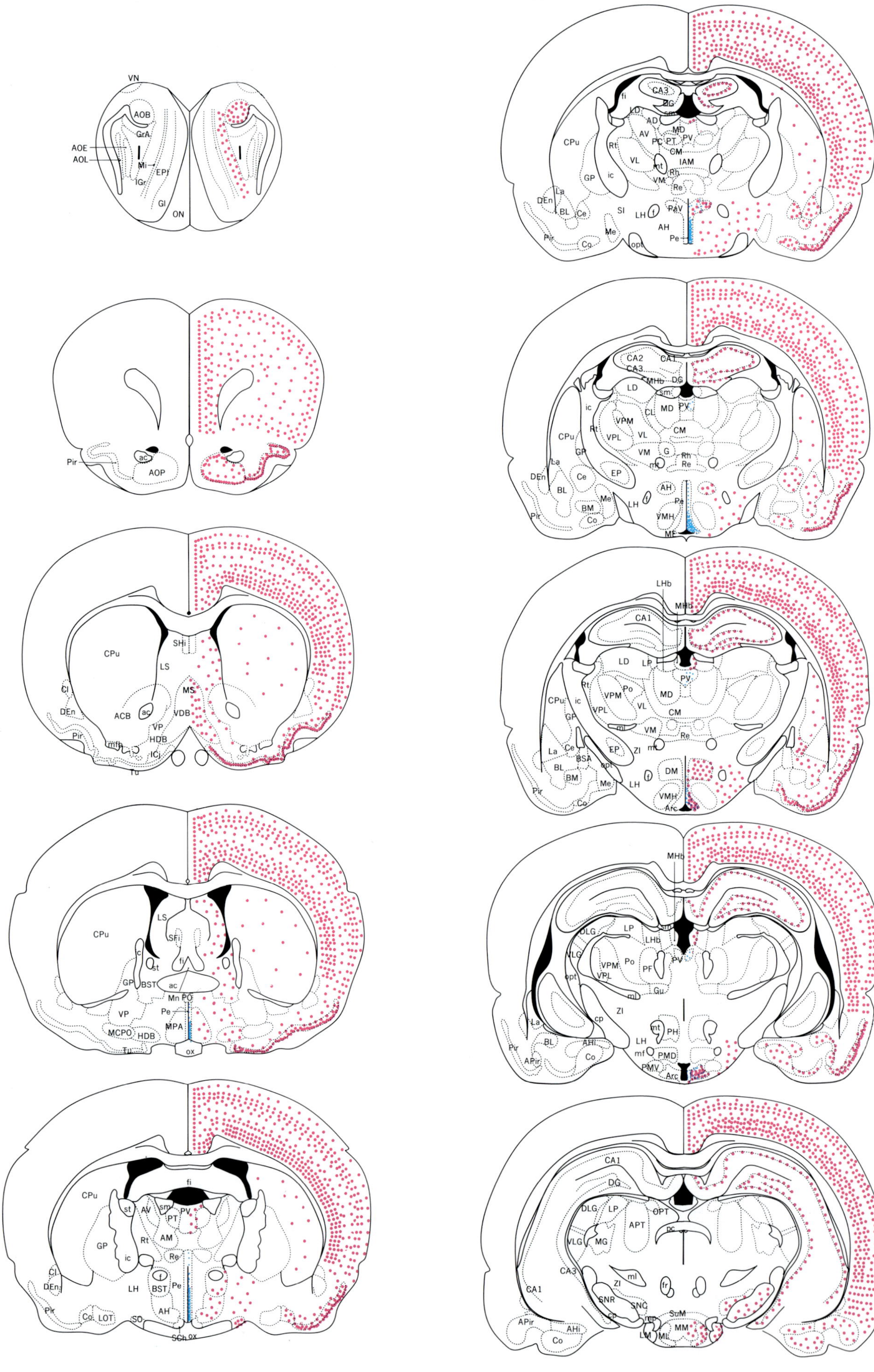

β-Adrenoceptor (immunocytochemistry)

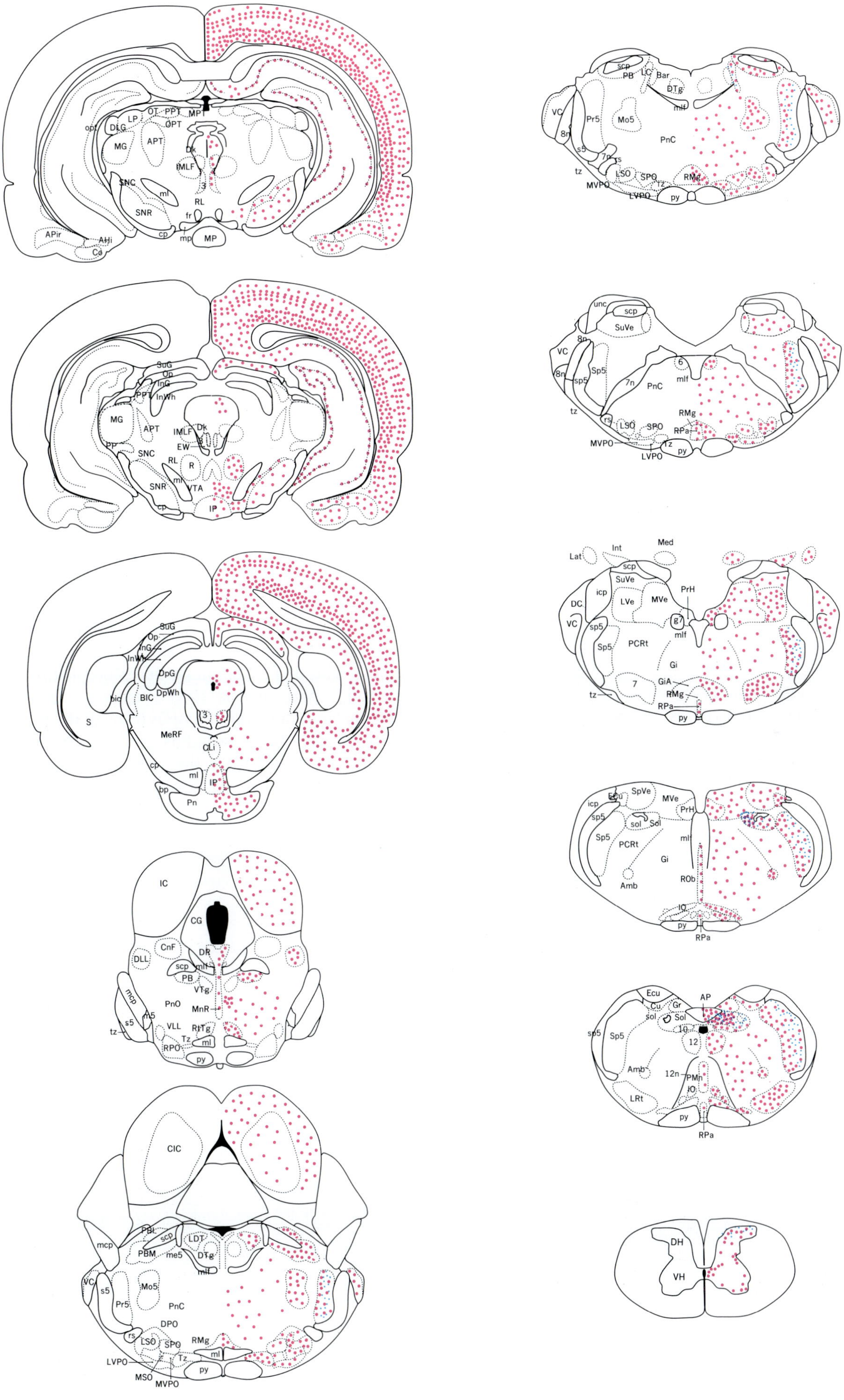

2 The serotonin system

Serotonin

Synthesis and degradation

Serotonin (5-hydroxytryptamine, 5-HT) is synthesized from tryptophan by the sequential actions of tryptophan hydroxylase and aromatic L-aminoacid dehydroxylase[3] (Fig. 2.1). There are two pathways which lead to its degradation; one by N-acetylase and the other by monoamine oxidase. The action sites of various pharmacological agents on serotonin are shown in Fig. 2.2.

Distribution and projection pathway

The distribution of 5-HT has been studied by histofluorescence autoradiography using radiolabelled 5-HT, and immunocytochemistry using synthesized enzyme-antibody. The most remarkable advance in this area was brought about by the introduction of the immunocytochemical method by Steinbusch *et al.*[2]. Their method made it possible to map the distribution of serotonin in more detail than the previously used methods. In the central nervous system, fibres which are immunoreactive to 5-HT are widely distributed. However, the cells which produce 5-HT are found only in restricted areas of the brain, namely in raphe nuclei groups which exist from the midbrain to the medulla oblongata. These areas are classified into nine regions (B1–B9) from caudal to rostal (Fig. 2.3).

B1 raphe pallidus
B2 raphe obscurus
B3 raphe magnus
B4 central grey matter of the medulla oblongata
B5 pontine medial raphe
B6 pontine dorsal raphe
B7 midbrain dorsal raphe
B8 caudal linear nucleus
B9 medial lemniscus.

Among these regions, B1–B3 have projection pathways mainly to the spinal cord and the others have those linking mainly to the forebrain (Fig. 2.3). The dorsal raphe nucleus (B7) has two ascending pathways to the cerebral cortex. One originates from the anterior part of the dorsal raphe nucleus, passing through the lateral part of medial forebrain bundle and external capsula, and reaching the cortex. The other pathway originates from the posterior part of the dorsal raphe nucleus, ascending within the median forebrain bundle, ascending dorsally at the diagonal band, and reaching the cingulate gyrus. Thus, different pathways from the dorsal raphe nucleus connect to different cortical areas, suggesting different functional significance for each pathway. There are projection pathways for other neurons in the dorsal raphe nucleus that connect to wide areas in the midbrain such as thalamus and hypothalamus to the lateral geniculate body among others. The 5-HT-containing fibres originating from the median raphe nucleus (B8) project to areas such as hypothalamus and hippocampus through the median forebrain bundle. The descending pathways from the raphe magnus (B3) and from the raphe obsculus and raphe pallidus (B2, B1) project mainly to the spinal cord, the

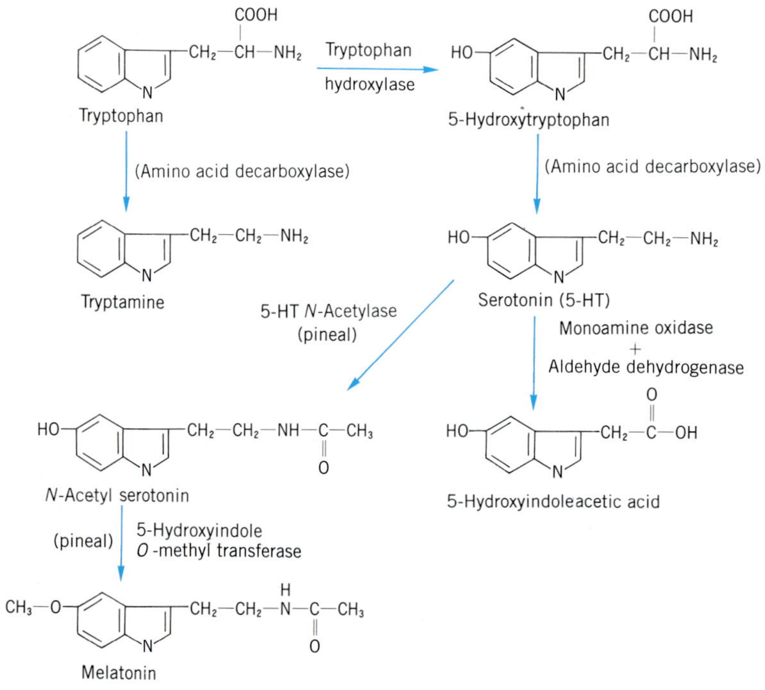

Fig. 2.1 Synthesis and metabolism of 5-HT.

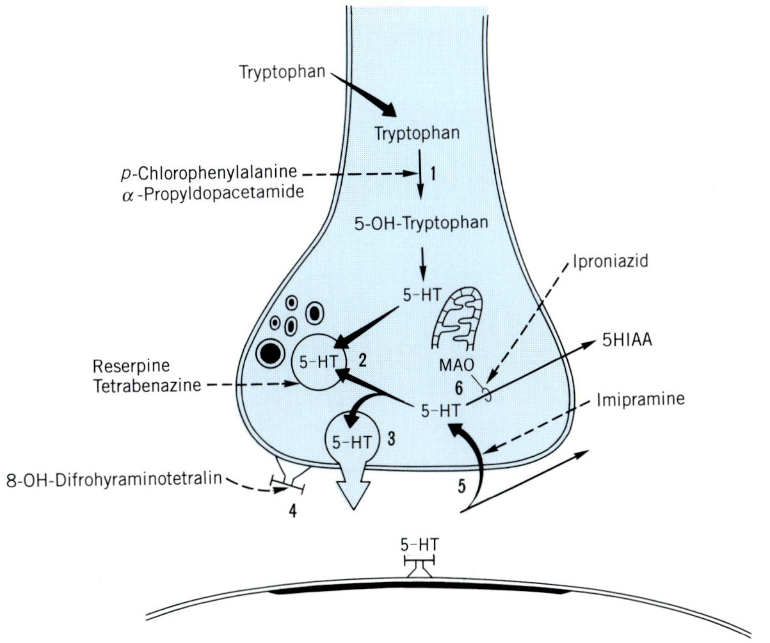

Fig. 2.2 Possible drug actions on the serotonergic neuron. 1, enzymatic synthesis; 2, storage; 3, release; 4, presynaptic regulation by autoreceptor; 5, uptake by serotonin transporter; 6, degradation of monoamine oxidase (MAO). 5HIAA: 5-hydroxy-indoleacetaldehyde (from ref. 1)

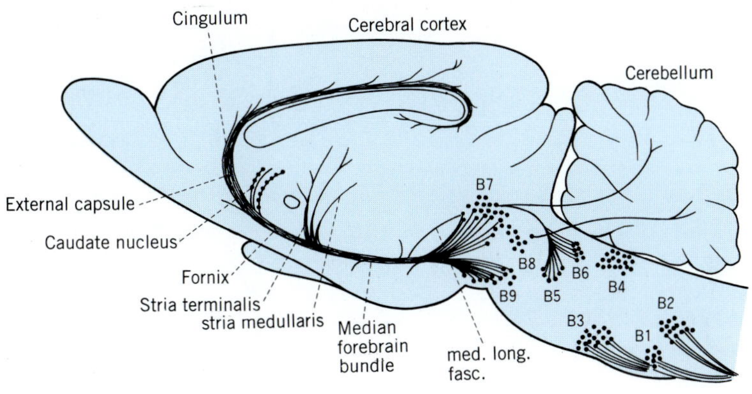

Fig. 2.3 Main serotonin-containing pathways in the rat CNS (from ref. 1).

former passing through the dorsal part of the lateral column and the latter through the ventral column. Thus, although the 5-HT system in the brain is being elucidated and its critical role in mental functions is being demonstrated pharmacologically, its definitive role is yet to be determined.

References

1. Cooper, J. R., Bloom, F. E., and Roth, R. E. (1991). *The biochemical basis of hemopharmacolocy.* Oxford University Press.
2. Steinbusch, H.W.M., Verhofstad, A.A.J., and Joosten, H.W.J. (1978). *Neuroscience*, **3**, 811–19.
3. Twarog, B.M. and Page, I.H. (1953). *J. Physiol. (Lond.)*, **175**, 157–61.

Serotonin (immunocytochemistry)

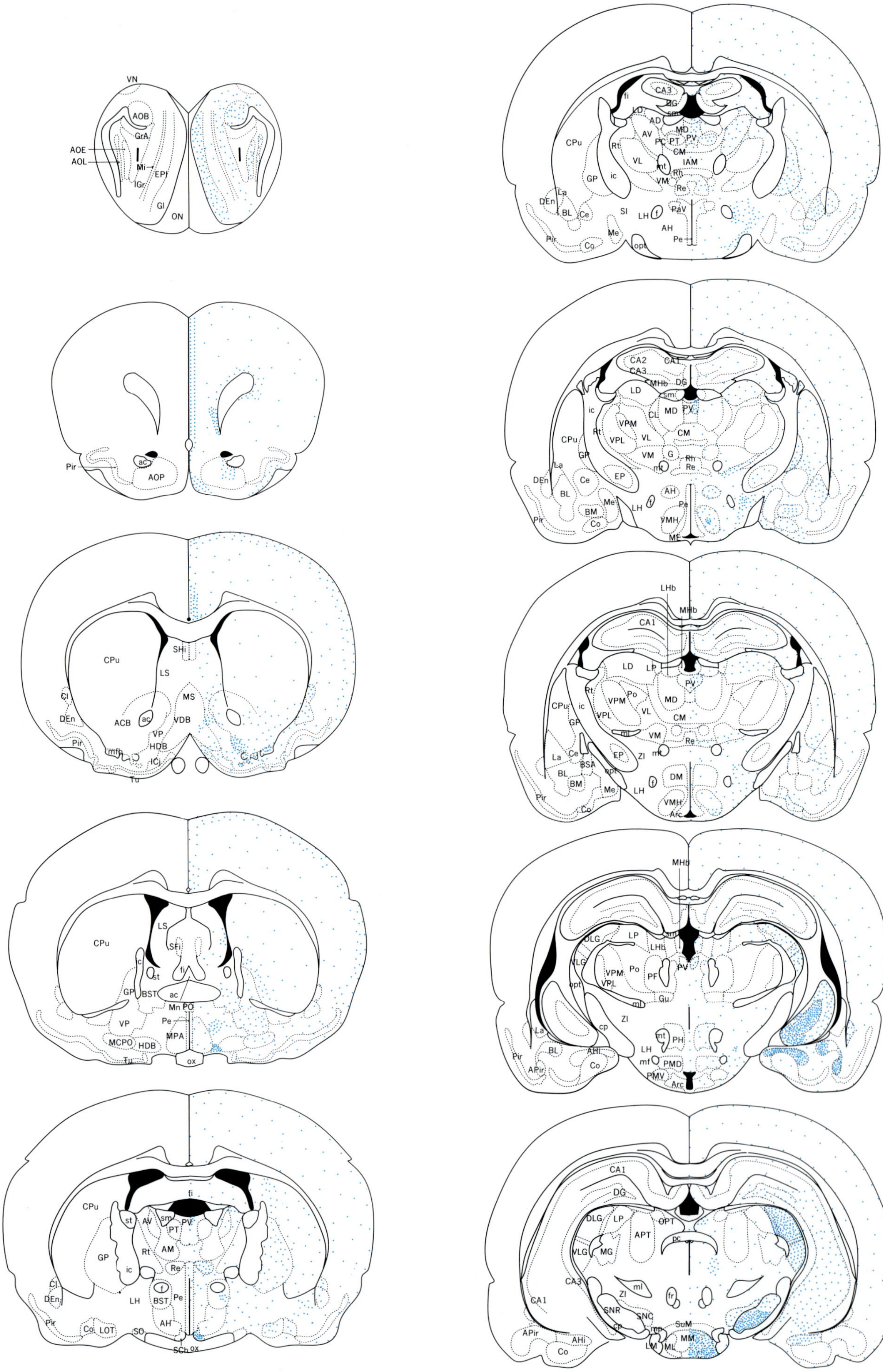

Serotonin (immunocytochemistry)

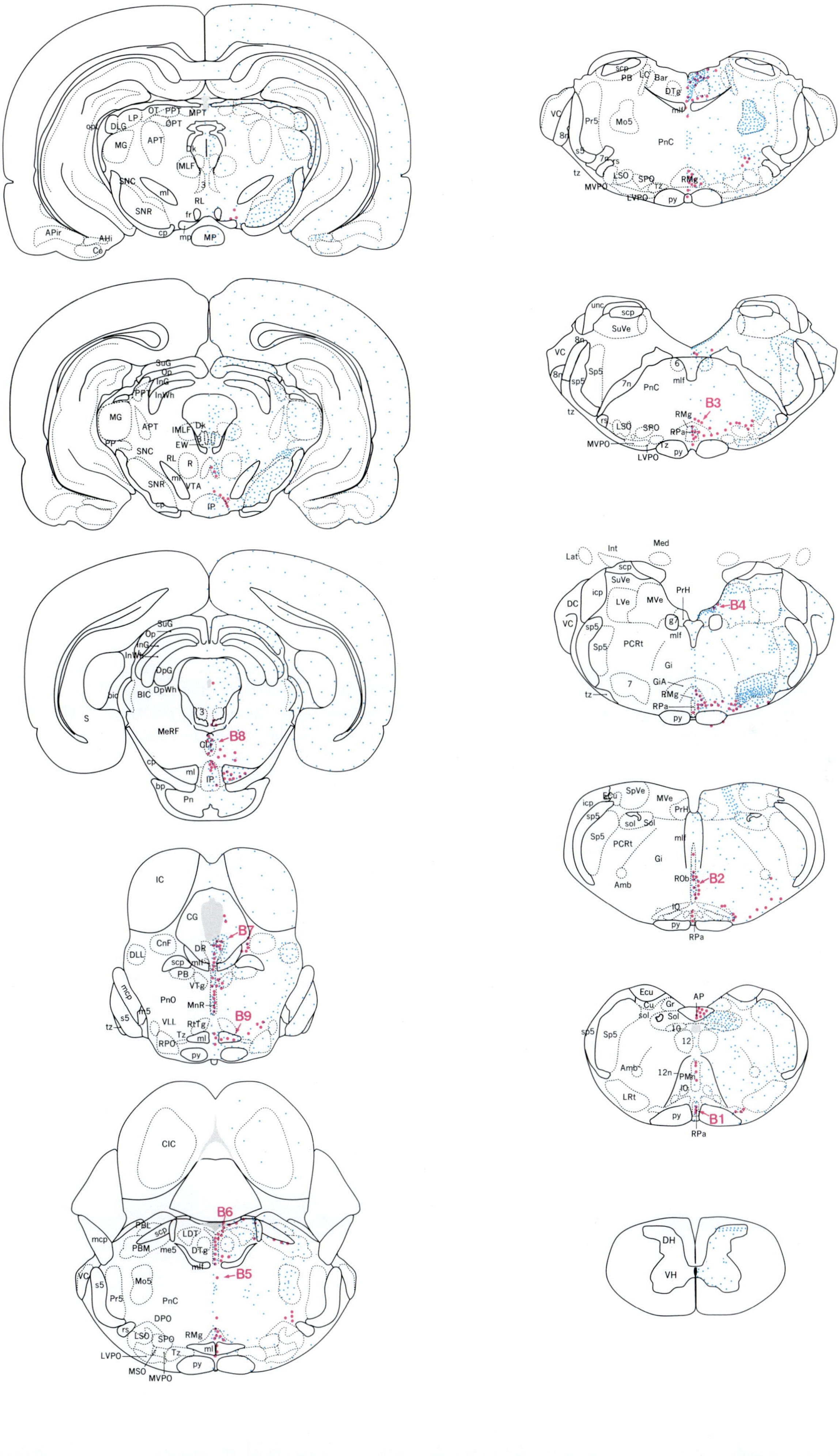

Serotonin receptors

Classification and structure

Guddum and Picarelli[2] showed in 1957 that the receptors for 5-HT exist in cholinergic nerves and are classified as M type, which is antagonized by morphine, and D type, which elicits contraction of smooth muscle and is antagonized by dibenamine and LSD. Since then, many ligands of high affinity have been developed, and they have been further subclassified into various subtypes. However, these subtypes have been based on response to ligands and not on their molecular structures. The advances in molecular biology have made cloning of 5-HT receptors possible, and a new classification and nomenclature have evolved.

The most recent consensus was reached at the International Union of Pharmacology, Classification of Receptors for 5-Hydroxytriptamine (Table 2.1).[5] The receptors for 5-HT are now classified into at least seven subtypes: 5-HT_1, 5-HT_2, 5-HT_3, 5-HT_4, 5-HT_5, 5-HT_6, and 5-HT_7. Among these, 5-HT_1 is further classified into subtypes A, B, C, and D. The previous 5-HT_{1C} is now known as 5-HT_{2C}. The original D type proposed in 1957 corresponds to 5-HT_2 in the current classification, while the M type corresponds to 5-HT_3. All the subtypes except 5-HT_4 have been cloned and their structures elucidated thus far. However, it is only for 5-HT_{1C} (5-HT_{2C}) that the mRNA for these receptors have been traced in detail in the central nervous system. Thus, only the distribution of (5-HT_{1C}/5-HT_{2C}) is shown in the atlas.

All types of receptors except 5-HT_3 are proteins with seven transmembrane sites coupled with G proteins, while 5-HT_3 is a protein with four transmembrane sites of ion channel type. The 5-HT_2 group is structurally and pharmacologically similar in increasing the intracellular level of IP_3/DG. 5-HT_1 receptors are known to decrease the level of cAMP. It is also known that the third transmembrane site of 5-HT_{1A} and the first transmembrane site of 5-HT_2 receptors have structures similar to the leucine zipper (Fig. 2.4), which is widespread for DNA-binding proteins and ion channels, but is rarely seen in receptors for neurotransmitters and makes the 5-HT receptors unique among them. This suggests that receptors for 5-HT_{1A} and 5-HT_{2A} have dimeric structure. The leucine zipper structure is not found either in 5-HT_{2C} or 5-HT_{1D} (which resembles 5-HT_{1A}), indicating that structurally different types do exist in G-protein coupled receptors which are classified into two groups on a physiological basis. The 5-HT_4 receptor is known to increase the level of cAMP.

Distribution

The most detailed study on the distribution of 5-HT receptors in the brain has been reported by Mengod *et al.*[4], who employed *in situ* hybridization for 5-HT_{2C} receptors. The information presented in this atlas is mainly based on their data. The characteristic of the distribution of 5-HT_{2C} receptors is that they exist most abundantly in the choroid plexus. Their mRNAs are distributed in anterior olfactory nuclei, septum, pyriform cortex, amygdaloid body, lateral habenula, C3 area of hippocampus, subthalamic nucleus, and substantia nigra. A moderate amount of mRNA was detected in the lower brain stem as well.

Future research directions

Recent efforts to clone 5-HT receptors have resulted in the elucidation of their structures, enabling the researchers to establish their classification based on structure. Until recently their distribution has been mapped only by the autoradiography of ligands; however, methods such as *in situ* hybridization for mRNA and immunocytochemistry using specific antibodies to receptors will make it possible to elucidate completely the distribution of 5-HT receptors in the brain. In addition, it is noteworthy that the recently cloned 5-HT transporter protein will be recognized as the third important protein, apart from the ligand and the receptor, in serotoninergic signal transmission system[1,3] (Fig. 2.5). It has been assumed that transporters exist in the presynaptic region and terminate the neurotransmission by actively re-uptaking the neurotransmitters released from the neuron. As they have just been cloned and their existence been demonstrated, the exact roles of 5-HT transporters remain to be clarified in the future.

References

1. Blakely, R.D., Berson, H.E., Fremeau, R.T. Jr. *et al.* (1991). *Nature*, **354**, 66–70.
2. Guddum, J.H. and Picalleri, Z.P. (1957). *Br. J. Pharmacol.*, **12**, 323–8.
3. Hoffman, B.J., Mezey, E., and Brownstein, M.J. (1991). *Science*, **254**, 779–80.
4. Mengod, G., Nguyen, H., Le, H. *et al.* (1990). *Neuroscience*, **35**, 577–91.
5. Hoyer, D., Clarke, D.E., Fozard J.R. *et al.* (1994). *Pharmacol. Rev.* **46**, 157–203.
6. Julius, D. (1991). *Annu. Rev. Neurosci.*, **14**, 335–60.

Table 2.1 5-Hydroxytryptamine receptors

Nomenclature	5-HT_{1A}	5-HT_{1B}	5-HT_{1D}	5-HT_{1E}	5-HT_{1F}	5-HT_{2A} (D, 5-HT_2)	5-HT_{2B} ($5HT_{2F}$)
Radioligands	8-OH-DPTA	5-CT	5-CT	5-HT	LSD	ketanserin	5-HT
Effectors	cAMP− K+ channel	cAMP−	cAMP−	cAMP−	cAMP−	IP3/DG	IP3/DG
Accession No.	(H)X57829 (R)J05276	(H)M75128 (R)M89954	(H)M89955	(H)L05597	(H)L05567 (R)L05596	(H)S71229 (R)M30705	(H)X77307 (R)X66842

Nomenclature	5-HT_{2C} (5-HT_{1C})	5-HT_3	5-HT_4	5-HT_{5A}	5-HT_{5B}	5-HT_6	5-HT_7
Radioligands	mesulergine	zacopride	GR113808	LSD	LSD	5-CT/LSD	Clozapine/LSD
Effectors	IP3/DG	Cation channel	cAMP+	—	—	cAMP+	cAMP+
Accession No.	(H)U49516	(H)D49394 (R)D49395	—	(H)X81412 (R)L10072	(R)L10073	(H)L41147 (R)L41146	(H)L21195 (R)L19654

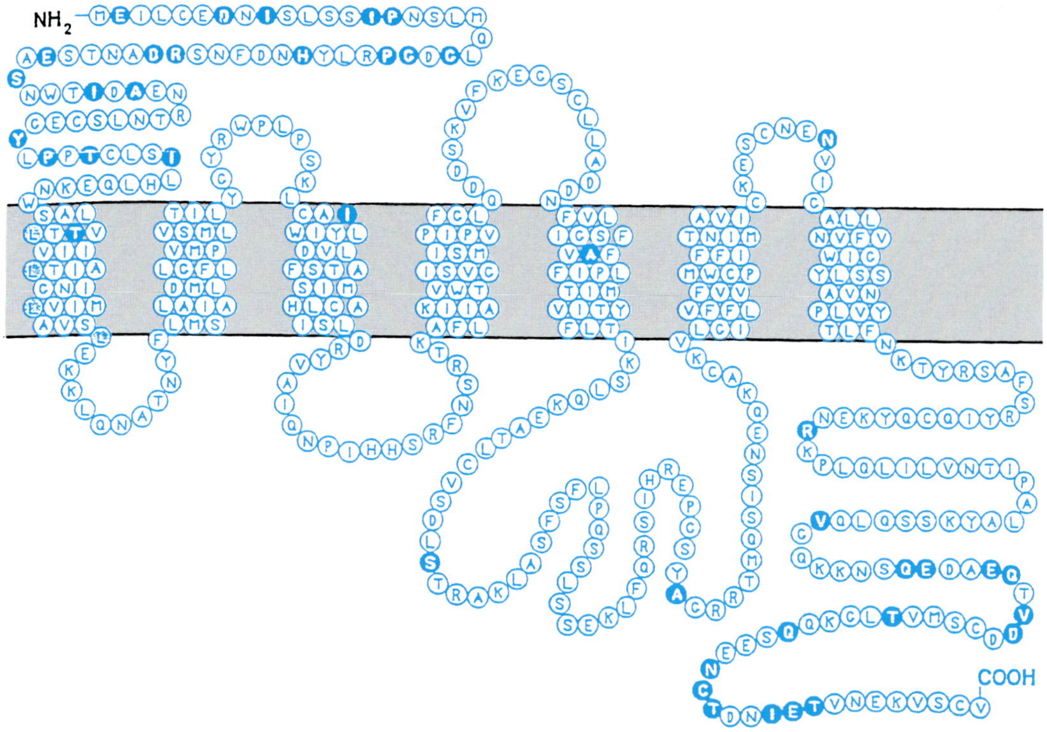

Fig. 2.4 Structure of rat serotonin 5-HT$_{2A}$ receptor. 5-HT$_{2A}$ receptor is a G-protein coupled type receptor, and a leucine zipper-like structure is found in the first transmembrane domain. Shaded residues, leucine zipper; dark coloured residues differ in human 5-HT$_2$ receptor sequence (from ref. 6).

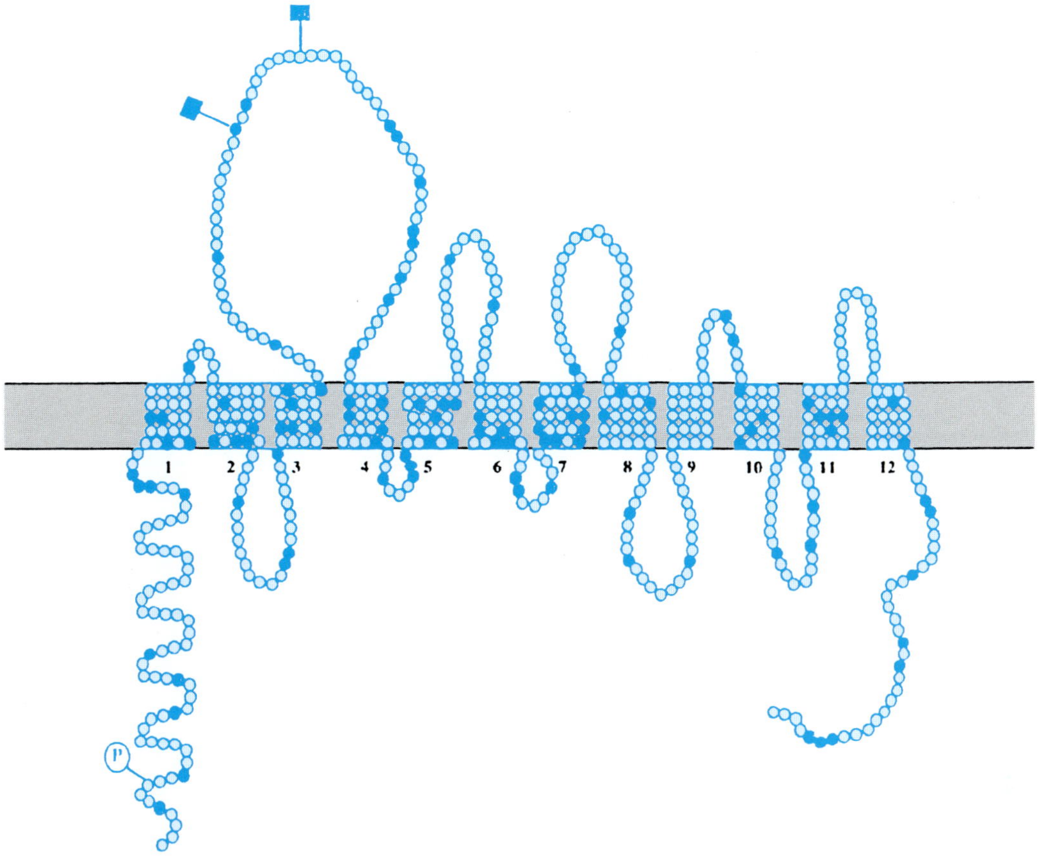

Fig. 2.5 Structure of 5-HT transporter. It has 12 transmembrane domains and is located on the presynaptic membrane. This transporter is a neuronal type, and another transporter has also cloned from platelet (from refs 1 and 3).

Serotonin receptor (binding sites)

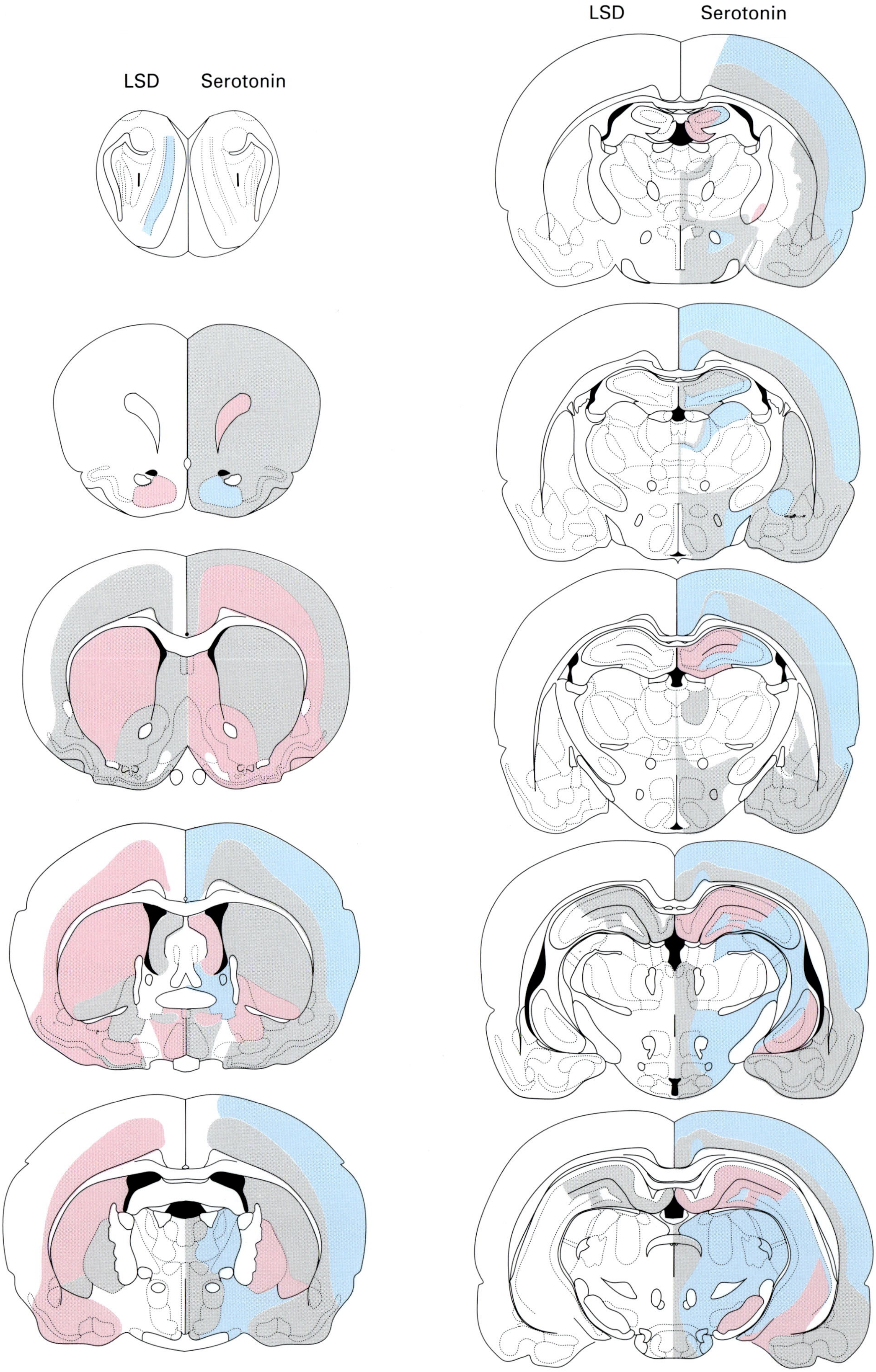

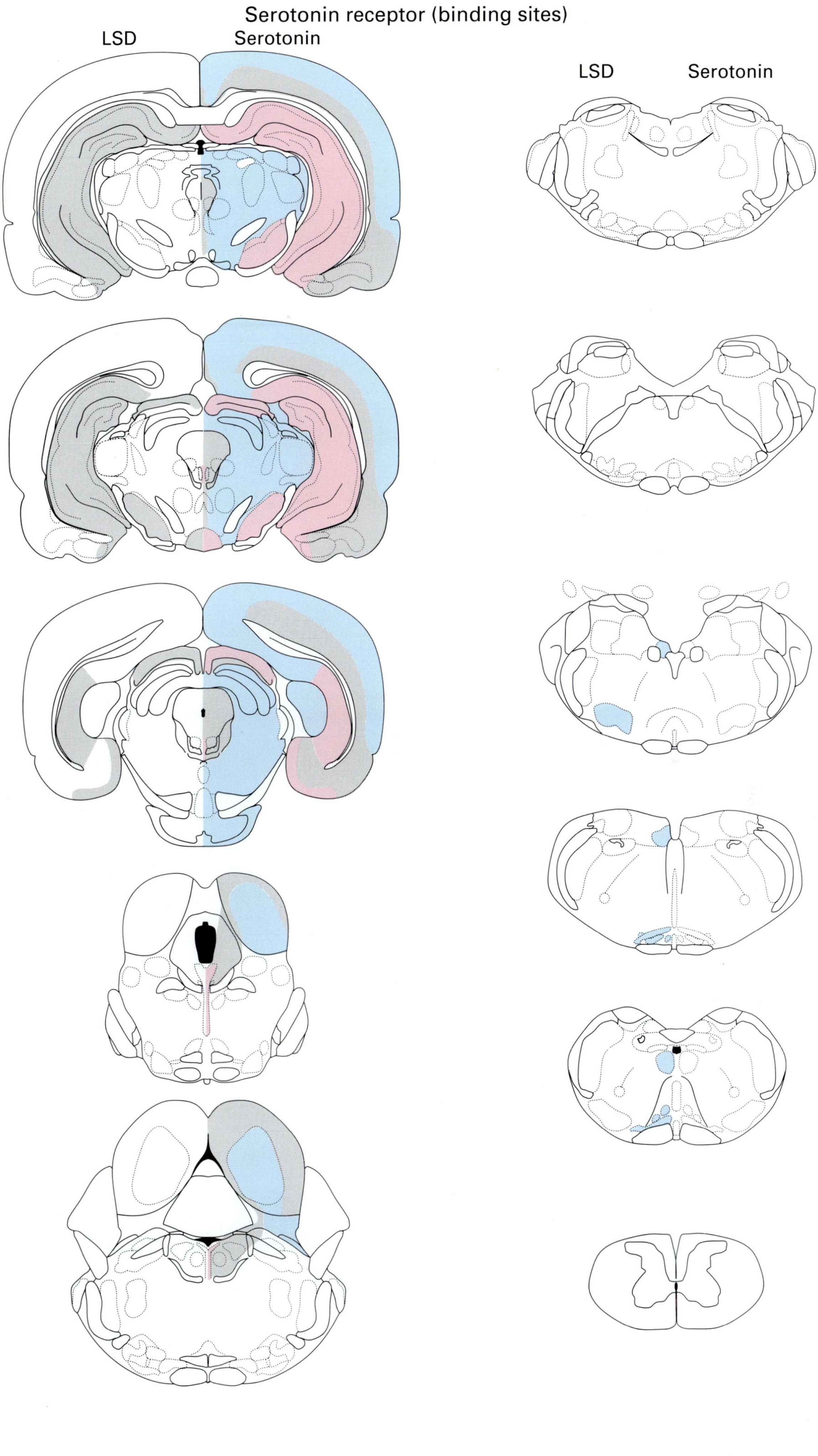

Serotonin receptor (binding sites)
LSD
Serotonin
LSD
Serotonin

Serotonin receptor (ISH)

Serotonin 2c

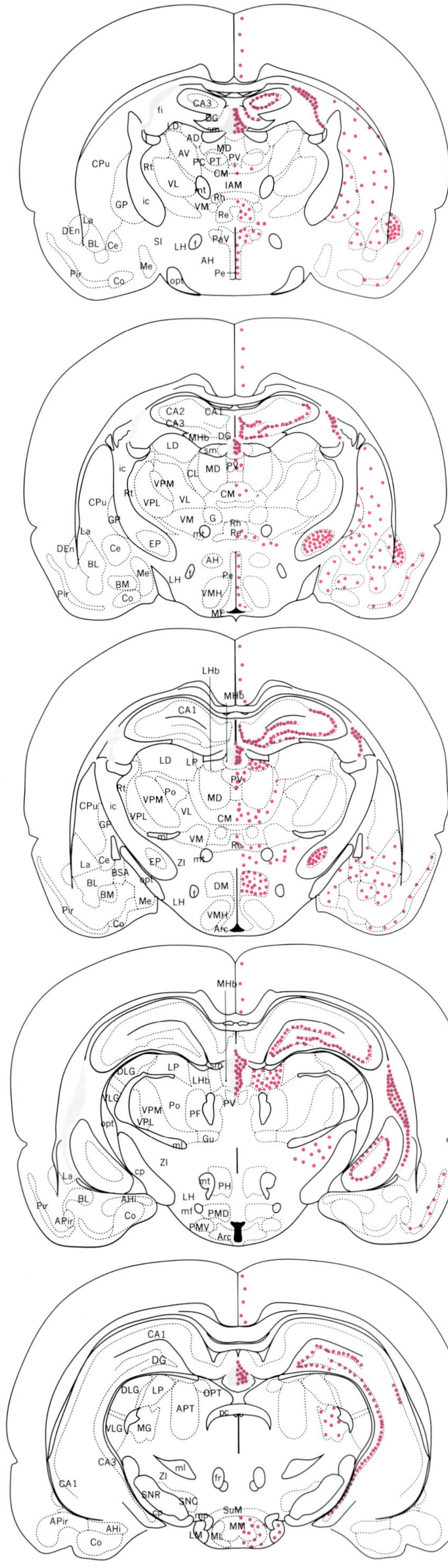

Serotonin 2c

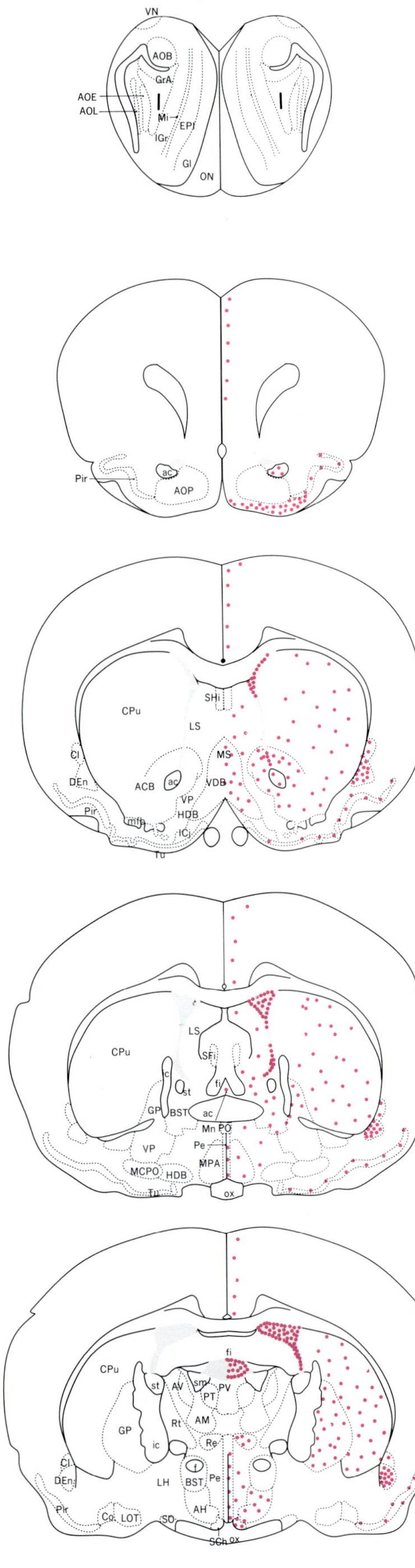

Serotonin receptor (ISH)

Serotonin 2c

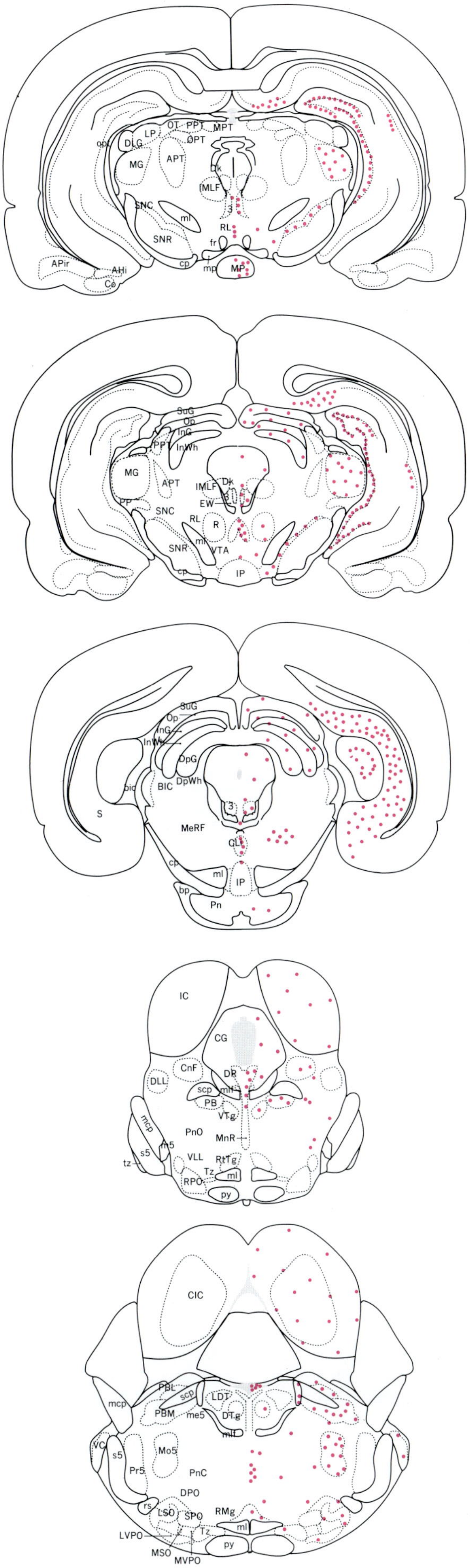

Serotonin 2c

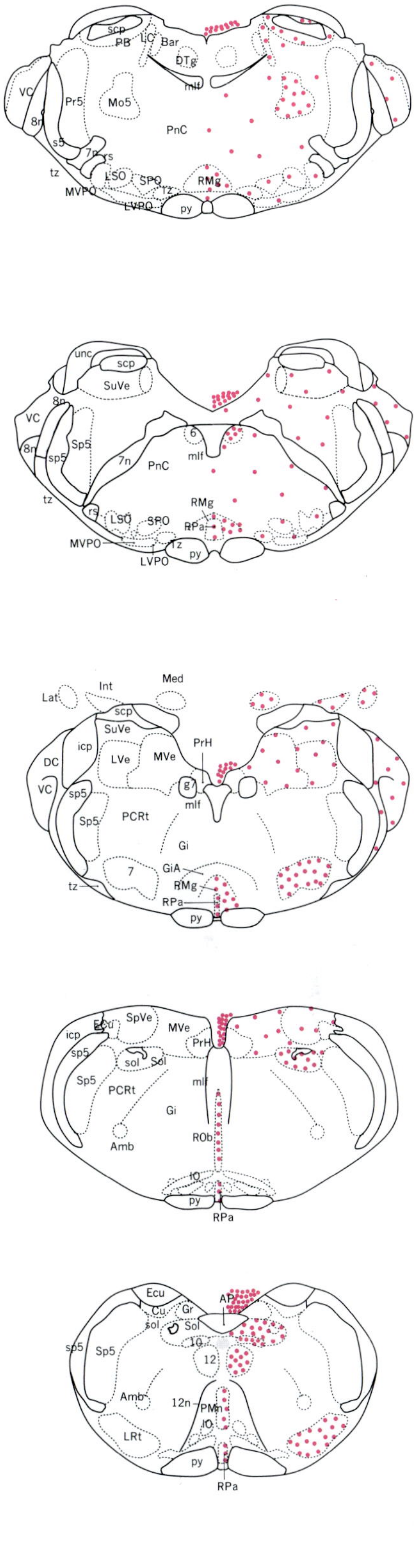

3 The histaminergic system

Histamine

Transmission of information

The transmission of information in the histaminergic system takes place in the following steps (Fig. 3.1). L-Histidine taken up in the histamine neuron is decarboxylated by L-histidine decarboxylase (HDC, EC 4.1.1.22) (1), resulting in the synthesis of histamine (2); histamine thus synthesized is stored in synaptic vesicles (3), and is released on stimulation, depending upon Ca^{2+}, into the synaptic junction (4); then it binds either to H_1 or H_2 receptors (5, 6), exerting its physiological actions. However, in contrast to other monoamines, histamine's re-uptake to presynaptic terminals has not been demonstrated. Thus it is assumed that histamine is inactivated through methylation of the N^τ site of its imidazole ring by *N*-methyltransferase (HMT; histamine *N*-methyltransferase, EC 2.1.1.8) which converts histamine to N^τ-methyl histamine (N^τ-MeHA) (8). The released histamine regulates the biosynthesis and release of histamine acting through the H_3 receptors on the presynaptic membrane (7). Therefore, H_3-receptor agonists inhibit, whereas the antagonists promote, the release of histamine.

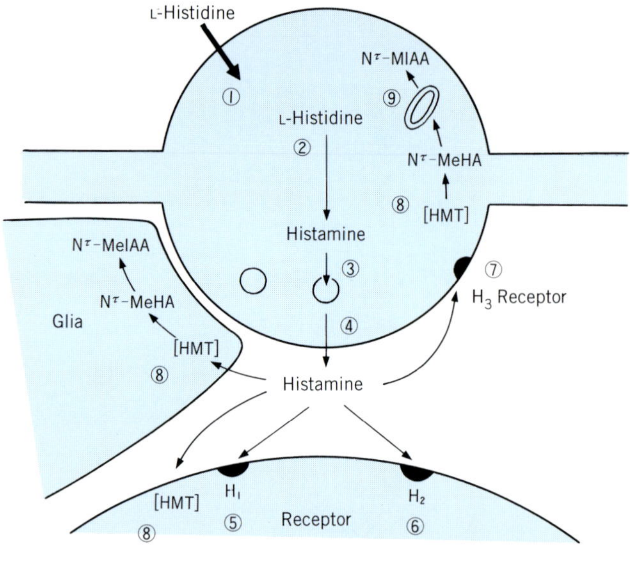

Fig. 3.1 Neurotransmission at the histaminergic varicosity. Few synapse formations are found between presynaptic and postsynaptic structures along the various axons of the histaminergic neurons. Histamine released from synaptic vesicles in the varicosity is therefore distributed in proximity to the site of release.

Properties of HDC and HMT

Both HDC and HMT have been cloned. The nucleotide sequence of their cDNA and the deduced amino acid sequences are shown in Fig. 3.2. The molecular weight of HDC as calculated from the cDNA is 74 kDa, while that of purified protein is 54 kDa, suggesting post-translational processing.

Distribution and functions

The histaminergic neuronal system locates its cell bodies at the tuberomammillary nucleus (TM) in the posterior hypothalamus-mammillary body (Fig. 3.3). It widely projects axons which have varicosities caudally and rostrally to almost all the brain regions. As shown in detail in the map, the fibres of this system are distributed densely in the hypothalamus, thalamus, cerebral cortex, basal nuclei, amygdala, and septum[1]. Such a diffuse distribution suggests that the histaminergic system does not act for one-to-one transmission of information but for a more 'universal' regulation of brain functions[3,4,6,7] Thus, histamine is more aptly called a neuromodulator or a neuroregulator, rather than a neurotransmitter. Its role in many functions can hardly be ignored.

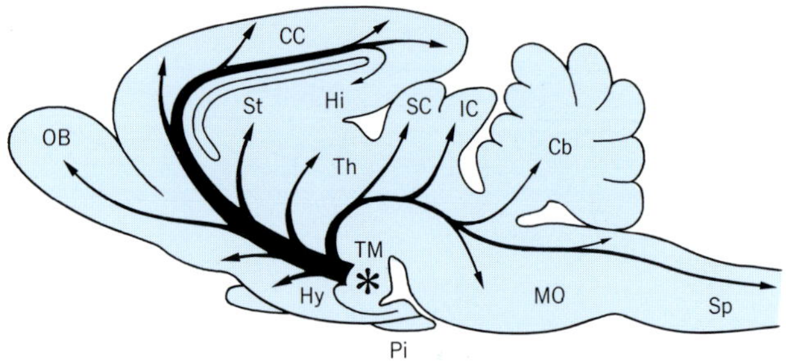

Fig. 3.3 Location and distribution of the histaminergic neurons[6]. The cell bodies of the histaminergic neurons in rat brain (shown in sagittal section) send nerve fibres to almost all regions of the brain (arrows). Abbreviations: Cb, cerebellum; CC, cerebral cortex; Hi, hippocampus; Hy, hypothalamus; IC, inferior colliculus; MO, medulla oblongata; OB, olfactory bulb; Pi, pituitary; SC, superior colliculus; Sp, spinal cord; Th, thalamus; TM, tuberomammillary nucleus.

References

1. Inagaki, N., Toda, K., Taniuchi, I. *et al.* (1990). *Exp. Brain Res.*, **80**, 374–80.
2. Mamure-Sato, R. Yamauchi, Tanno, Y. *et al.* (1992). *Eur. J. Biochem.*, **209**, 533–39.
3. Onodera, K., Yamatodani, A., Watanabe, T., and Wada, H. (1994). *Progr. Neurobiol.*, **42**, 685–702.
4. Schwartz, J.C., Arrang, J.-M., Garbarg, M. *et al.* (1991). *Physiol. Rev.*, **71**, 1–51.
5. Takemura, M., Tanaka, T., Taguchi, Y. *et al.* (1992). *J. Biol. Chem.*, **267**, 12687–15691.
6. Wada, H., Inagaki, N., Yamatodani, A., and Watanabe, T. (1991). *Trends Neurosci.*, **14**, 415–18.
7. Watanabe, T. and Wada, H. (1991). *Histaminergic neurons: morphology and function.* CRC Press, Boca Raton.

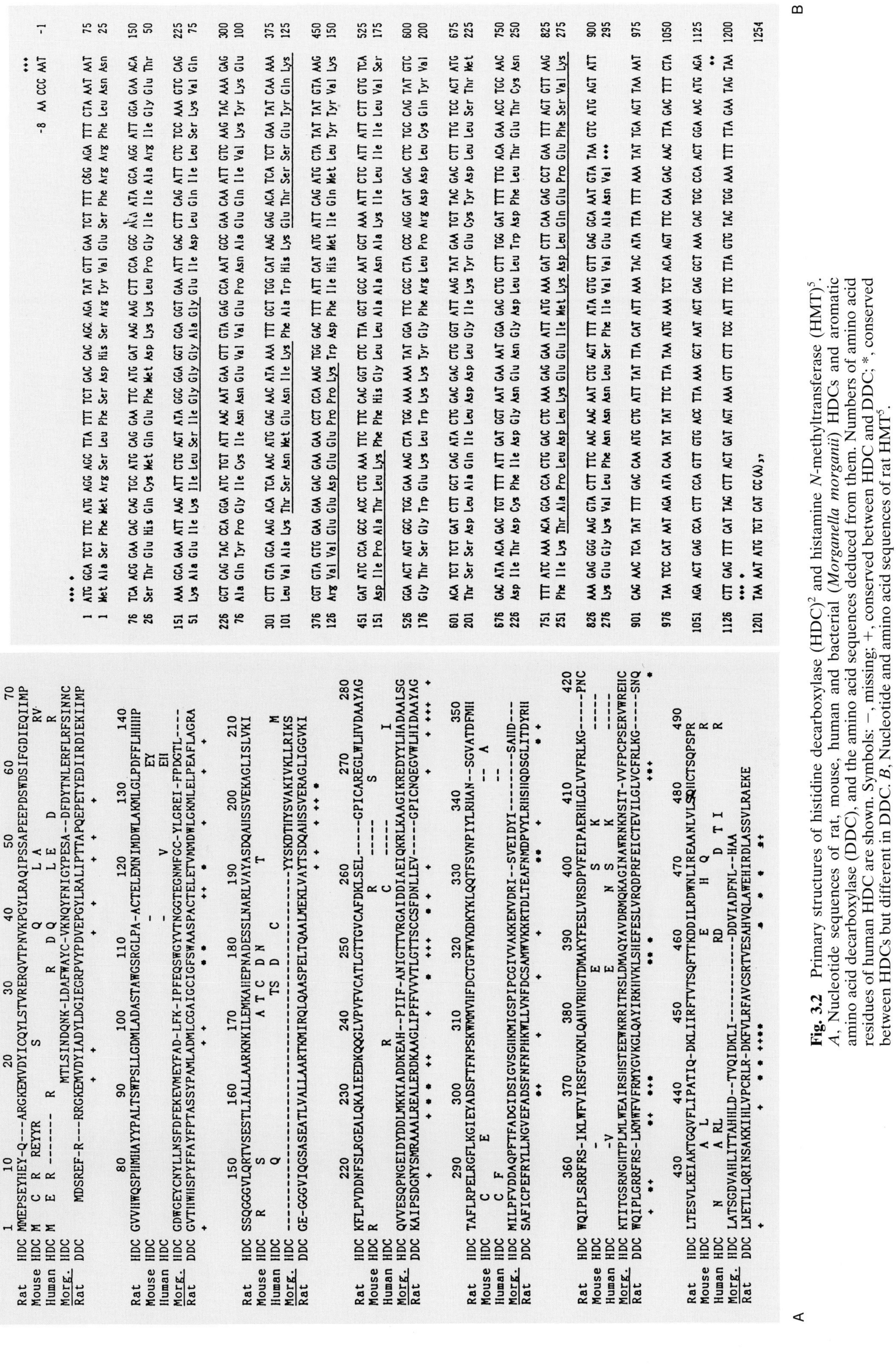

Fig. 3.2 Primary structures of histidine decarboxylase (HDC)[2] and histamine *N*-methyltransferase (HMT)[5]. *A*, Nucleotide sequences of rat, mouse, human and bacterial (*Morganella morganii*) HDCs and aromatic amino acid decarboxylase (DDC), and the amino acid sequences deduced from them. Numbers of amino acid residues of human HDC are shown. Symbols: −, missing; +, conserved between human HDC and DDC; *, conserved between HDCs but different in DDC. *B*, Nucleotide and amino acid sequences of rat HMT[5].

Histamine (immunocytochemistry)

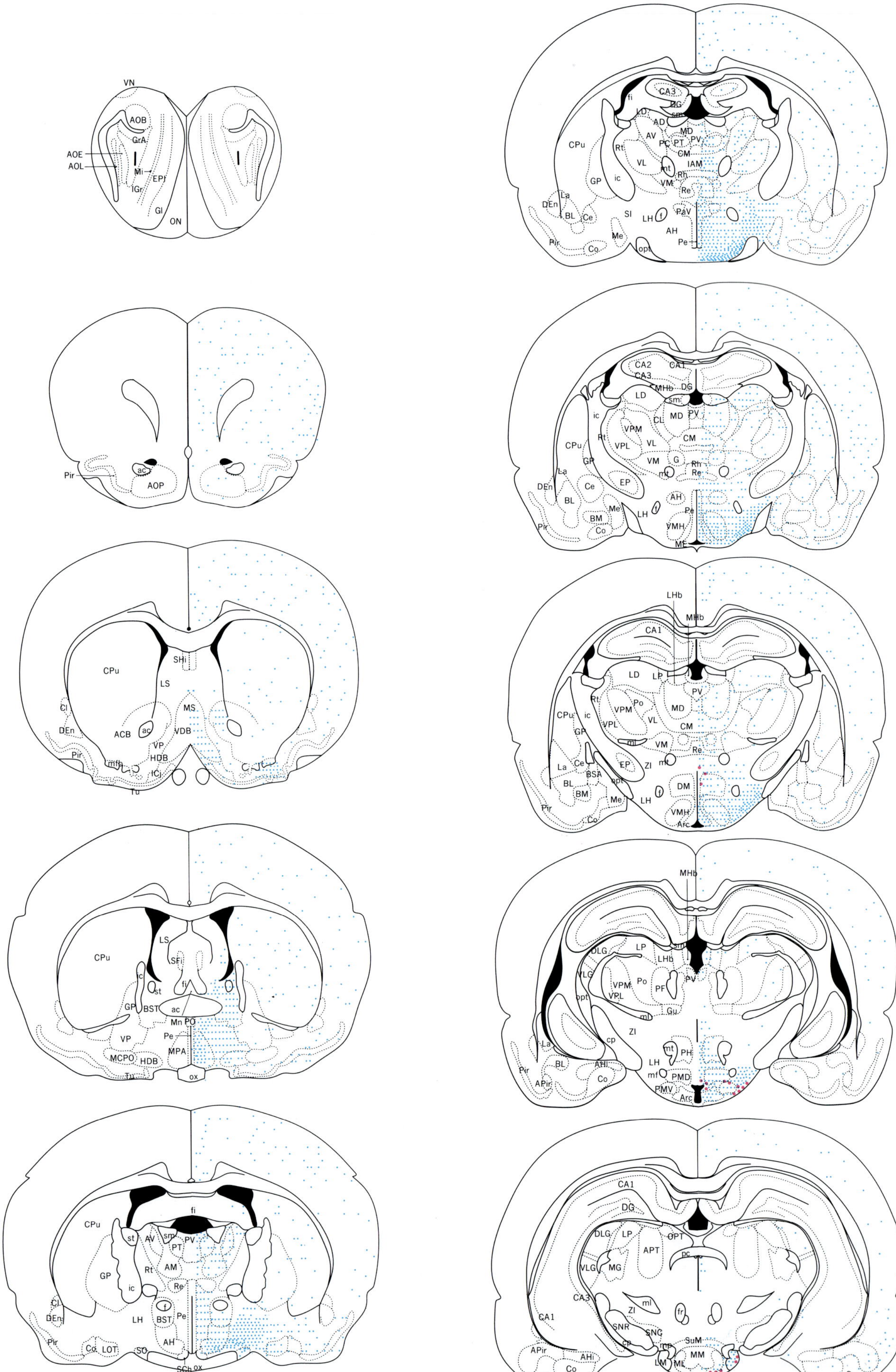

Histamine (immunocytochemistry)

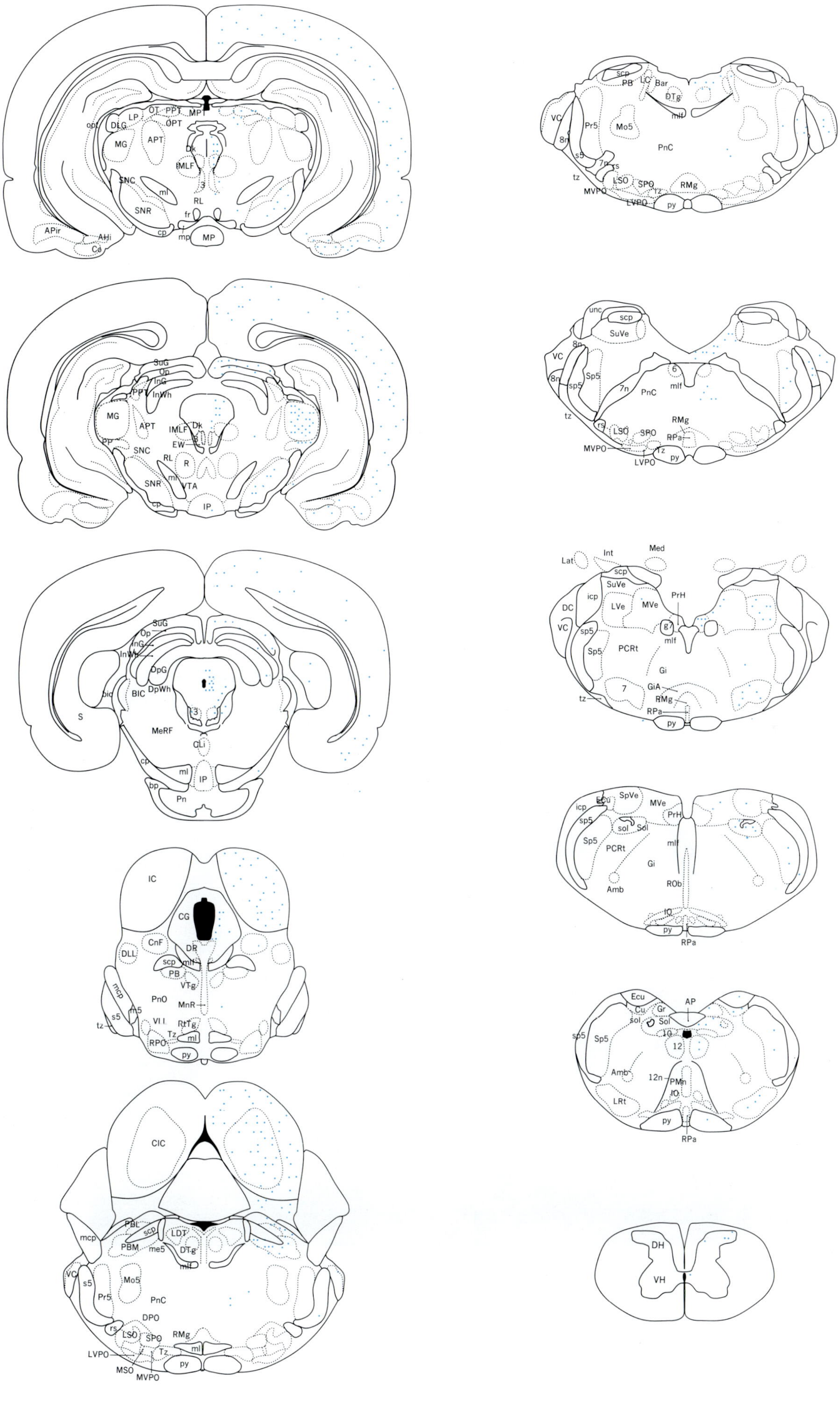

Histamine receptors

Subtypes

The subtypes of histamine receptors have been studied by the development of specific antagonists. The so-called classical antihistamine agents demonstrated the existence of H_1 receptors in 1937. Subsequently, H_2 receptors and H_3 receptors were identified in 1972 and 1987, owing to the development of cimetidine and thioperamide, respectively. It can be forseen that newer subclasses in the subtypes (e.g. H_{1a}, H_{1b} and H_{1c}) and even newer subtypes (e.g. H_4) will be identified, as has been the case for other receptors. Table 3.1 summarizes the agonists and antagonists for each identified receptor. H_1 and H_2 receptors have recently been cloned and their amino acid sequences have been determined[1,2] (Fig. 3.4). Both are seven-transmembrane proteins, i.e. the G-protein coupled receptors. H_1 receptors resemble M_1, α_1, 5-HT_{1C}, and D_2 much more than H_2 receptors.

Table 3.1. Agonists and antagonists of histamine receptors

Subtype	Agonist	Antagonist
H_1	Betahistine 2-Methylhistamien 2-Pyridylhistamine Thiazolylethylamine	*d*-Chlorpheniramine Cyproheptadine Diphendydramine Promethadine Astemizole[a] Mequitazine[a] Terfenadine[a]
H_2	Dimaprit Impromidine 4-Methylhistamine	Burimamide Cimetidine Famotidine Ranitidine Zolantidine[b]
H_3	(*R*)-α-Methylhistamine Imetit Immepip	Thioperamide Clobenpropit

[a] Non-sedative antihistaminics.
[b] Blood–brain barrier penetrating.

Distribution

The distribution of histamine receptors corresponds to the distribution of the terminals of histamine neurons. It should be noted that the distribution, especially of H_1 receptors, has widespread interspecies difference among rats, guinea pigs, and humans. Imaging of the H_1 receptors by positron emission tomography (PET) in the human brain is shown in Fig. 3.5.[3]

Second messengers

The stimulation of H_1 receptors results in the activation of phospholipase C, promotion of PI turnover, eliciting the increase in Ca^{2+} and activation of C kinase by diacylglycerol. On the other hand, H_2 receptors increase intracellular cAMP and activate A kinase. The second messenger for H_3 receptors has not been identified.

References

1. Gantz, I., Schaffer, M., DelValle, J. *et al.* (1991). *Proc Natl Acad. Sci. USA,* **88**, 429–33.
2. Yamashita, M., Fukui, H., Sugama, K. *et al.* (1991). *Proc. Natl Acad. Sci. USA*, **88**, 11515–19.
3. Yanai, K., Watanabe, T., Yokoyama, H. *et al.* (1992). *Neurosci. Lett.*, **137**, 145–8.

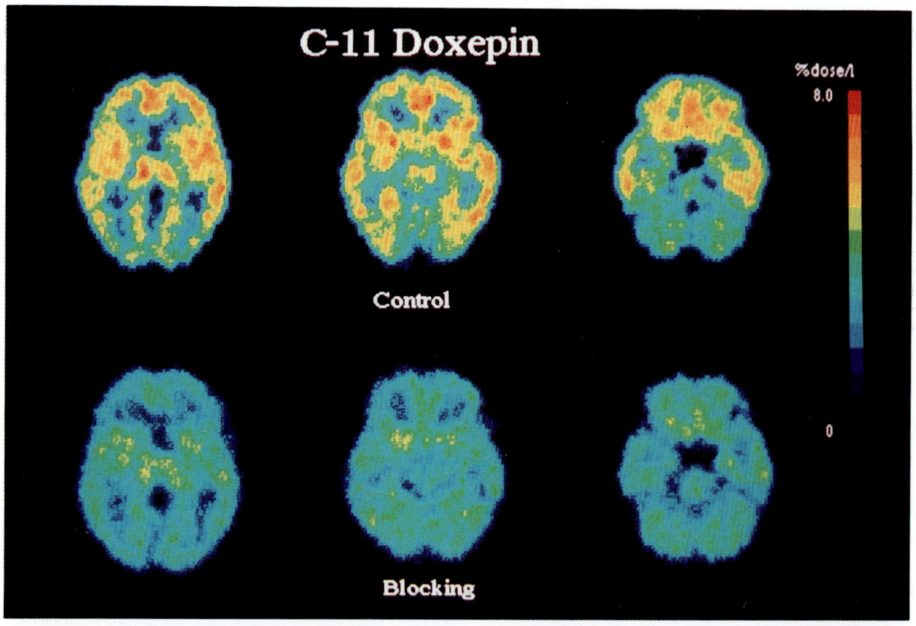

Fig. 3.5 Distribution of histamine H_1 receptors in the living human brain as shown by positron emission tomography (PET). From left to right, the figures represent coronal sections of human brain at the levels of OM+48, 34, and 20 mm, respectively (OM, orbitomeatal line). The upper and lower panels are total and non-specific bindings, respectively.

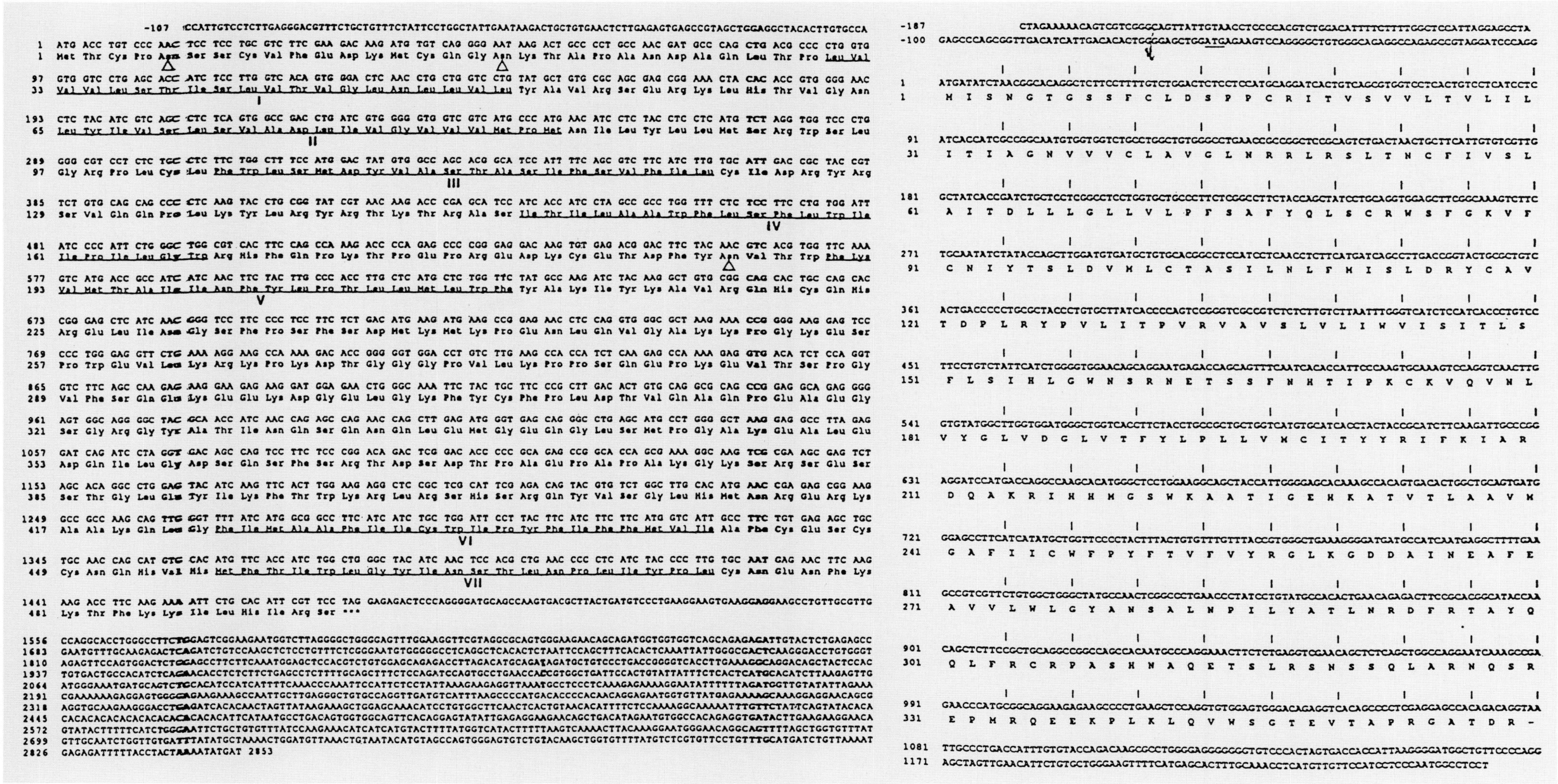

A B

Fig. 3.4 Structures of histamine H_1 and H_2 receptors. *A*, Nucleotide sequence of histamine H_1 receptor of bovine adrenal medulla and the amino acid sequence deduced from it; *B*, nucleotide sequence of histamine H_2 receptor of dog gastric mucosa and the amino acid sequences deduced from it.

Histamine receptor (binding sites)

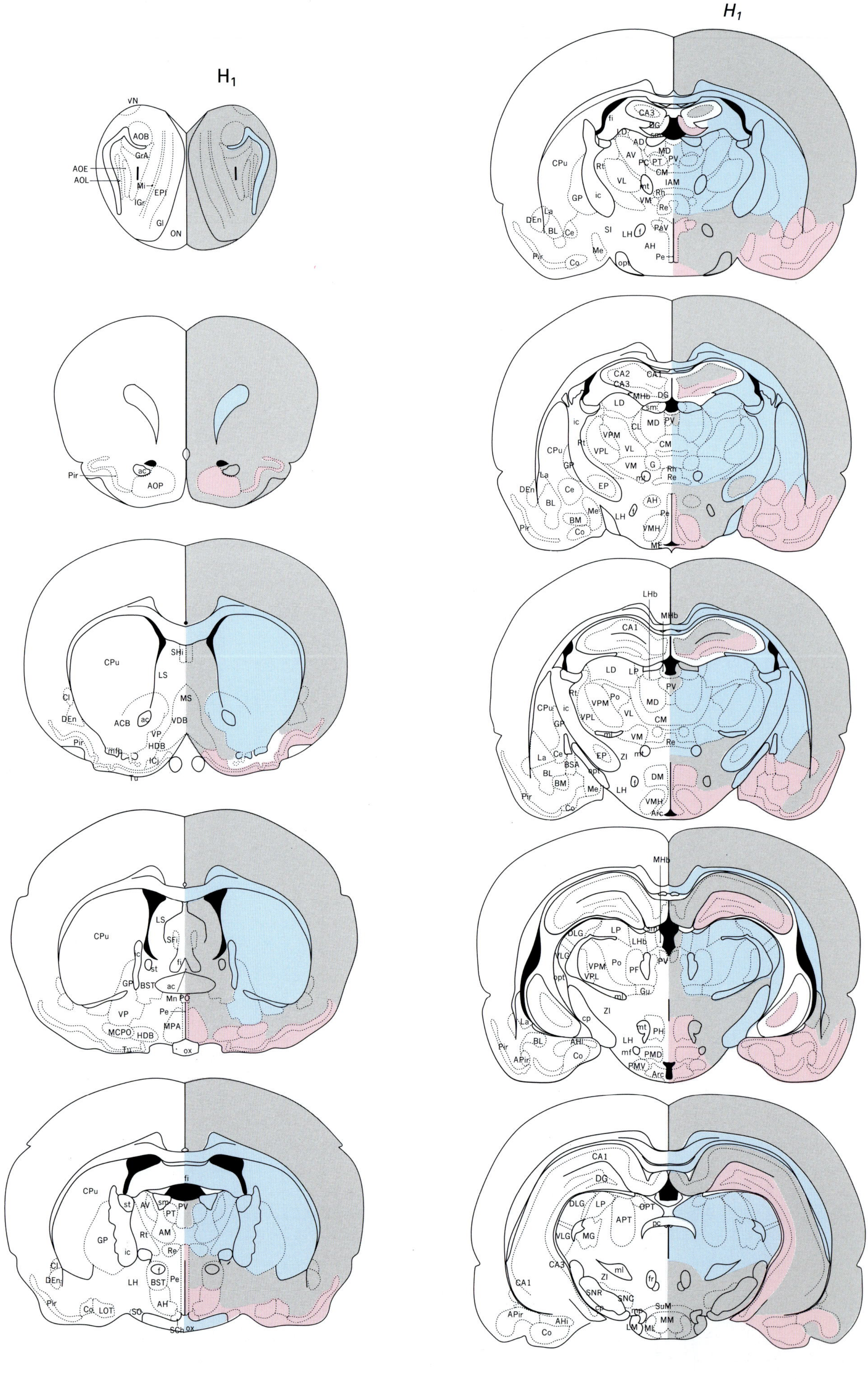

Histamine receptor (binding sites)

H_1

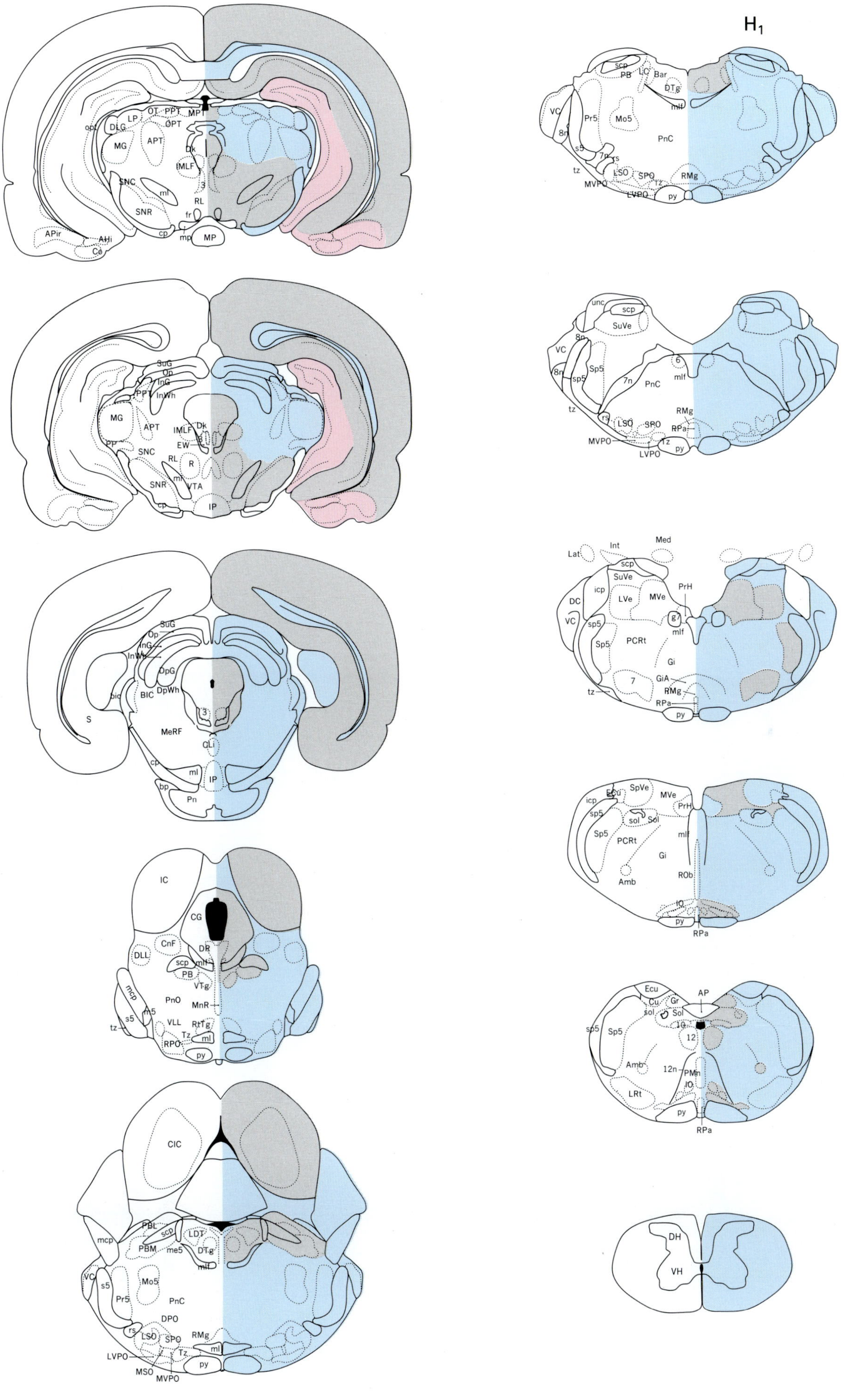

Histamine receptor (binding sites)

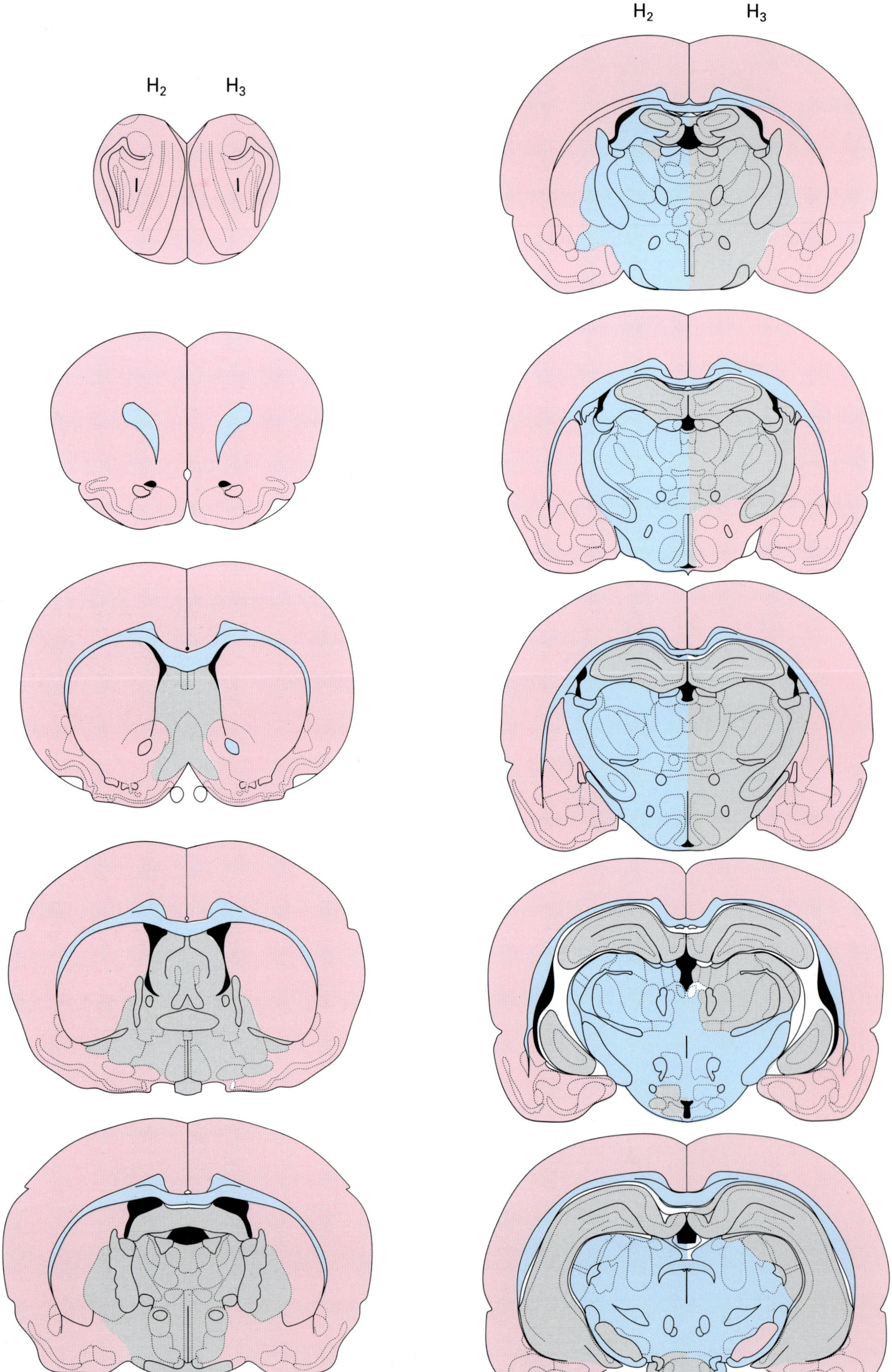

Histamine receptor (binding sites)

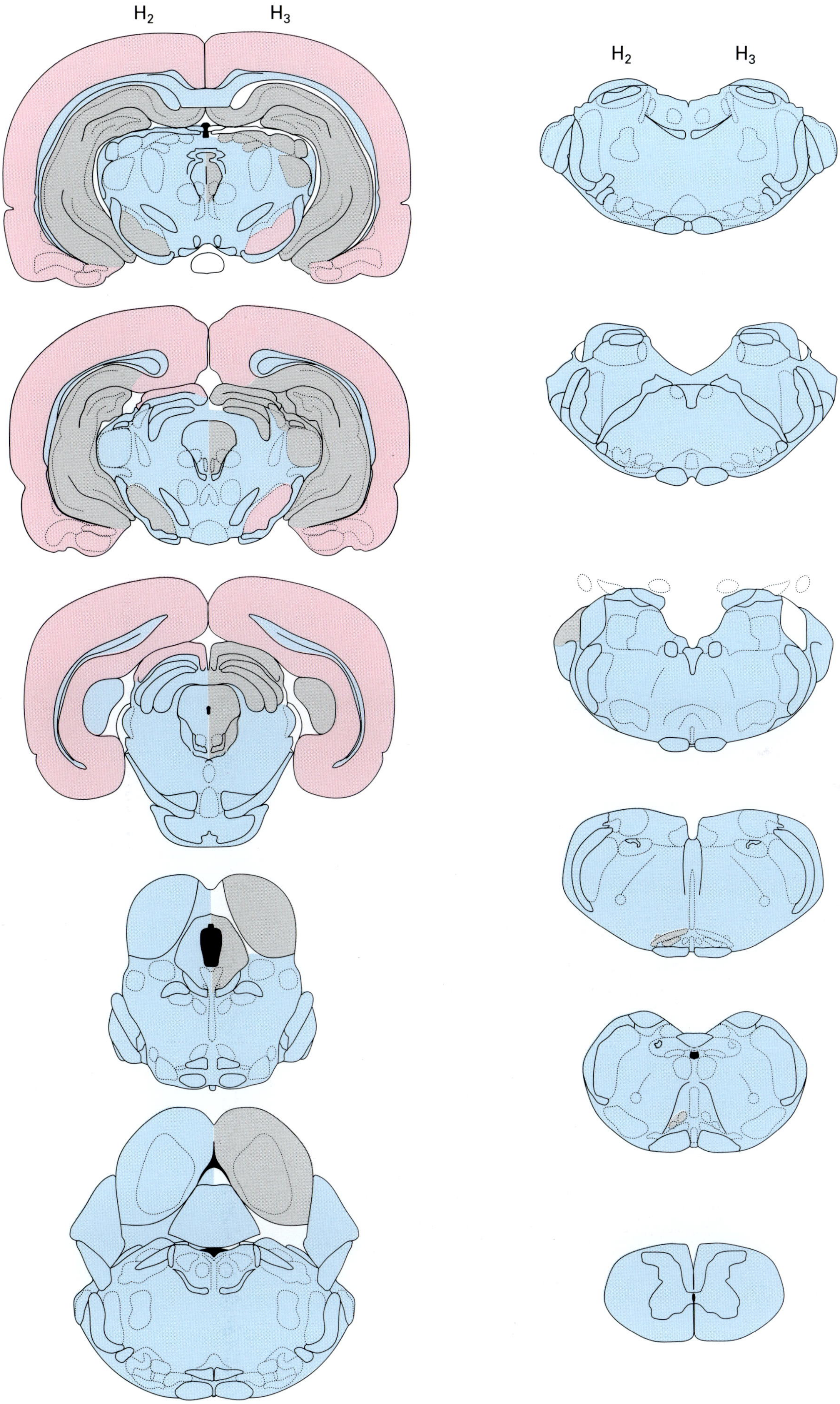

Acetylcholine

4. The acetylcholinergic system

Acetylcholine

Biosynthesis and release

Acetylcholine (ACh) is synthesized from acetyl-CoA and choline by choline acetyl transferase (ChAT) (Fig 4.1, 4.2). It is converted into acetate and choline by acetylcholine esterase (AChE). ACh is mainly synthesized in the cell body and then transported to the terminal, but a certain amount is synthesized at the terminal as well. At the terminal, ACh is taken up to synaptic vesicles and is released to the synaptic junction on stimulation. The released ACh either binds to the receptor or is degraded by AChE. The resulting choline is transferred back into the terminal site by choline transporter and is utilized again in the synthesis of ACh (Fig. 4.2).

Distribution and projection pathway

The distribution of ACh was first studied by enzyme histochemical methods. However, these methods were far from specific, leaving quite a number of uncertainties. The immunocytochemical method, utilizing antisera to ChAT, has made it possible to clarify the outline of the distribution of cholinergic neurons and fibres[1,5–9,11,12,14–16,19,22,25,26]. ACh-containing neurons are further subclassified into Ch1–6,[1,7,8,15,16,19]. Table 4.1 summarizes the localization and projection pathway of these neurons. Undoubtedly many other ACh neurons must exist within the brain.

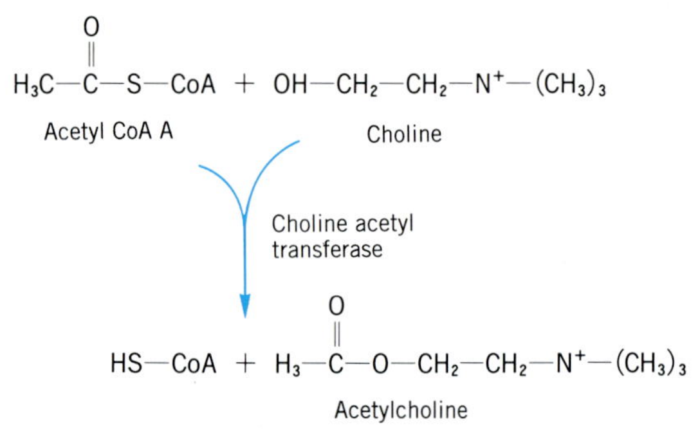

Fig. 4.1 Synthesis of acetylcholine.

Coexistence

It is well known that vasoactive intestinal peptides (VIP) coexist as neurotransmitters in the parasympathetic ganglion (Fig. 4.3) with ACh[10,17]. Thus, it is sometimes wrongly assumed that ACh always coexists with VIP. There are many neurotransmitters (neuromodulators) which coexist with ACh, as shown in Table 4.2[10]. The combination of coexisting neurotransmitters differs from neuron to neuron.

For example, ACh present in the cerebral cortex is often described generally as 'ACh in the cortex'. However, there are at least three types of ACh neuron systems, according to the classification based on projections and coexistence. These are:

- ACh neurons containing VIP, which exist in the cerebral cortx and whose axons terminate within the cortex[7,8]

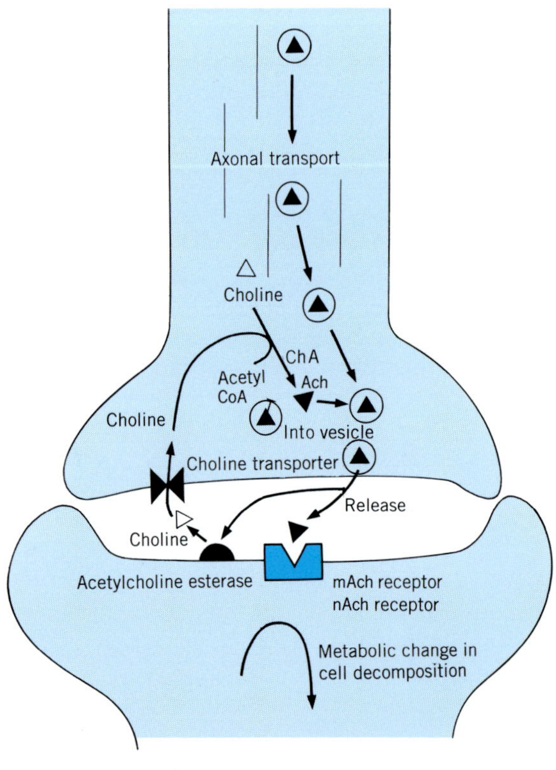

Fig. 4.2 Acetylcholinergic transmission at the synaptic site.

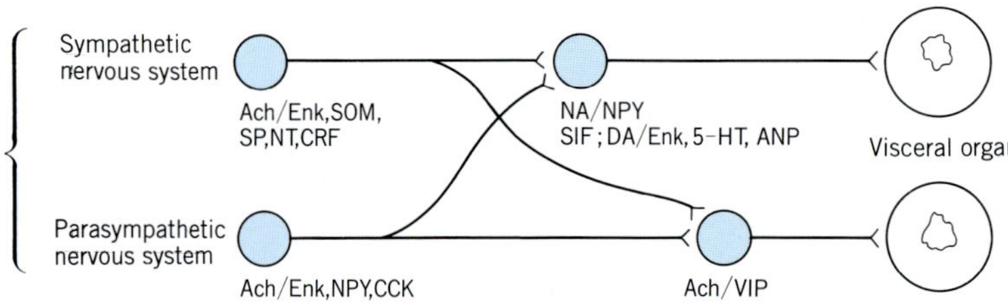

Fig. 4.3 Neurotransmitters or neuromodulators in the autonomic nervous system.

Table 4.1. Ch1–6 neurons and their projections

	Locatin of cholinergic nervous	**Projecting area**
Ch1	Medial septal area	Hippocampus
Ch2	Vertical limb of the diagonal band of Broca	Hippocampus
Ch3	Horizontal limb of the diagonal band of Broca	Olfactory bulb, Thalamic reticular nucleus
Ch4	Basal nucleus of Mynert	Cerebral cortex, Amygdala Thalamic reticular nucleus
Ch5	Nucleus pedunculopontinus	Thalamus
Ch6	Laterodorsal tegmental nucleus	Cerebral cortex, Thalamus and many others

- those containing substance P (SP) which originate from the pontine laterodorsal tegmentum and terminate within the cerebral cortex[23,26]
- those containing galanin which originate from the basal nucleus of Meynert and terminate within the cerebral cortex.[18,19]

Even for those belonging to the same neuron system, the coexistence pattern is different. For example, the vestibular efferent pathway containing Ach originates from the reticular formation, areas lateral and medial to the nucleus of the abducent nerve. Some of the neurons lateral to the nucleus of the abducent nerve contain CGRP in addition to ACh, while others lack CGRP.[21]

The efferent nerve from the superior olivary nucleus terminates in the cochlea. This olivocochlear bundle, which is divided into lateral and medial fibres, contains ACh[2,3]. In addition, the bundle contains enkephalin, CGRP, dynorphin, and GABA[2,3,20,24].

What is the significance of such coexistence? This problem has been studied in postgranglionic neurons of the parasympathetic system and motor neurons. In salivary glands, ACh is released on weak stimulation, while VIP is released only on strong stimulation[10,17].

In bronchial smooth muscles[17], while ACh exhibits contractive actions, VIP shows dilatatory actions[17]. Therefore it is postulated that under physiological conditions these two neurotransmitters, released from the same nerve terminal, cooperate in a coordinated fashion in controlling the tonus of the bronchus. In pathological situations such as bronchial asthma such coordination may be disturbed. It is hypothesized that under such conditions, the activity of VIP is decreased by proteinases released from accumulating neutrophils, eosinophils, and mast cells in the bronchus, resulting in its hypersensitivity due to relative dominance of ACh over VIP[4]. In motor neurons, CGRP potentiates the contraction caused by ACh[13]. This is known to be elicited by the activation of adenylate cyclase.

Thus, in the peripheral nervous system, ACh seems to play a major role while other coexisting components function as neuromodulators. However, such a proposition cannot be extrapolated to the central nervous system, since the role of ACh differs from region to region.

Table 4.2. Examples of neurotransmitters/neuromodulators coexisting with acetylcholine, and their locations

Enkephalin	Superior olivary nucleus, Sympathetic preganglionic neurons
Substance P	Laterodorsal tegmental nucleus, Sympathetic postganglionic neurons
VIP	Cerebral cortex, Parasympathetic postganglionic neurons
Galanin	Basal nucleus of Mynert
CGRP	Motor neuron, Superior olivary nucleus
GABA	Superior olivary nucleus
Somatostatine	Sympathetic preganglionic neurons
Neurotensin	Sympathetic preganglionic neurons
ACTH	Sympathetic preganglionic neurons
Neuropeptide Y	Parasympathetic preganglionic neurons
Cholecystokinin	Parasympathetic preganglionic neurons

References

1. Armstrong, D.M., Saper, C.B., Level, A.I. *et al.* (1983). *J. Comp. Neurol.*, **216**, 53–68.
2. Altschuler, R.A., Hoffman,, D.W., Reeks, D.M., and Fex, J. (1985). *Hearing Res.*, **17**, 249–58.
3. Altschuler, R.A., Parakkal, M.H., and Fex, J. (1983). *Neuroscience*, **9**, 621–30.
4. Barnes, P. J. (1987). *Trends Pharmacol. Sci.*, **8**, 24–7.
5. Carlsen, J. and Heimer, L. (1986). *J. Comp. Neurol.*, **244**, 121–36.
6. Carlsen, J., Zaborsky, L., and Heimer, L. (1985). *J. Comp. Neurol.*, **234**, 155–65.
7. Eckenstein, F. and Baughman, R.W. (1984). *Nature (Lond.)*, **309**, 153–5.
8. Eckenstein, F.P., Baughman, R.W., and Quinn, J. (1988). *Neuroscience*, **25**, 457–74.
9. Hallanger, A.E., Levey, A.I., Lee, H.J. *et al.* (1987). *J. Comp. Neurol.*, **2632**, 105–24.
10. Hokfelt, T., Holets, V.R., and Stains, W. (1987). *Progr. Brain Res.*, **68**, 33–70.
11. Houser, C.R., Crawford, G.D., Barber, R.P. *et al.* (1983). *Brain Res.*, **266**, 97–119.
12. Ichikawa, T. and Hirata, Y. (1986). *Neuroscience*, **6**, 281–92.
13. Ishida-Yamamoto, A. and Tohyama, M. (1989). *Progr. Neurobiol.*, **33**, 335–86.
14. Kimura, H., MacGeer, P.L., and Peng, J.H. (1984). In *Handbook of chemical neuroanatomy*, Vol. 3. (eds. T. Hökfelt and A. Björklund), pp. 51–67. Elsevier, Amsterdam.
15. Levey, A.I., Wainer, B.H., Rye, D.B. *et al.* (1984). *Neuroscience*, **13**, 341–53.
16. Levey, A.I., Hallanger, A.E., and Wainer, B.H. (1987). *J. Comp. Neurol.*, **257**, 317–32.
17. Lundberg, J.M. and Hökfelt, T. (1987). *Progr. Brain Res.*, **68**, 241–62.
18. Melander, T., Staines, W.A., Hökfelt, T. *et al.* (1985). *Brain Res.*, **360**, 130–8.
19. Mesulam, M.-M., Mufson, E.J., Wainer, B.H., and Levey, A.L. (1983). *Neuroscience*, **10**, 1185–201.
20. Ohno, K., Takeda, N., Kiyama, H. *et al.* (1994). *BrainRes.*, **644**, 135–43.
21. Ohno, K., Takeda, N., Yamano, M. *et al.* (1991). *Brain Res.*, **566**, 103–7.
22. Rao, Z.R., Shiosaka, S., and Tohyama, M. (1987). *J. Hirnforsch.*, **28**, 553–60.
23. Sakanaka, M., Shiosaka, S., Takatsuki, K. *et al.* (1983). *Brain Res.*, **259**, 123–6.
24. Takeda, N., Kitajiri, M., Girgis, S., Hillyard, C.J. *et al.* (1986). *Exp. Brain Res.*, **61**, 575–8.
25. Tatehata, T., Shiosaka, S., Wanaka, A. *et al.* (1987). *J. Hirnforsch.*, **28**, 707–16.
26. Vincent, S.R., Satoh, K., Armstrong, D.M., and Fibigerm, H.C. (1983). *Nature (Lond.)*, **306**, 688–91.

Acetylcholine (immunocytochemistry)

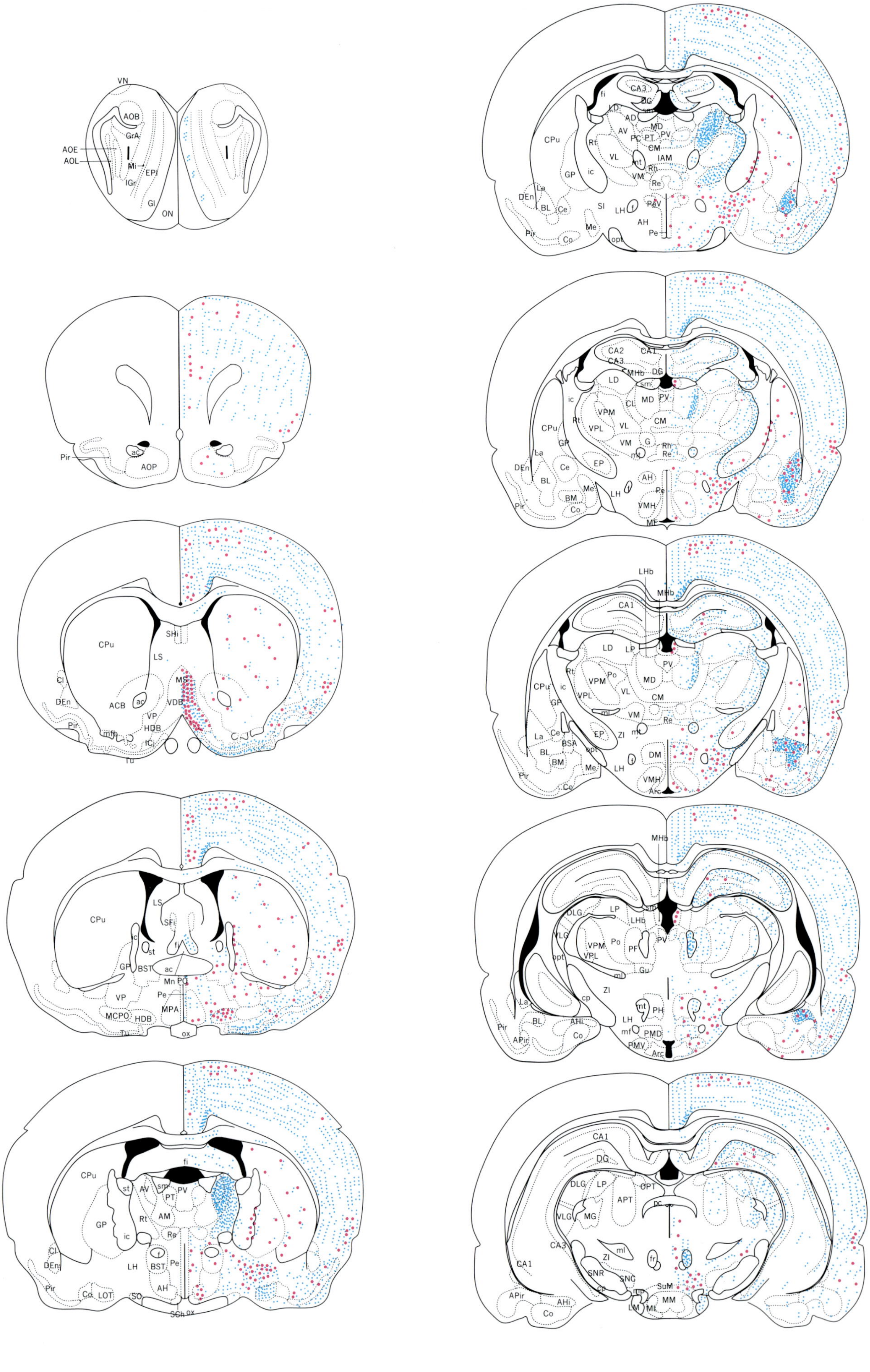

Acetylcholine (immunocytochemistry)

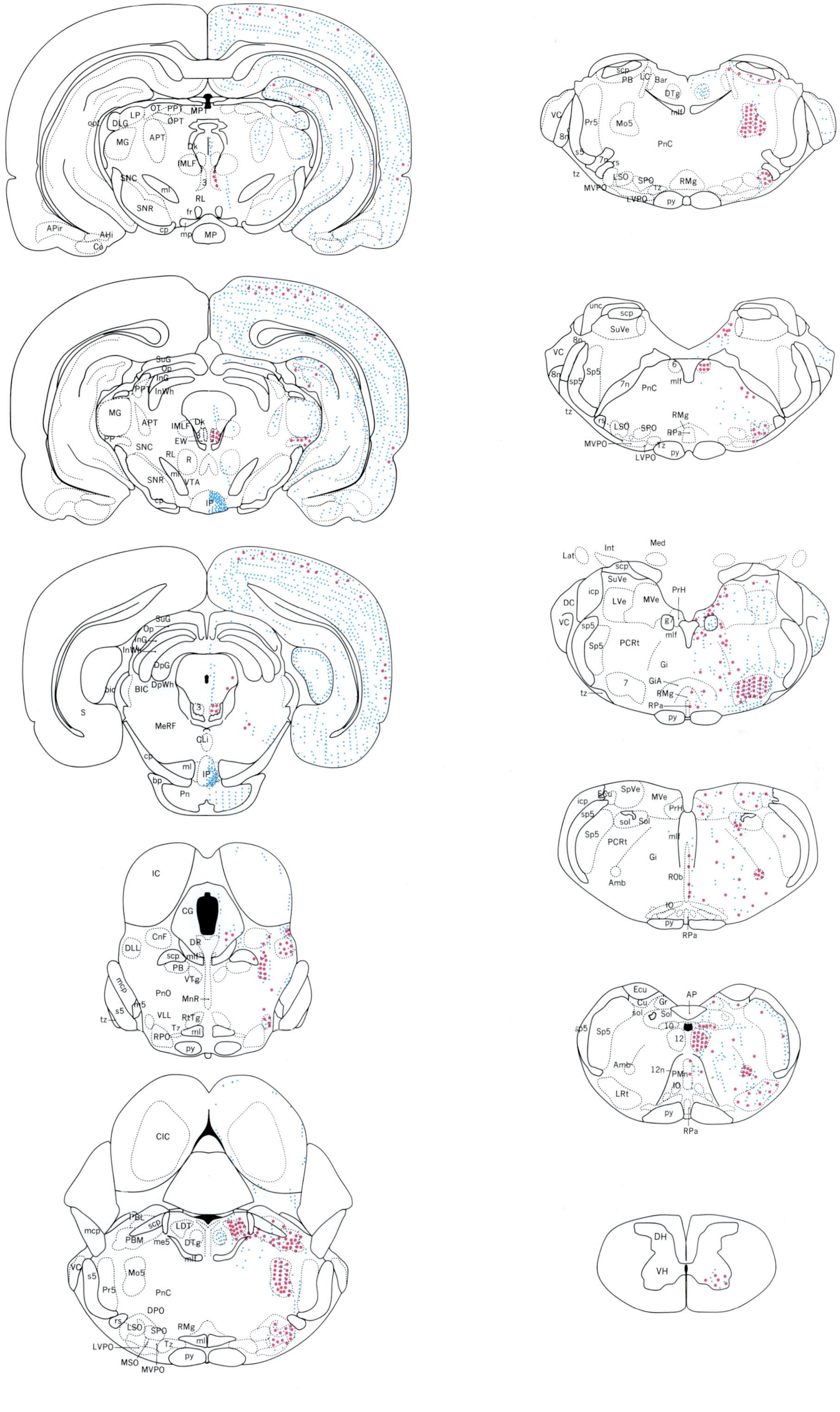

Acetylcholine receptors

Subtypes

The receptors for ACh are classified into muscarinic (mACh) and nicotinic (nACh) receptors. While the mACh receptor is G-protein coupled, the nACh receptor is ionotropic (Na^+ channel).

The nACh receptor of the electric eel is known to be a 500 kDa protein, composed of α_1, β_1, γ, and δ subunits[6] (Table 4.3). Two α_1 subunits and one of each of the other subunits assemble to form a monomer (L-type), constituting a dimer by bridging between δ chains (Table 4.3). All the subunits have four transmembrane perforation sites, like those for other ion-channel proteins. Genes for nACh receptors in the central nervous system have been cloned and their structures elucidated. Those in the central nervous system are different from those in the peripheral nervous system in being composed of only α and β subunits. In addition, their structures are slightly different from that of peripheral nervous system, in that the α subunit has four variants (α_2–α_5) and the β subunit has three variants (β_2–β_4) (Table 4.3).

Distribution

The distribution of nACh receptors has been studied by autoradiography utilizing $[^{125}]\alpha$-bungarotoxin, $[^3H]$nicotine, $[^3N]$ACh, and $[^3H]$methyl carbamylcholine[4]. It should be noted that mACh receptors are to be inhibited when $[^3H]$ACh is used as a ligand. Immunocytochemical methods utilizing antibodies to nACh receptors have also been employed.[5,14] Recently, localization of the mRNAs for each subunit was elucidated by *in situ* hybridization.[16,17]

The mACh receptors were initially classified into M_1 and M_2 based on their affinities for pirenzepine. The M_1 type has high affinity, and the M_2 type showed low affinity. Subsequently, the M_2 type was further subclassified into cardiac type M_2—which accelerates the turnover of inositol phosphate, though it has considerably higher affinity for pirenzepine (Table 4.4).

Recently, advances in molecular biological techniques have made it possible to identify subtypes HM_1–HM_4[10,11] in humans and m_1–m_5 in rats (Table 4.4)[1–3,7]. All these are G-protein coupled transmembrane proteins with seven perforation sites. The distribution of mACh receptors has been studied by autoradiography using quinuclidinyl benzilate [QNB] as ligand[8,9,12]. However, it is difficult to differentiate M_1 from M_2 by this method. $[^3H]$pirenzepine and $[^3H]$AF-DX 116 are widely employed as the specific ligands for M_1 and M_2 respectively[8,12]. There is another method of identifying them. In this method, the samples are first reacted with pirenzepin or then reacted with $[^3H]$pirenzepine; the sites thus visualized are identified as M_2; on the other hand, samples are first reacted with M_2 sites by carbachol and then visualized by $[^3H]$QNB; the sites thus visualized are identified as M_1[5,12]. Immunocytochemical methods utilizing anti-mACh receptor antibody have also been employed[14]. Recently the *in situ* hybridization technique has elucidated the distribution of each subtype of mACh receptors[3,11,15].

The *in situ* hybridization method has made it possible to clarify the diversity of ACh receptors by localization of mRNA of each subtype of ACh receptor or that of each subtype of nACH receptor. For example, subunits for nACh receptors vary in their distribution depending upon the sites they exist in the brain, indicating the possibility of difference and diversity in their constitution (Fig. 4.4). In addition, there are neurons showing multiple mRNA receptor subtypes. It is even possible that one neuron expresses both nACh and mACh receptors simultaneously. It remains to be discovered if each subunit and subtype are expressed on the same postsynaptic membrane or if they are expressed on different sites (Fig. 4.5). Identification of the localization of the receptor in the peripheral autonomic nervous system is very important, because the nervous system here is a 'synapse en passent' and the targets of the neurotransmitter are not clear. Using *in situ* hybridization histochemistry, the analysis for determination of the cells expressing ACh receptors has been done in the peripheral system from this point of view.

Table 4.3. Subunits of nAch receptor in the peripheral and central nervous system

Peripheral	α_1, β_1, γ, δ
Central	$\alpha_{2\text{–}5}$, $\beta_{2\text{–}4}$

References

1. Bonner, T.I., Buckley, N.J., Young, A.C., and Brann, M.R. (1987). *Science*, **237**, 527–31.
2. Bonner, T.I., Young, A.C., Brann, M.R., and Buckley, N.J. (1988). *Neuron*, **1**, 403–10.
3. Buckley, N.G., Bonner, T.I., and Brann, M.R. (1988). *J. Neurosci.*, **8**, 4646–52.
4. Clarke, P.B.S., Schwartz, R.D., Paul, S.M. *et al.* (1985). *J. Neurosci.*, **5**, 1307–15

Table 4.4. Subtype and function of receptor

Subtype of human mAchR	Subtype rat mAchR	Subtype based on pharmacological reaction	Affinity for ligands		Cyclase activity, PI turnover, and permeability
HM_1	m_1	M_1	Pirenzepine AD-DX116	High Low	K^+ Permeability ↓, Depolarization PI turnover ↑
HM_2	m_2	Heart type M_2	Pirenzepine AF-DX116	Low High	K^+ outflow, Hyperpolarization, Cyclase ↓
HM_3	m_3	M_3	Pirenzepine AF-DX116	Relatively high Low	PI turnover ↑
HM_4	m_4	Lobular M_2	Pirenzepine AD-DX116	Relatively high Relatively low	Cyclase ↑
	m_5		Pirenzepine AF-DX116	Medium Low	

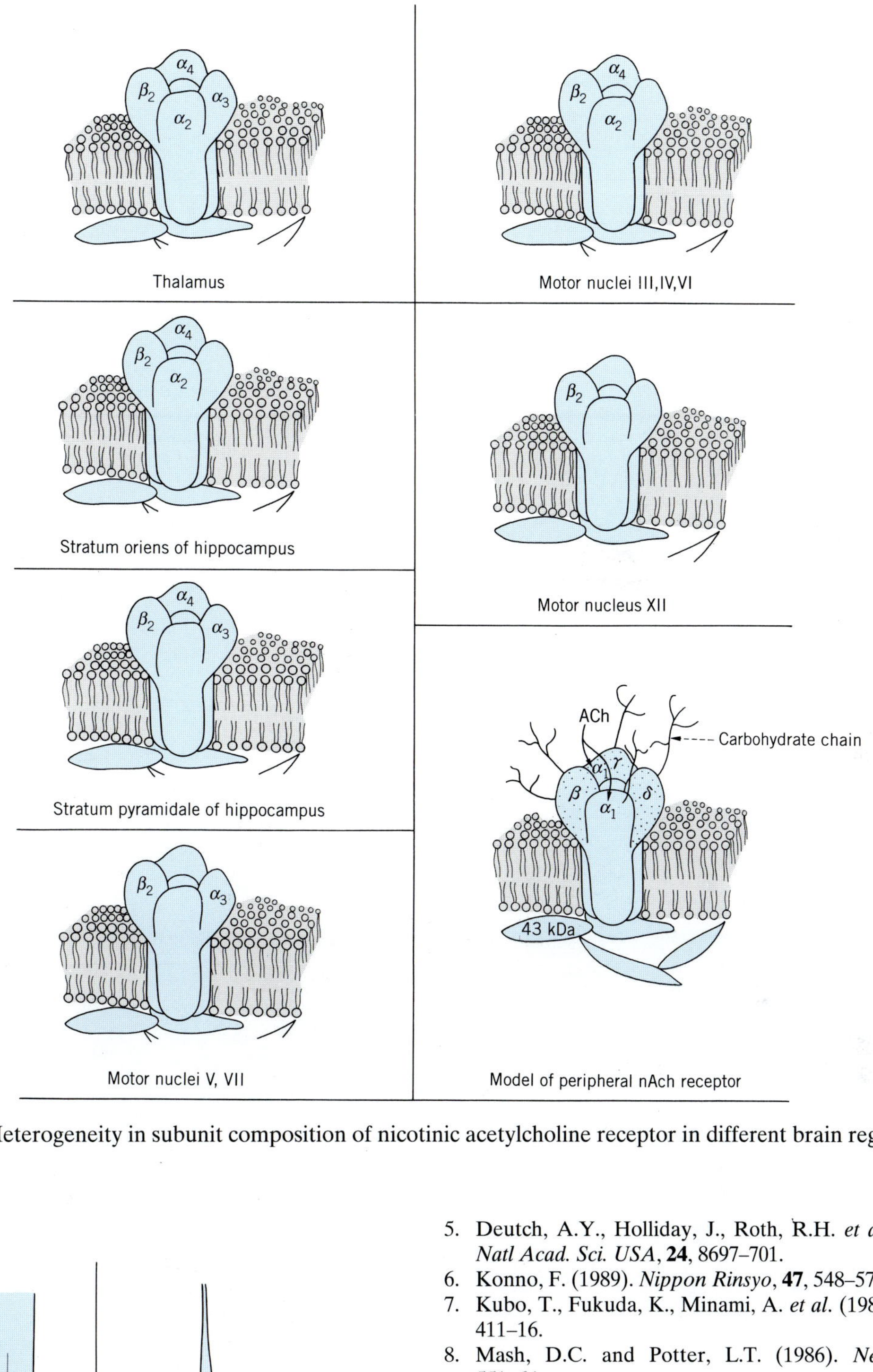

Fig. 4.4 Heterogeneity in subunit composition of nicotinic acetylcholine receptor in different brain regions.

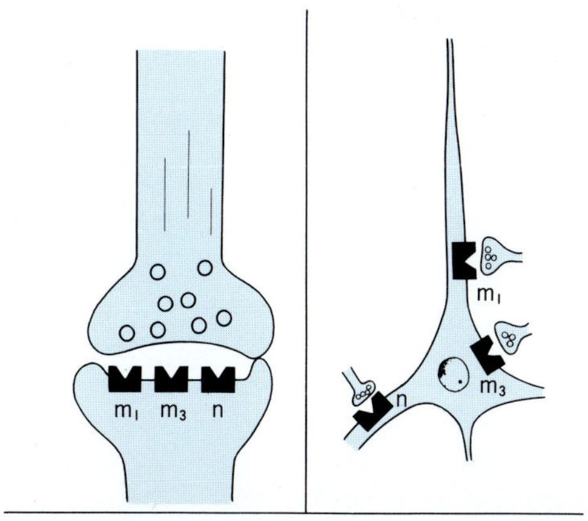

Fig. 4.5 Mode of expression of the cholinergic receptors.

5. Deutch, A.Y., Holliday, J., Roth, R.H. *et al.* (1987). *Proc. Natl Acad. Sci. USA*, **24**, 8697–701.
6. Konno, F. (1989). *Nippon Rinsyo*, **47**, 548–57.
7. Kubo, T., Fukuda, K., Minami, A. *et al.* (1986). *Nature*, **323**, 411–16.
8. Mash, D.C. and Potter, L.T. (1986). *Neuroscience*, **19**, 551–64.
9. Neustadt, A., Frostholm, A., and Rotter, A. (1988). *J. Comp. Neurol.*, **272**, 317–30.
10. Perlta, E.G., Ashkenazi, A., Winslow, J.W. *et al.* (1987). *EMBO J.*, **6**, 3923–9.
11. Perlta, E.G., Winslow, J.W., Peterson, G.L. *et al.* (1987). *Science*, **236**, 600–5.
12. Regenold, W., Araujo, D.M., and Quirion, R. (1989). *Synapse*, **4**, 115–25.
13. Swanson, L.W., Simmons, D.M., Whiting, P.J., and Lindstrom, J. (1987). *J. Neurosci.*, **7**, 3334–42.
14. Van der Zee, E.A., Matsuyama, T., Strosberg, A.D. *et al.* (1989). *Histochemistry*, **92**, 4646–57.
15. Vilaro, M.T., Palacios, J.M., and Mengod, G. (1990). *Neurosci. Lett.*, **114**, 154–9.
16. Wada, E., Wada, K., Boutler, J. *et al.* (1989). *J. Comp. Neurol.*, **284**, 314–35.
17. Wada, E., Mckinnon, D., Heinemann, S., Patrick, J., and Swanson, L.W. (1990). *Brain Res.*, **526**, 45–53.

Nicotinic acetylcholine receptor (binding sites)

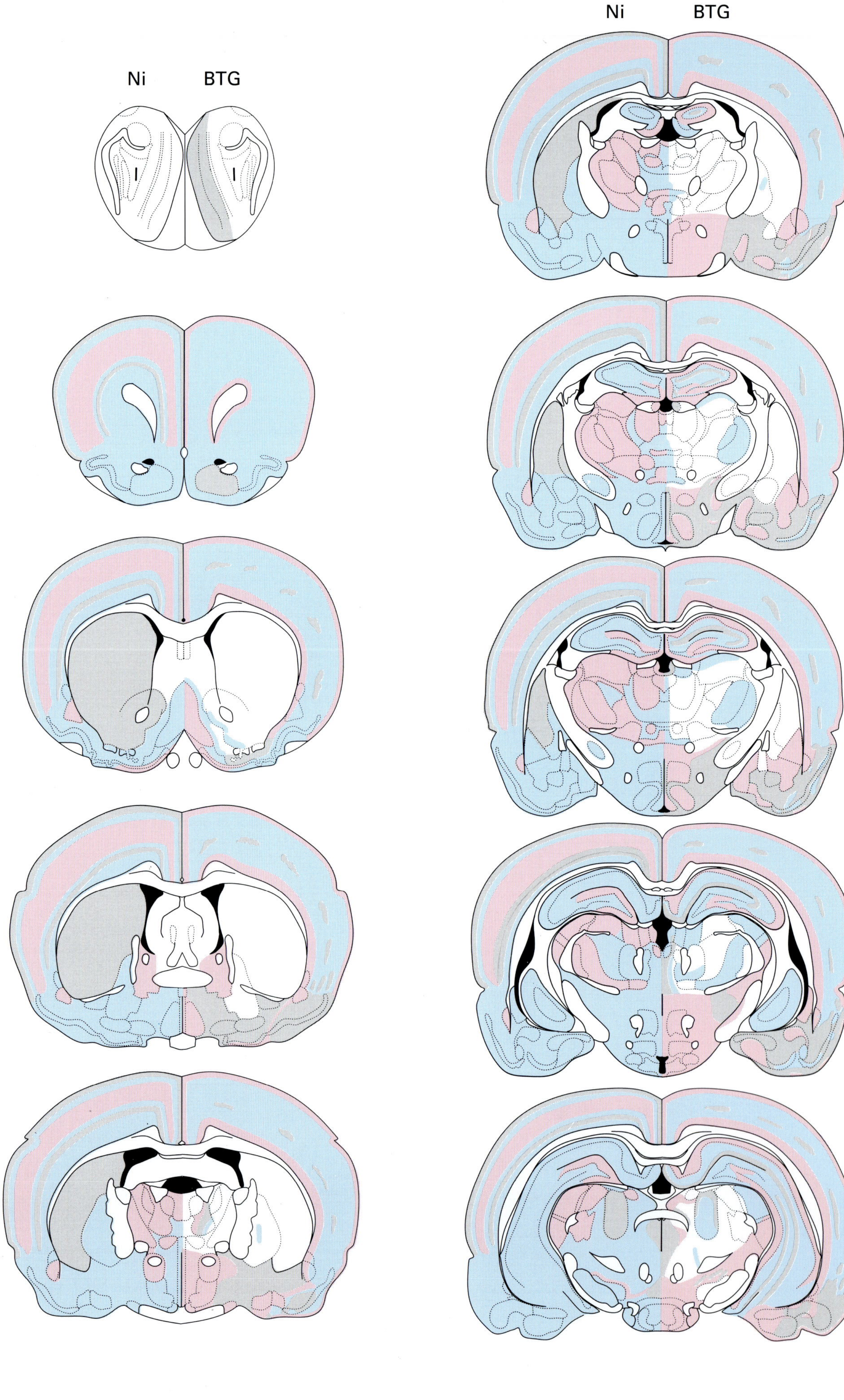

Nicotinic acetylcholine receptor (binding sites)

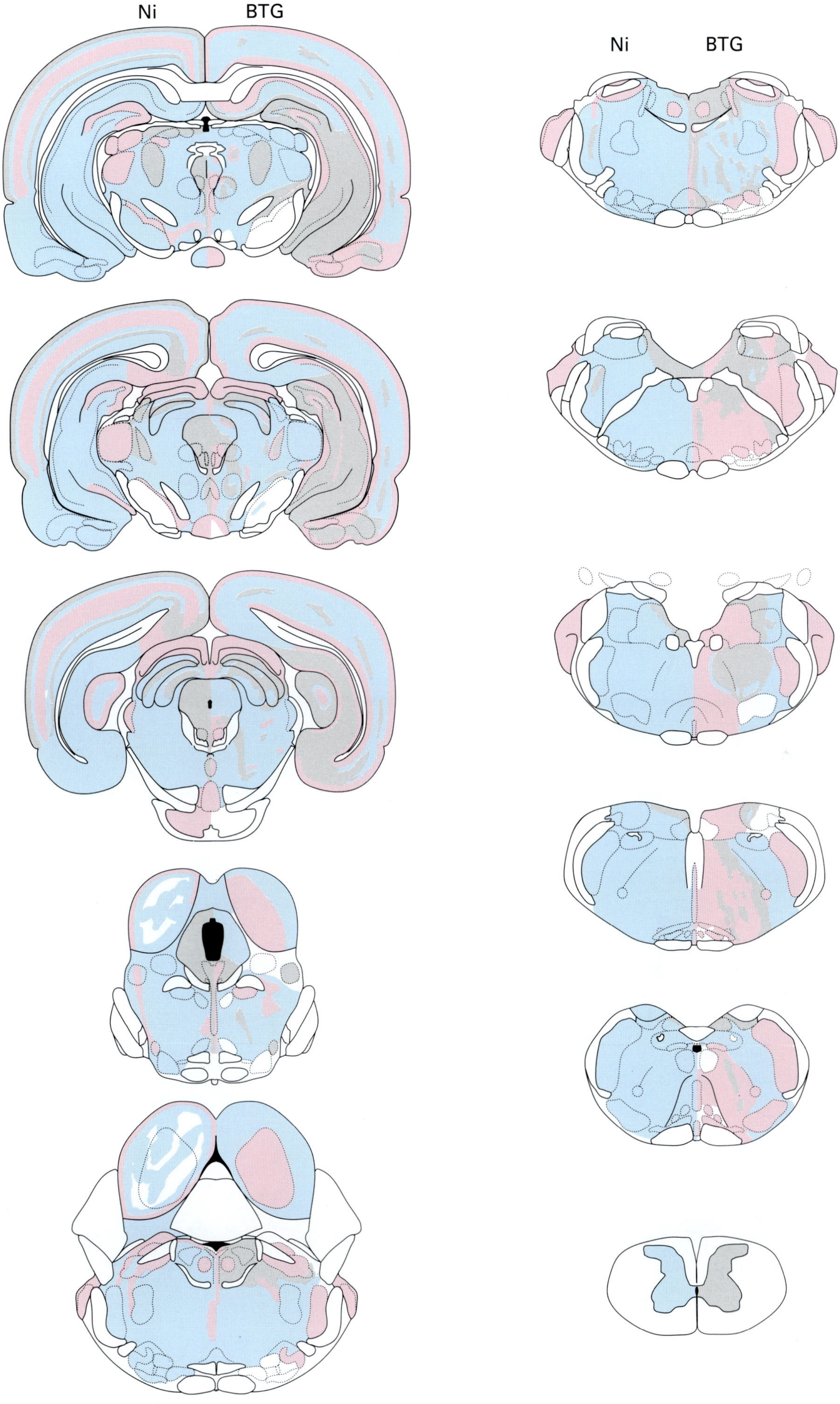

Nicotinic acetylcholine receptor (immunocytochemistry)

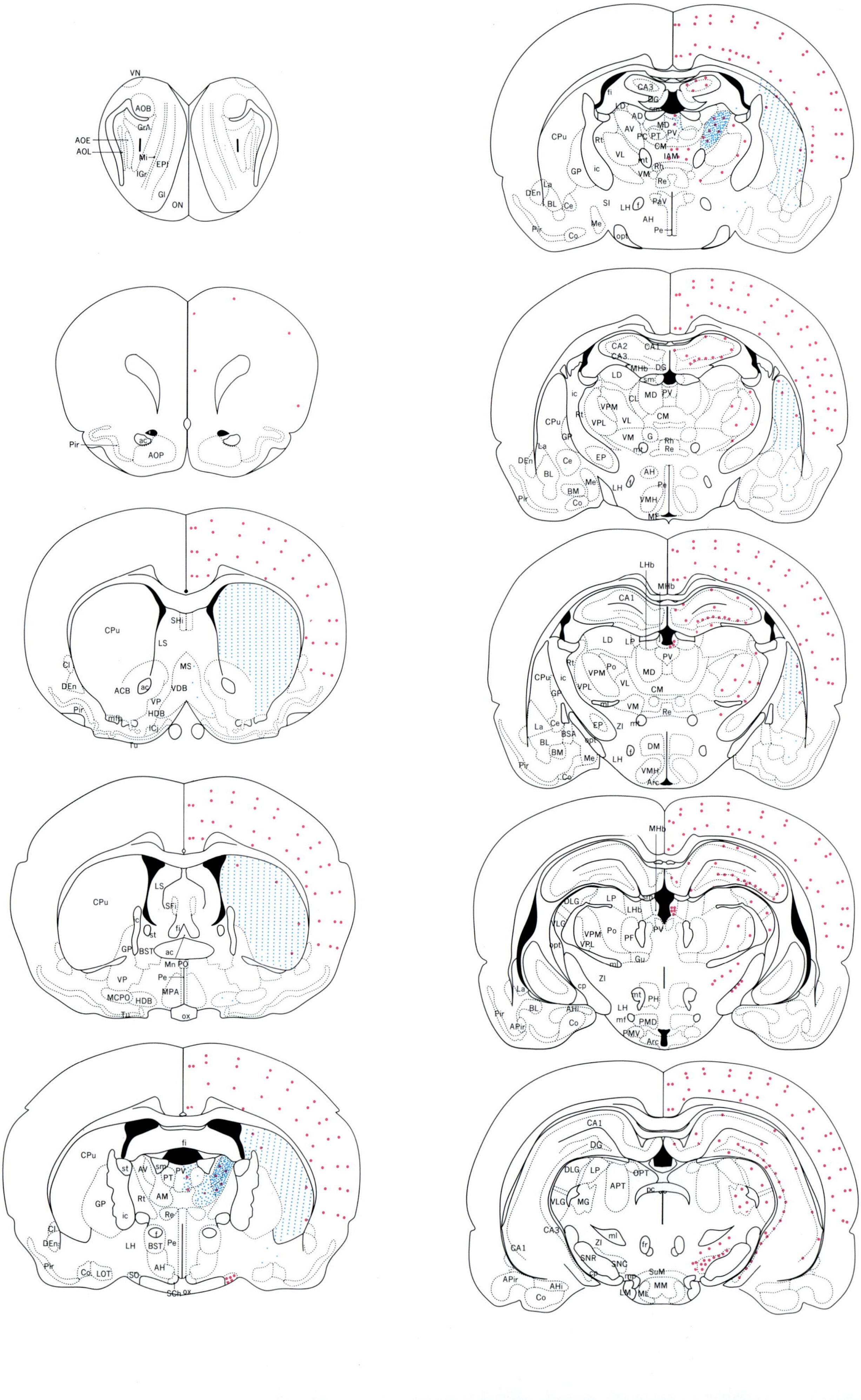

Nicotinic acetylcholine receptor (immunocytochemistry)

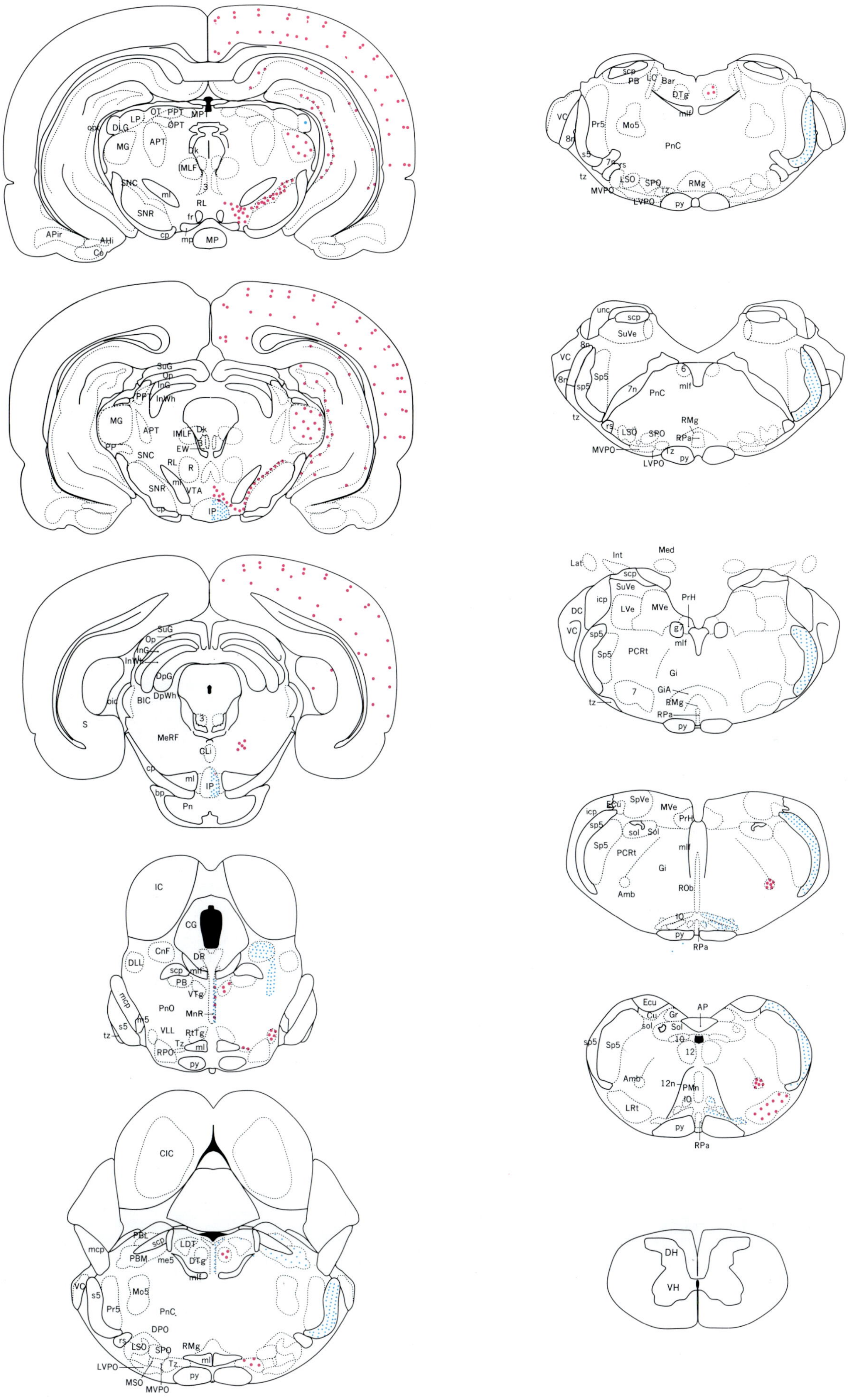

Nicotinic acetylcholine receptor (immunolabelling)

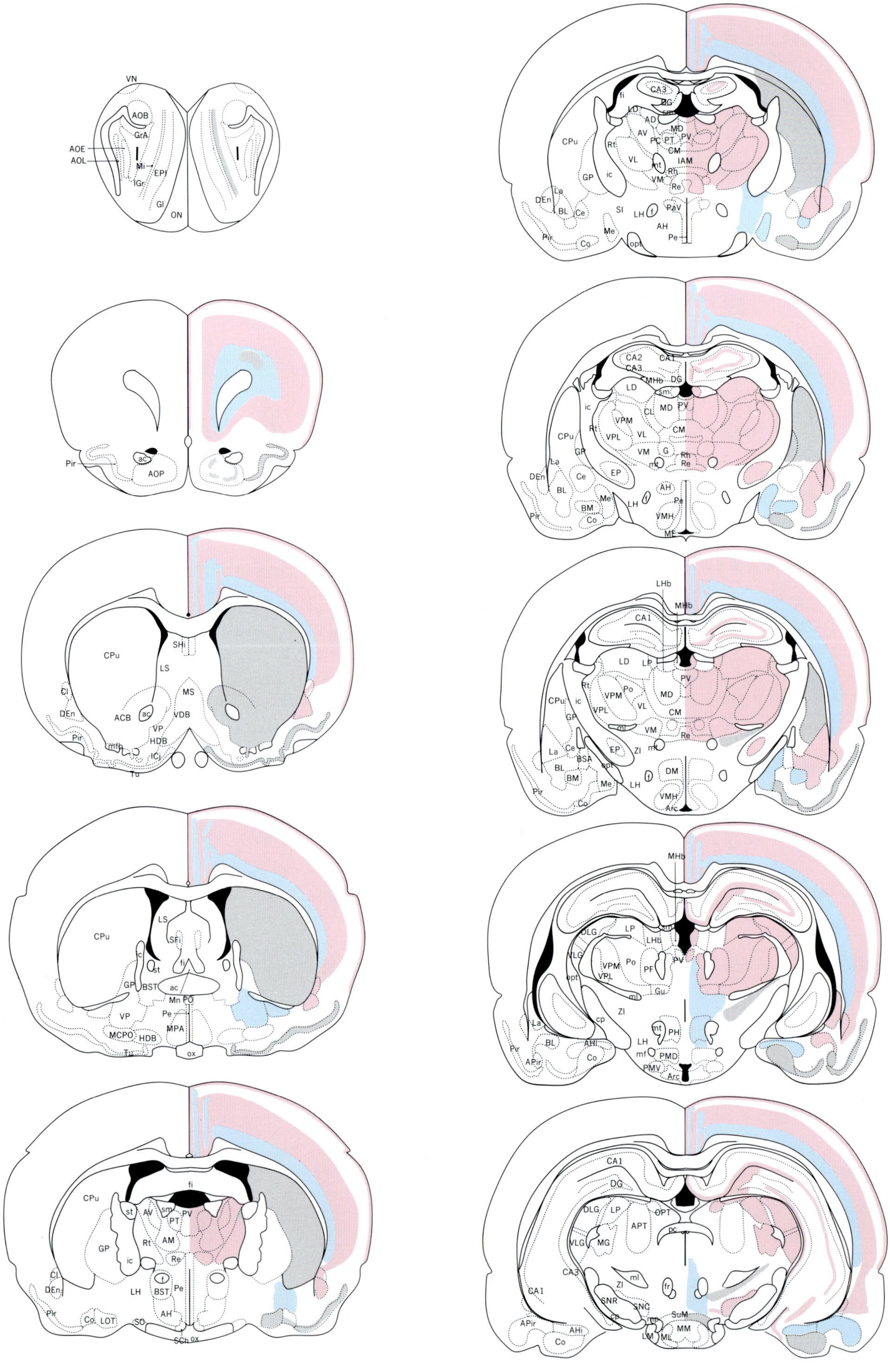

Nicotinic acetylcholine receptor (immunolabelling)

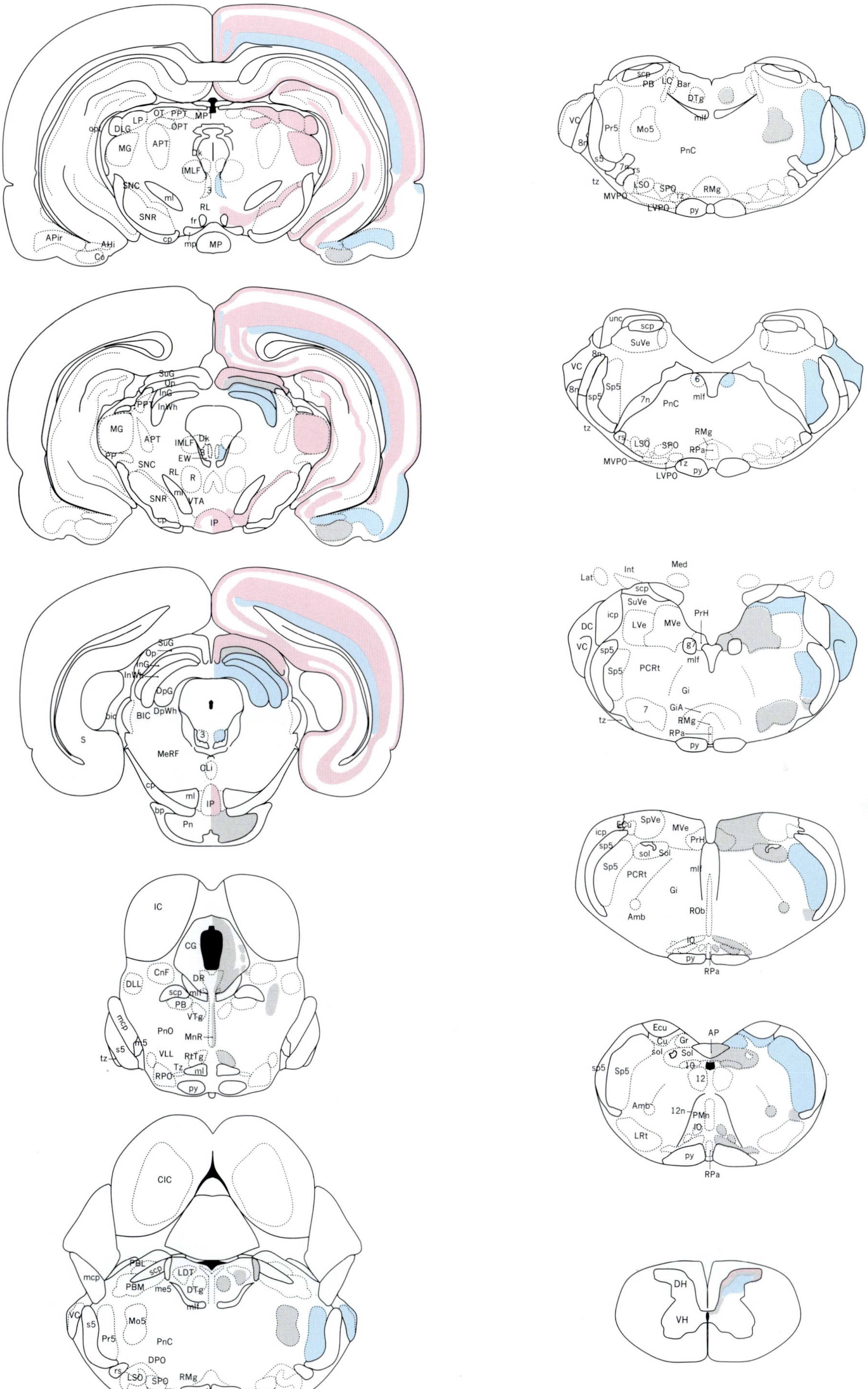

Nicotinic acetylcholine receptor (ISH)

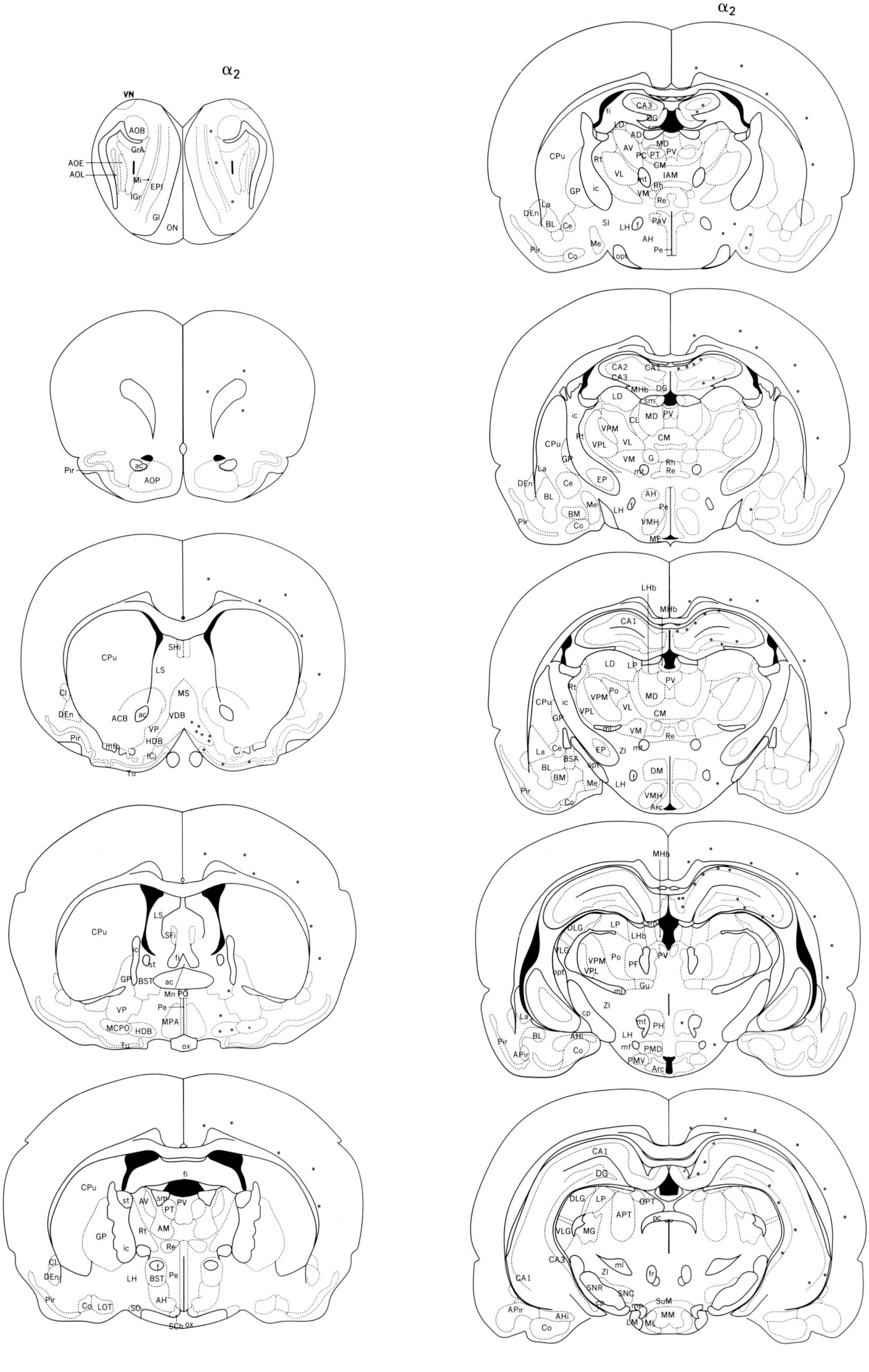

Nicotinic acetylcholine receptor (ISH)

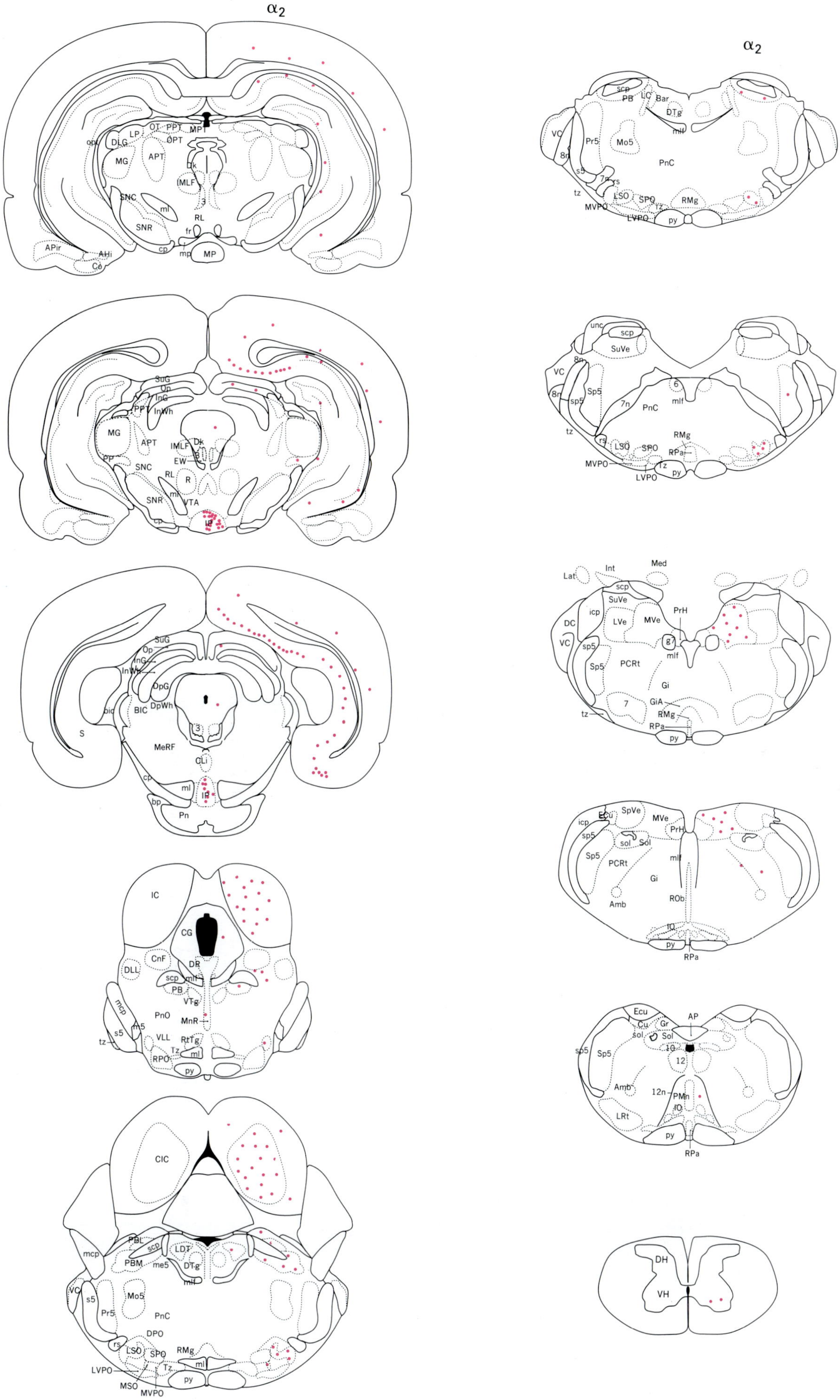

Nicotinic acetylcholine receptor (ISH)

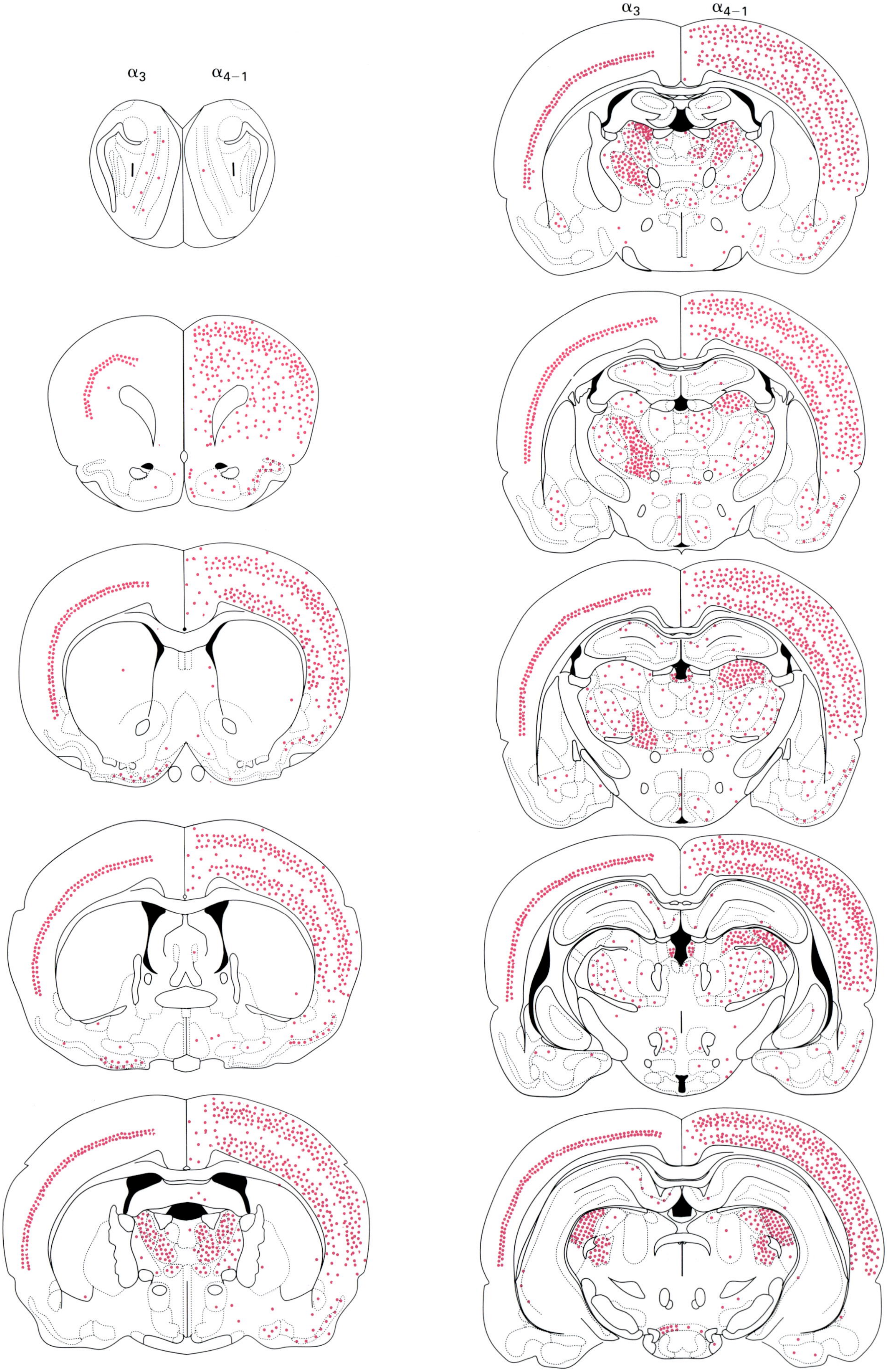

Nicotinic acetylcholine receptor (ISH)

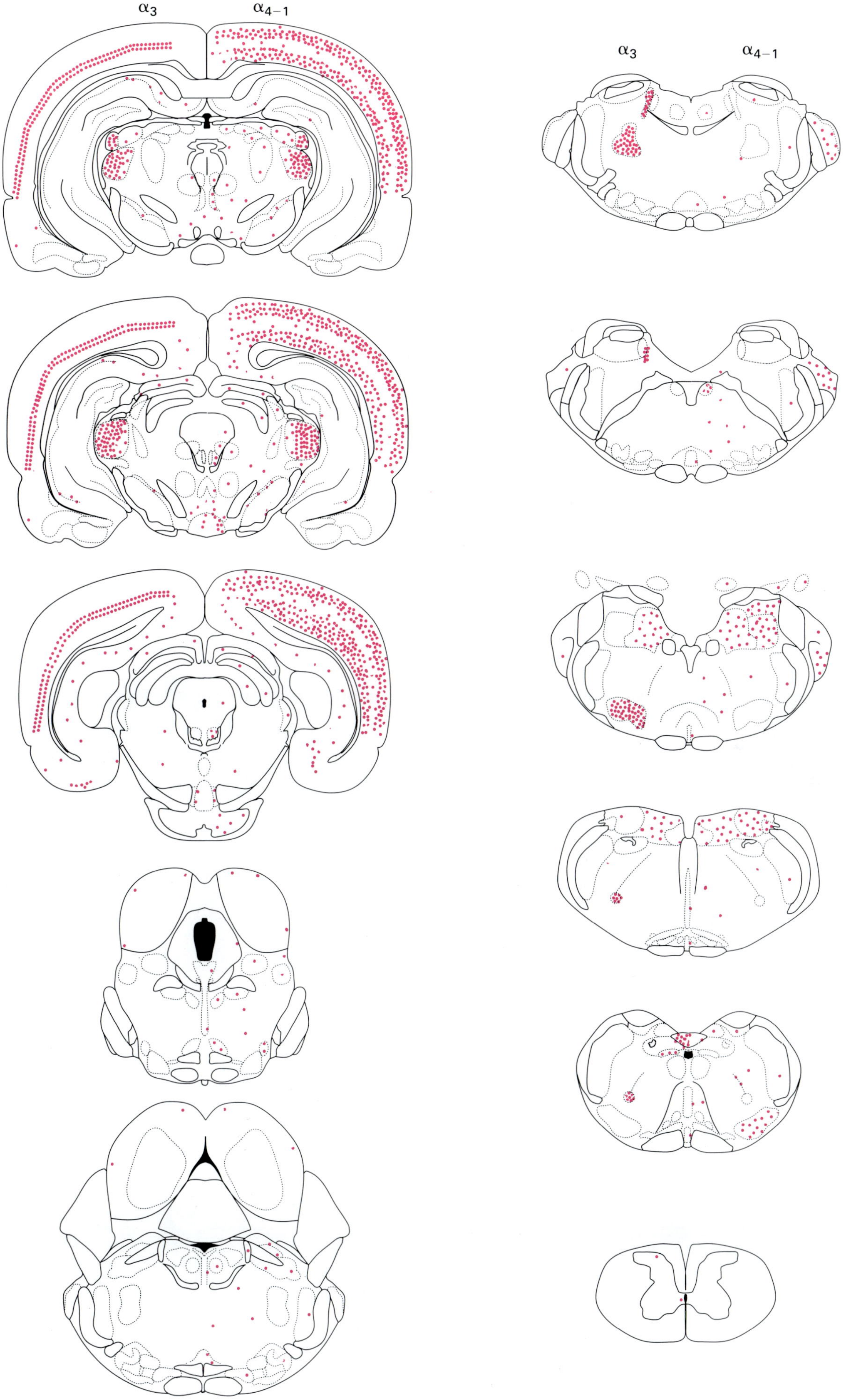

Nicotinic acetylcholine receptor (ISH)

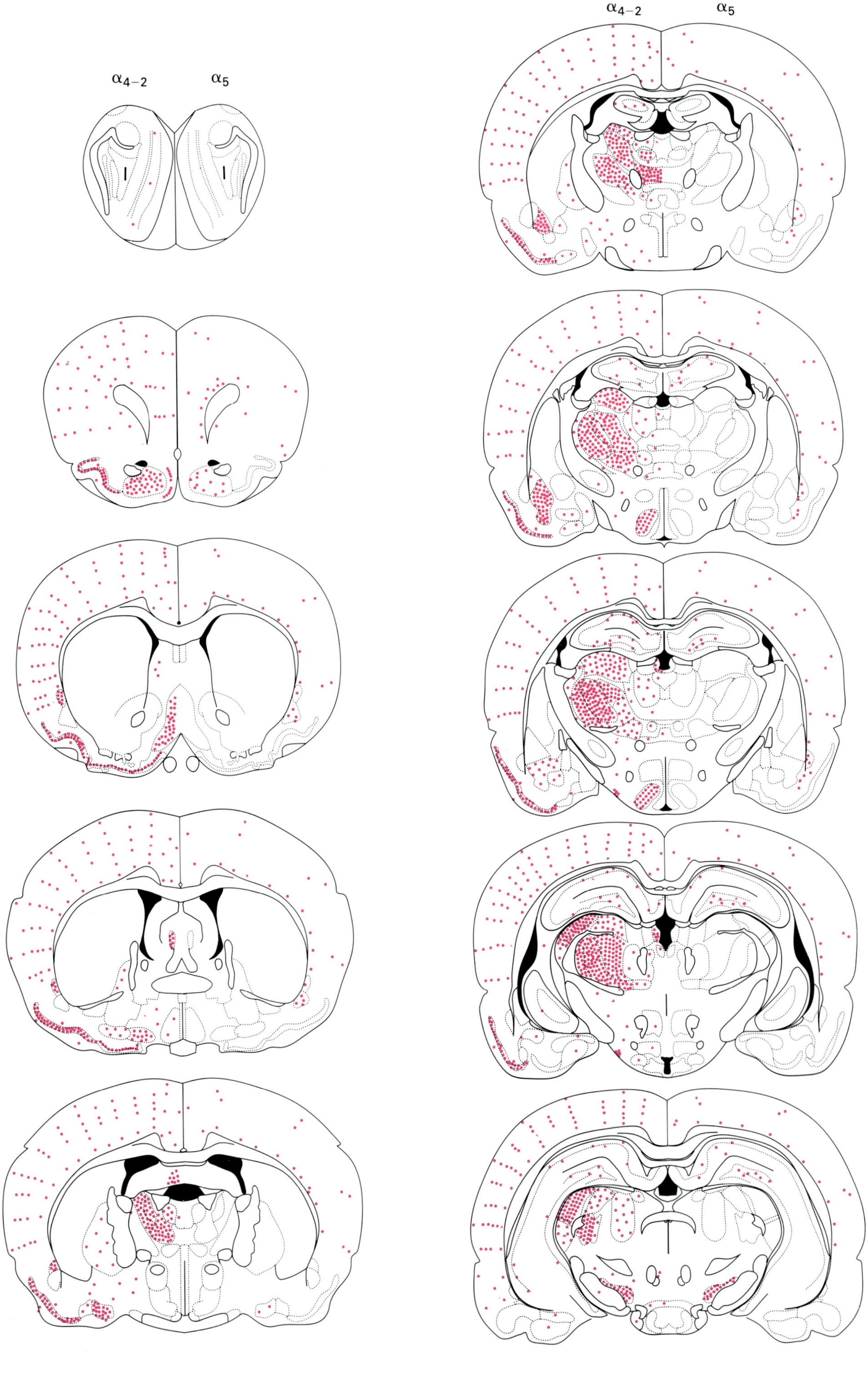

Nicotinic acetylcholine receptor (ISH)

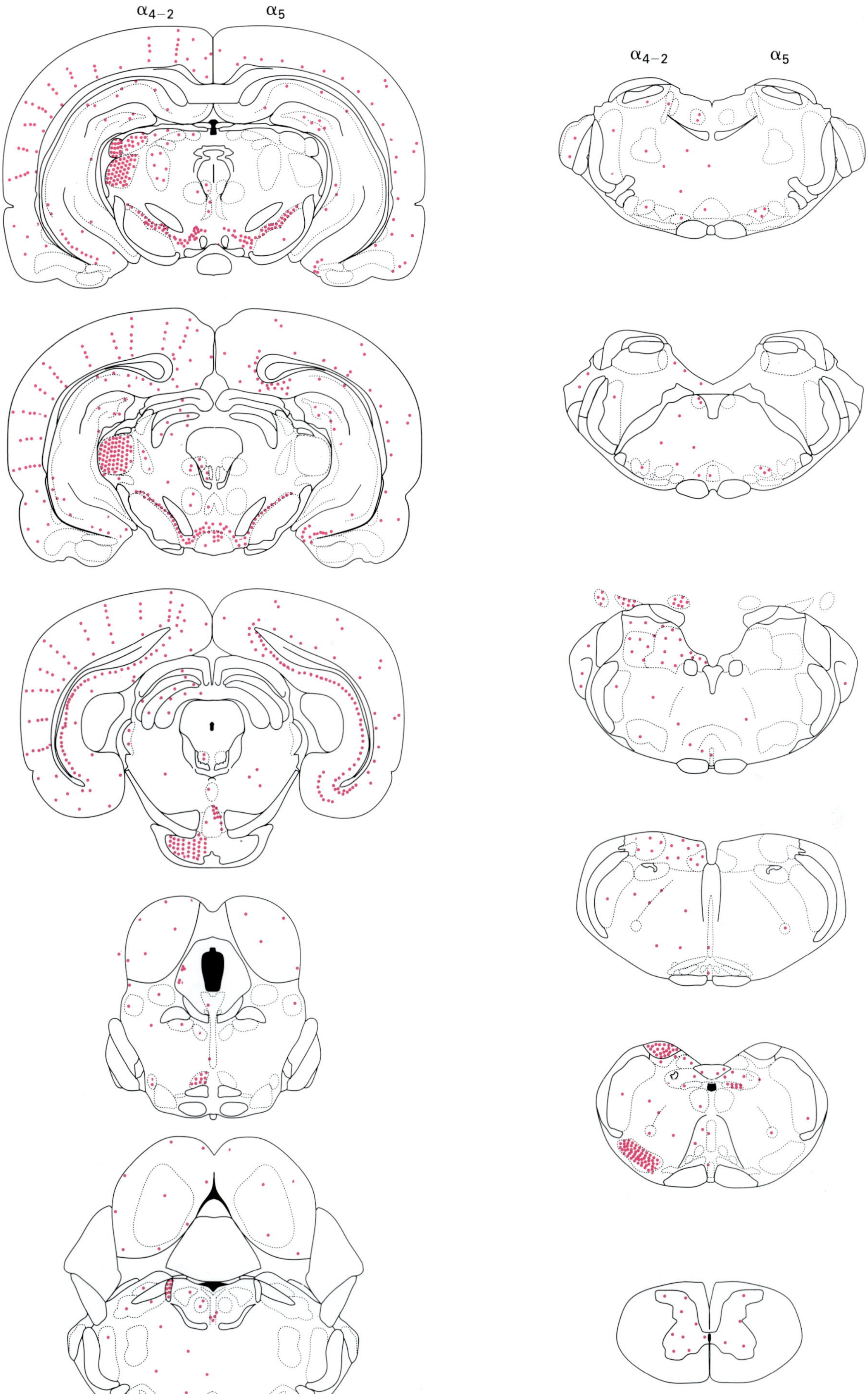

Nicotinic acetylcholine receptor (ISH)

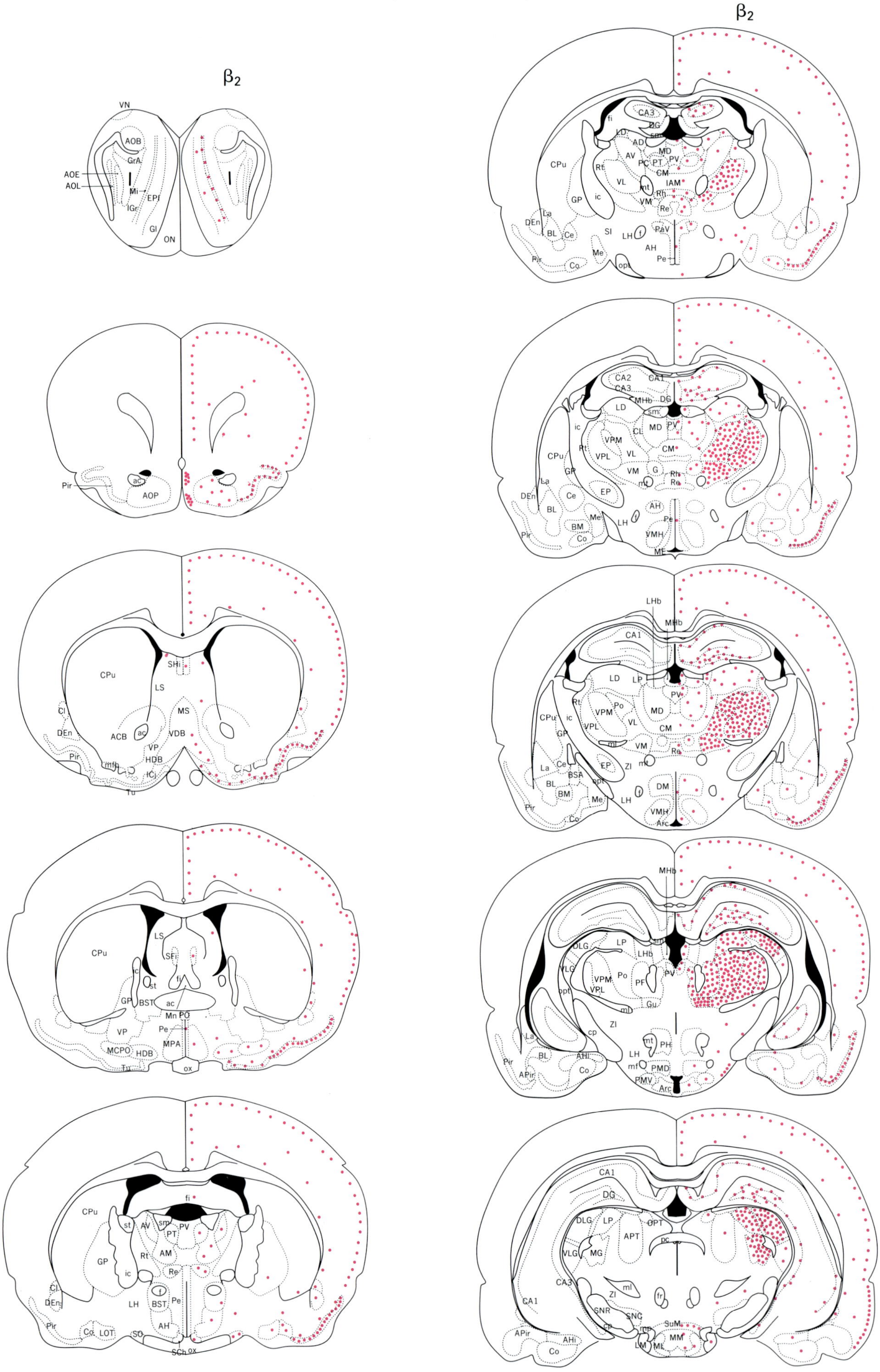

Nicotinic acetylcholine receptor (ISH)

β_2

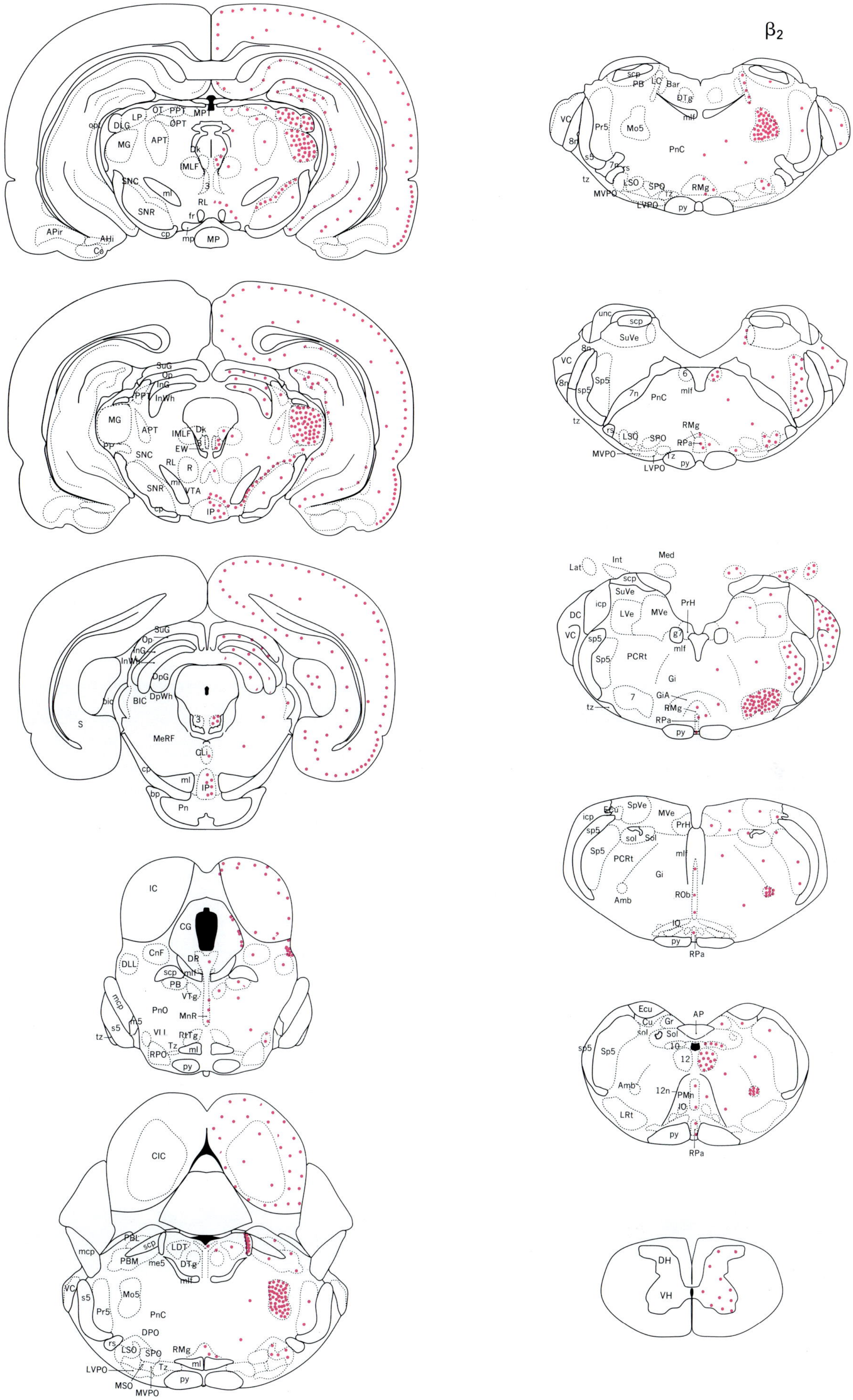

Muscarinic acetylcholine receptor (binding sites)

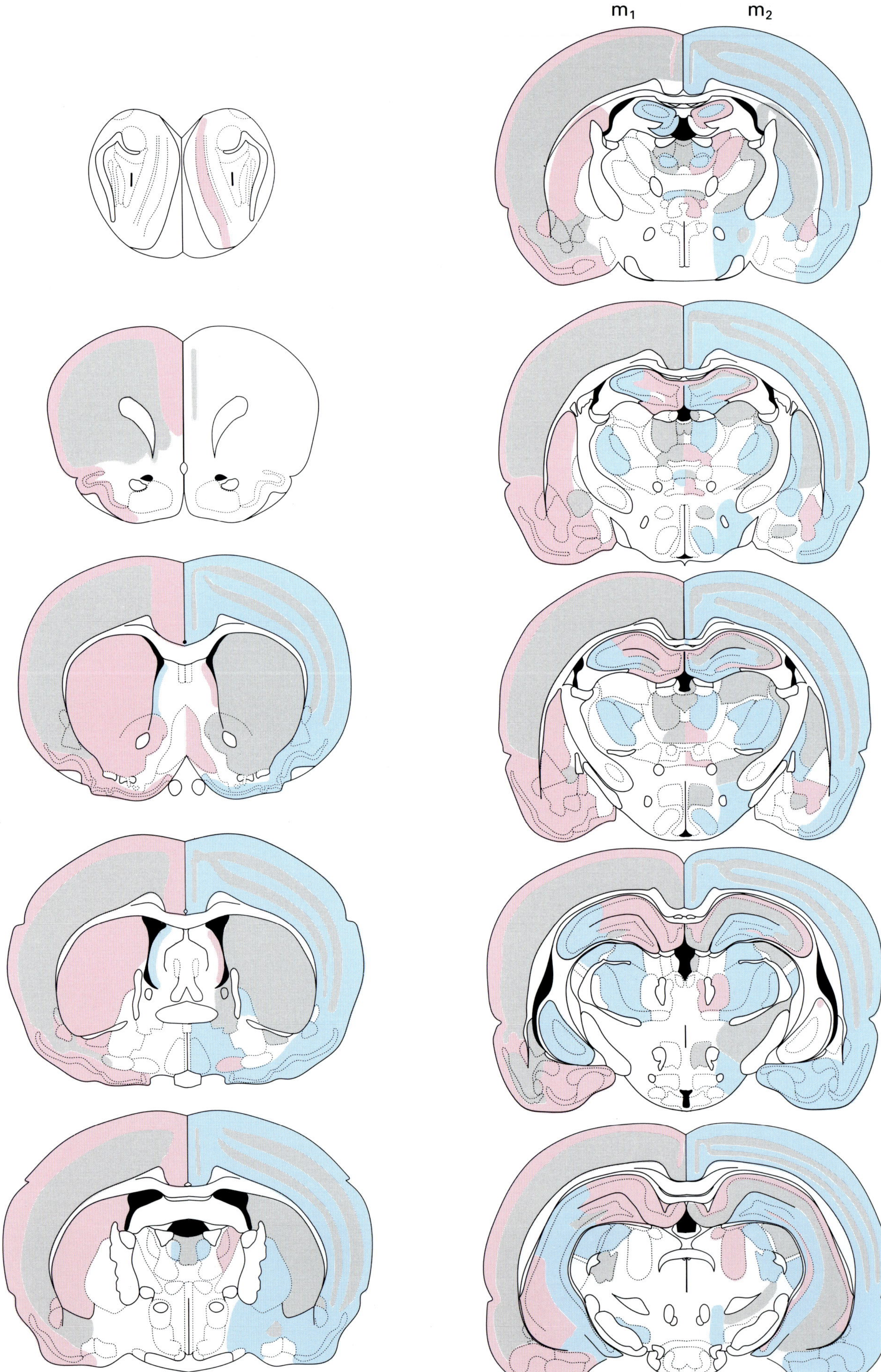

Muscarinic acetylcholine receptor (binding sites)

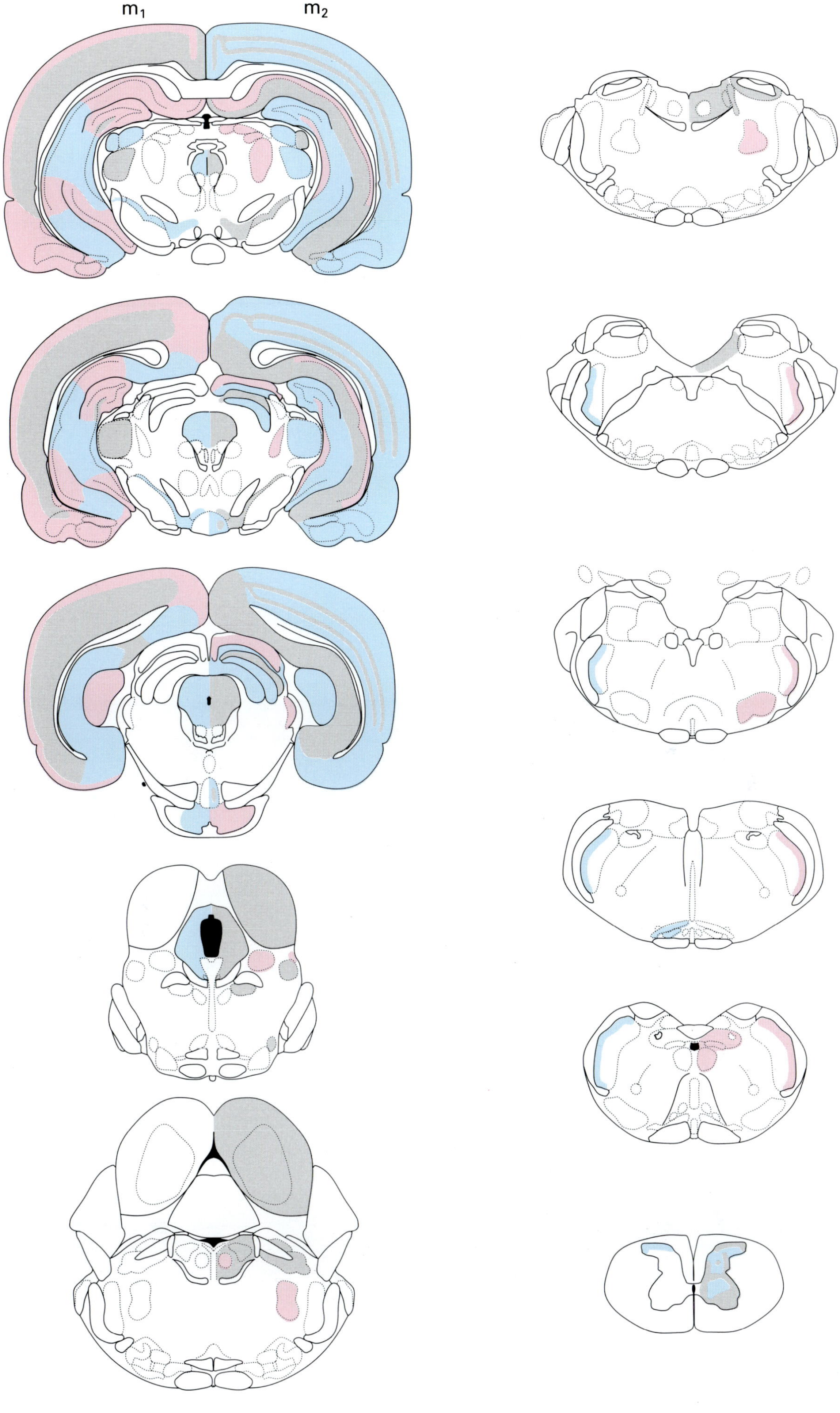

Muscarinic acetylcholine receptor (immunocytochemistry)

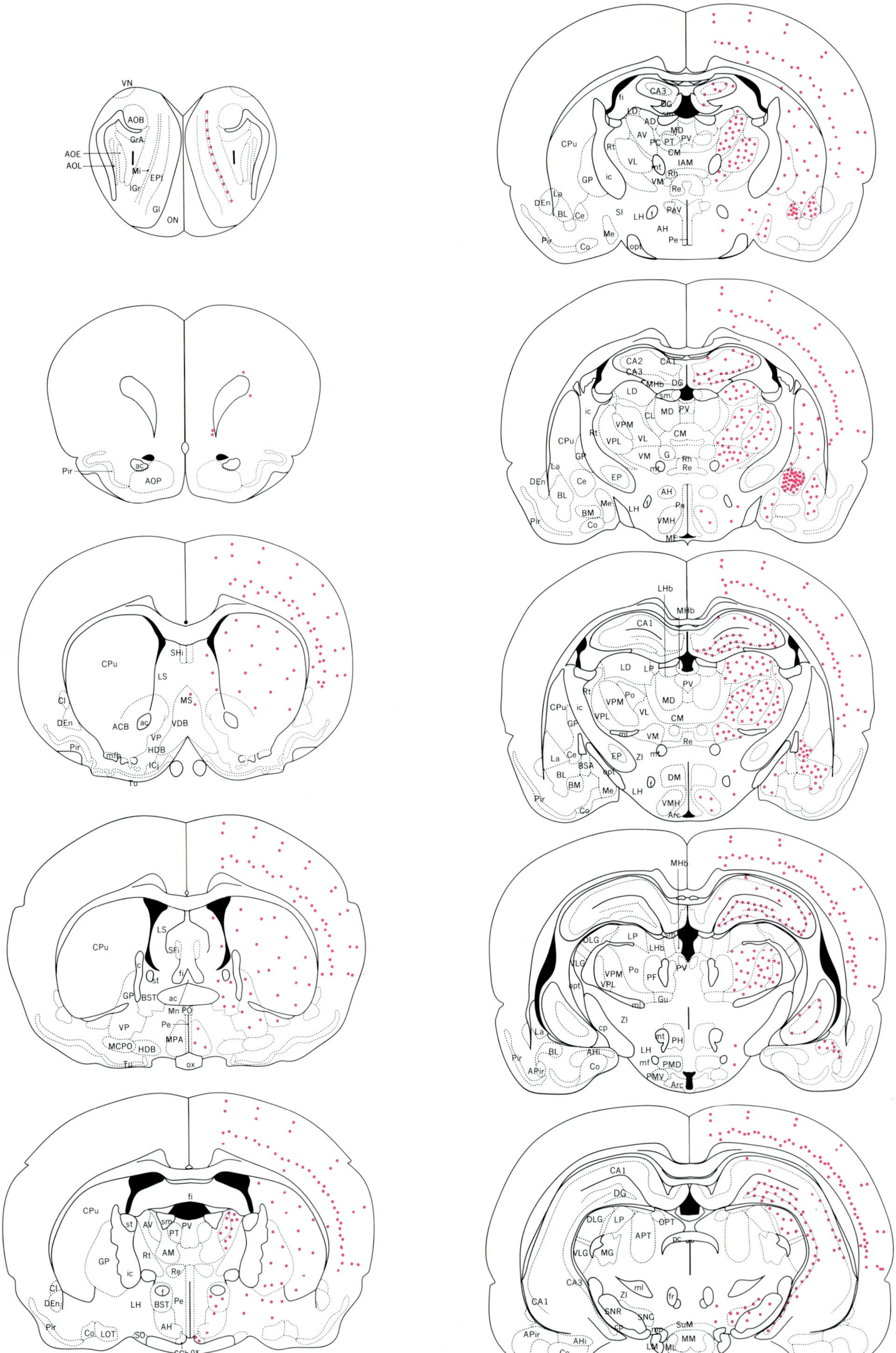

Muscarinic acetylcholine receptor (immunocytochemistry)

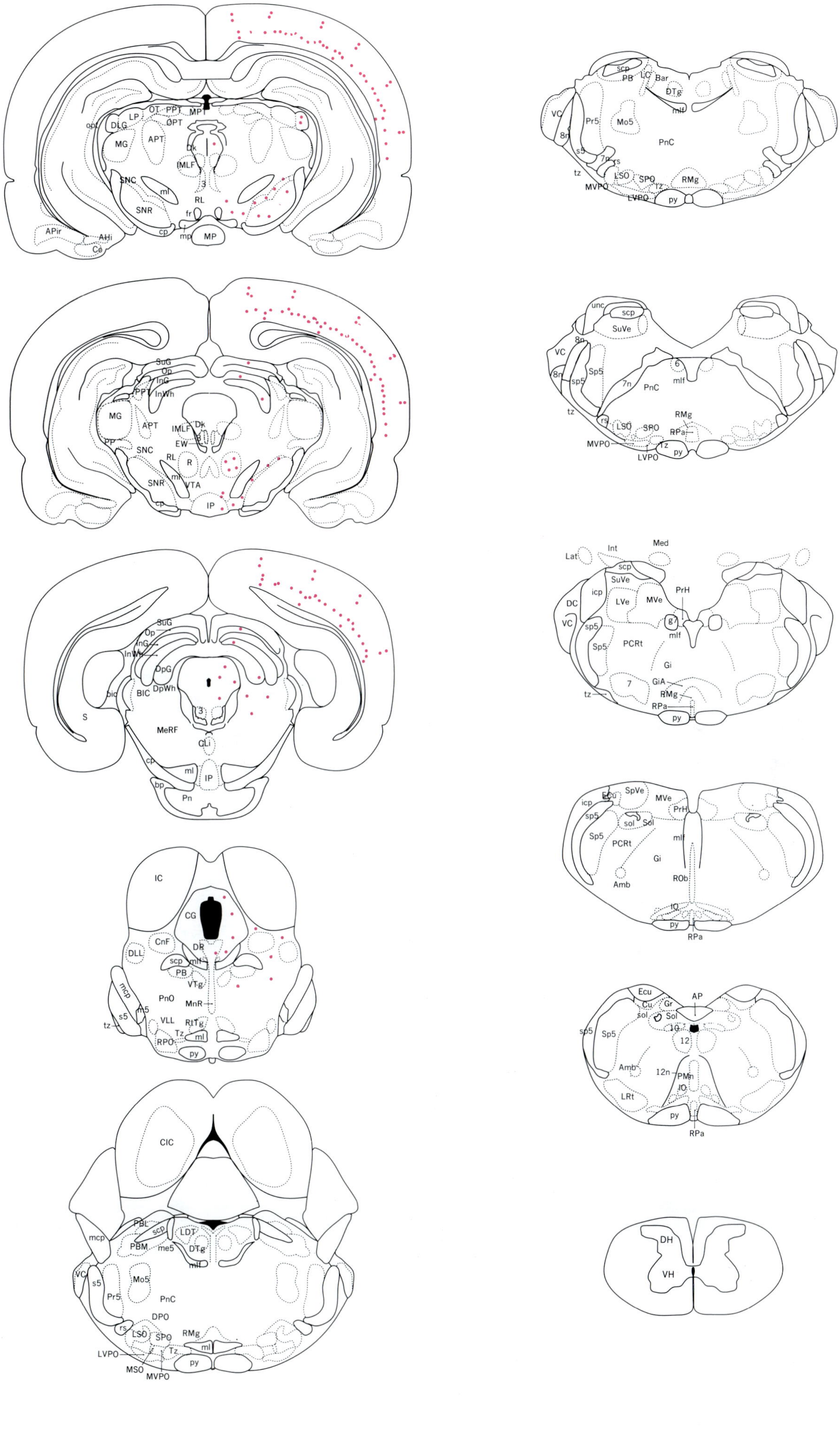

Muscarinic acetylcholine receptor (ISH)

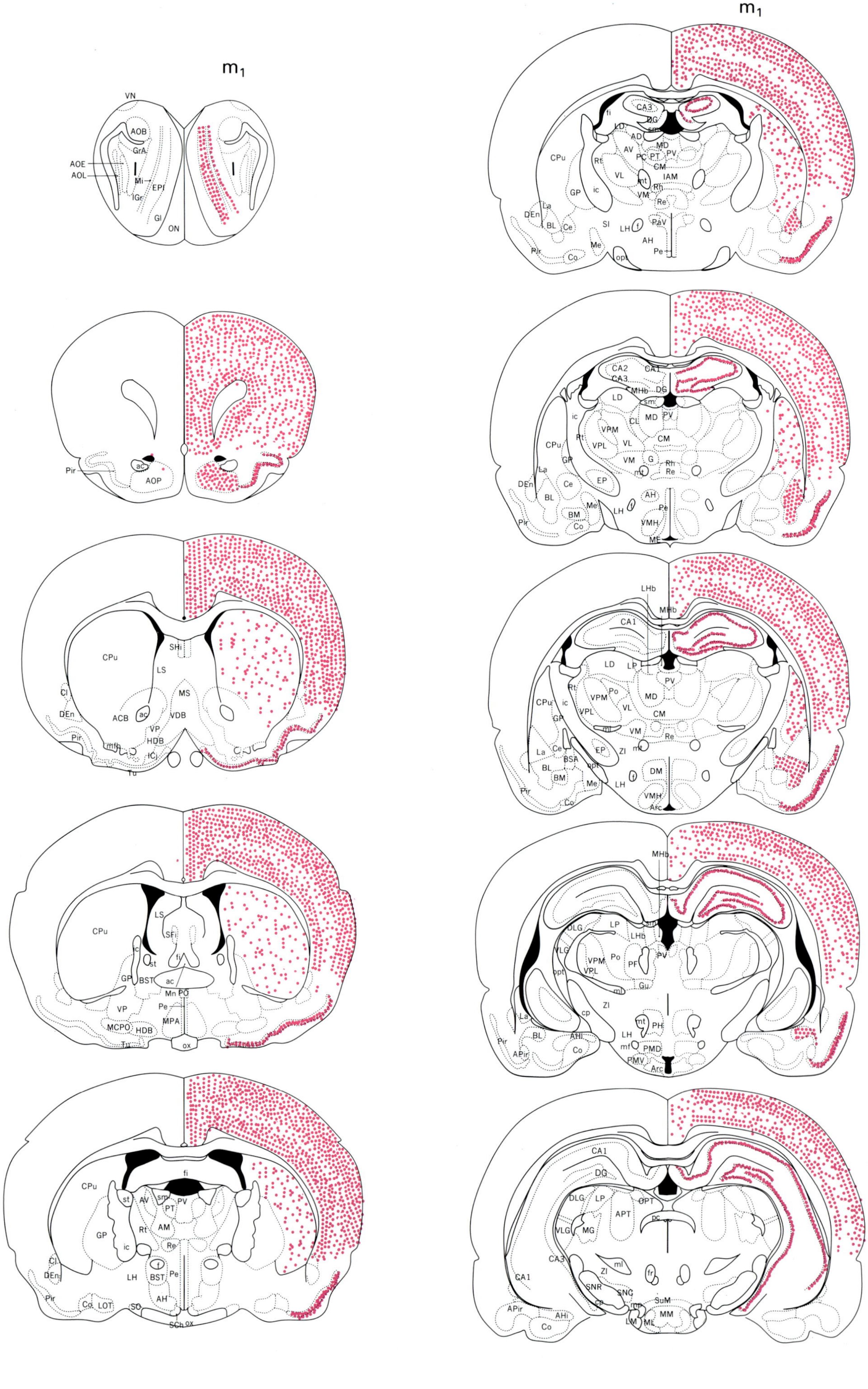

Muscarinic acetylcholine receptor (ISH)

m_1

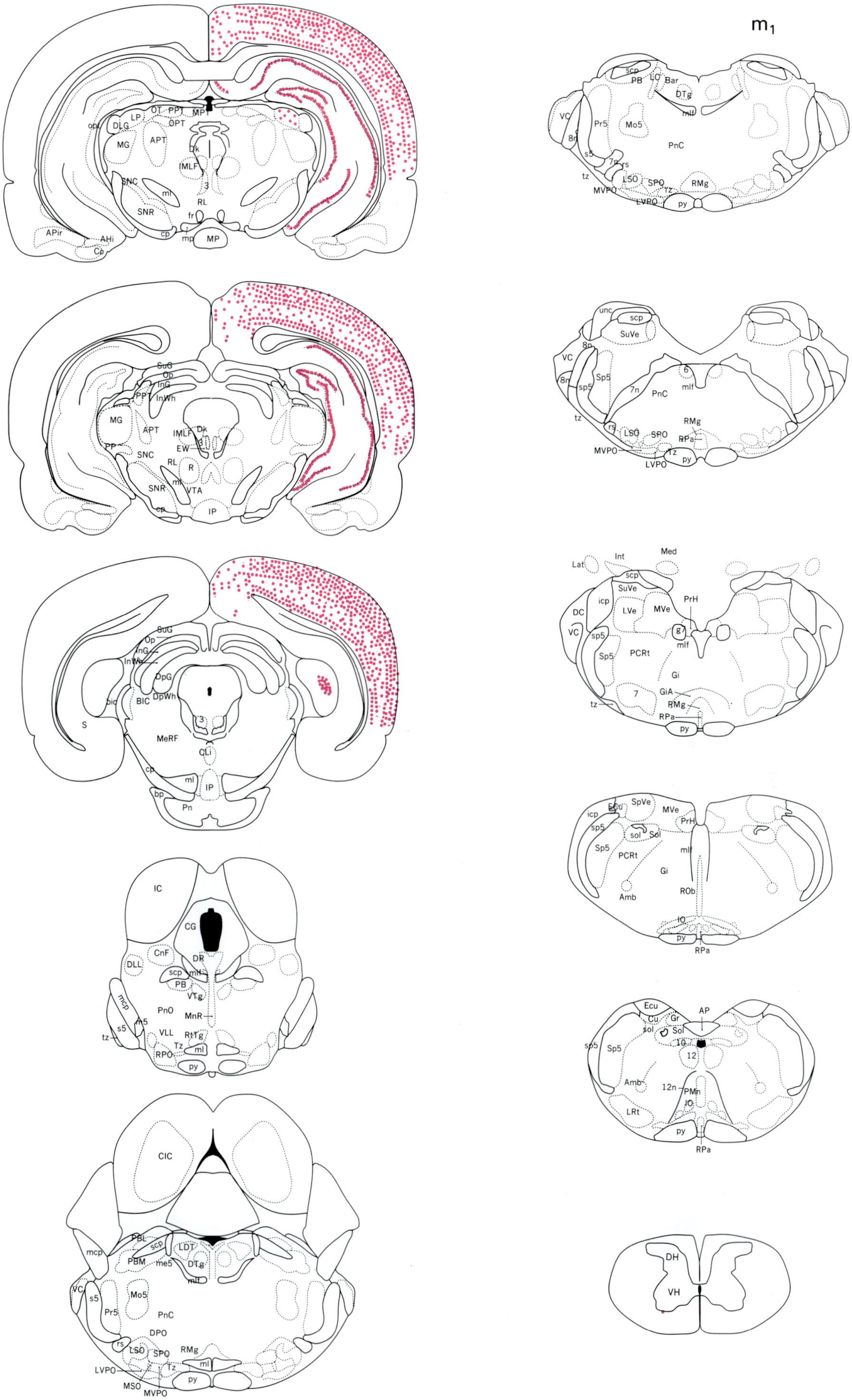

Muscarinic acetylcholine receptor (ISH)

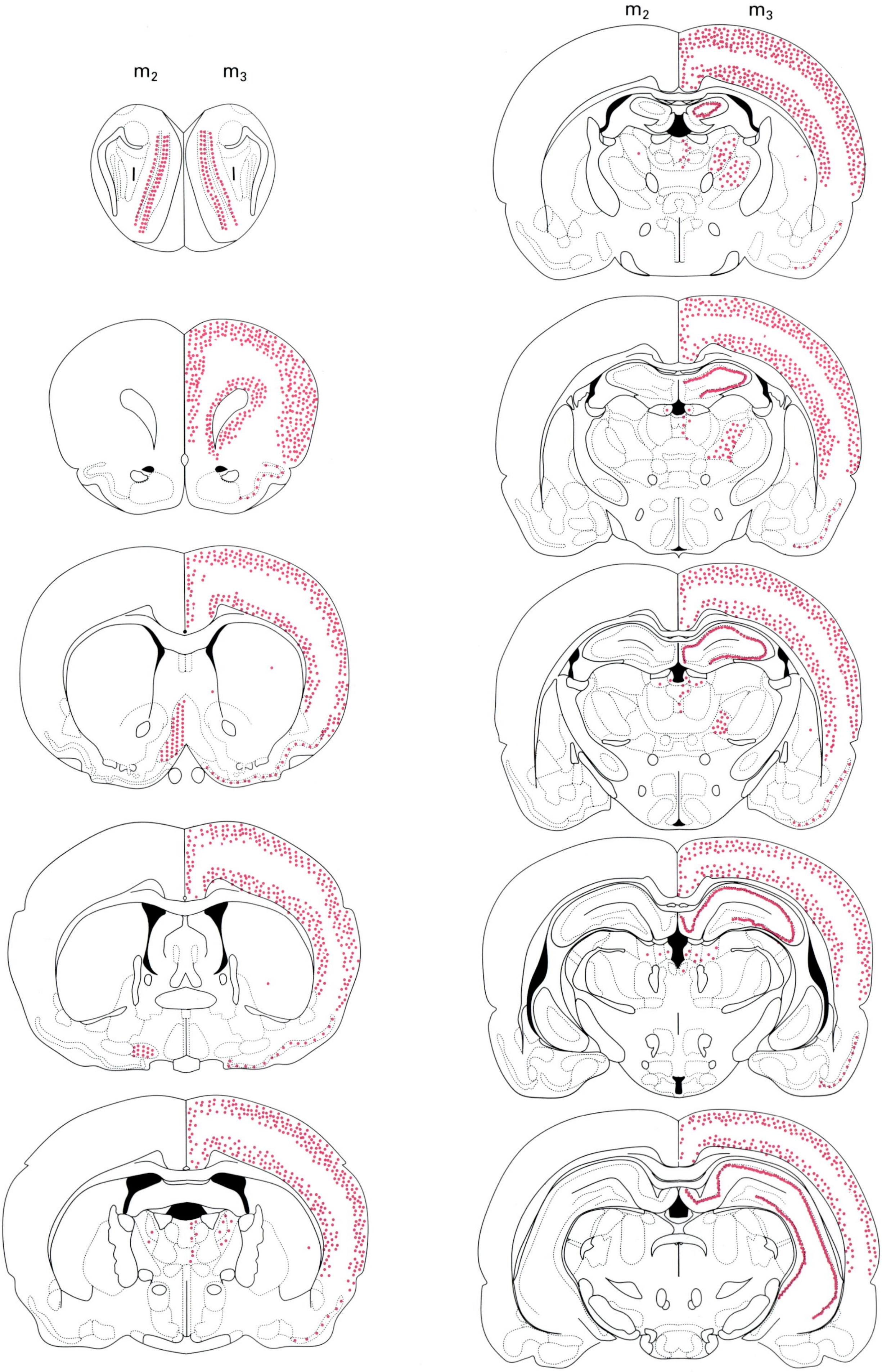

Muscarinic acetylcholine receptor (ISH)

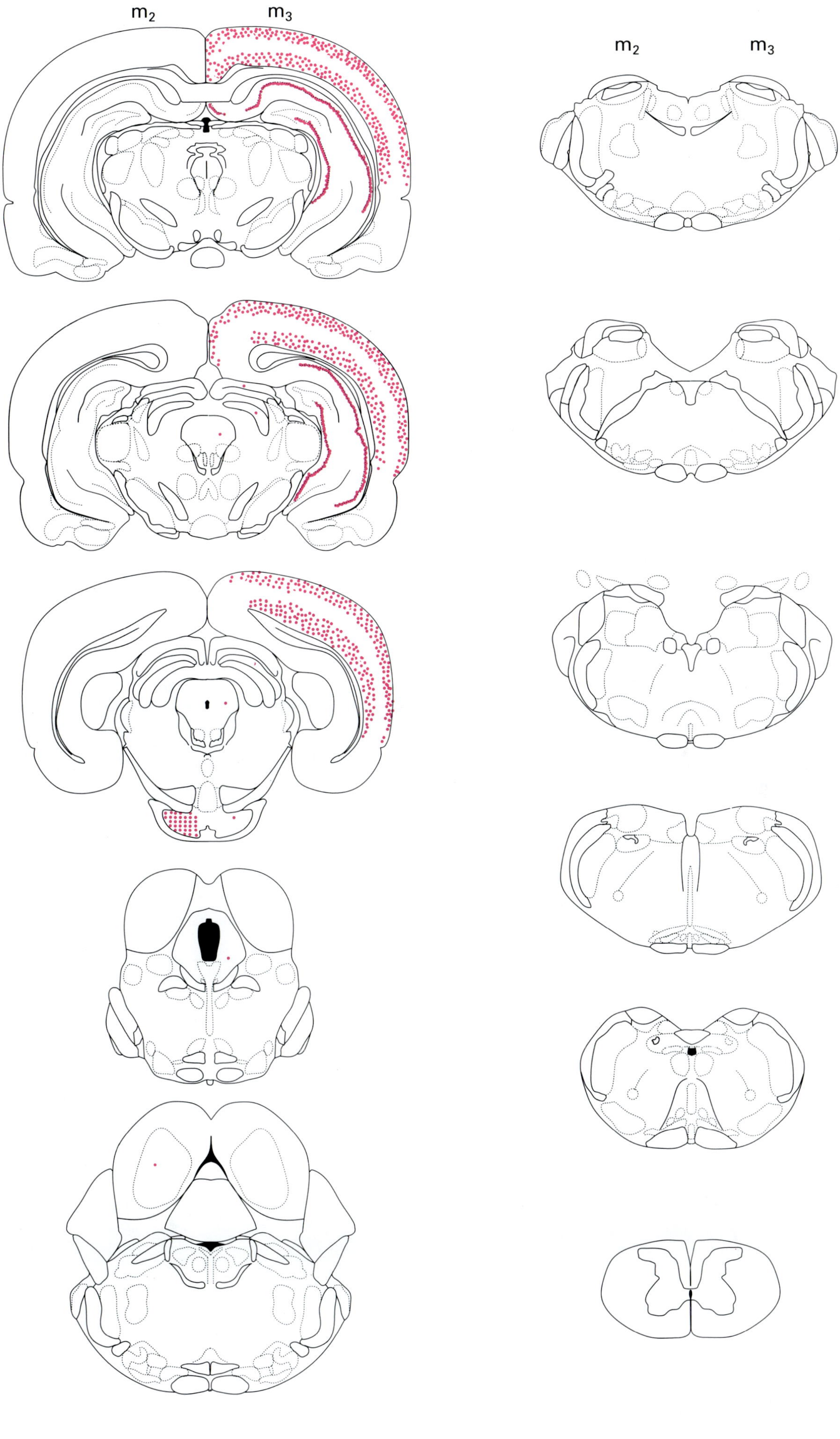

Muscarinic acetylcholine receptor (ISH)

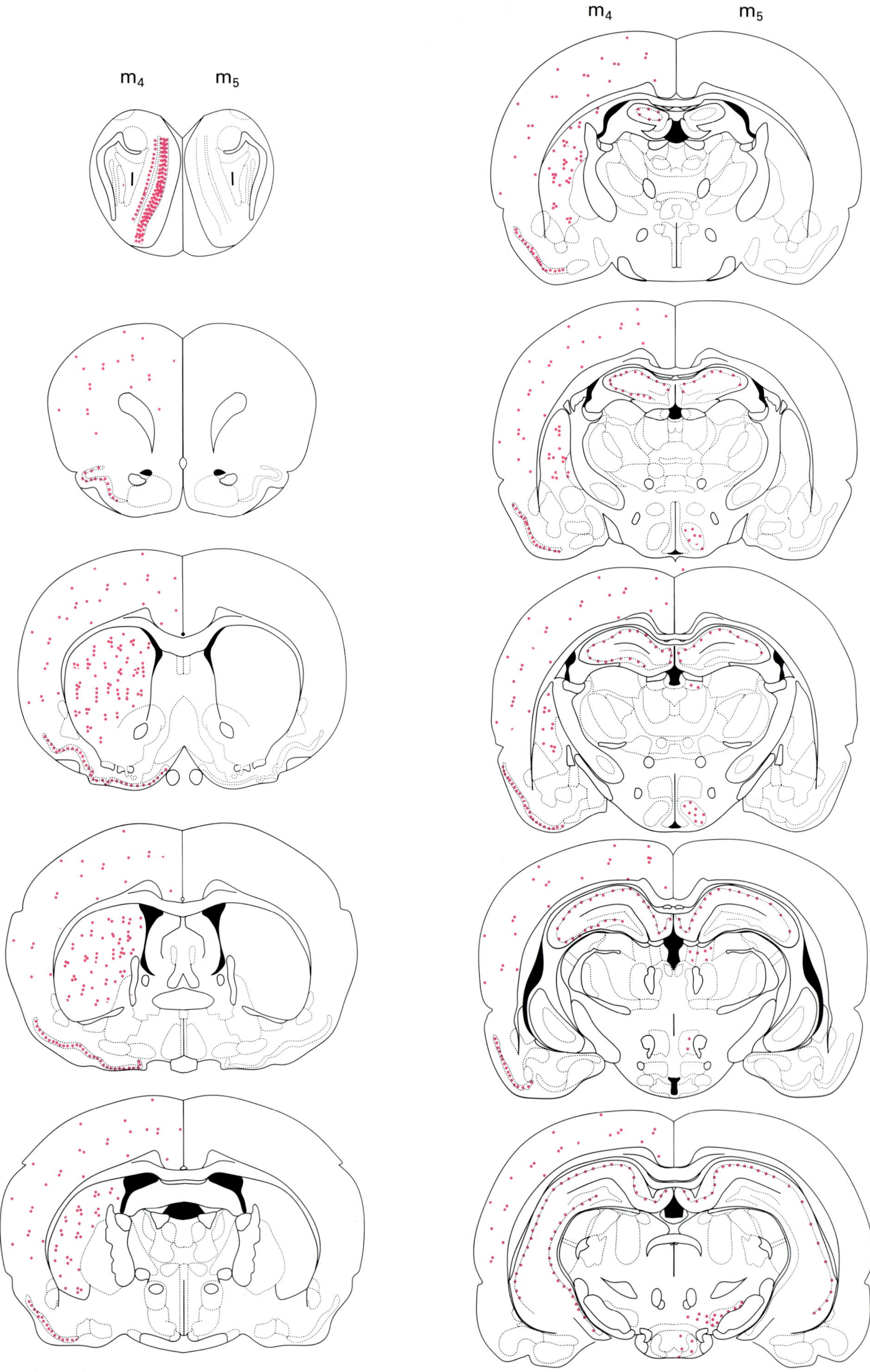

Muscarinic acetylcholine receptor (ISH)

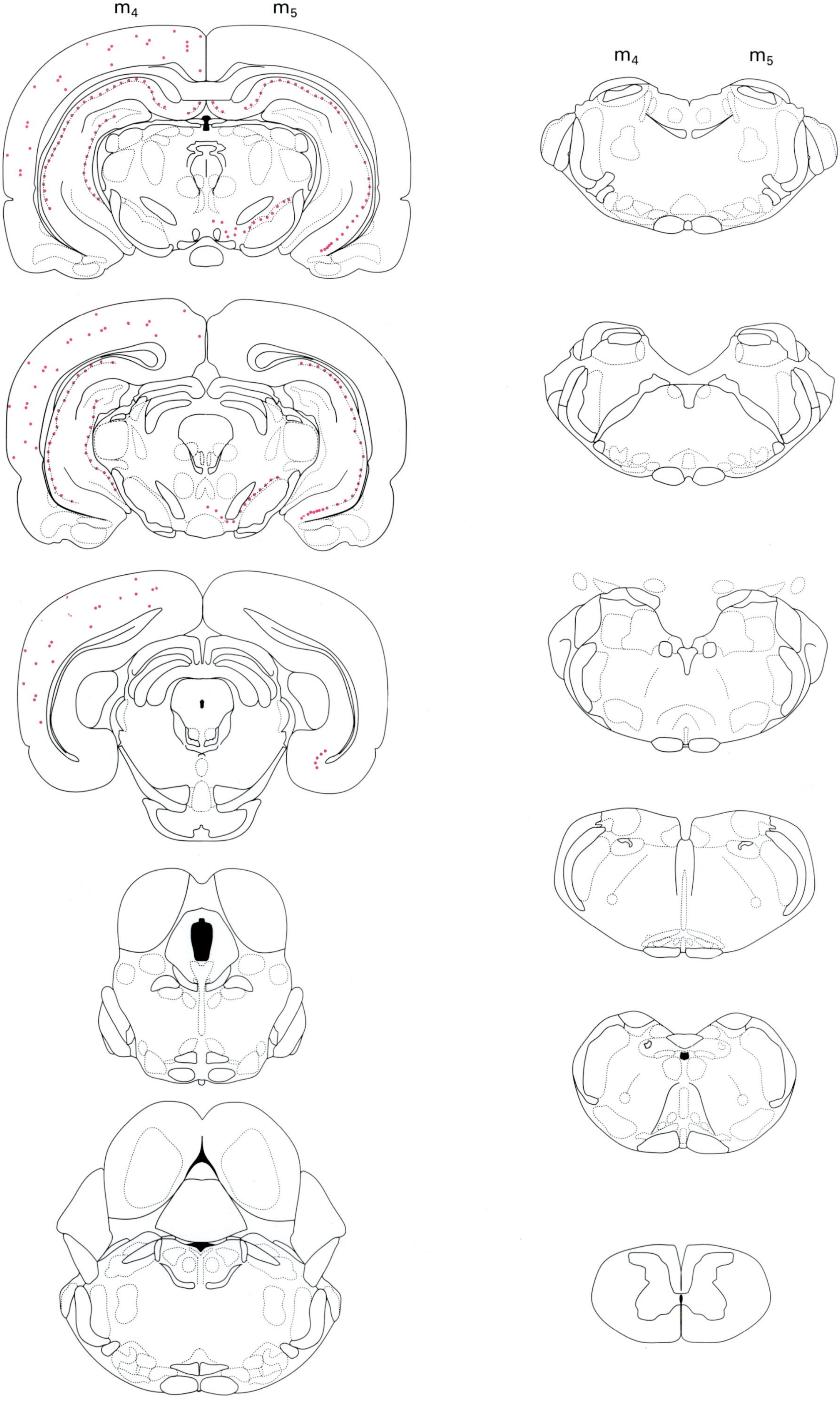

Amino acids

5 The γ-aminobutyric acid activation system

γ-Aminobutyric acid

Synthesis

γ-Aminobutyric acid (GABA) is synthesized by decarboxylation of glutamic acid. The enzyme which catalyses this reaction (shown in Fig. 5.1) is glutamic acid decarboxylase (GAD; EC 4.1.1.15)[5]. It has been shown that there are two genes which code for GAD, which correspond to proteins of molecular weight 65 kDa (585 amino acids) and 67 kDa (593 amino acids)[7]. Both show 65 per cent homology in amino acids, with low degree of resemblance at the N-terminal end. *In situ* hybridization has shown that the expression sites of mRNAs for them largely overlap in the brain, though the quantities expressed at the same sites differ. The 67 kDa GAD is abundant in the cerebral cortex, medial septum, inferior colliculus, and cerebellum, while 65 kDa GAD is expressed notably in the olfactory tubercle, substantia innominata, ventral part of lateral geniculate body, and nucleus of Darkschewitsch.

GABA is catabolized mainly by amino acid transfer reactions and then brought into the tricarboxylic acid cycle. GABA-T is the enzyme which catalyzes this reaction. This enzyme can catalyse the reaction towards the production of GABA; however, it cannot produce GABA *in vivo* and instead, succinic hemialdehyde is rapidly oxidized to succinate by succinic hemialdehyde dehydrogenase. Whereas GABA-T is localized in the mitochondrial fraction, GAD is localized in the synaptosome fraction (see Fig. 5.1 for GABA metabolism).

Inactivation of neuroactive substances released to the synaptic junction occurs by two mechanisms: by enzymatic inactivation, and by energy dependent uptake by neurons and glial cells. The GABA transporter, responsible for the second mechanism, was first cloned in 1990 (GAT1)[9], and GAT2–4 were subsequently cloned on the basis of the structural homology to GAT1[3,4,15]. These transporters have 12 putative transmembrane domains but do not show sequence homology to other previously identified transporters, indicating that GABA transporters form a distinct gene family.

The transporter-mediated GABA uptake is dependent on concentrations of extracellular sodium and chloride ions. A molecule of GABA is co-transported with two sodium and one chloride ion molecules, by utilizing electric potentials formed by different concentrations of these two ions[17].

Among four GABA transporters identified thus far, GAT1 and GAT4 are expressed in the mouse nervous system. GAT1 is located throughout the CNS, while GAT4 is mainly observed in the thalamus, pons, and brain stem[11,13,16]. By immunoelectronmicroscopic analysis, GAT1 is mainly observed in the presynaptic terminals, while GAT4 is observed in the glial cells. The different pharmacological characteristics of these two types of GABA transporters match well with previous pharmacology of GABA uptake.

Distribution

The distribution of GAD overlaps with that of GABA,

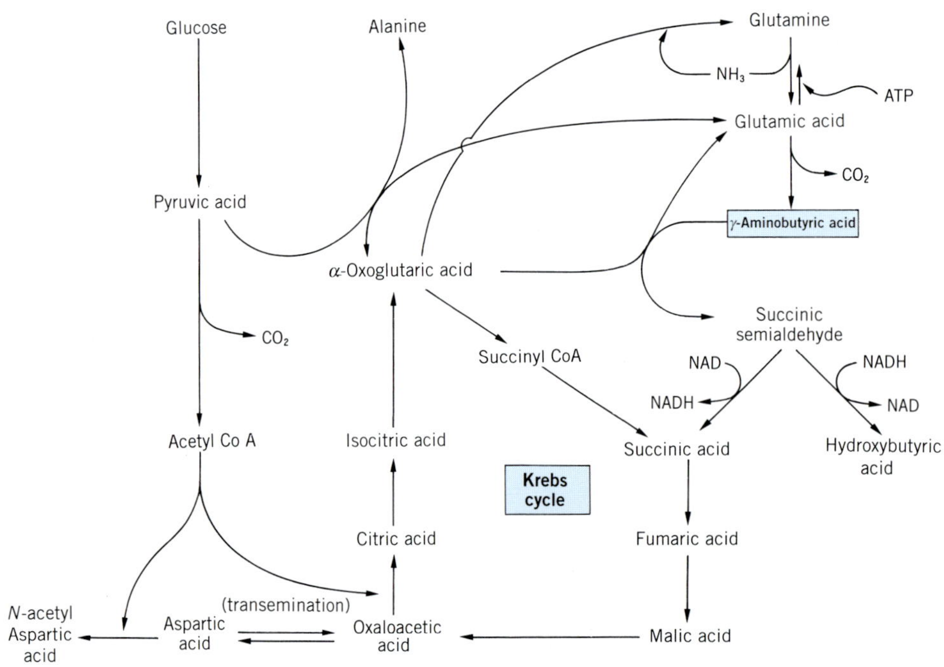

Fig. 5.1 GABA metabolism.

and it has therefore been widely used to localize GABA cytochemically[14]. Recently, *in situ* hybridization has been applied to localize GAD by detecting its mRNA[8]. Although the results from these methods agree in general, there are slight differences in specific details. Immunocytochemical methods show a high density of GAD positive cells in striatum and globus pallidus, though the expression of mRNA by the *in situ* hybridization technique seems only moderate. Conversely, a higher density of positive cells is seen in bed nucleus of the stria terminalis and granular layers at Ammon's horn of hippocampus by *in situ* hybridization.

Physiological actions

GABA was first discovered in nervous tissue in 1950. Subsequent physiological and pharmacological experiments have established that it is an inhibitory neurotransmitter and is distributed widely in the nervous systems of invertebrates and vertebrates. Additional functions of GABA include participation in the secretion of prolactin and growth factors, actions on cardiovascular system, and synergism with the action of anesthetic drugs. It has been speculated that GABA also plays important roles in the pathogenesis of diseases such as Huntington's disease, Parkinson's disease, epilepsy, schizophrenia, senile dementia, and hepatic encephalopathy.

Recently GAD has been implicated to be directly related to two human disorders: stiff-man syndrome and insulin-dependent diabetes[6]. Stiff-man syndrome is a rare neurological disorder, and is characterized by painful spasms and marked rigidity of the skeletal muscles. In the majority of stiff-man sydrome cases, autoimmune pathogenesis is responsible for the disease, and GAD is implicated as a major antigen. However, some stiff-man syndrome cases may result from other non-immune etiologic processes, because some patients lack autoimmune characteristics. Insulin-dependent diabetes is a chronic endocrine disorder that results from an autoimmune destruction of the insulin-producing β cells in the pancreas. It has been demonstrated that autoantibodies to a 64 kDa islet cell protein are present in the sera of approximately 80 per cent of new-onset insulin-dependent diabetes patients and arise long before the onset of clinical symptoms. This 64-kDa autoantigen was identified as 65 kDa isoform of GAD, which is the dominant form of GAD in human islet cells.

Targetted disruption of GAD subtypes has been recently reported. Selective elimination of GAD65 resulted in normal GABA contents and normal animal behaviour, except for a slight increase in susceptability to seizures[2]. On the hand, GAD67-null mutant mice have cleft palate, which results in neonatal death, and a reduction of GABA in the absence of major defects in brain[1]. It has been proposed that GAD65 plays a more active role in nervous system function than GAD67, but these results indicate that GAD67 substantially controls GABA levels.

References

1. Asada, H., Kawamura, Y., Maruyama, K. *et al.* (1997). *Proc. Natl. Acad. Sci. USA*, **94**, 6496–9.
2. Asada, H., Kawamura, Y., Maruyama, K., Kume K. *et al.* (1996). *Biochem. Biophys. Res. Commun.*, **229**, 891–5.
3. Borden, L.A., Smith, K.E., Hartung, P.R. *et al.* (1992). *J. Biol. Chem.*, 21105–11.
4. Clark, J.A., Deutch, A.Y., Gallipoli, P.Z. *et al.* (1992). *Neuron*, **9**, 337–48.
5. Cooper, J.R., Bloom, F.E., and Roth, R.H. (1991). *The biochemical basis of pharmacology*, 6th edn, pp. 133–66. Oxford University Press.
6. Ellis, T.M. and Atkinson, M.A. (1996). *Nat. Med.*, **2**, 148–53
7. Erlander, M.G., Tilakaratne, N.J., Feldblum, S. *et al.* (1991). *Neuron*, **7**, 91–100.
8. Ferraguti, F., Zoli, M., Aronsson, M. *et al.* (1990). *J. Comp. Neurol.*, **3**, 377–96.
9. Gustella, J., Nelson, N., Czyzyk, L. *et al.* (1990). *Science*, **249**, 1303–6.
10. Ikegaki, N., Saito, N., Hashima, M. *et al.* (1994). *Mol. Brain Res.*, **26**, 47–54.
11. Itouji, A., Tanaka, C., and Saito, N. (1996). *Mol.Brain Res.*, **37**, 309–16.
12. Madtes, P.J. and Redburn, D.A. (1983). *Life Sci.*, **33**, 979–84.
13. Minelli, A., Brecha, N.C., Karschin, C. *et al.* (1995). *J. Neorosci.*, **15**, 7734–46.
14. Mugnaini, E. and Oetel, W.H. (1985). *In Handbook of chemical neuroanatomy* Vol. 3., (ed. A. Björklund and T. Hoekfeld) pp. 436–595, Elsevier, Amsterdam.
15. Liu, Q.R., Lopez, C.B., Mandiyan, S. *et al.* (1993). *J. Biol. Chem*, **268**, 2106–12.
16. Ribak, C.E., Tong, W.M.Y., and Brecha, N.C. (1996). *J. Comp. Neurol.*, **367**, 595–606.
17. Risso, S., Defelice, L.J., and Blakely, R.D. (1996). *J. Physiol.*, **490**, 691–702.
18. van Eden, C.G., Mrzljak, L., Voorn, P., and Uylings, H.B. (1989). *J. Comp. Neurol.*, **289**, 213–27.

GABA (immunocytochemistry)

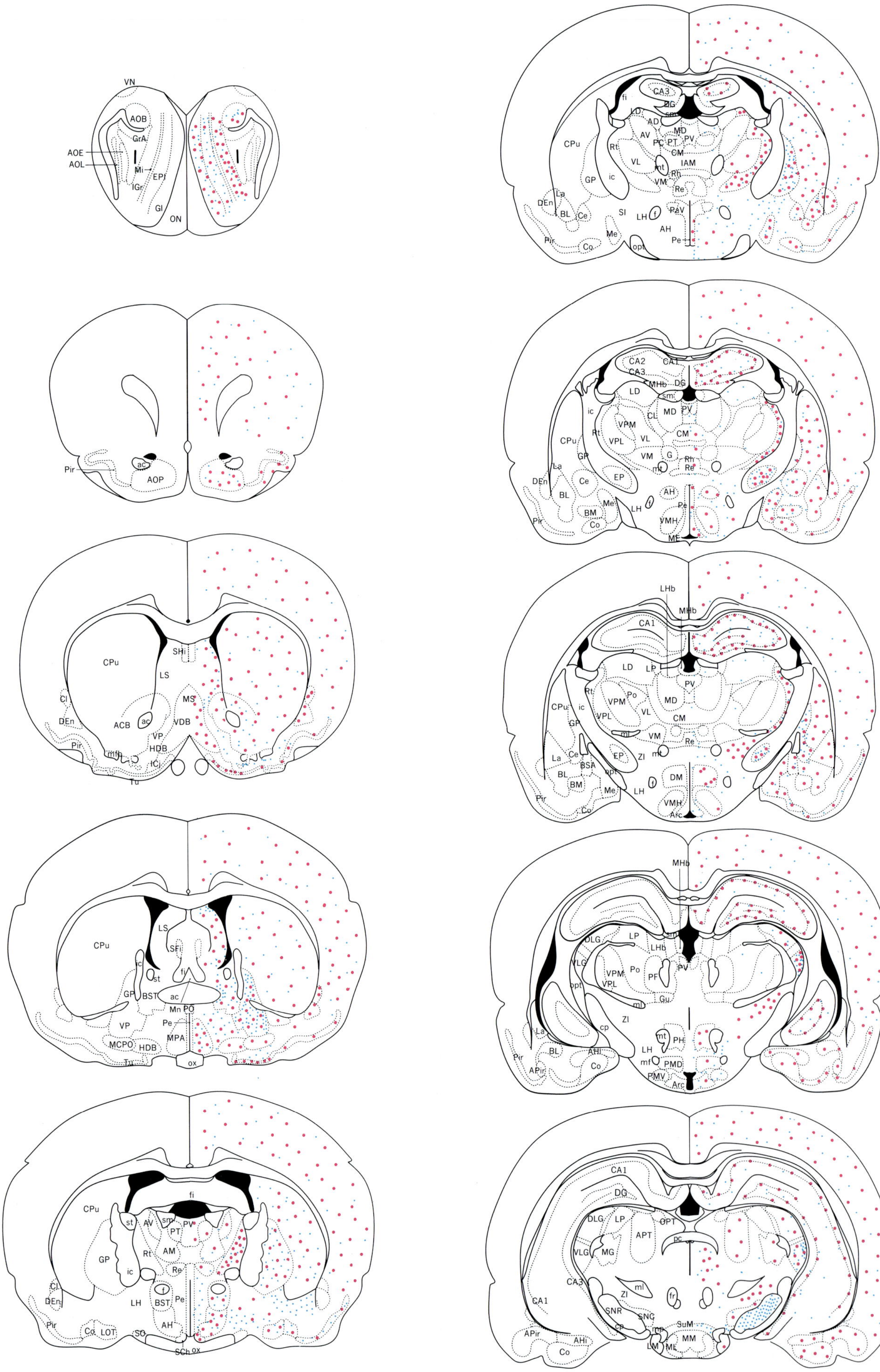

GABA (immunocytochemistry)

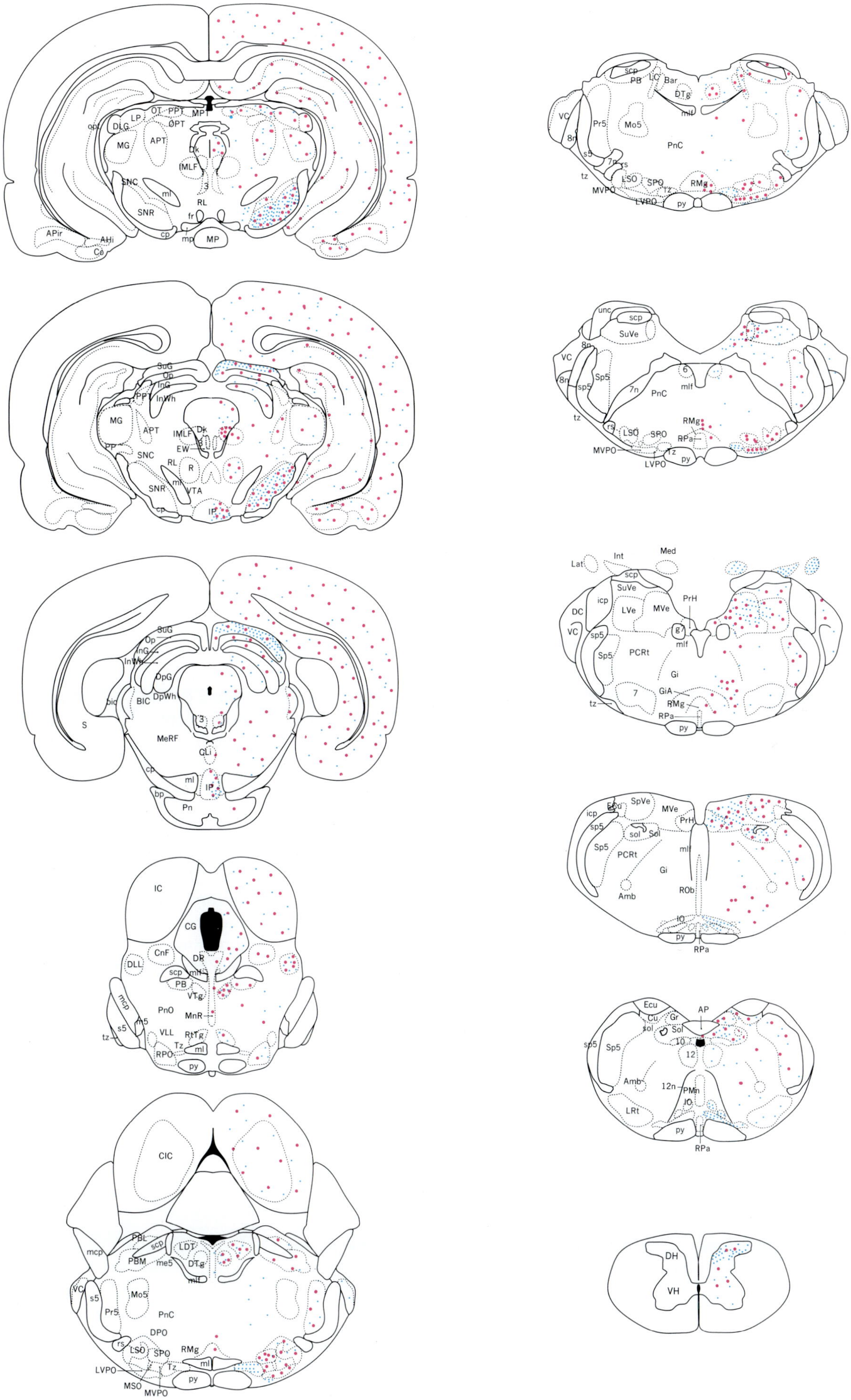

γ-Aminobutyric acid receptors

Classification

Transmission of information through the GABAergic system depends on the protein receptors in the plasma membrane. As shown in Table 5.1, GABA receptors are classified into three distinct subtypes: $GABA_A$, $GABA_B$, and $GABA_C$ receptors. $GABA_A$ and $GABA_C$ receptors constitute Cl^- channels, and $GABA_B$ receptors are negatively coupled to adenylate cyclase[7,14,18].

Significance of diversity and subunit structure

The $GABA_A$ receptor was purified in 1987[29], and the primary sequence of DNA coding the protein indicated that it is a member of the ion channel superfamily which possesses four transmembrane sites. Initially, $GABA_A$ receptor was thought to be a tetramer, consisting of α and β subunits. However, subsequent examination using molecular biological techniques has revealed its diversity, and many subunits have been cloned whose sequences resemble each other, especially in the transmembrane regions[19]. Moreover, γ_2 subunit has two subtypes, γ_{2L} and γ_{2S}; the former has an eight-amino-acid insertion absent in the latter, which constitutes the site phosphorylated by cAMP-dependent protein kinase[32]. The diversity of subunits elucidated so far is reported to have the following functional significance:

- regulation of opening time of ion channel, affinity to ligands and permeability of membranes[30,31]
- relation to the action of alcohol on the central nervous system (α_6, γ_{2L})[18,32]
- difference in response to drugs such as benzodiazepines which influence the binding of substrates to the receptor[26]
- possibly constituting autoreceptors (γ_2)[4]
- possibly regulating information transmission at some special neural circuits (γ_1)[1].

Naturally occuring and genetically manipulated mutants of $GABA_A$ receptor subunits have been described[11,20]. The pink-eyed cleft-palate mutant mice, characterized by hypopigmentation associated with cleft palate, neurological disorders and runting, have a deletion of $GABA_A$ receptor α_5, β_3, and γ_3 subunits which are clustered on mouse chromosone 7. The human homologue of the region deleted in this mutation is associated with Angelman syndrome. Targetted disruption of $GABA_A$ receptor β_3 subunit, which is a candidate gene of this genetic disease, results in hyperactivity and hyperresponsiveness to sensory stimuli, lack of motor coordination, and seizures[16].

The existing sites of $GABA_A$ receptors have been reported as the binding sites of receptor ligands, which have been classified as follows:

- ligands which bind to GABA binding sites (muscimol, isoguvacine, etc.)[25]
- ligands which bind to the Cl^- channel (t-butyl bicyclophospholothionate, t-butyl bicycloorthobenzoate, etc.)[13,23]
- ligands which bind to the benzodiazepine binding site (flunitazepam, etc.)[21].

The binding sites of ligands differ.

Following the finding of the diversity of subunits in the $GABA_A$ receptor, their expression sites have been studied by cytochemical method using *in situ* hybridization[1,4,15,33,34]. Comparison of the results of these studies and earlier study on the binding sites of receptor ligands has revealed that the latter only partially reflects the distribution of various subunits. It should be noted that no ligand has been reported to bind specifically to any subunit. When the results of studies using *in situ* hybridization are compared, each subunit has its unique distribution and thus it is impossible to comprehend the subunit structure of $GABA_A$ receptors solely on the basis of these results. However, some of the possible roles listed above can be explored by examining the expression sites.

There are some reports on the localization of receptors employing antibodies to receptors[12,27]. Although advances have been made in producing specific antibody to each subunit, it is usually difficult to visualize their binding sites in tissues and only a few groups have succeeded in obtaining definitive results[5,6]. This atlas shows distribution of receptors of $GABA_A$ and $GABA_B$ detected as binding sites by autoradiography using ^{3}H-GABA as ligand which is blocked by baclofen ($GABA_A$ receptors) or isoguvacine ($GABA_B$ receptors)[8] and distribution of subunits $\alpha_{1,3,4}$, $\beta_{1,3}$, $\gamma_{1,2}$ in the central nervous system[1,3,4,33,34]. Data from the studies describing an immunocytochemical method using antibodies to receptors was adopted in this atlas from the report using bd-17 antibody, which is considered to recognize the β subunit specifically[12]. However, several sites are uncertain since the original reports do not mention if they were immunopositive. Though there are

Table 5.1. GABA receptors

Receptors	$GABA_A$		$GABA_B$	$GABA_C$
Agonists	Competitive site	Benzodiazepine modulatory site		
	Muscimol isoguvacine	Flunitrazepam Zolpidem CGS9896	L-bacrofen 3-Aminopropyl-phosphonic acid	Muscimol *trans*-4-aminocrotonic acid *cis*-4-aminocrotonic acid
Antagonists	Bicuculline	Flumazenil	Sacrofen CGP35348 2-Hydroxysacrofen CGP64213 CGP71872	Imidazol-4-acetic acid SR95531

Table 5.2. Multiplicity of $GABA_A$ and $GABA_C$ receptor subunits and subunit-specific antibodies

α subunit	α_{1-6}	bd 24 antibody
β subunit	β_{1-4}	bd 17 antibody
γ subunit	γ_{1-3} (γ_2L, γ_2S)	
δ subunit	δ	
ρ subunit	ρ_{1-3}	

reports which have used monoclonal antibody bd-24 to recognize the α subunit[12] and polyclonal antibodies to recognize single subunits[5,6], these have not been adopted here since they fail to show detailed distribution of the sites where those antibodies are presumed to be present (Table 5.2).

A third class of ionotropic GABA receptor, which is distinct from bicuculline-sensitive $GABA_A$ receptors and baclofen-sensitive $GABA_B$ receptors, has been reported to be observed predominantly in the retina of vertebrates, and is referred to as $GABA_C$ receptors. $GABA_C$ receptors are insensitive to drugs that modulate $GABA_A$ and $GABA_B$ receptors and are activated selectively by *cis*-4-aminoacrotonic acid. $GABA_C$ receptors are more sensitive to GABA than $GABA_A$ receptors. $GABA_C$ receptors are composed of ρ subunits. ρ_{1-3} subunits have been identified[7,22]. In the retina, ρ subunits are expressed in rod bipolar cells. Expression of ρ_2 receptor is also demonstrated in other parts of the CNS, most notably in the hippocampus and cortex.

$GABA_B$ receptors, which are a metabotropic receptor system, modulate synaptic transmission by presynaptic inhibition of transmitter release by increasing a K^+-conductance responsible for long-lasting inhibitory postsynaptic potentials. Presynaptic $GABA_B$ receptors have been reported to control GABA, whereas $GABA_B$ postsynaptic receptors regulate the release of *L*-glutamate, noradrenaline, dopamine, 5-hydroxytryptamine, substance P, cholesystokinin, and somatostatin.

The long-persued primary structure of $GABA_B$ receptor was revealed in 1997 by expression cloning using a new high-affinity $GABA_B$ receptor anatagonist[17]. Two $GABA_B$ receptor proteins, $GABA_BR1_a$ and $_{-b}$ are detected as 130 and 100 kDa proteins respectively in the CNS, and comprise a gene family together with metabotropic glutamate receptors. The cloned $GABA_BR1_a$ receptor negatively couples to adenylate cyclase when stably expressed in HEK293 cells.

Northern blot analysis revealed RNAs of 4.4 kb and 1.8 kb for $GABA_BR1_a$, and RNAs of 4.3 kb and 3 kb for $GABA_BR1_b$ in brain. A transcript of 3 kb was also identified in testis, but not in any other organs. *In situ* hybridization studies using a probe detecting 4.4, 4.3, and 3 kb transcripts showed that $GABA_BR1$ transcripts are abundant in all cerebral cortex layers, in the pyramidal cell layers of the hippocampus, the granular cell layer of the dentate gyrus, and in the basal ganglia, including caudate putamen, nucleus accumbens, and olfactory tubercle. In the cerebellum, abundant transcripts are found in the Purkinje cells, and a moderate level in the granular layer. The distribution of $GABA_BR1$ mRNA and $GABA_B$ receptor binding sites are closely similar in the medial habenula medial geniculate nucleus, and interpeduncular nucleus, but in other CNS areas the regional abundance of transcripts and binding sites differs. $GABA_BR1$ mRNAs are expressed in neuronal but not in glial cells.

References

1. Araki, T., Kiyama, H., and Tohyama, M. (1992). *Mol. Brain Res.*, **15**, 121–32.
2. Araki, T., Sato, M., Zhang, J.-H. *et al.* (1992). *Neuroscience*, **47**, 45–61.
3. Araki, T. and Tohyama, M. (1992). *Mol. Brain Res.*, **12**, 293–314.
4. Araki, T., Kiyama, H., and Tohyama, M. (1991). *J. Neurosci. Res., Abstr. Suppl.*, **14**, S138.
5. Benke, D., Mertens, S., Trezciak, A. *et al.* (1990). *Eur. J. Pharmacol.*, **189**, 337–40.
6. Benke, D., Mertens, S., Trezeciak, A. *et al.* (1991). *FEBS Lett.*, **283**, 145–9.
7. Bormann, J., and Feigensen, A. (1995). *Trends Neurosci.*, **18**, 515–19.
8. Bowery, N.G., Hudson, A.L., and Price, G.W. (1987). *Neuroscience*, **20**, 365–83.
9. Bowery, N. (1989). *Trends Pharmacol. Sci.*, **10**, 401–11.
10. Burt, D.R. and Kamatchi, G.L. (1991). *FASEB J.*, **5**, 2916–23.
11. Culiat, C.T., Stubbs, L., Montgomery, C.S. *et al.* (1994). *Proc. Natl. Acad. Sci USA*, **91**, 2815–18.
12. de Blas, A.L., Vitorica, J., and Friedrich, P. (1988). *J. Neurosci.*, **8**, 602–14.
13. Edgar, P.P. and Schwartz, R.D. (1990). *J. Neurosci*, **10**, 603–12.
14. Fagg, G.E. and Foster, A.C. (1983). *Neuroscience,* **9**, 701–19.
15. Hironaka, T., Morita, Y., Hagihira, S. *et al.* (1990). *Mol. Brain Res.*, **7**, 335–45.
16. Homanics, G.E., DeLorey, T.M., Firestone, L.L. *et al.* (1997). *Proc. Natl. Acad. Sci USA*, **94**, 4143–48.
17. Kaupmann, K., Huggel, K., Heid, J. *et al.* (1997). *Nature*, **386**, 239–46.
18. Lüddens, H., Protchett, D.B., Köhler, M. *et al.* (1990). *Nature*, **346**, 648–51.
19. Lüddens, H. and Wisden, W. (1991). *Trends Pharmacol. Sci.*, **12**, 49–51.
20. Nakatsu, Y., Tyndale, R.F., DeLorey, T.M. *et al.* (1993). *Nature*, **364**, 448–50.
21. Niddam, R., Dubois, A., Scatton, B. *et al.* (1987). *J. Neurochem.*, **49**, 890–9.
22. Ogurusu, T. and Samgathingai, R. (1996). *Biochem. Biophys. Acta*, **1305**, 15–18.
23. Olsen, R.W. and J., T.A. (1990). *FASEB J.*, **4**, 1469–80.
24. Olsen, R.W., McCabe, R.T., and Wamsley, J.K. (1990). *J. Chem. Neuroanat.*, **3**, 59–76.
25. Palacios, J.M., Wamsley, J.K., and Kuhar, M.J. (1981). *Brain Res.*, **222**, 285–307.
26. Puia, G., Vicini, S., Seeburg, P.H., and Costa, E. (1991). *Mol. Pharmacol.*, **39**, 691–6.
27. Richards, J.G., Schoch, P., Häring, P., Takacs, B. and Möhler, H. (1987). *J. Neurosci.*, **7**, 1866–86.
28. Roberts, E. (1986). In *Biochemistry and Methology* Vol. 5 (ed. R. W. Olsen and J. C. Venter), pp. 1–39.
29. Schofield, P.R., Darlison, M.G., Fujita, N. *et al.* (1987). *Nature*, **328**, 221–7.
30. Siegel, E., Baur, R., Trube, G. *et al.* (1990). *Neuron*, **5**, 703–11.
31. Stephenson, F.A. (1988). *Biochem. J.*, **249**, 21–32.
32. Wafford, K.A., Burnett, D.M., Dunwiddie, T.V., and Harris, R.A. (1990). *Science*, **249**, 291–3.
33. Zhang, J.-H., Araki, T., Sato, M., and Tohyama, M. (1991). *Mol. Brain Res.*, **11**, 239–47.
34. Zhang, J.-H., Sata, M., and Tohyama, M. (1991). *J. Comp. Neurol.*, **303**, 637–57.

GABA receptor (binding sites)

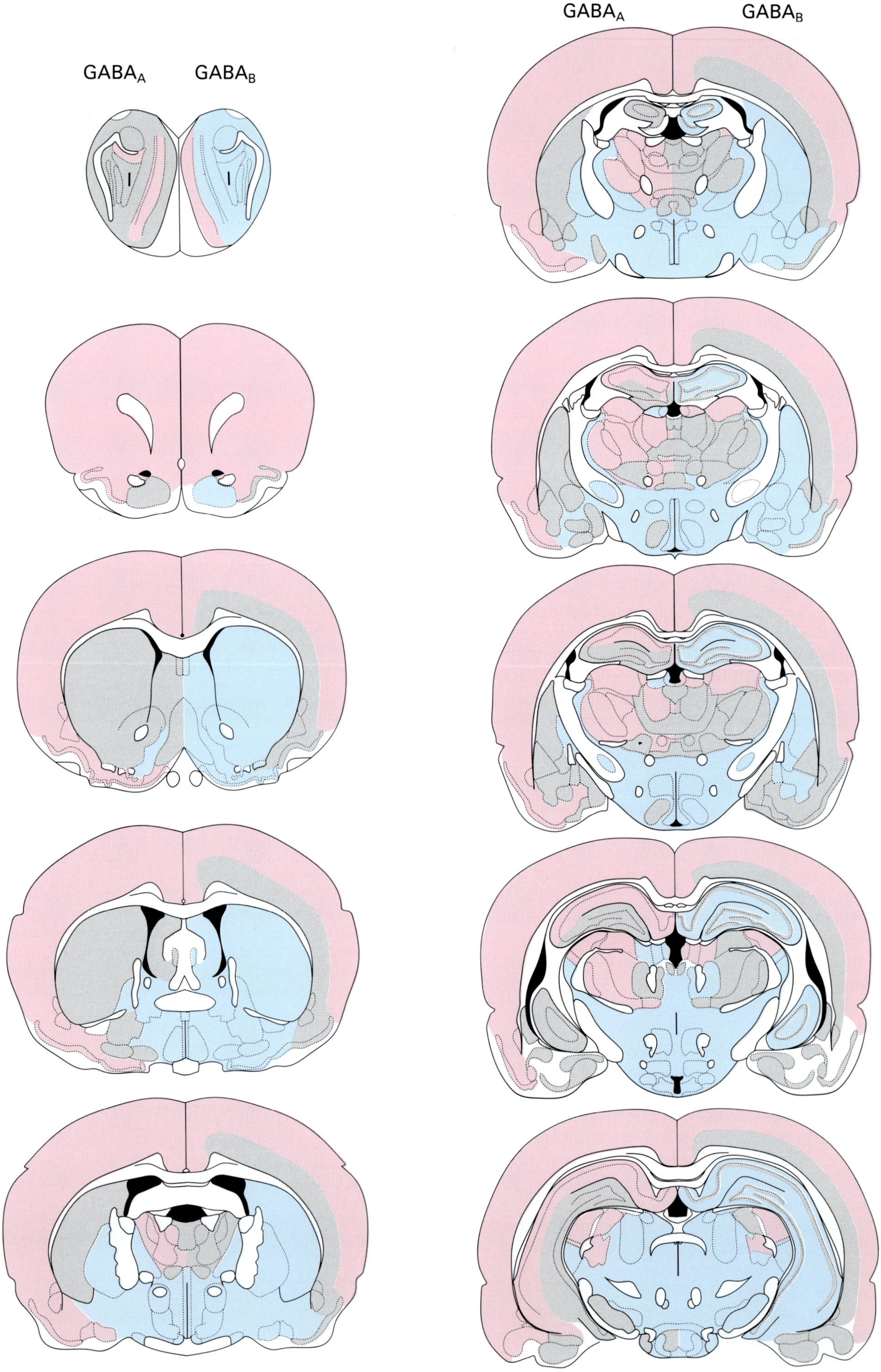

GABA receptor (binding sites)

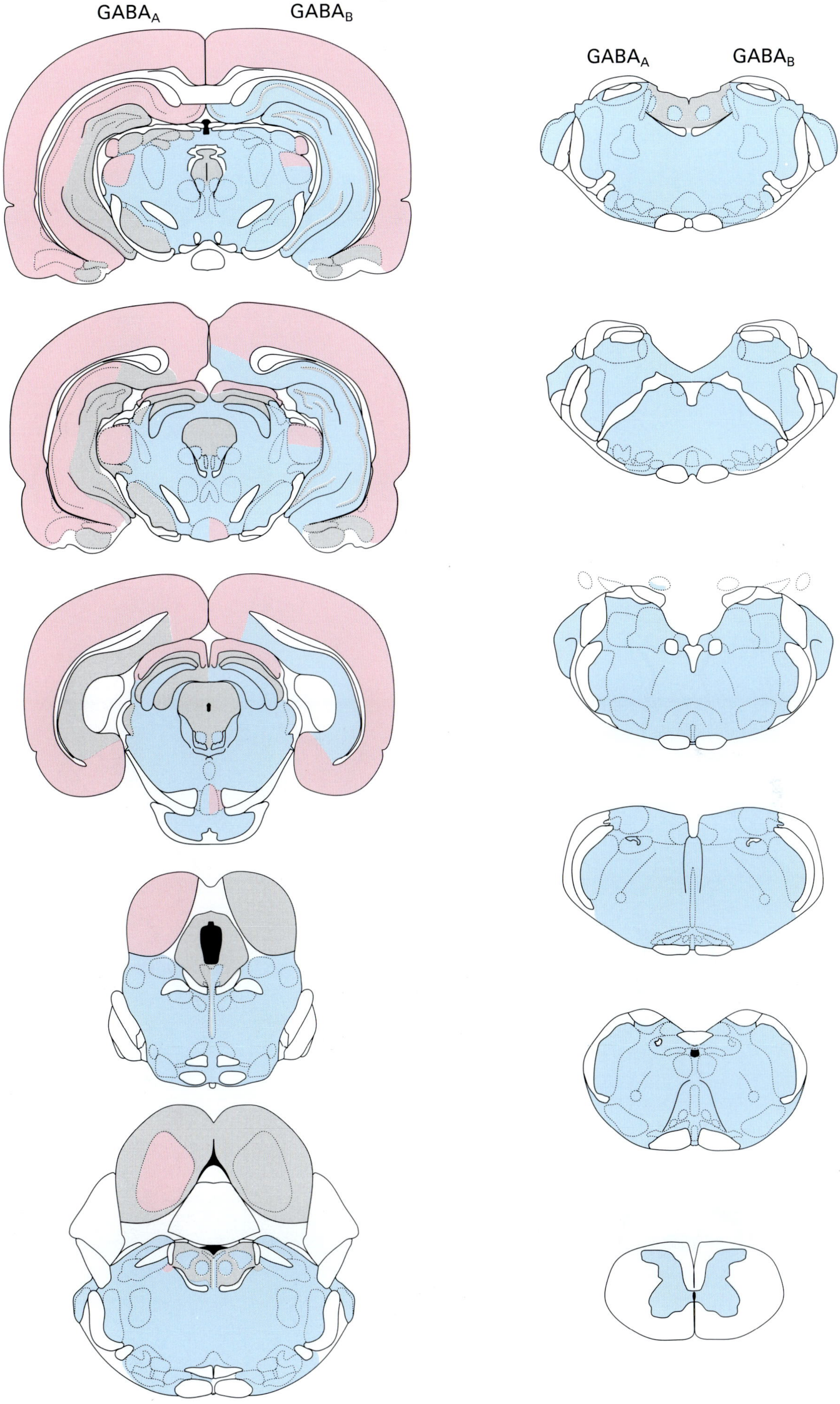

GABA receptor (immunocytochemistry)

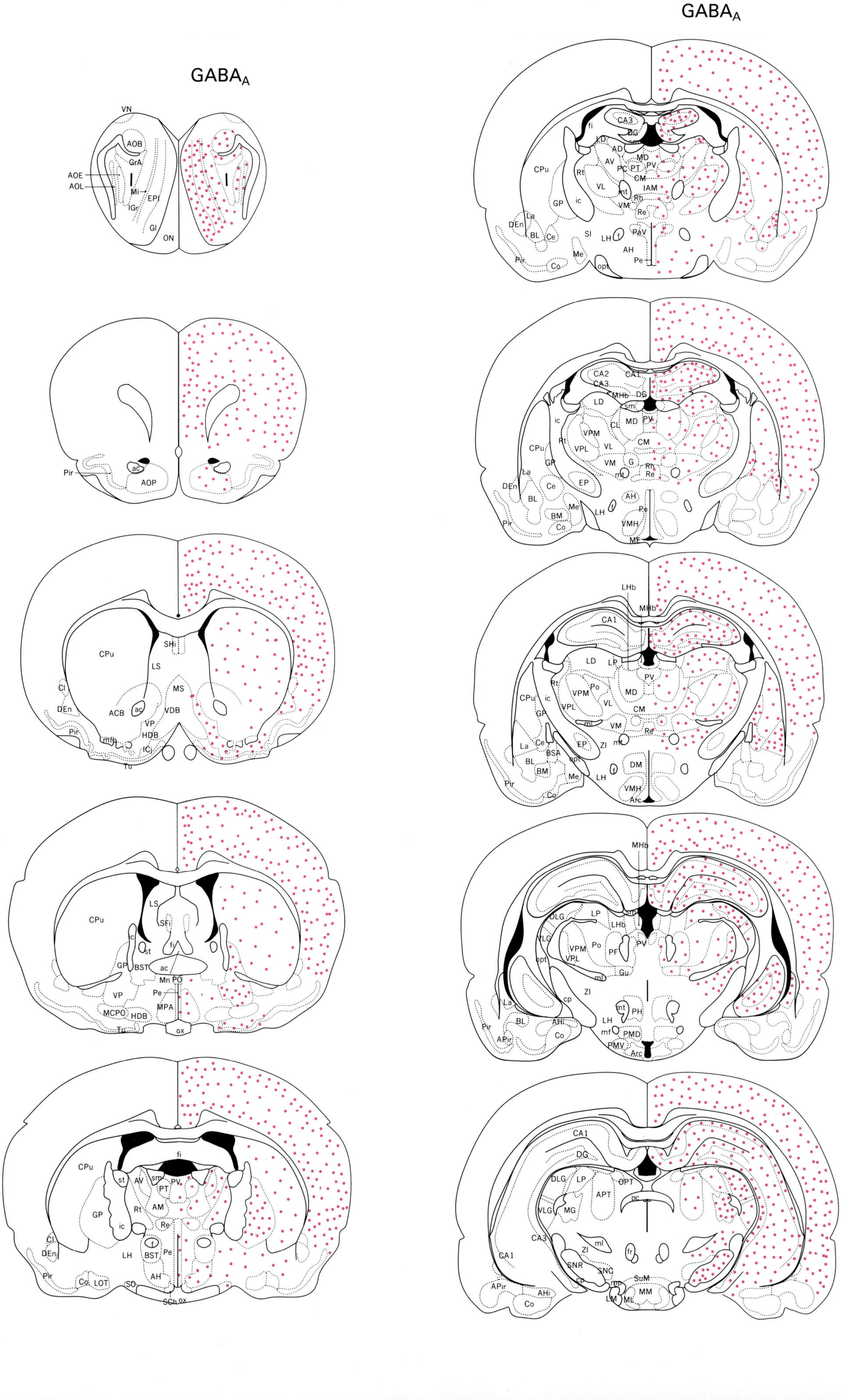

GABA receptor (immunocytochemistry)

$GABA_A$

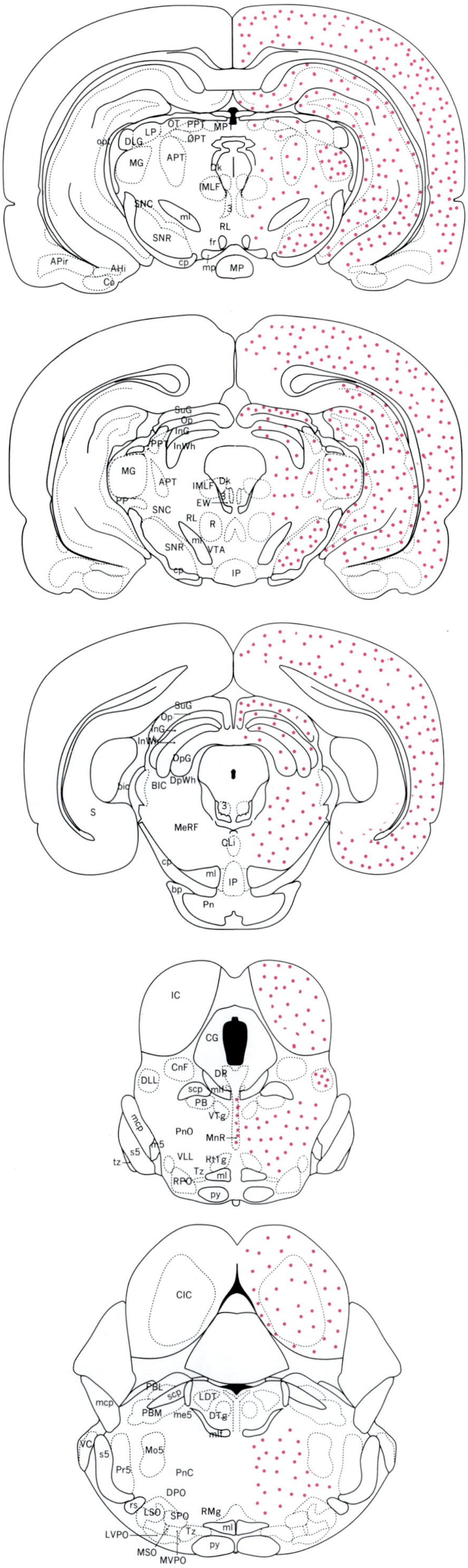

$GABA_A$

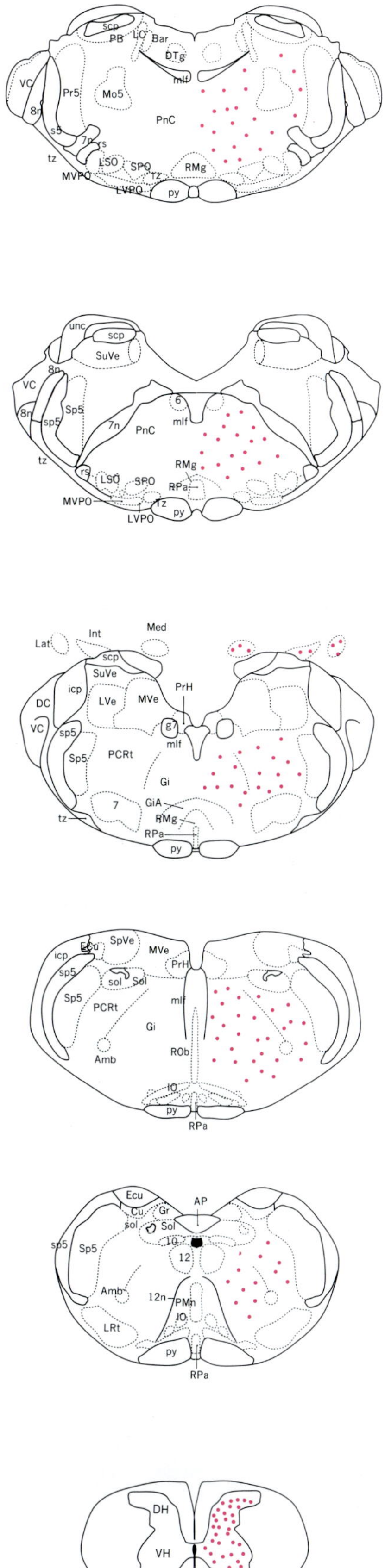

GABA$_A$ receptor (ISH)

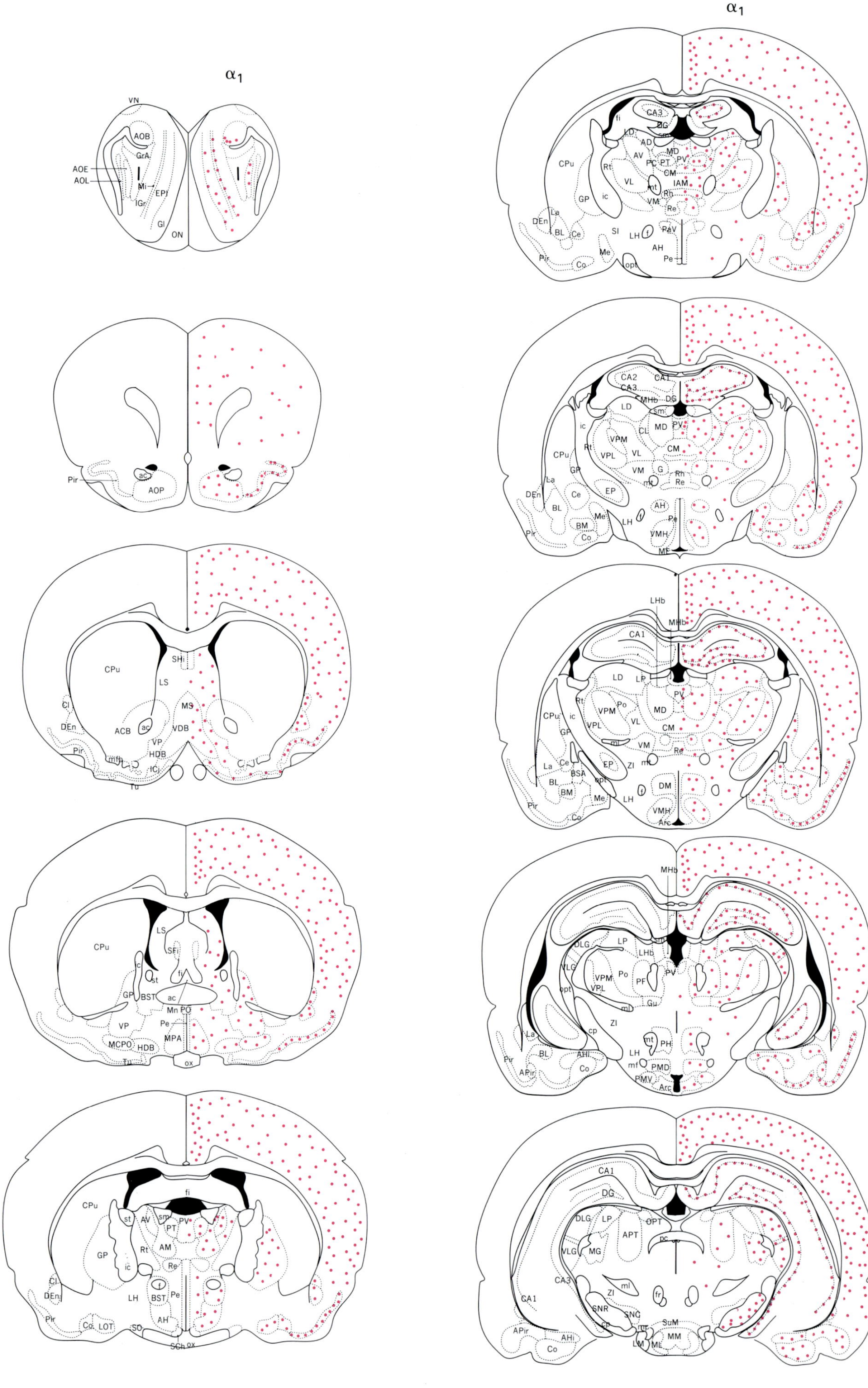

$GABA_A$ receptor (ISH)

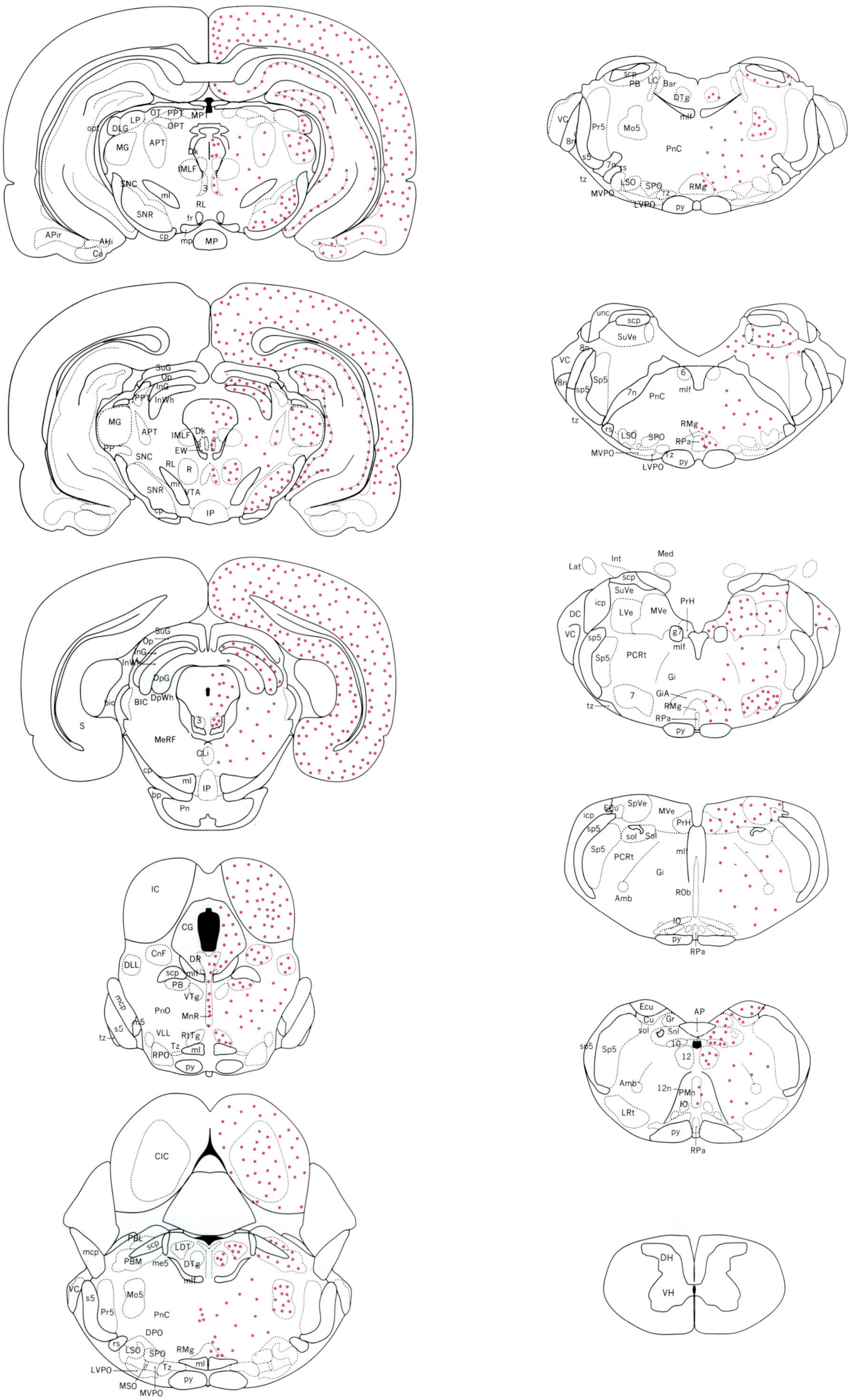

$GABA_A$ receptor (ISH)

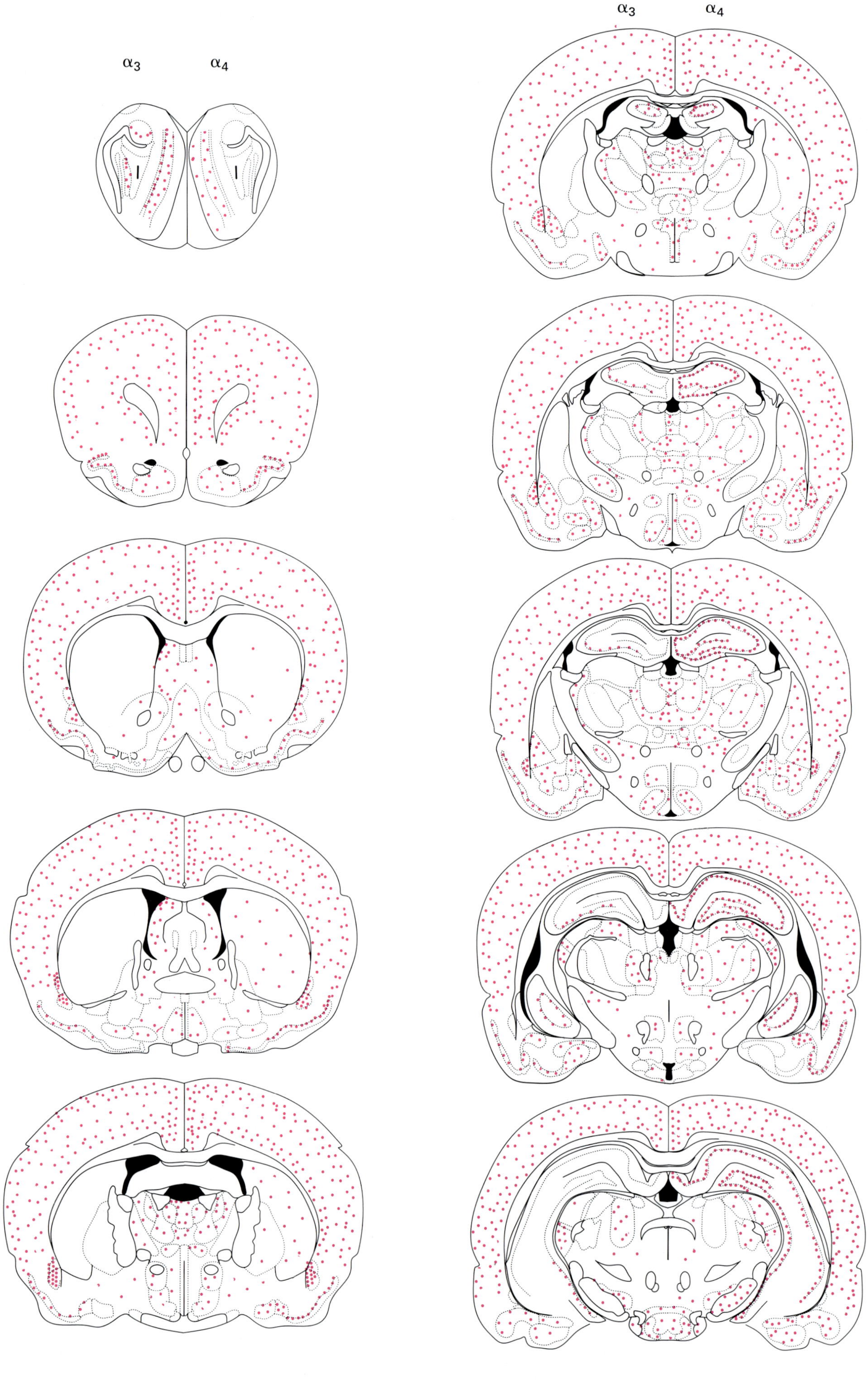

$GABA_A$ receptor (ISH)

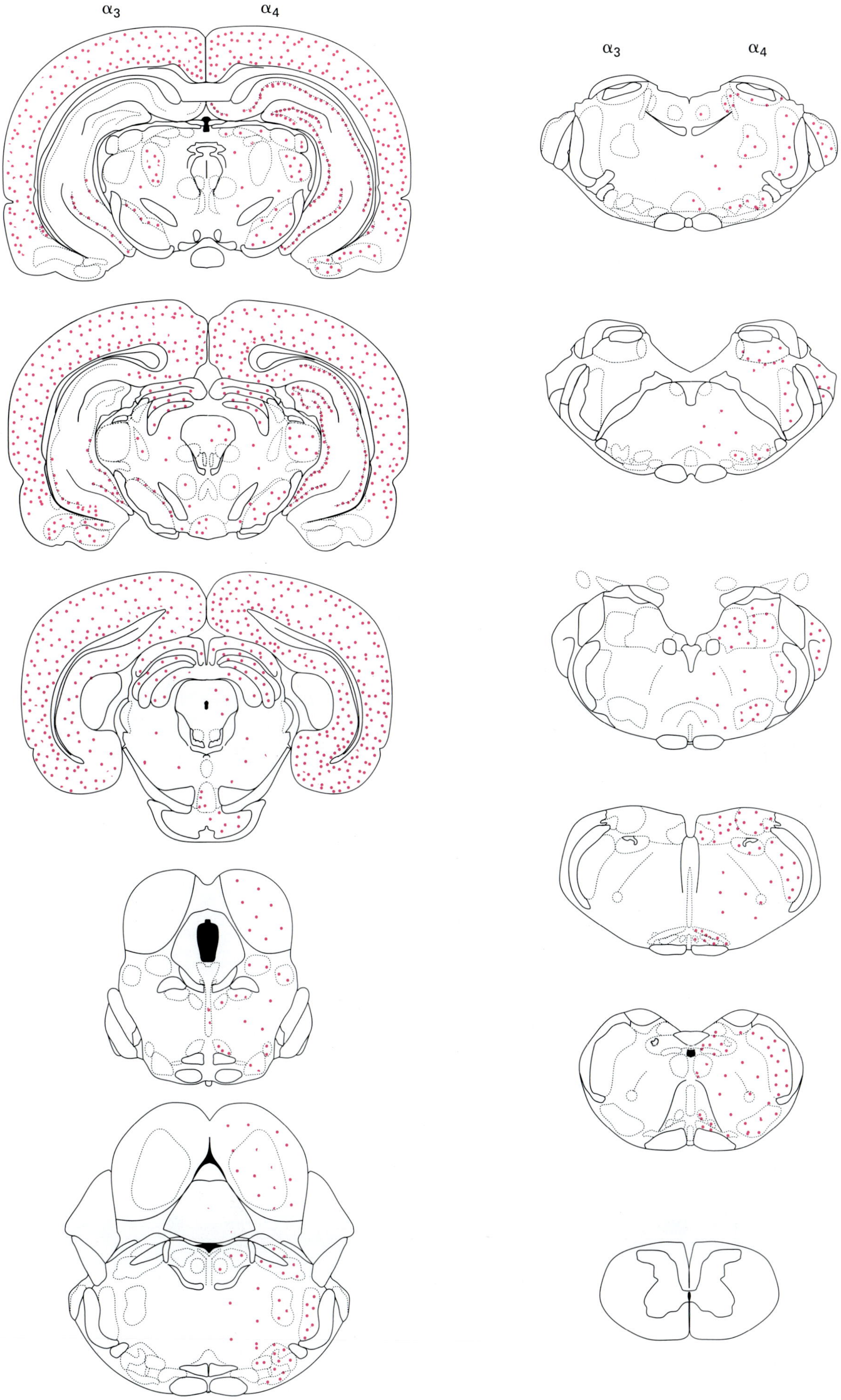

GABA$_A$ receptor (ISH)

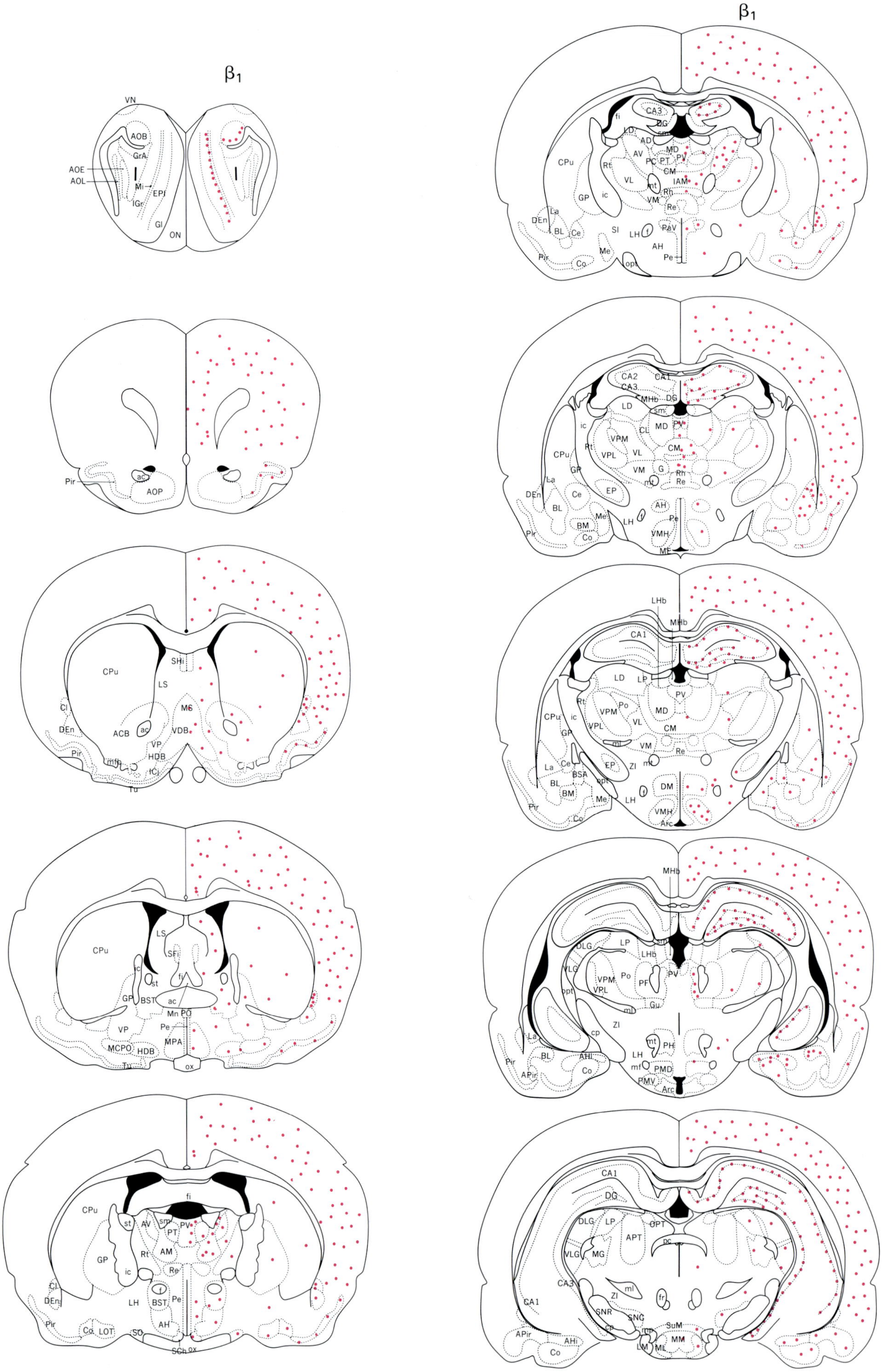

$GABA_A$ receptor (ISH)

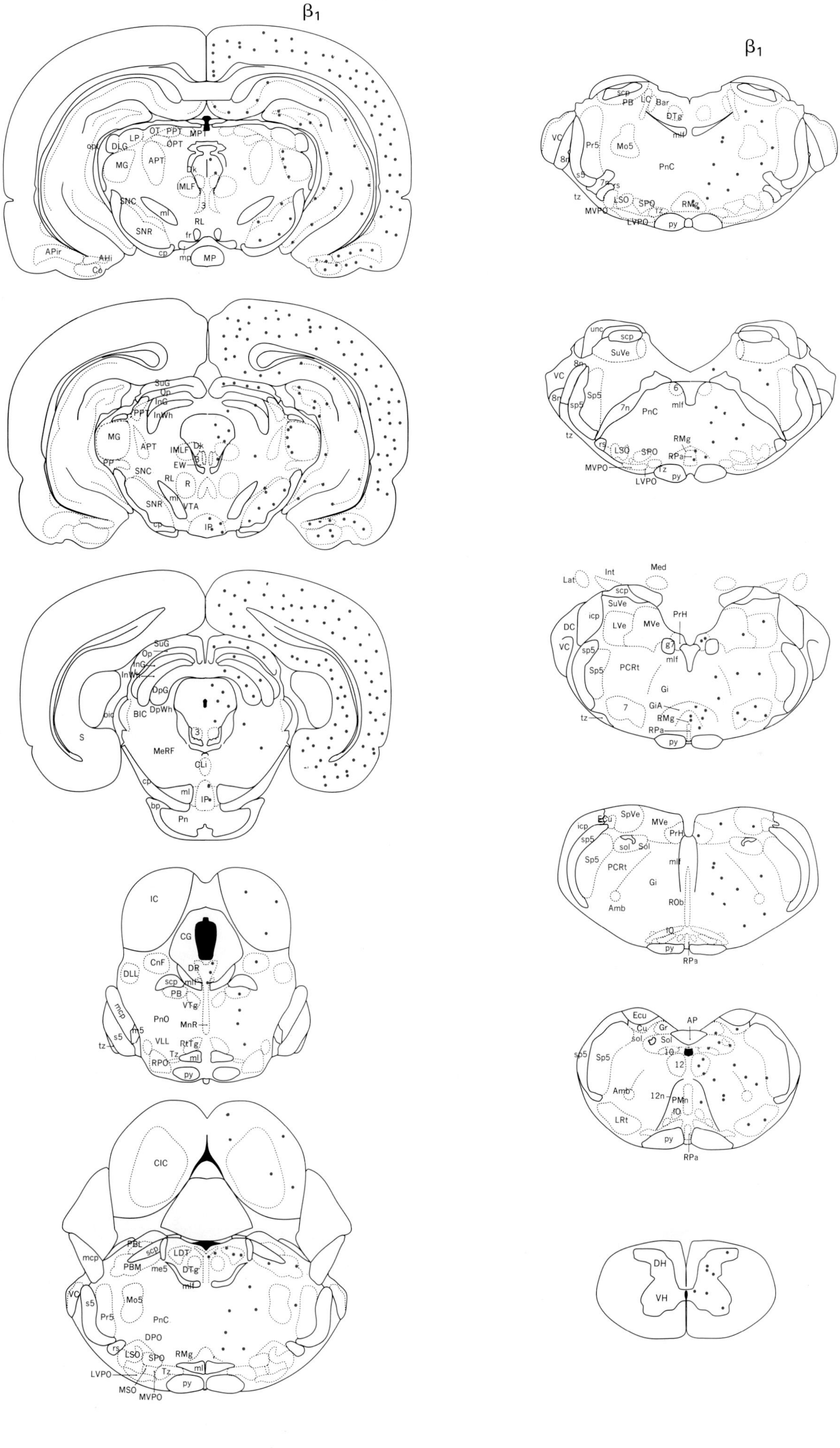

GABA$_A$ receptor (ISH)

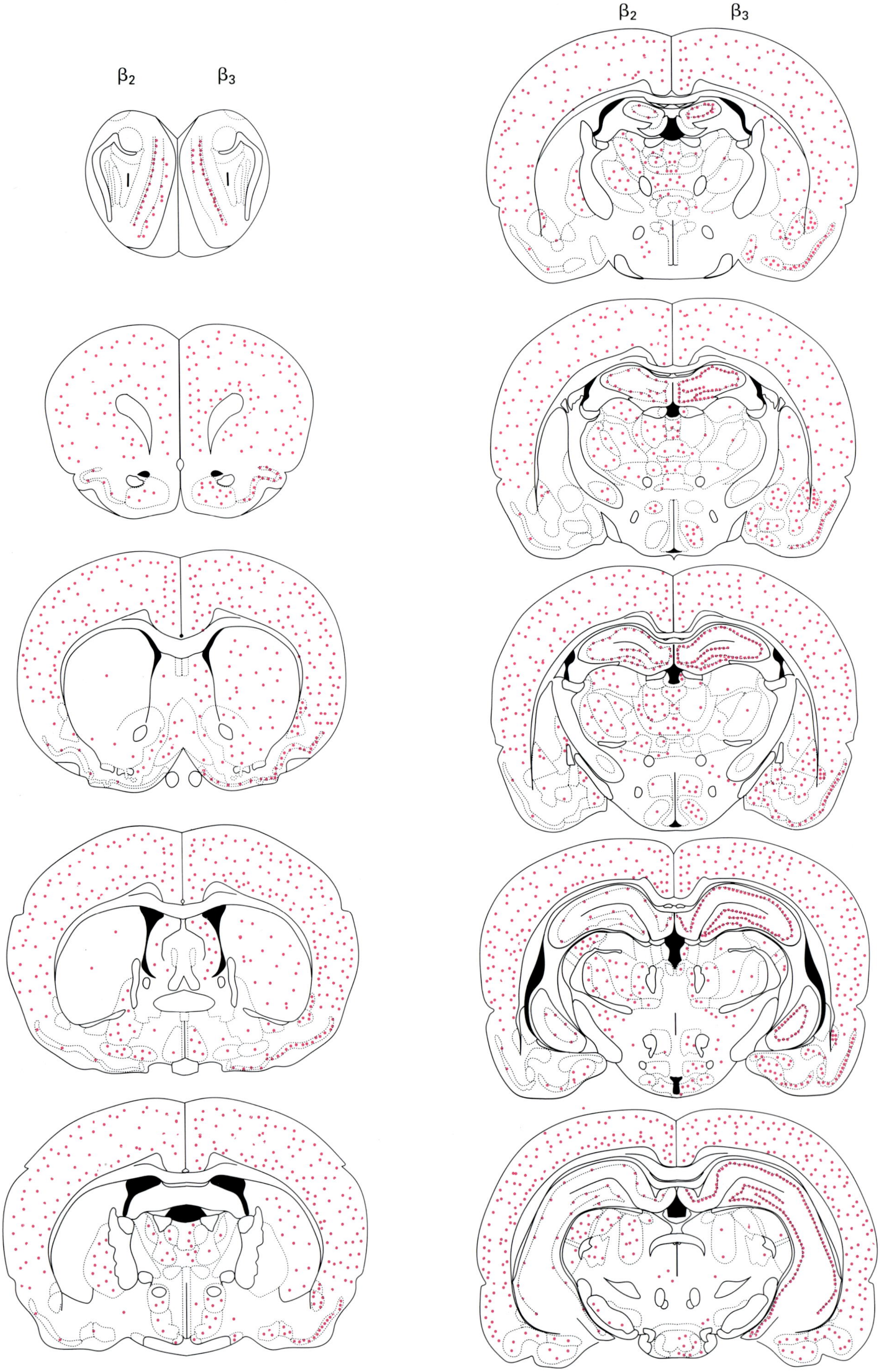

GABA$_A$ receptor (ISH)

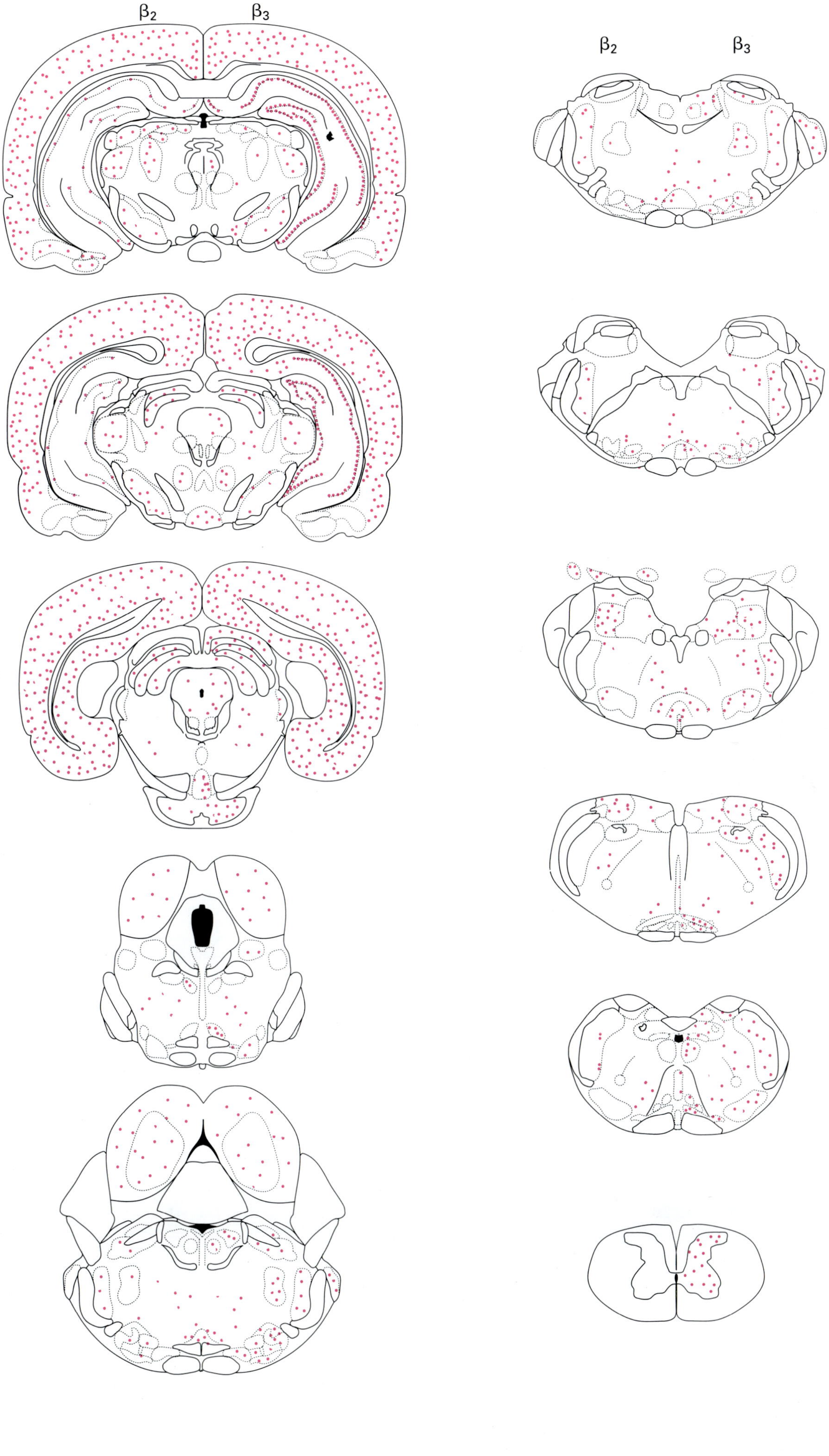

GABA$_A$ receptor (ISH)

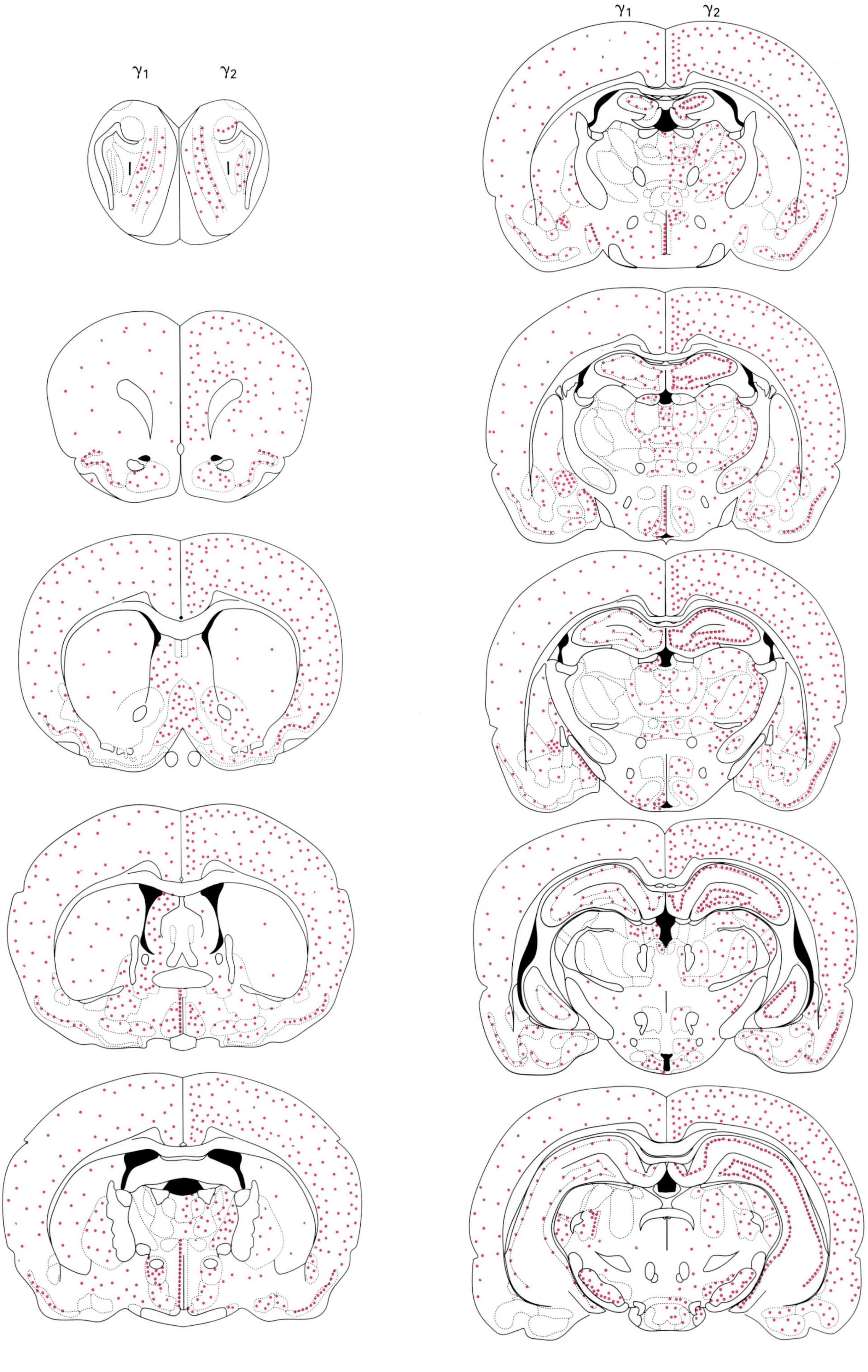

$GABA_A$ receptor (ISH)

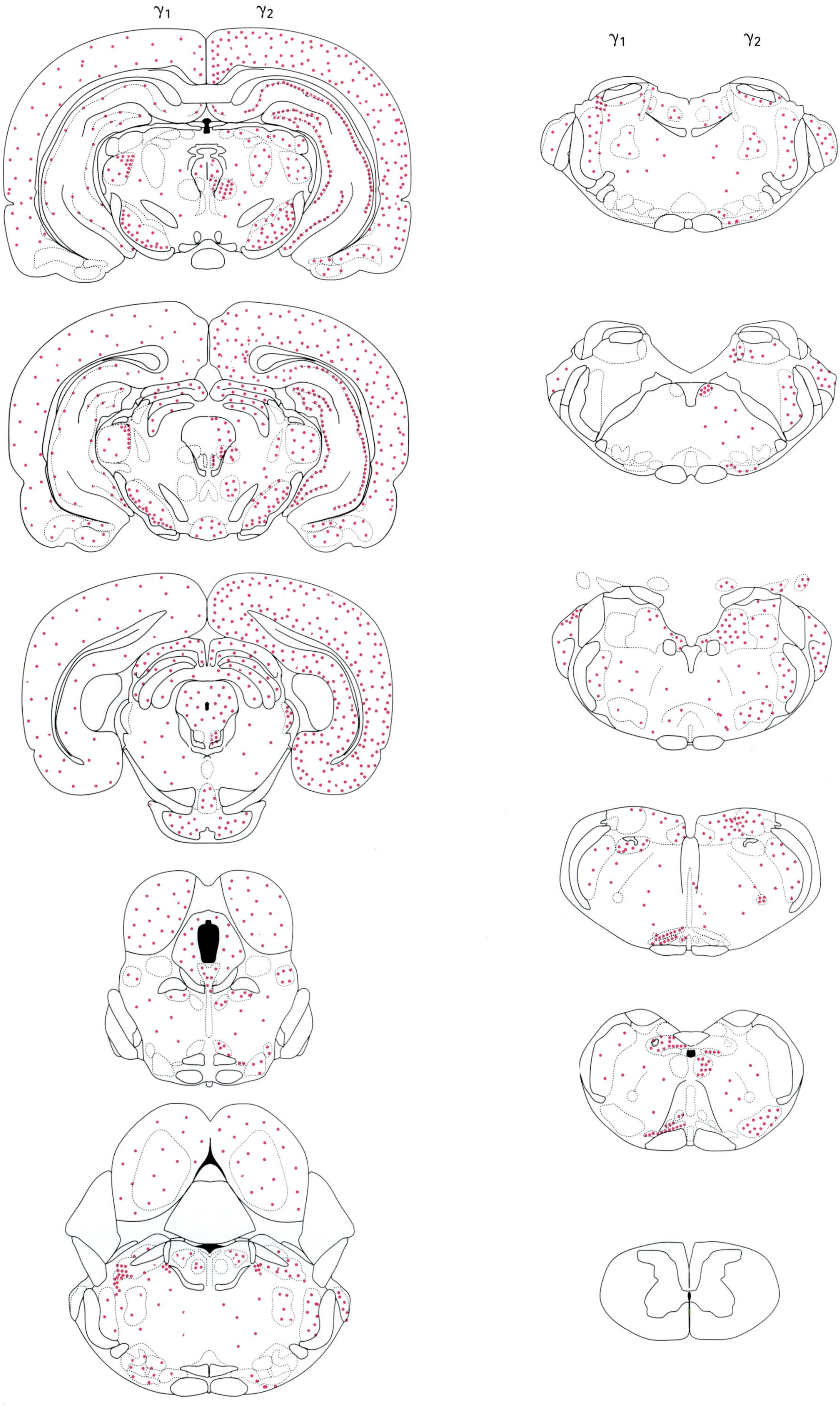

6 The glycinergic system

Glycine

Biosynthesis and decomposition

Glycine is the amino acid with the lowest molecular weight (Fig. 6.1). In the nervous system, it is thought to be synthesized mainly from glucose via glycolysis. There are several pathways for such synthesis, as shown in Fig. 6.2, the dominant one being through serine. Some reports have indicated that the activity of serine hydroxymethyltransferase (Fig. 6.2; 17) is positively correlated with the concentration of glycine. However, no specific synthetic pathway has been demonstrated that can be used as the indicator to identify the glycine-producing neurons[2]. Glycine has dual roles in the nervous system, as a neurotransmitter and as a component of the metabolic pool. Glycine released into the synaptic cleft from the axon terminal is re-uptaken either into the presynaptic terminal or

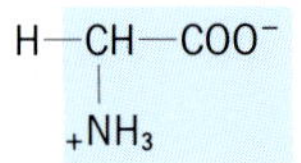

Fig. 6.1 The structure of glycine.

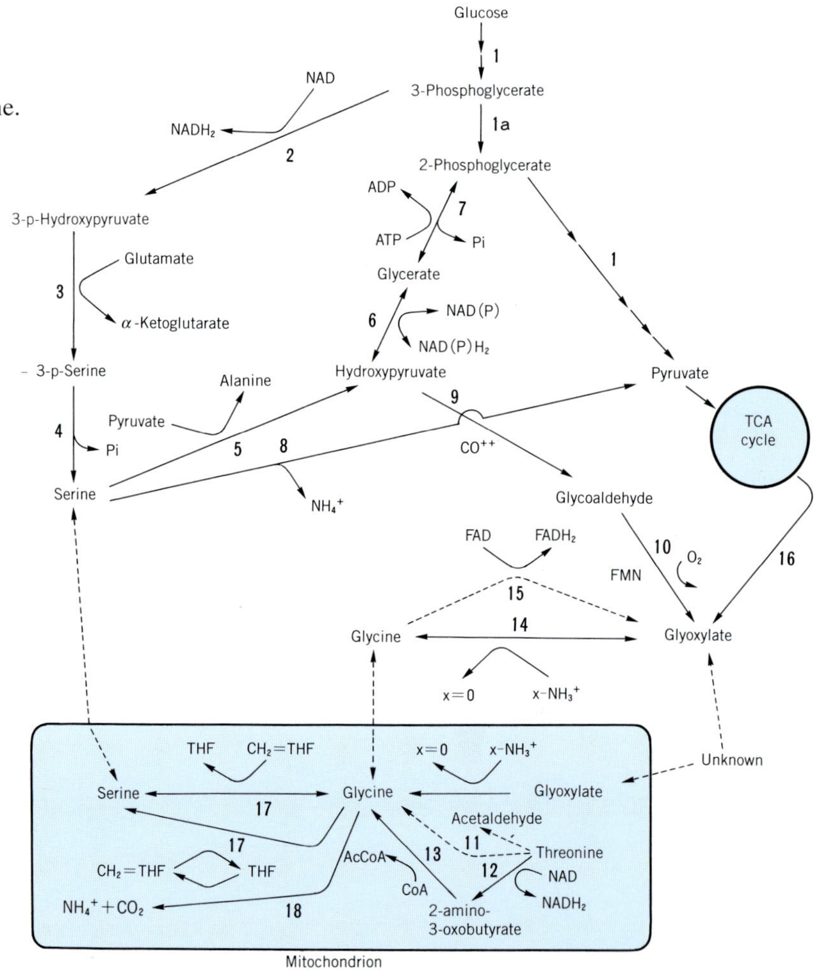

Fig. 6.2 Pathways of glycine synthesis. Abbreviations: THF, tetrahydrofolate; CH_2=THF, N^5,N^{10}-methylene THF; CoA, coenzyme A; AcCoA, acetyl coenzyme A; x-NH_3^+, amino acid transaminase donor; x=O, α-ketoacid transaminase acceptor; P_i, inorganic phosphate. Enzymes: (1) glycolytic pathway; (1a) phosphoglyceromutase; (2) D-3-phosphoglycerate dehydrogenase; (3) 3-phosphoserine:2-oxoglutarate aminotransferase; (4) 3-phosphoserine phosphatase; (5) L-serine:pyruvate aminotransferase; (6) D-glycerate dehydrogenase; (7) D-glycerate kinase; (8) L-serine dehydratase; (9) hydroxypyruvate decarboxylase; (10) glycolate oxidase; (11) 'threonine aldolase'; (12) threonine dehydrogenase; (13) 2-amino-3-oxobutyrate CoA-ligase; (14) glycine transaminase (amino acid:glyoxylate aminotransferase); (15) D-amino acid oxidase; (16) isocitrate lyase; (17) serine hydroxymethyltransferase (SHMT); (18) glycine cleavage system (coupled to SHMT). (From ref. 2)

Table 6.1. Contents of free amino acids in eight areas of the rat CNS in μmol/g tissue (from ref. 7)

	Glutamate	Glutamine	Aspartate	GABA	Glycine	Taurine
Telecephalon	11.24	4.35	2.39	1.98	0.67	7.51
Diencephalon	8.62	4.22	2.40	3.45	0.93	4.51
Mesencephalon	7.06	3.51	2.53	3.59	1.63	2.98
Cerebellum	9.68	4.74	1.99	1.58	0.63	6.18
Pons	6.24	2.80	2.53	1.74	2.93	2.42
Medulla	6.19	3.07	2.70	1.64	4.02	2.15
Spinal cord (grey)	6.36	3.27	3.45	1.40	5.53	1.58
Spinal cord (white)	3.69	3.15	1.26	0.47	3.19	1.88

into astrocytes where it is inactivated. In mammals, its inactivation is carried out by the glycine cleavage system located specifically in astrocyte mitochondria[6] (Fig. 6.3). In humans, it is known that an abnormality in this enzyme system leads to an abnormal increase in the concentration of glycine in body fluids, which in turn results in the manifestation of severe neurological symptoms. This pathological condition is well known as congenital hyperglycinemia[9]. Recently, the gene for the glycine transporter, which plays the major role in the uptake of glycine into glial cells, has been cloned and attracted much attention[8].

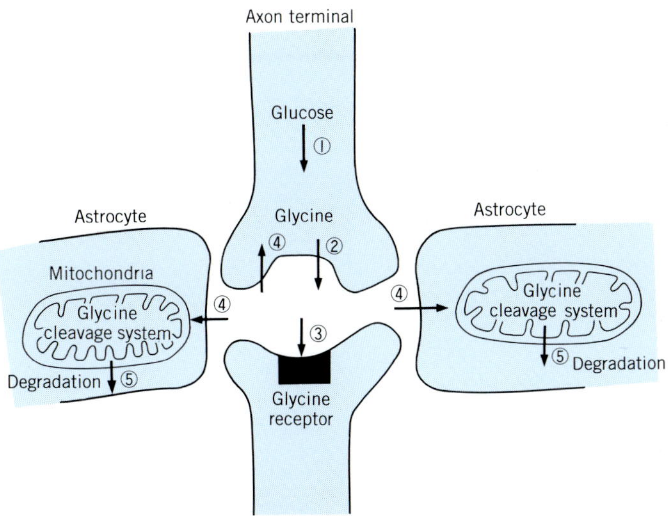

Fig. 6.3 Glycine neurotransmission. (1) synthesis of glycine from glucose; (2) release of glycine from axon terminals; (3) glycine receptors are activated by glycine; (4) uptake of glycine from synaptic cleft; (5) degradation of glycine by glycine cleavage system in astrocytes.

Neurotransmitter functions

Though glycine is generally known as an inhibitory neurotransmitter, it also has excitatory actions.

Inhibitory actions

The inhibitory action of glycine is due to its binding to the ion-channel type glycine receptor, causing an increase in the permeability of Cl^- in the membrane. As a result, the membrane is hyperpolarized and an inhibitory postsynaptic potential is elicited. This effect is inhibited by strychnine, an alkaloid. It is well known that administration of strychnine induces muscle cramp and spasm[1]. Recently the existence of a strychnine insensitive glycine receptor has also been reported[4].

Excitatory actions

Glycine potentiates the action of the excitatory NMDA receptor[3]. This action is manifested by glycine at a concentration far below the concentration at which it activates glycine receptors and is not inhibited by strychnine.

Distribution

The distribution of glycine has been studied by an antibody to glycine, which is linked with a carrier protein[5,10,11]. There are many glycine-positive fibres in the whole area of grey matter in the spinal cord of rat. Scattered positive cells are seen in the anterior and posterior horn. In the lower brain stem, positive fibres and cells are found in the reticular formation, cochlear nucleus, superior olivary nucleus, medial trapezoid nucleus, and the central grey matter of the midbrain. Positive fibres are also found in the oculomotor nucleus, trochlear nucleus, and abductor nucleus. At the upper level, they are found in the external plexiform layer and mitral cell layer of the olfactory bulb, hypothalamus, and hippocampus. In the cerebellum, positive fibres are seen in the granular layer and the molecular layer, while a few positive cells exist in the granular layer.

Biochemical studies indicate that the concentration of glycine tends to be high at the lower brain and is low in the upper brain[2] (Table 6.1). This tendency is characteristic of glycine and may reflect the difference in its role at the two levels.

References

1. Betz, H. and Becker, C.M. (1988). *Neurochem. Int.*, **13**, 137–46.
2. Daly, E.C. (1991). In *Glycine neurotransmission* (ed. O.P. Ottersen, and J. Storm-Mathisen) pp. 25–66. Wilcy, Ncw York.
3. Johnson, J.W. and Ascher, P. (1987). *Nature*, **325**, 529–31.
4. Kuhse, J., Schmieden, V., and Betz, H. (1990). *Neuron*, **5**, 867–73.
5. Ottersen, O.P., Davanger, S., and Storm-Mathisen, J. (1987). *Exp. Brain Res.*, **66**, 211–21.
6. Sato, K., Yoshida, S., Fujiwara, K. *et al.* (1991). *Brain Res.*, **567**, 64–70.
7. Shank, R.P. and Aprison, M.H. (1970). *J. Neurochem.*, **17**, 1461–75.
8. Smith, K.E., Borden, L.A., Hartig, P.R. *et al.* (1992). *Neuron*, **8**, 927–35.
9. Tada, K. and Hayasaka E. (1987). *Eur. J. Pediatr.*, **146**, 221–7.
10. Van den Pol, A.N. and Gorcs, T. (1988). *J. Neurosci.*, **8**, 472–92.
11. Wenthold, R.J., Huie, D., Altschuler, R.A., and Reeks, K.A. (1987). *Neuroscience*, **22**, 897–912.

Glycine (immunocytochemistry)

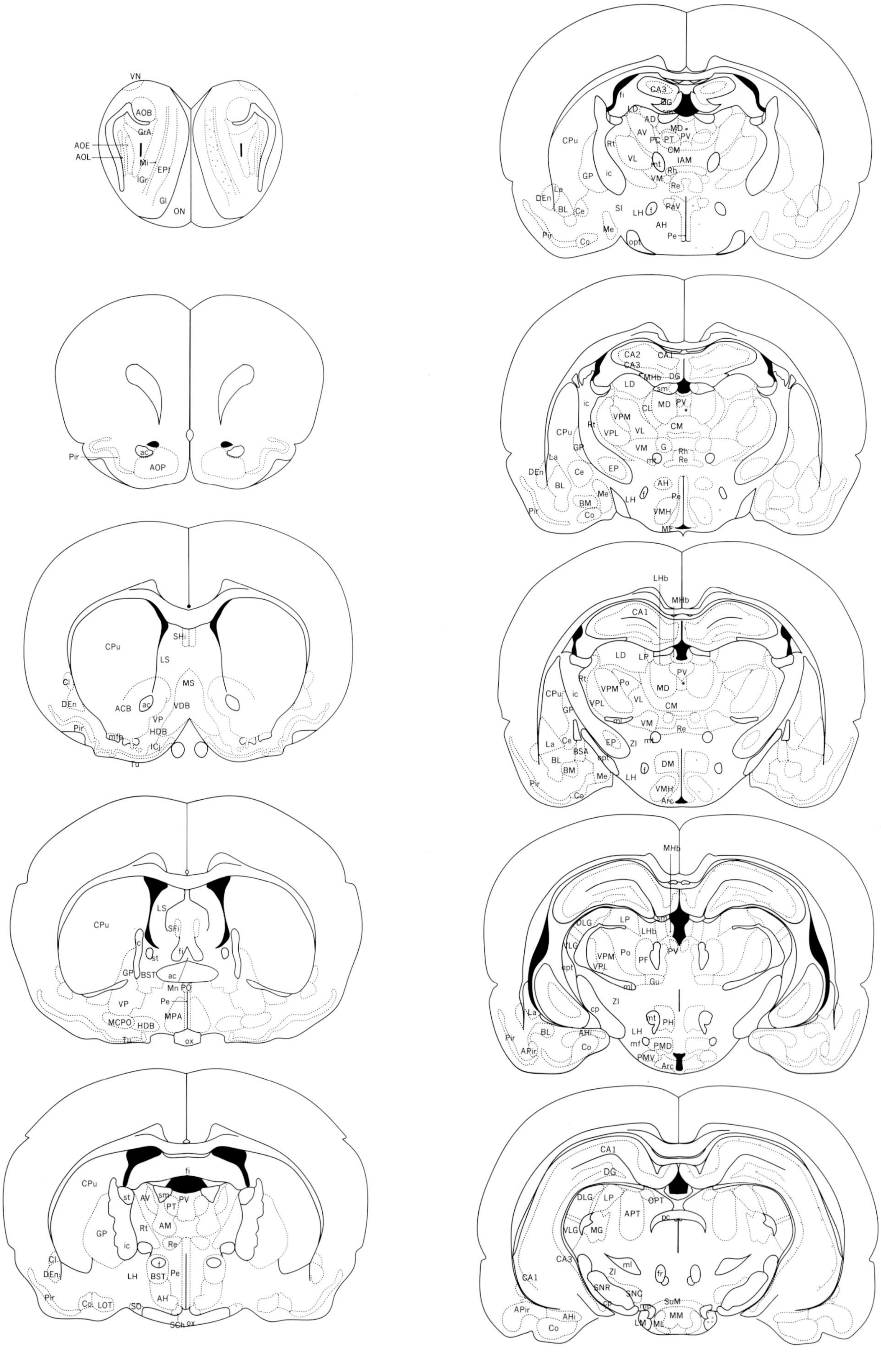

Glycine (immunocytochemistry)

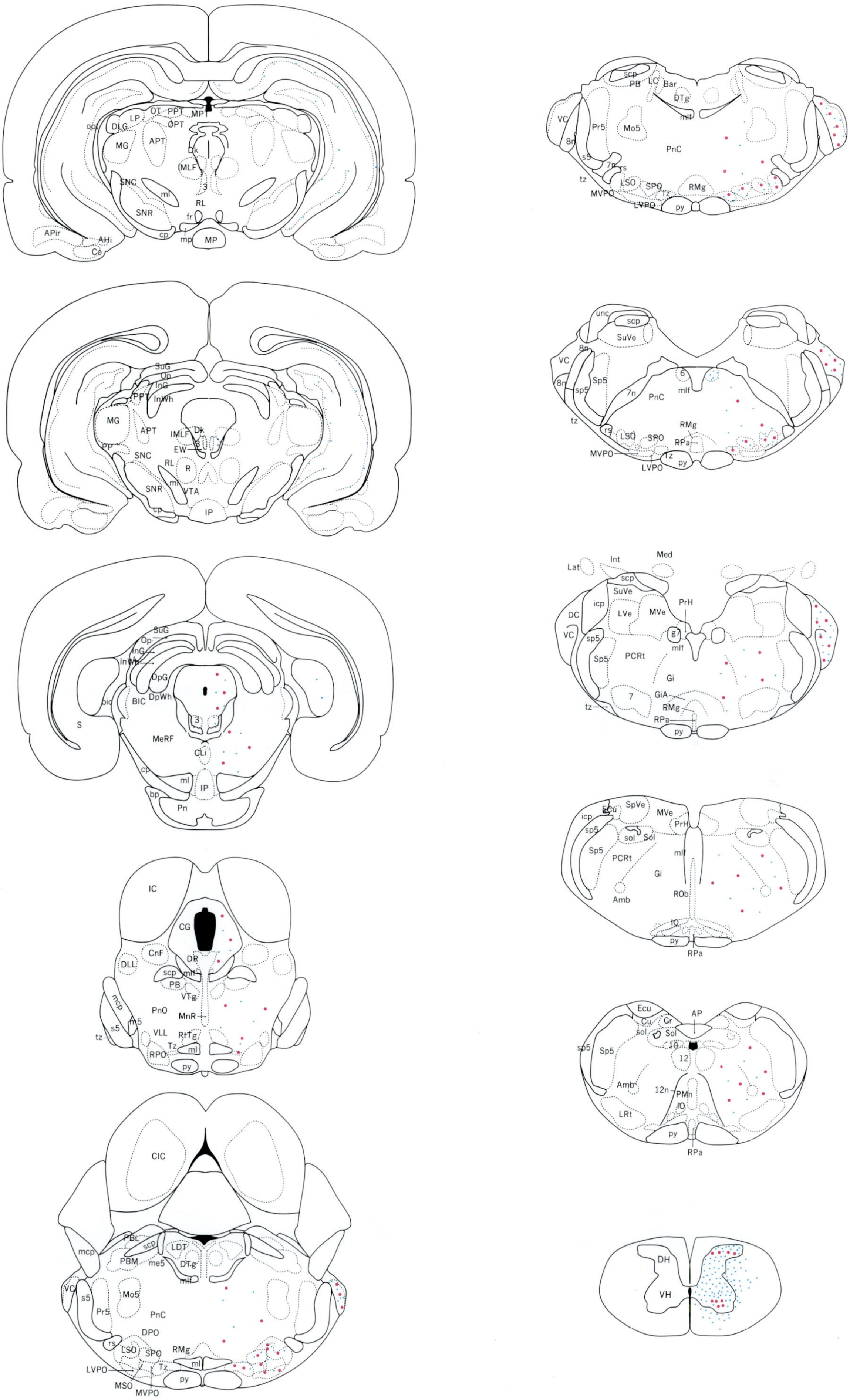

Glycine receptor

Structure and subunits

The adult type glycine receptor was purified by the affinity column method using 2-aminostrychnine as a ligand and was found to be composed of three subunits; α_1 subunit (48 kDa), β subunit (58 kDa), and one with a molecular weight of 93 kDa (Fig. 6.4). The amino acid and base sequences of the α_1 and β subunits have been determined[3,4]. These are thought to have four hydrophobic portions which are assumed to be transmembrane sites, with their N-terminal facing the extracellular space (Fig. 6.5) which is highly analogous to the nicotinic ACh receptor protein and $GABA_A$ receptor protein. The α_1 subunit is assumed to have glycine and strychnine binding sites near the N-terminal which is exposed to the extracellular surface. The adult type glycine receptor is thought to comprise three α_1 subunits, two β subunits (which constitute the Cl^- channel), and the 93 kDa protein which underlies them at the intracellular surface (Fig. 6.4).

Betz *et al.*[1] have reported various subunits, which are shown in Table 6.2. One of the remarkable findings is the presence of an α_2 subunit which is expressed specifically in infants. It appears around day 14 of gestation, reaches its peak around birth, and diminishes afterwards[11]. It can be considered as the precursor to the α_1 subunit and is assumed to play an important role in the development of glycinergic system. The 93 kDa protein has been cloned and named gephyrin[13]. It has high affinity for tubulin and thus is believed to take part in linking the glycine receptor and the cytoskeleton.

Action mechanism

When glycine binds to its receptor, the latter changes its structure so that permeability to Cl^- is increased. As a result, the membrane becomes hyperpolarized and the neuron is inhibited[1]. This effect is inhibited by strychnine, though the α_2* subunit in rat is insensitive to it[7]. An explanation has been put forward for the difference in sensitivity among the subunits[7]. It has been elucidated that in the rat, whereas the α_2* subunit (which is insensitive to strychnine) has glutamic acid at position 167, in other α subunits the corresponding amino acid is glycine. As this site is thought to constitute the strychnine binding site, the sensitivity to strychnine is determined by the difference in one amino acid at this site.

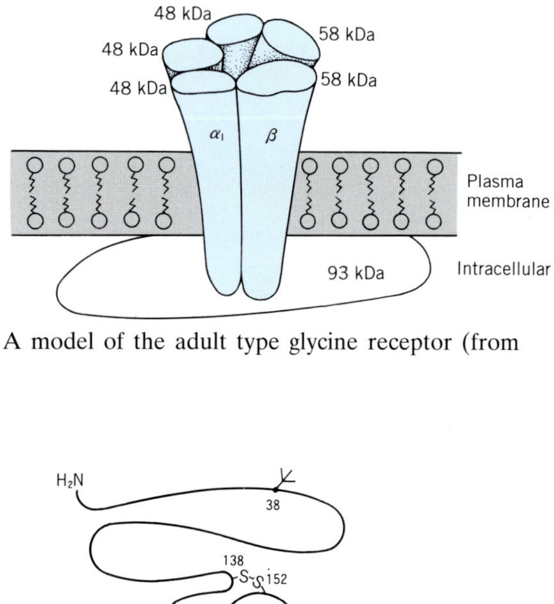

Fig. 6.4 A model of the adult type glycine receptor (from ref. 6).

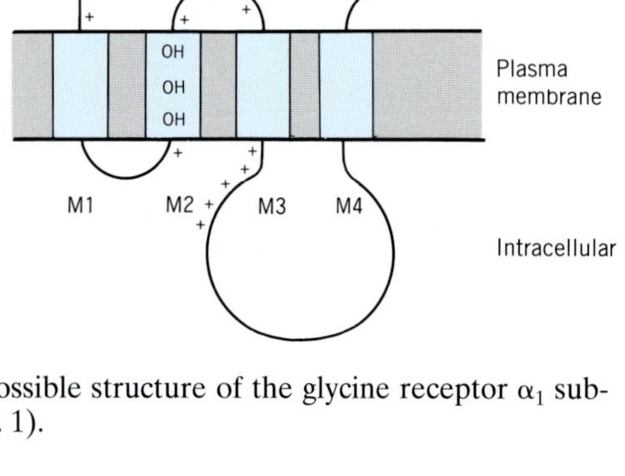

Fig. 6.5 A possible structure of the glycine receptor α_1 subunit (from ref. 1).

Table 6.2. Glycine receptor subunits

α subunit (glycine binding subunit)	
α (adult type)	
Rat α_1[3]	Glycine binding subunit of the rat adult type glycine receptor, 48 kDa, strychnine sensitive
Rat α_1 ins[10]	A variant of rat α_1, with eight amino acids insertion by alternative splicing, strychnine sensitive
Human α_1[5]	Glycine binding subunit of human adult type glycine receptor, 48 kDa, six amino acids are different from α_1, strychnine sensitive, autosomal chromosome
α_2 (infant type)	
Rat α_2*[7]	Glycine binding subunit of rat infant type glycine receptor, 49 kDa, strychnine insensitive
Rat α_{2A}[9]	A variant of rat α_2*. Only one amino acid is different, strychnine sensitive
Rat α_{2B}[9]	A variant of rat α_2*. Two amino acids are different
Human α_2[5]	Glycine binding subunit of human infant type glycine receptor, 49 kDa. One amino acid is different from rat α_2*, strychnine sensitive, X chromosome
α_3 (adult type)	
Rat α_3[8]	Weak expression in the olfactory bulb, hippocampus, etc., in the adult rat brain
β subunit	
Rat β[4]	Makes the glycine receptor by coupling with α subunit. Homo-oligomer of β subunit is also weakly reactive to glycine
93 kDa anchoring protein	
Gephyrin[12]	93 kDa, involved in the association between cytoskeletons and glycine receptors

Distribution

The binding sites for ^{3}H-glycine are distributed abundantly in the upper brain, such as cerebral cortex, basal ganglia, and hippocampus. However, they are sparse in the lower brain, where previous studies have assumed that glycine receptors exist abundantly[11]. This suggests that there are binding sites whose affinity is higher than glycine receptors. At present, it is believed that the binding sites detected by ^{3}H-glycine represent the glycine binding sites of the NMDA receptor. Therefore, in order to specifically detect the glycine receptor it is mandatory that strychnine is used as its specific inhibitor. The binding sites for ^{3}H-strychnine are distributed abundantly in the spinal cord and the lower brain stem, and are sparse in the upper brain[16].

In situ hybridization for mRNA of each subunit of the glycine receptor showed that cells positive to mRNA for the α_1 subunit exist mainly in the spinal cord and the lower brainstem and are sparse in the upper brain. These results correspond perfectly with those obtained from ^{3}H-strychnine autoradiography[14].

Positive cells for mRNA of the β subunit exist in the spinal cord, lower brainstem, cerebral cortex, hippocampus, cerebellum, and most other regions in the brain. The role it plays in the upper brain still remains obscure, though it is suggested that it may couple with an unknown α subunit to constitute glycine receptors[2].

The mRNA for the α_2 subunit is expressed in large amounts around birth and then gradually decreases. In the atlas, the distribution of the α_2 subunit is shown at 7 days after birth[15]. It is distributed more widely than that for the α_1 subunit, suggesting that the α_2 subunit not only plays a role as precursor to the α_1 subunit, but also possesses other important functions in wider areas of the brain.

Monoclonal antibodies to the 93 kDa protein have shown that the positive cells and fibres to this protein are distributed widely in the brain.

References

1. Betz, H. and Becker, C.M. (1988). *Neurochem. Int.*, **13**, 137–46.
2. Fujita, M., Sato, K., Sato, M. *et al.* (1991). *Brain Res.*, **560**, 23–7.
3. Grenningloh, G., Rienitz, A., Schmitt, B. *et al.* (1987). *Nature*, **328**, 215–20.
4. Grenningloh, G., Pribilla, I., Prior, P. *et al.* (1990). *Neuron*, **4**, 963–70.
5. Grenningloh, G., Schmieden, V., Schofield, P.R. *et al.* (1990). *EMBO J.*, **9**, 771–6.
6. Langosch, D., Betz, H., and Becker, C. M. In *Glycine neurotransmission* (ed. O. P. Ottersen and J. Storm-Mathisen), pp. 67–82. Wiley, New York.
7. Kuhse, J., Schmieden, V., and Betz, H. (1990). *Neuron*, **5**, 867–73.
8. Kuhse, J., Schmieden, V., and Betz, H. (1990). *J. Biol. Chem.*, **265**, 22317–20.
9. Kuhse, J., Kuryatov, A., Maulet, Y. *et al.* (1991). *FEBS Lett.*, **283**, 73–7.
10. Malosio, M.L., Grenningloh, G., Kuhsek, J. *et al.* (1991). *J. Biol. Chem.*, **266**, 2048–53.
11. Malosio, M.L., Marqueze-Pouey, B., Kuhse, J., and Betz, H. (1991). *EMBO J.*, **10**, 2401–9.
12. McDonald, J.W., Penney, J.B., Johnston, M.V., and Young, A.B. (1990). *Neuroscience*, **35**, 653–68.
13. Prior, P., Schmitt B., Grenningloh, G. *et al.* (1992). *Neuron*, **8**, 1161–70.
14. Sato, K., Zhang, J.H., Saika, T. *et al.* (1991). *Neuroscience*, **43**, 381–95.
15. Sato, K., Kiyama, H., and Tohyama, M. (1992). *Brain Res.*, **590**, 95–108.
16. Kirsch, J. and Betz, H. (1993). *Brain Res.*, **621**, 301–10.
17. Zarbin, M.A., Wamsley, J.K., and Kuhar, M.J. (1981). *J. Neurosci.*, **1**, 532–47.

Glycine receptor (binding sites)

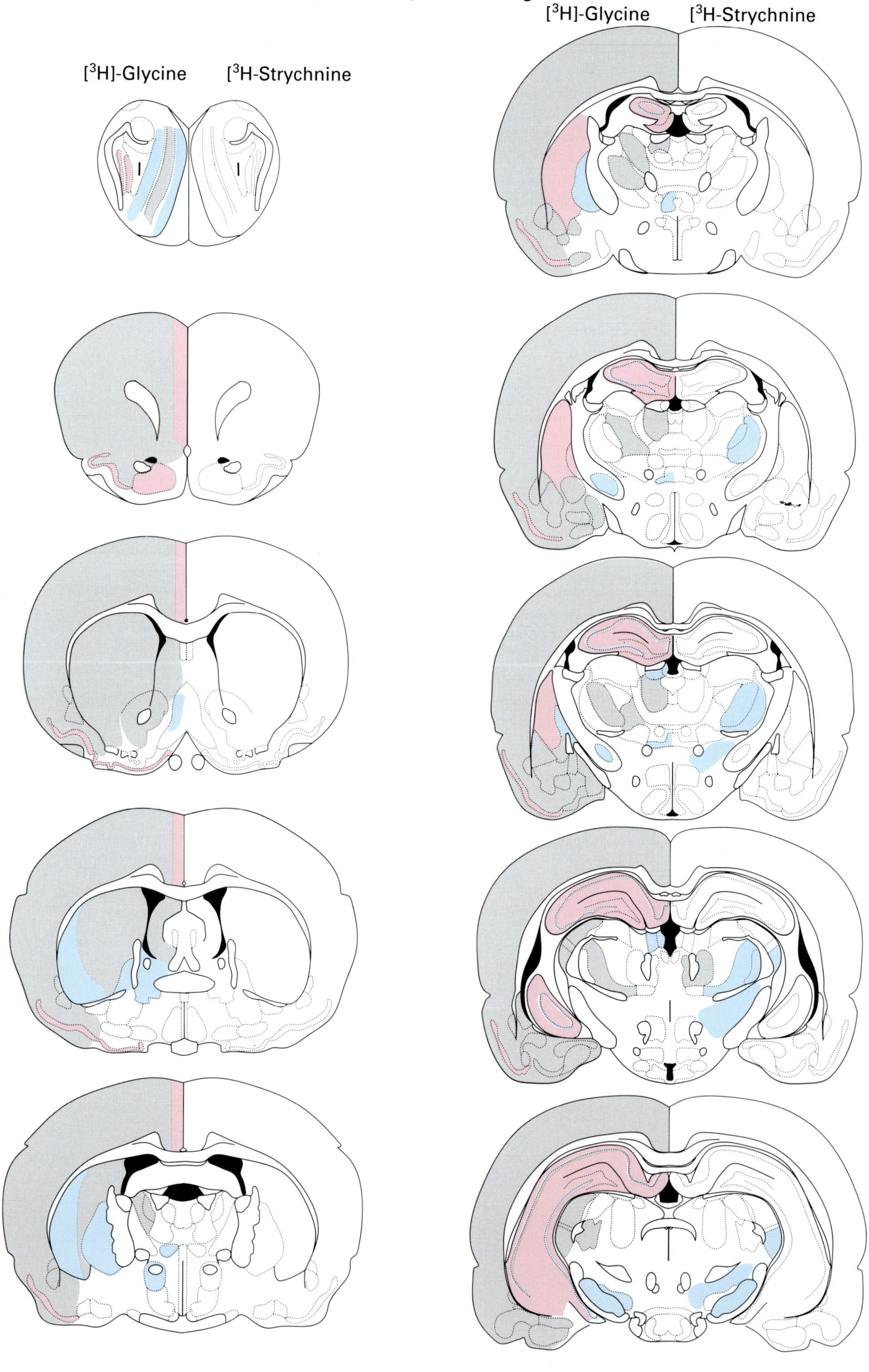

Glycine receptor (binding sites)

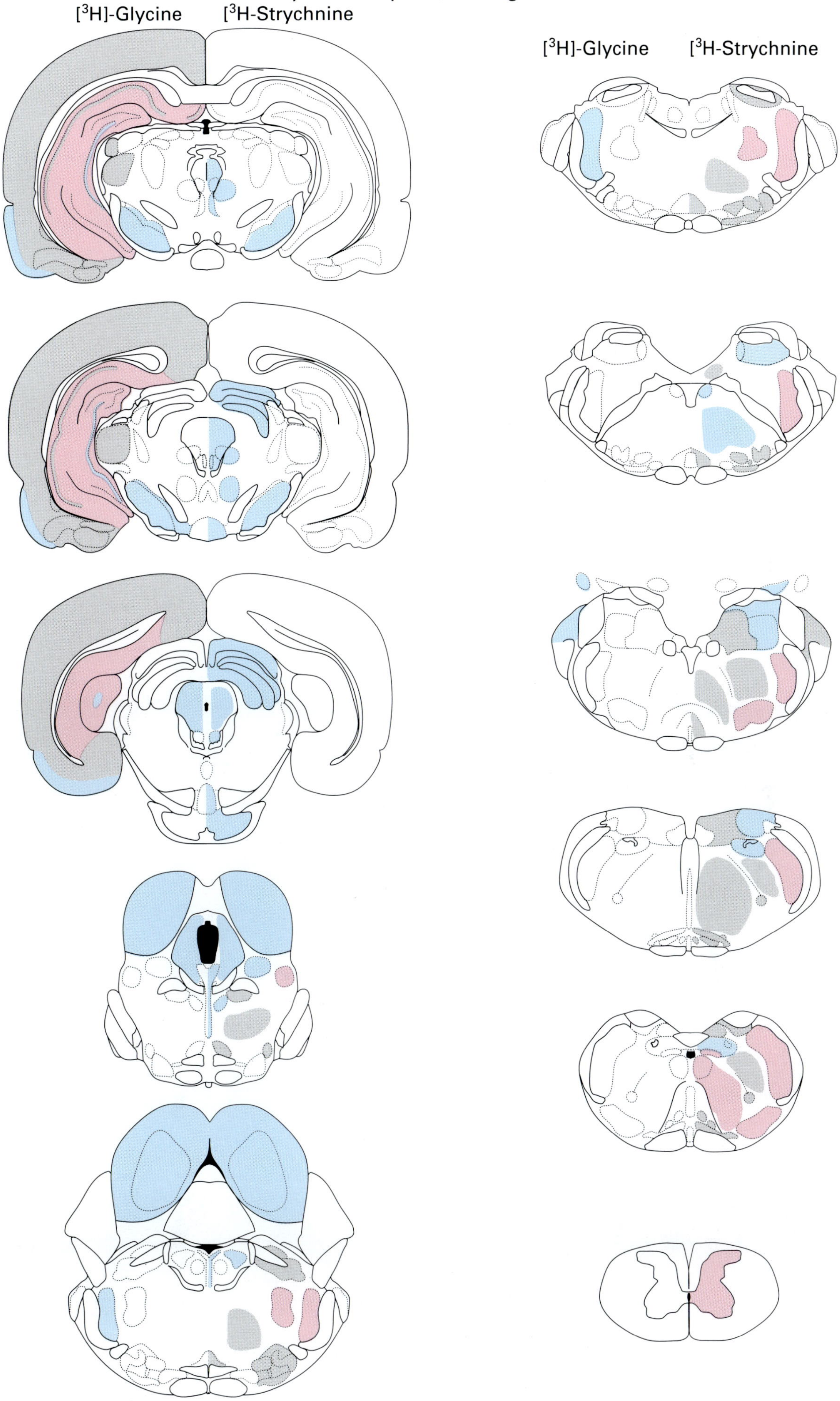

Glycine receptor (ISH)

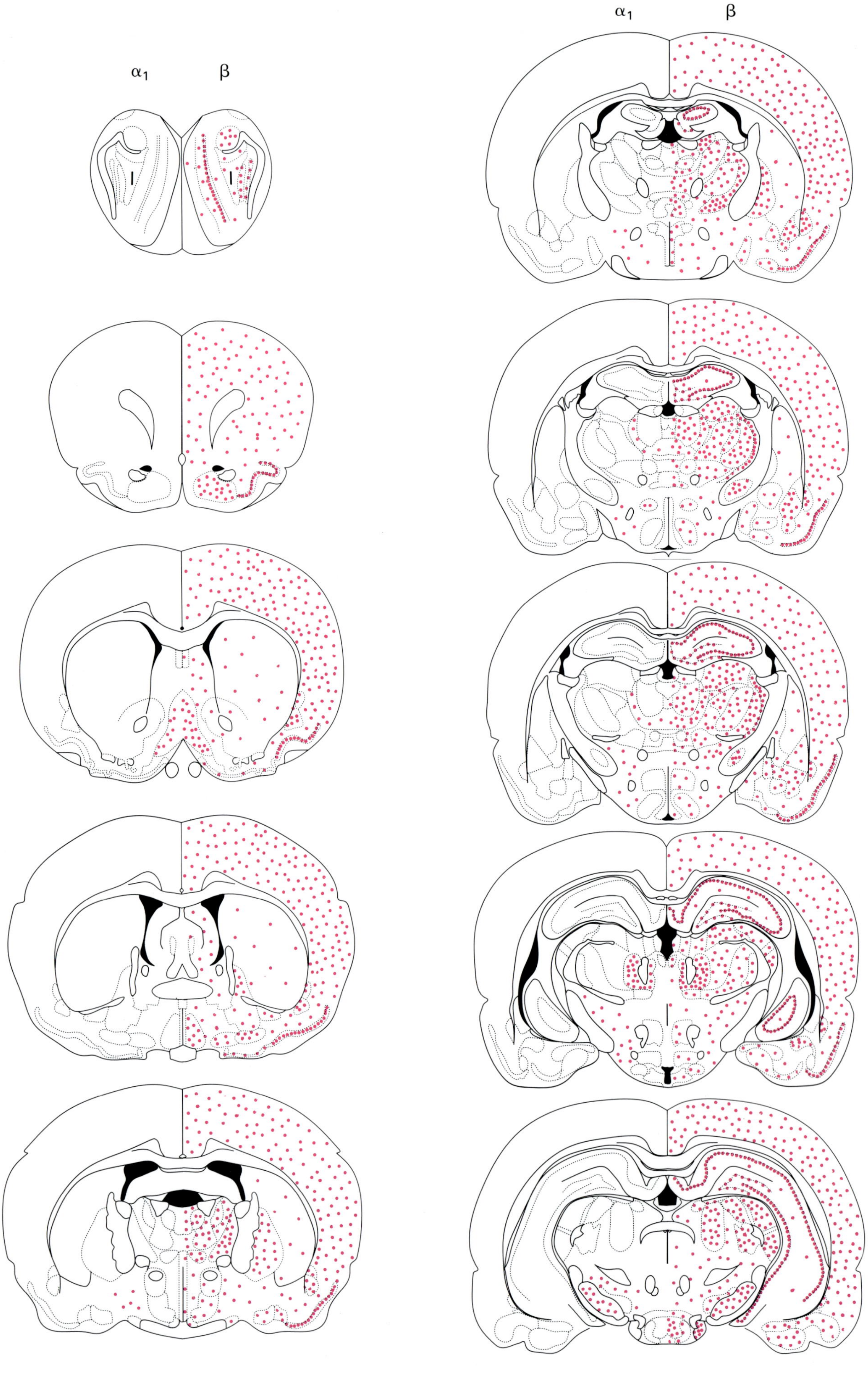

Glycine receptor (ISH)

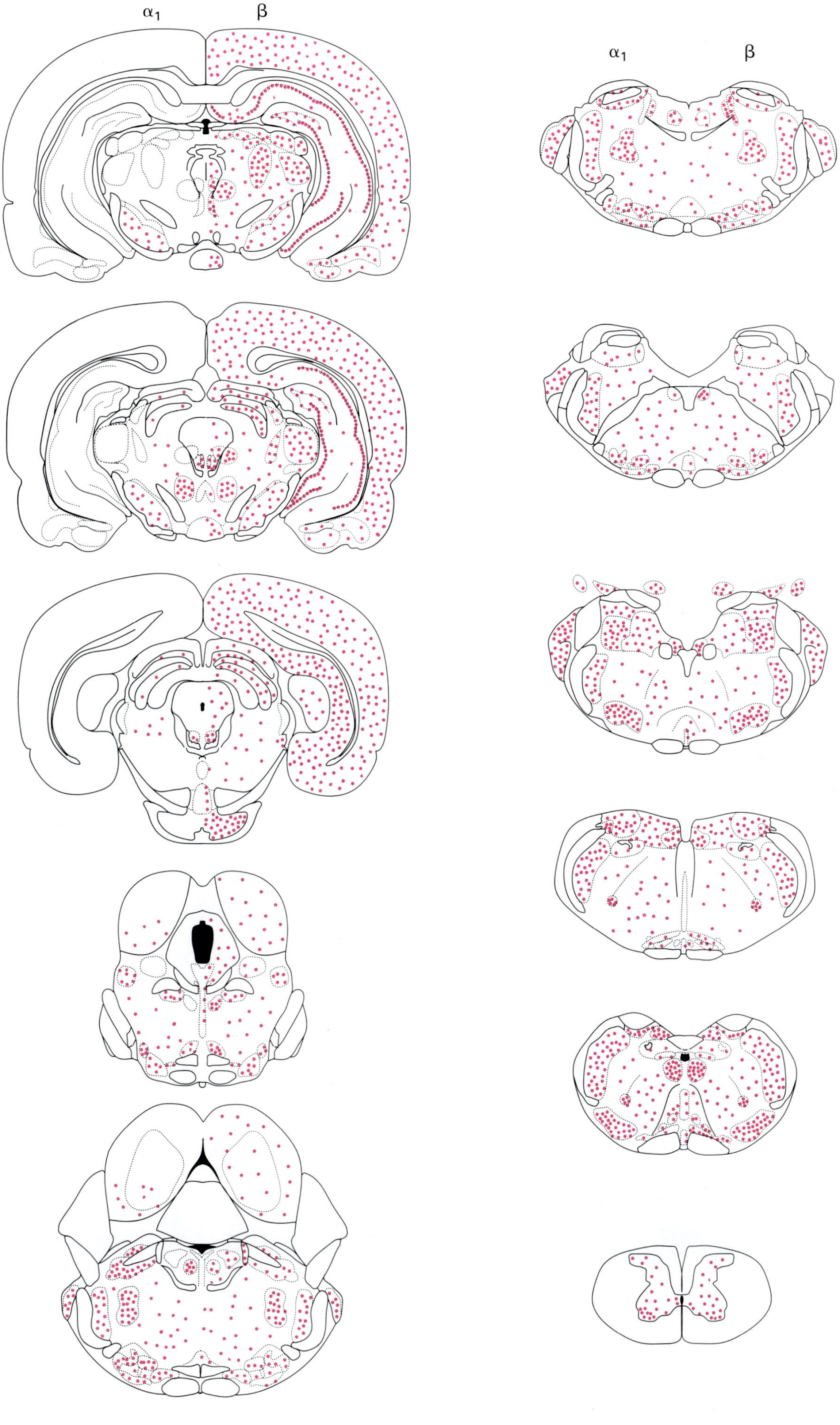

Glycine receptor (ISH and immunocytochemistry)

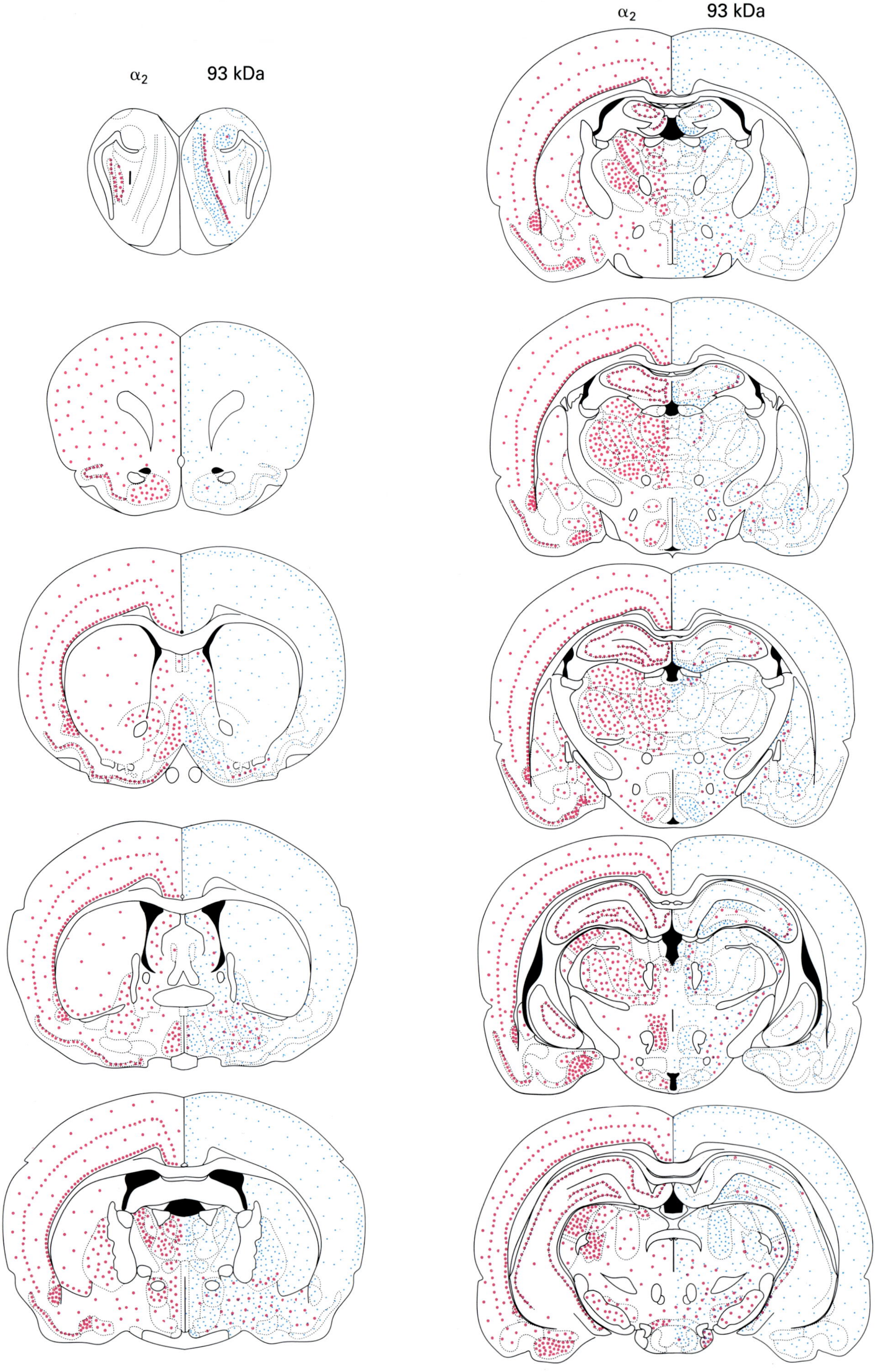

Glycine receptor (ISH and immunocytochemistry)

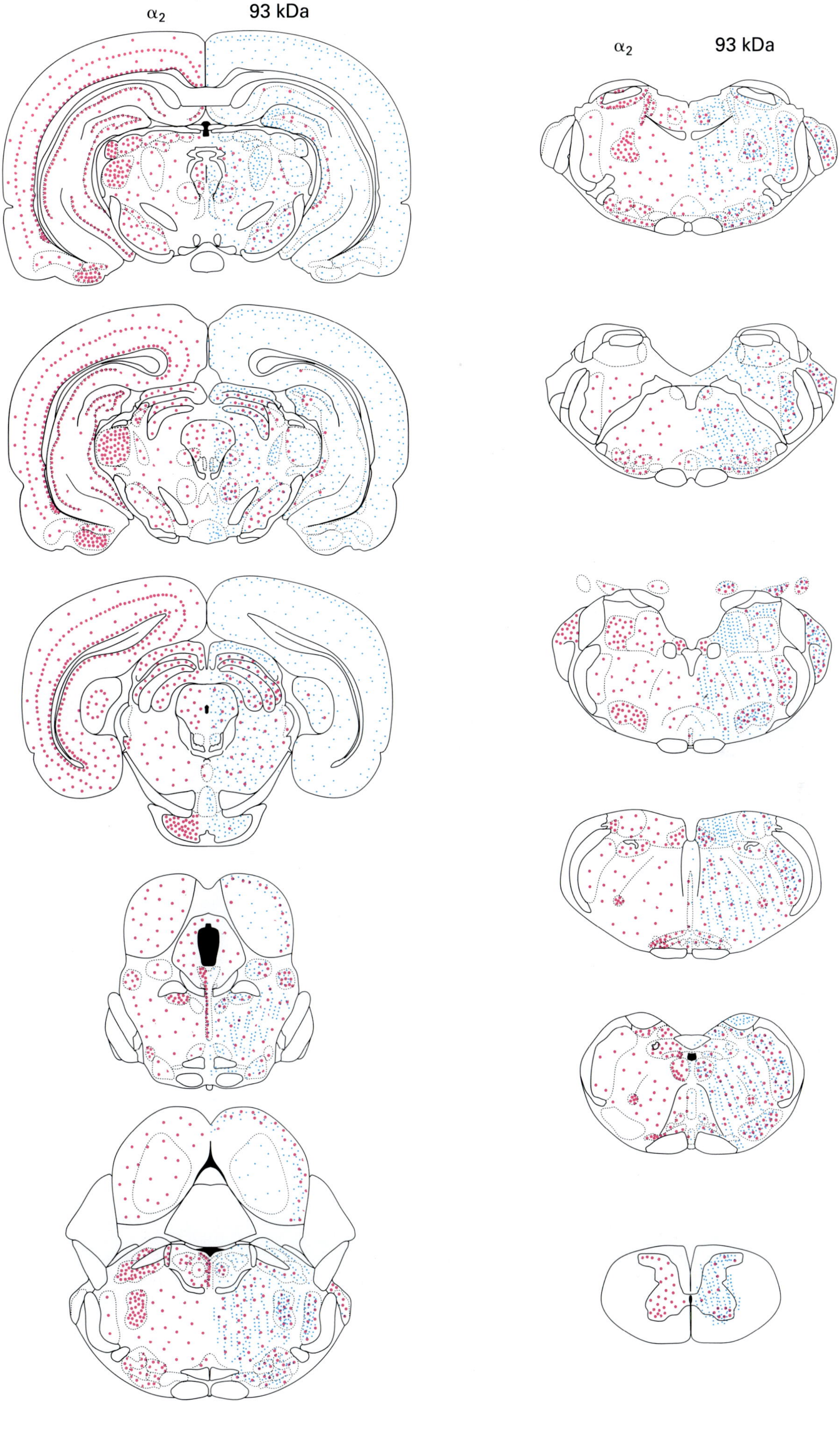

7 The glutamatergic system

Glutamic acid

Biosynthesis and decomposition

Glutamic acid is an amino acid which is present in abundance in the central nervous system (Fig. 7.1) and plays an important role in neurotransmission and metabolism[2,3]. Since glutamic acid cannot pass through the blood–brain barrier easily, it is believed to be synthesized within the central nervous system.

Glutamic acid is usually synthesized by two pathways: from glucose through α-ketoglutaric acid (Fig. 7.2; 8), and from ornithine through glutamic acid-γ-semialdehyde (Fig. 7.2; 2,4). However, the glutamic acid which functions as a neurotransmitter is derived from the glutamic acid–glutamine circuit (Fig. 7.2; 6, Fig. 7.3)[2,3,9]. Glutaminase, which is highly localized at the nerve endings, acts on glutamine to release glutamic acid as a neurotransmitter. Glutamic acid released into the synaptic cleft binds with the glutamate receptor and expresses its function; subsequently it is re-uptaken into the presynaptic region and is re-used or is moved into astrocytes and converted into glutamine by the glutamine synthetase that is located there. Finally, it is recycled in the neuron[7–9] (Fig. 7.3). Glutamic acid is also an important intermediate metabolite in the biosynthesis of GABA (Fig. 7.2; 5) and in the biosynthesis of proline (Fig. 7.2; 3,4).

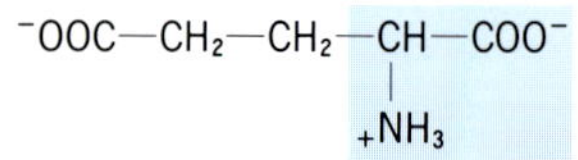

Fig. 7.1 The structure of glutamate.

Action mechanism

It has been known since the 1950s that intracerebral injection of glutamic acid leads to convulsion[4]. Later studies showed that glutamate acts through the glutamate receptor existing in postsynaptic cells[1]. Glutamic acid receptors have been classified into NMDA receptor, AMPA/kainate (non-NMDA) receptor, and metabotropic receptor.

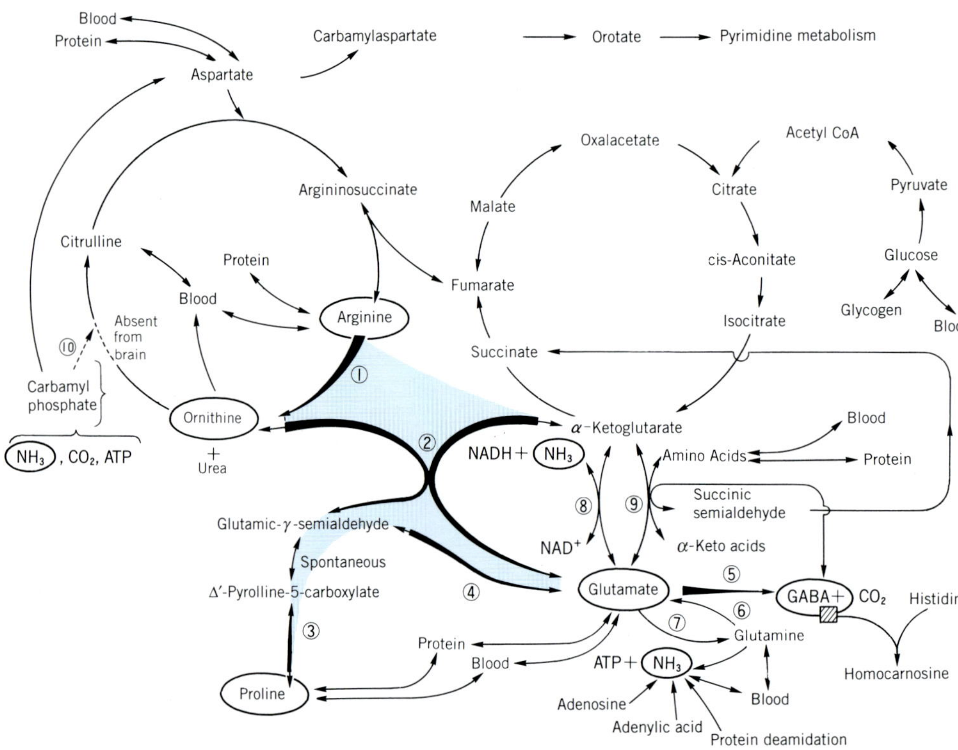

Fig. 7.2 The metabolic pathway of glutamate. (1) arginase; (2) ornithine-β-ketoglutarate-ω-aminotransferase; (3) proline dehydrogenase; (4) glutamate semialdehyde dehydrogenase; (5) glutamate decarboxylase; (6) glutaminase; (7) glutamine synthetase; (8) glutamate dehydrogenase; (9) GABA α-oxoglutarate transaminase, (10) ornithine transcarbamoylase (from ref. 9).

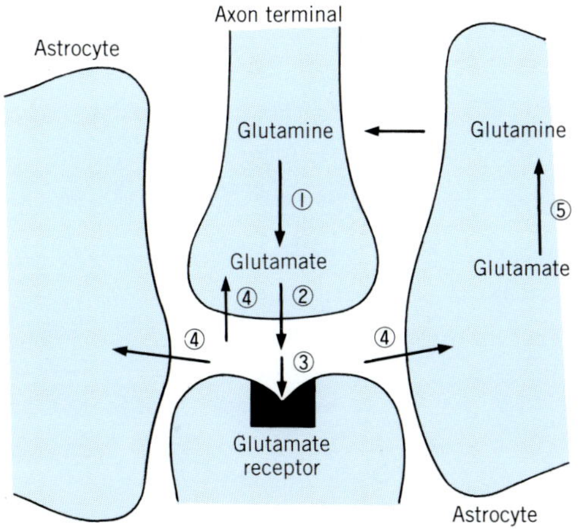

Fig. 7.3 Glutamate neurotransmission. (1) synthesis of glutamate from glutamine by glutaminase; (2) release of glutamate from axon terminals; (3) glutamate receptors are activated by glutamate; (4) uptake of glutamate from synaptic cleft; (5) synthesis of glutamine by glutamine synthetase in astrocytes.

Distribution

Since glutamic acid exists in abundance in the central nervous system, anti-glutamic acid antibody cannot distinguish the metabolic pool component from the neurotransmitter component. Glutaminase has therefore been used as a marker to study the glutamatergic system[5,6]. Glutaminase-positive cells and fibres exist throughout the brain. In cerebral cortex, layer V and the deep part of layer VI show strong distribution of glutaminase-positive cells; the surface layer of layer VI contains distribution of glutaminase positive cells in the intermediate range; but layers II–IV reveal a weak distribution. Glutaminase-positive fibres exist widely in the cortex. Furthermore, hippocampus CA3 contains strongly positive cells. Glutaminase-positive fibres also exist abundantly in the molecular layer and polymorphic layer of CA3. Glutaminase-positive cells and fibres exist in thalamic nuclei. These results suggest that the glutamatergic system shows wide range of functions in the brain.

References

1. Fagg, G.E., Foster, A.C., and Ganong, A.H. (1986). *Trends Pharmacol. Sci.*, **7**, 357–63.
2. Fonnum, F. (1984). *J. Neurochem.*, **42**, 1–11.
3. Hamberger, A.C., Chiang, G.H., Nylen, E.S. *et al.* (1979). *Brain Res.*, **168**, 513–30.
4. Hayashi, T. (1954). *Keio J. Med.*, **2**, 23–38.
5. Kaneko, T. and Mizuno, N. (1988). *J. Comp. Neurol.*, **267**, 590–602.
6. Kaneko, T., Itoh, K., Shigemoto, R., and Mizuno, N. (1989). *Neuroscience*, **32**, 79–98.
7. Martinez-Hernandez, A., Bell, K.P., and Norenberg, M.D. (1977). *Science*, **195**, 1356–8.
8. Norenberg, M.D. (1979). *J. Histochem. Cytochem.*, **27**, 756–62.
9. Roberts, E. (1981). In *Glutamate as a neurotransmitter*, pp. 71–102. Raven Press, New York.

Glutamate (immunocytochemistry)

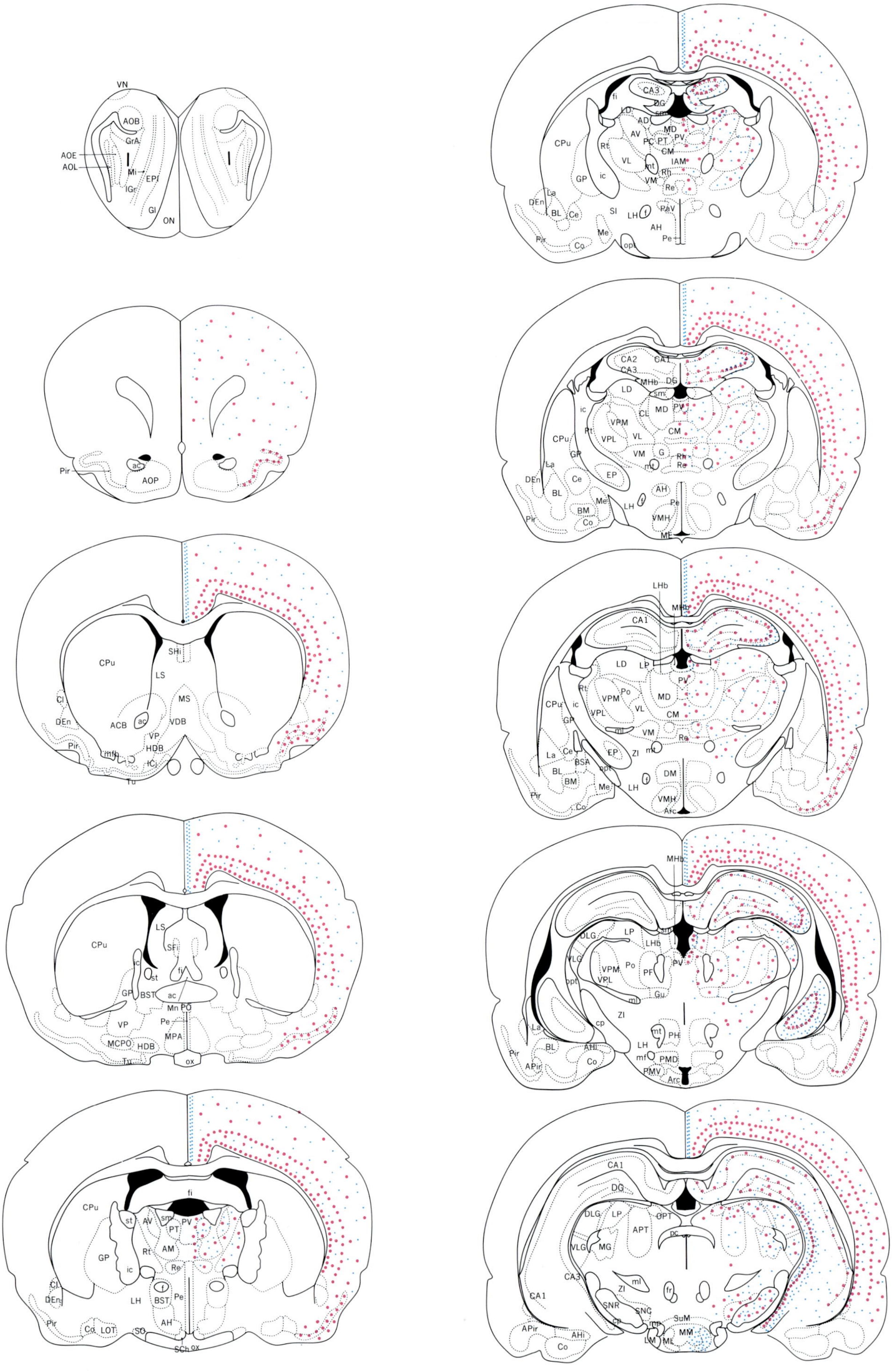

Glutamate (immunocytochemistry)

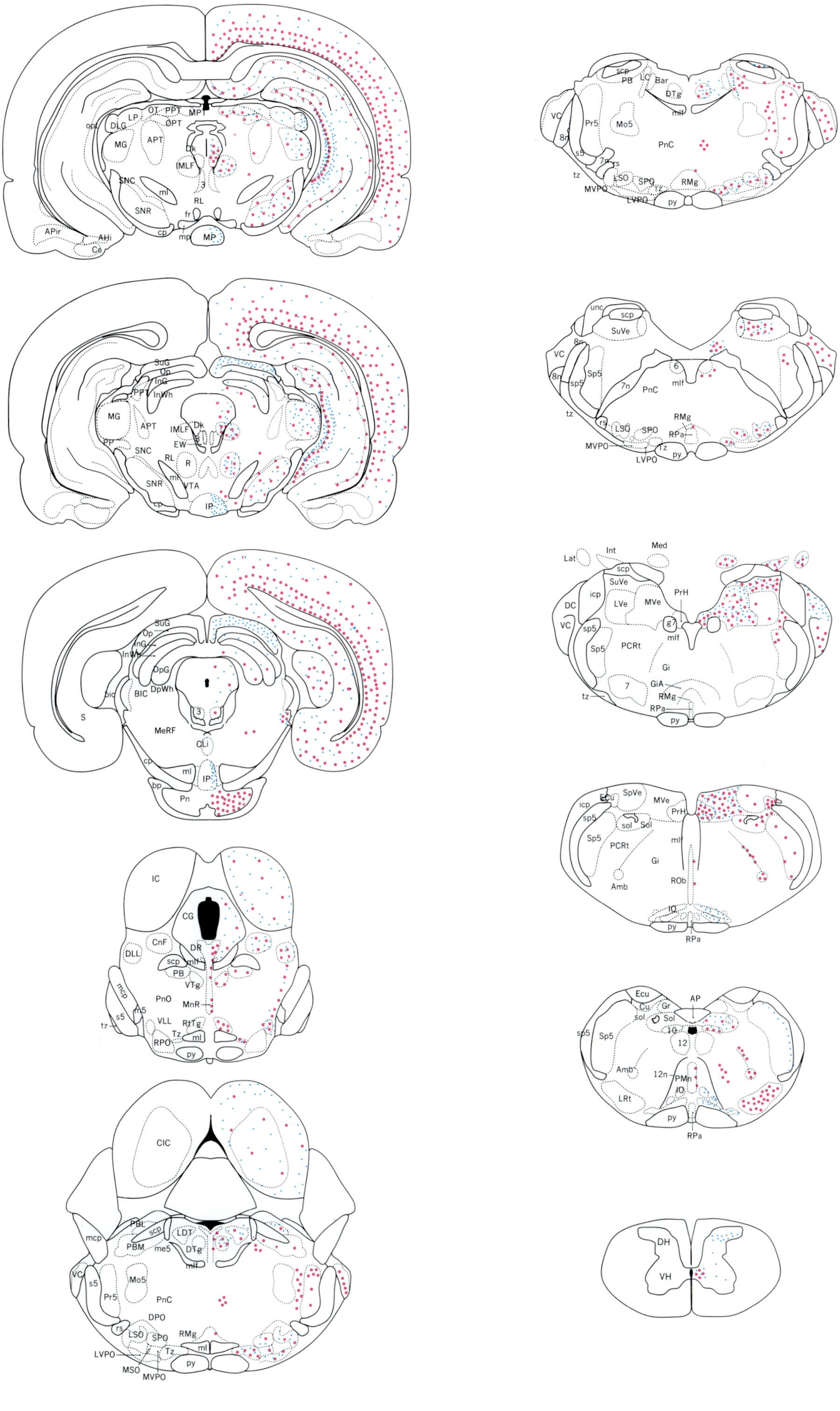

Glutamic acid receptor

Classification

At present, glutamate receptors are classified into three types; NMDA receptor, AMPA/kainate (non-NMDA) receptor, and metabotropic receptor (Table 7.1). The NMDA and AMPA/kainate (non-NMDA) receptors belong to the ion-channel receptor type, with a long N-terminal segment at the extracellular space and four transmembrane domains (Fig. 7.4). This type of structure profile is common to many ion-channel receptors. However, the metabotropic receptor is a G-protein conjugated type with seven transmembrane domains, and both ends have a large hydrophilic region[10] (Fig. 7.5).

Action mechanism and subunit subtypes

A distinguishing characteristic of a NMDA receptor is its permeability to Ca^{2+}. This inflow of Ca^{2+} is believed to initiate the second messenger system which in turn leads to neuroplasticity and a long-term potentiation effect. But, being inhibited by Mg^{2+} in membrane potential dependence, the NMDA receptor functions only in the depolarized state. Glycine is known to enhance the action of the NMDA receptor. Recently, many subunits of this receptor in the rat and mouse have been cloned[9,11,15,16].

Many receptor subunits have been reported for the AMPA/kainate (non-NMDA) receptor[2–4,6–8,17,19,21] and these are categorized arbitrarily into subunits with affinity to AMPA and subunits with affinity to kainate. The AMPA/kainate (non-NMDA) receptor is constituted with various combinations of these subunits. Pharmacological properties of the receptor therefore differ, based on the combination of subunits. Basically, this receptor produces excitatory postsynaptic potential by raising the permeability to univalent cations. Also, in combination with the NMDA receptor, it is believed that the AMPA/kainate receptor participates in the expression of long-term potentiation.

The existence of many subtypes of metabotropic receptor has also been reported[10,20]. These are associated with inositol phospholipid/Ca^{2+} signal transduction and with the cAMP cascade. It has been suggested that the metabotropic receptor also participates in long-term potentiation and long-term depression.

Table 7.1. The glutamate receptors

	Number of amino acids	Molecular weight (kDa)	Function
NMDA receptor			
NR1[16]	938	105.5	Homo-oligomer of NR1 subunit shows the full characteristics of the NMDA receptor, but the combination of NR1 and NR2A–C leads to a more functional expression
NR2A[15]	1445	n.d.	
NR2B[15]	1456	n.d.	
NR2C[15]	943	n.d.	
AMPA/kainate (non-NMDA) receptor			
Glu-R1[3,7,8,17]	889	99.8	AMPA type receptors consist of the combination of Glu-R1–4. If the combination has Glu-R2, it becomes Ca^{2+} impermeable. Kainate-type receptor consists of the combination of Glu-R5–7 and KA-1, 2. KA-2, Glu-R5 and Glu-R6 make a functional receptor
Glu-R2[3,8,17]	862	96.4	
Glu-R3[3,8,17]	855	98.0	
Glu-R4[8]	881	98.4	
Glu-R5[1,19]	890	100.9	
Glu-R6[4]	853	96.2	
Glu-R7[2]	888	n.d.	
KA–1[21]	956	105.0	
KA–2[6]	979	109	
Metabotropic receptor			
mGlu-R1[10]	1199	133.2	mGluR1 is involved in the inositol phosphate/ Ca^{2+} system. mGlu-R2 is involved in the cAMP cascade
mGlu-R2[20]	872	95.8	
mGlu-R3[20]	879	99.0	
mGlu-R4[20]	912	101.8	

n.d., not determined

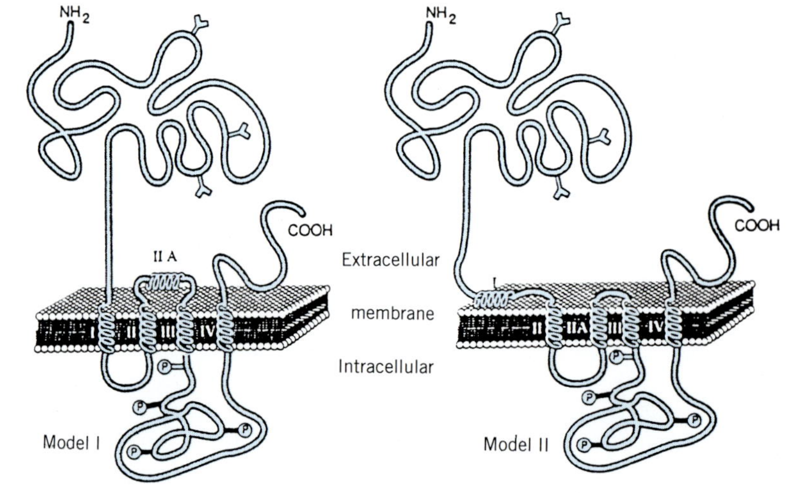

Fig. 7.4 Two models of the possible transmembrane arrangements in GluR1–4, assuming four transmembrane spanning domains. The Y symbols in the extracellular domain represent possible glycosylation sites. The P symbols between III and IV represent possible casein kinase II (open bar) and C kinase (solid bar) sites (from ref. 5).

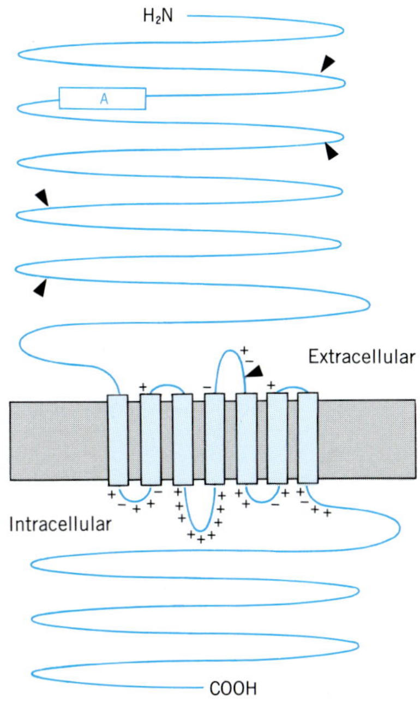

Fig 7.5 The structural model of metabotropic receptor 1. The black triangles mark *N*-glycosylation sites (from ref. 10, copyright 1991 Macmillan Magazines Ltd.).

Distribution

^{3}H-glutamic acid binding sites with NMDA sensitivity are distributed widely in the brain and seen in abundance in structures such as cerebral cortex, hippocampus, and cerebellum[4,14]. The existence of NMDA receptor in the radiate layer and polymorphic layer of CA1 region of hippocampus, which shows the strongest association with long-term potentiation, has attracted much attention. *In situ* hybridization results show that NR2B mRNA is expressed strongly in forebrain and NR2C mRNA in cerebellum. Neurons of the whole brain express NR1 mRNA strongly.

Similarly to the NMDA receptor, ^{3}H-AMPA binding sites and ^{3}H-kainate binding sites are distributed widely in the brain[12,13]. Whereas ^{3}H-AMPA binding sites are strong in the surface layers of cerebral cortex, ^{3}H-kainate binding sites show strong localization in the deep layers of the cortex. In hippocampus, ^{3}H-AMPA binding sites are strongest in the CA1 region and ^{3}H-kainate binding sites are predominantly seen in the molecular layers where mossy fibres from granule cells form the ends. Cloning and *in situ* hybridization of glutamate receptor 1–4 (Glu-R 1–4) has also been carried out[18]. Expression of four types of receptor mRNA appears widely in the brain, but Glu-R4 is only weakly expressed in all the regions. In the cerebral cortex, strong expression of Glu-R2 is observed in layers II–IV, though Glu-R1,3,4 expression is weak in the intermediary layers. The mRNA of four types of receptors are expressed strongly in the pyramidal cell layer and granular cell layer of hippocampus.

Metabotropic receptor is expressed in many neurons. mGlu-R1 is strongly observed in the granular cells, CA4 region of hippocampus, Purkinje cells in cerebellum, and mitral cells of olfactory bulb[10].

References

1. Bettler, B., Boulter, J., Hermans-Borgmeyer, I. *et al.* (1990). *Neuron*, **5**, 583–95.
2. Bettler, B., Egebjerg J., Sharma, G. *et al.* (1992). *Neuron*, **8**, 257–65.
3. Boulter, J., Hollmann, M., O'Shea-Greenfield, A. *et al.* (1990). *Science*, **249**, 1033–7.
4. Egebjerg, J., Bettler, B., Hermans-Borgmeyer, I., and Heinemann, S. (1991). *Nature*, **351**, 745–8.
5. Gasic, G.P. and Hollmann, M. (1992). *Annu. Rev. Physiol.*, **54**, 507–36.
6. Herb, A., Burnashev, N., Werner, P. *et al.* (1992). *Neuron*, **8**, 775–85.
7. Hollmann, M., O'Shea-Greenfield, A., Rogers, S.W., and Heinemann, S. (1989). *Nature*, **342**, 643–8.
8. Keinanen, K., Wisden, W., Sommer, B. *et al.* (1990). *Science*, **249**, 556–60.
9. Kutsuwada, T., Kashiwabuchi, N., Mori, H. *et al.* (1992). *Nature*, **358**, 36–41.
10. Masu, M., Tanabe, Y., Tsuchida, K. *et al.* (1991). *Nature*, **349**, 760–5.
11. Megro, H., Mori, H., Araki, K. *et al.* (1992). *Nature*, **357**, 70–4.
12. Monaghan, D.T., Yao, D., and Gotman, C.W. (1984). *Brain Res.*, **324**, 160–4.
13. Monaghan, D.T. and Cotman, C.W. (1982). *Brain Res.*, **252**, 91–100.
14. Monaghan, D.T. and Cotman, C.W. (1985). *J. Neurosci.*, **5**, 2909–19.
15. Monyer, H., Sprengel R., Schoepfer R. *et al.* (1992). *Science*, **256**, 1217–21.
16. Moriyoshi, K., Masu, M., Ishii, T. *et al.* (1991). *Nature*, **354**, 31–7.
17. Nakanishi, N., Schneider, N.A., and Axel, R. (1990). *Neuron*, **5**, 569–81.
18. Sato, K., Kiyama, H., and Tohyama, M. (1993). *Neuroscience* **52**, 515–39.
19. Sommer, B., Burnashev, N., Verdoorn, T.A. *et al.* (1992). *EMBO J.*, **11**, 1651–6.
20. Tanabe, Y., Masu M., Ishii T. *et al.* (1992). *Neuron*, **8**, 169–79.
21. Werner, P., Voigt, M., Keinanen, K. *et al.* (1991). *Nature*, **351**, 742–4.

Glutamate receptor (binding sites)

[^{3}H]-Glutamate

[^{3}H]-Glutamate

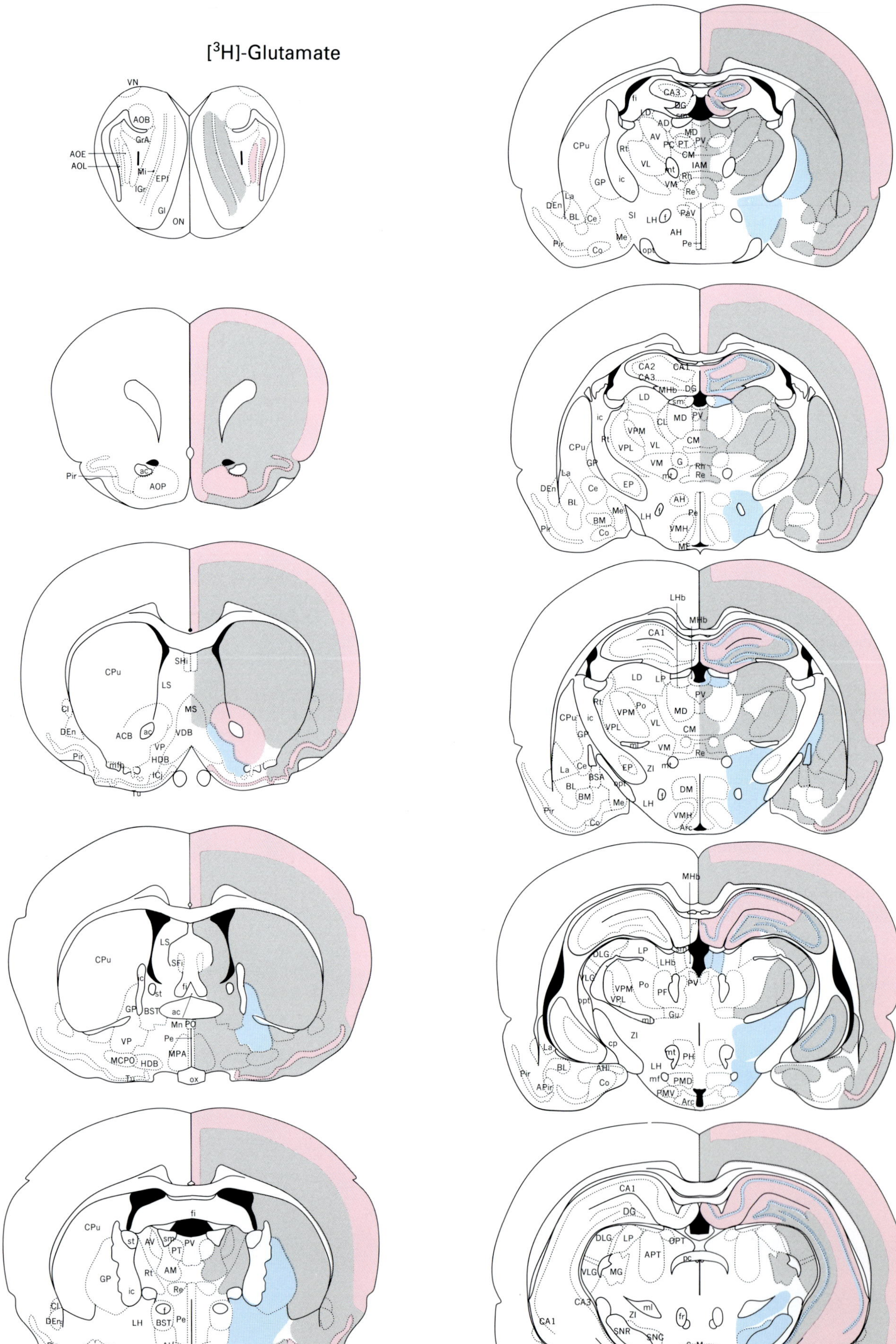

Glutamate receptor (binding sites)

[^{3}H]-Glutamate

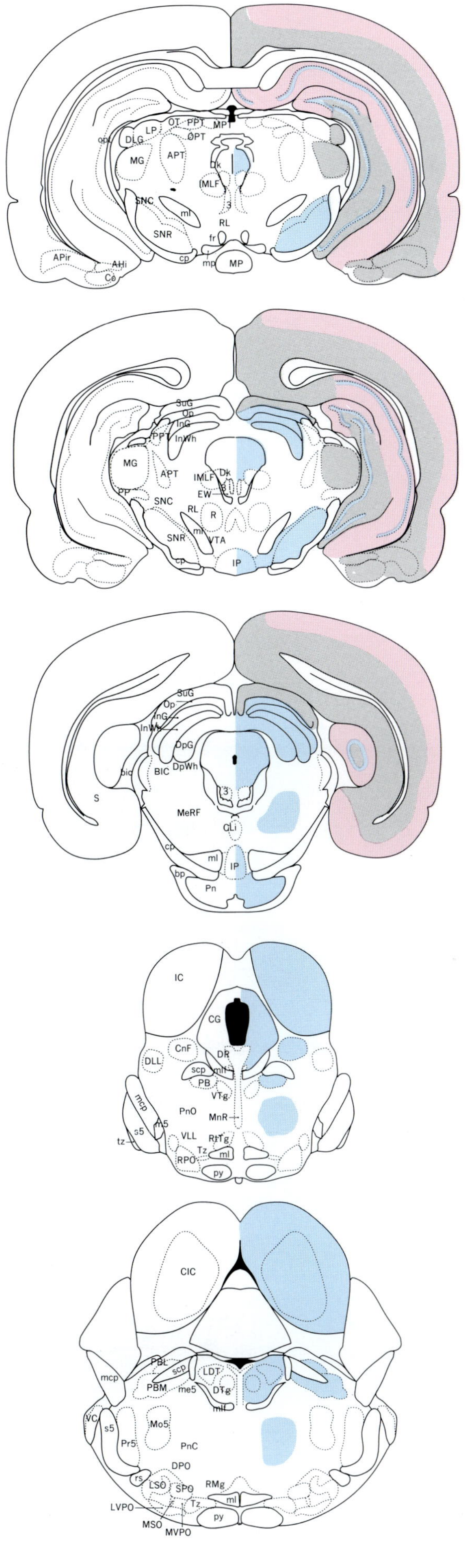

[^{3}H]-Glutamate

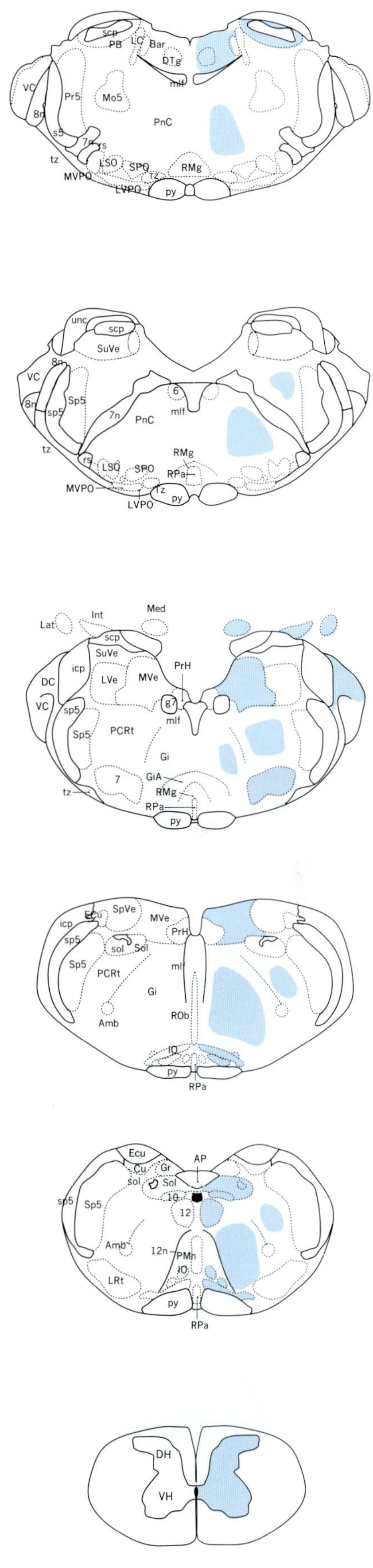

Glutamate receptor (binding sites)

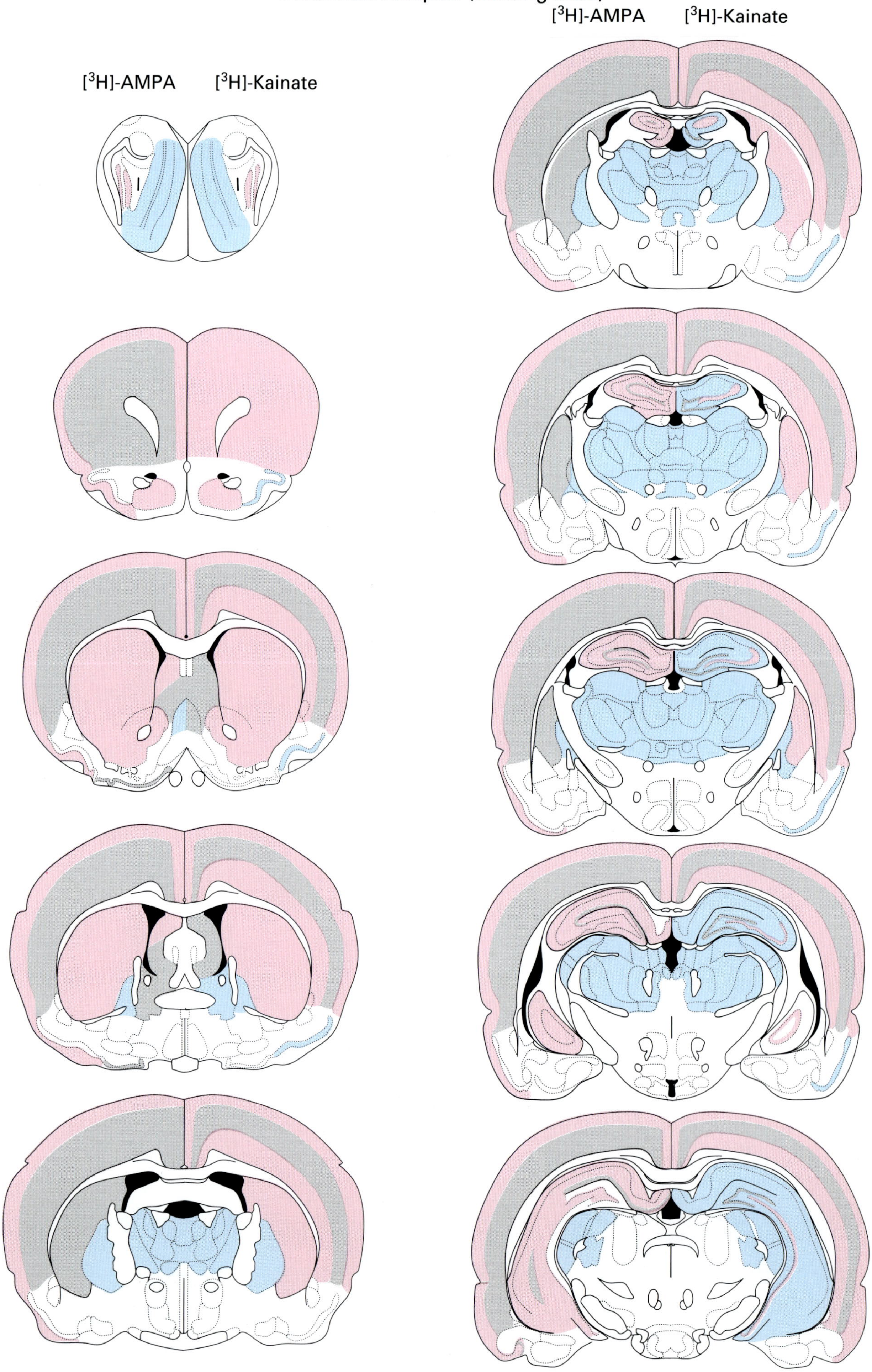

Glutamate receptor (binding sites)

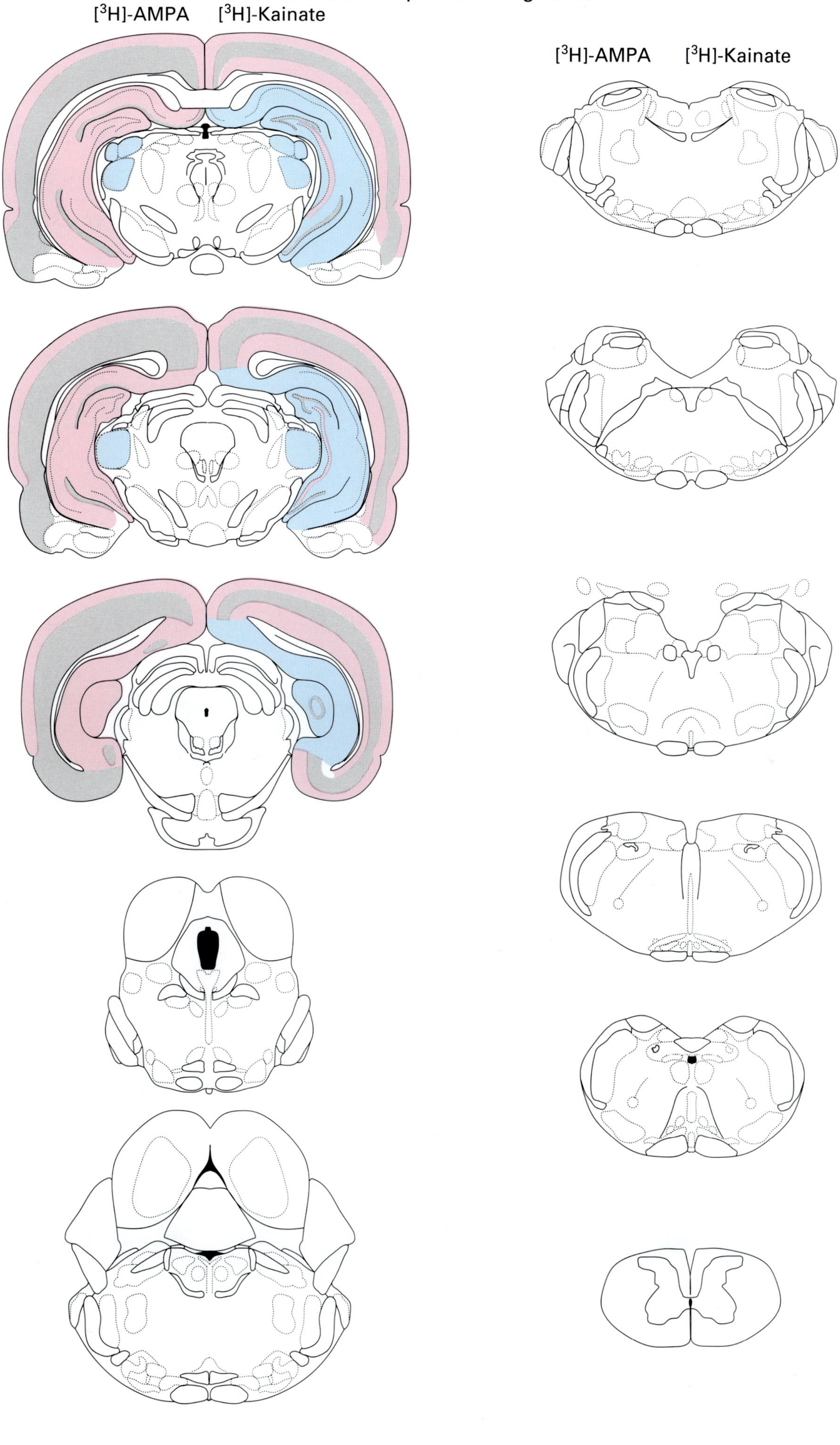

Glutamate receptor (ISH)

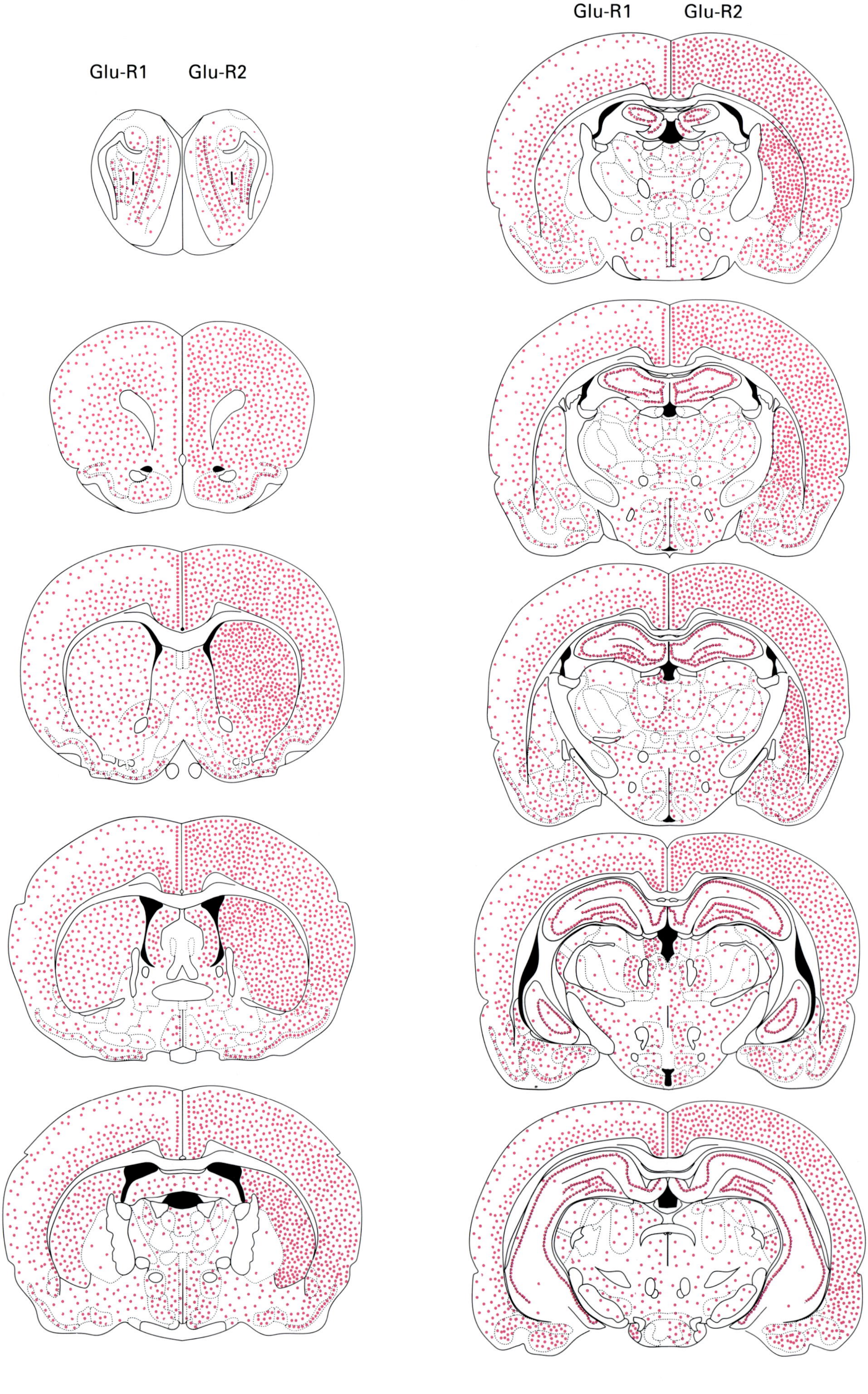

Glutamate receptor (ISH)

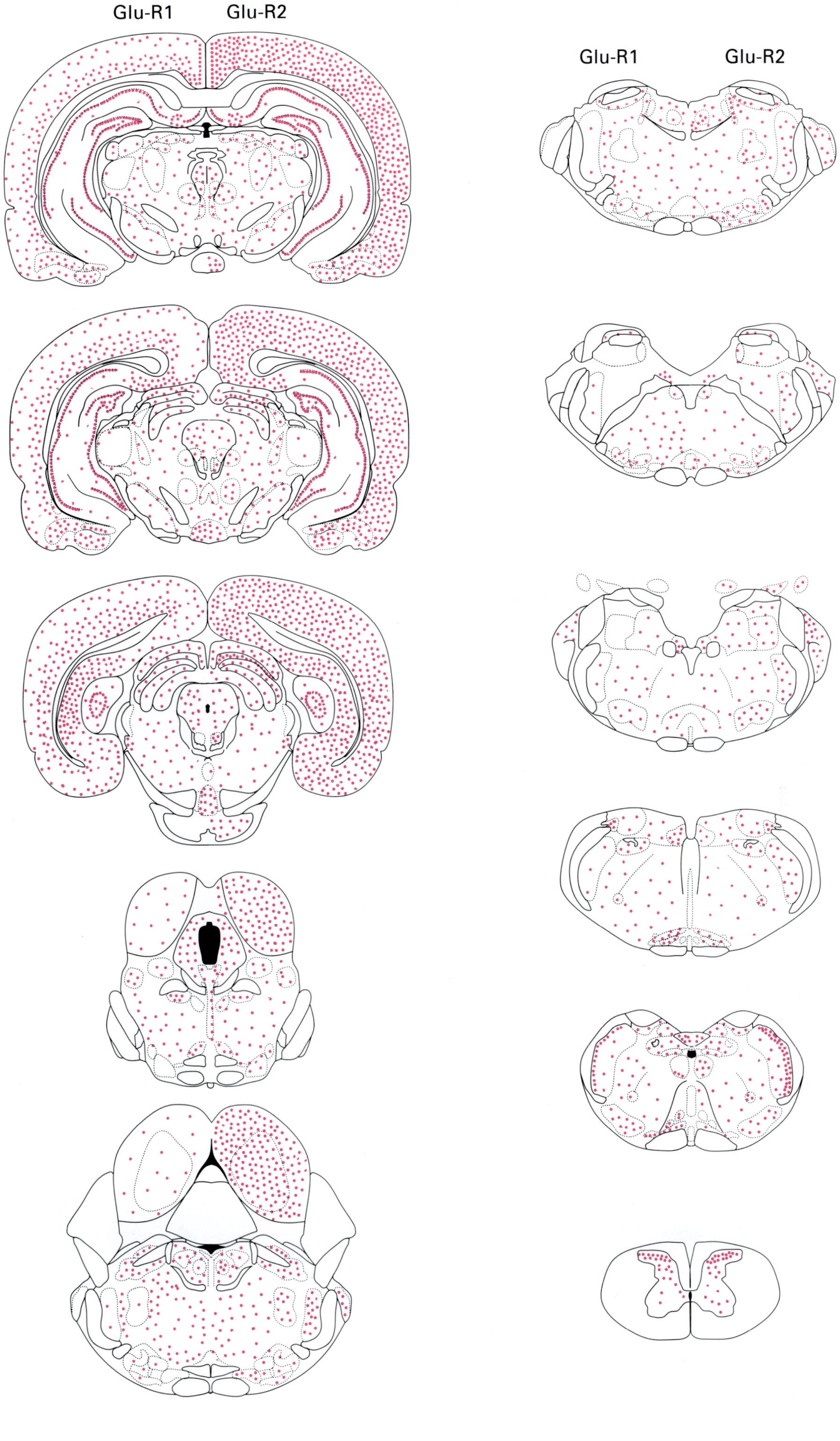

Glutamate receptor (ISH)

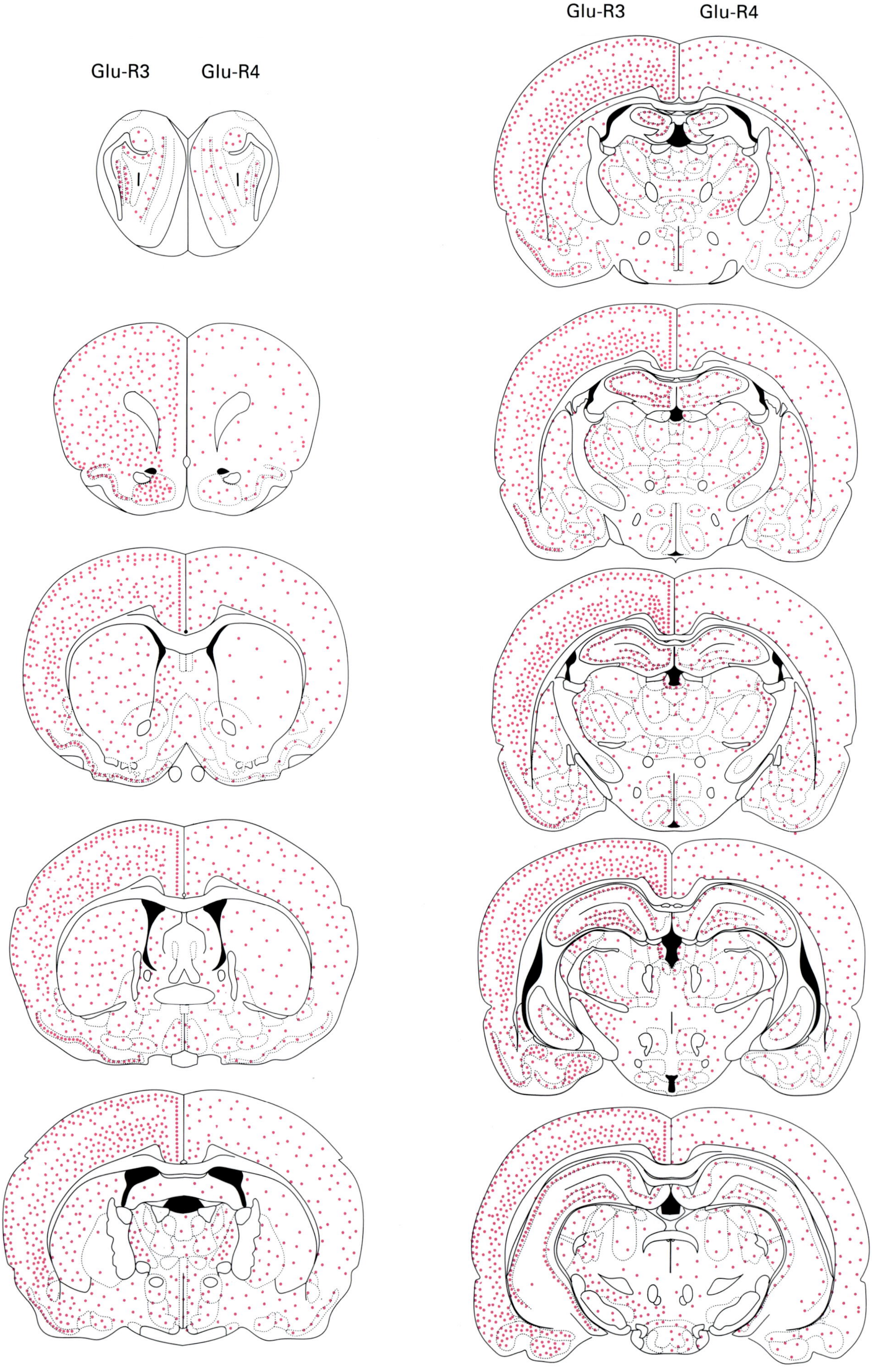

Glutamate receptor (ISH)

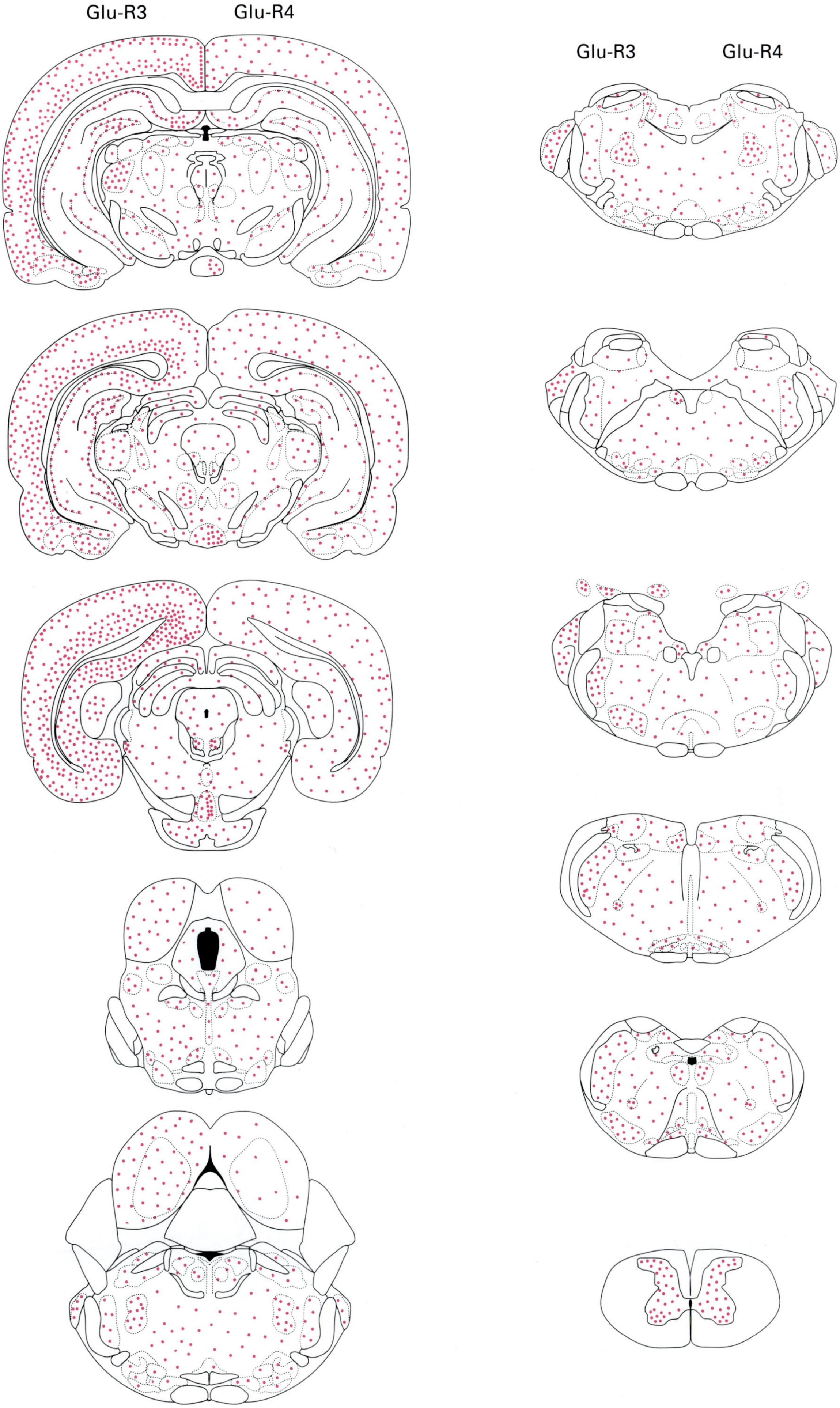

8 The aspartate system

Aspartate

Neurotransmitter function

Uneven and abundant uptake of [^{3}H]Asp into the intracerebral tissues, and the release of this [^{3}H]Asp from the nerve ending in response to a stimulus, is claimed as evidence for the action of aspartate as a neurotransmitter[7]. L-Aspartate also shows an excitatory action on the spinal cord neuron.

Asparaginergic innervation

Most studies on the intracerebral functions of Asp have been carried out in tandem with glutamic acid (Glu). At present, therefore, it is not easy to distinguish clearly the aspartate system from the glutamate one. Tracts that are believed to show the aspartate (or Asp/Glu) system are shown in Figs 8.1–8.7.

Localization of Asp-containing neurons and nerve endings has not been clarified in detail. Though a question of antibody specificity remains, studies on these lines have been reported using immunohistochemistry[1,3,5,6,9] and [^{3}H]Asp uptake[2,4,8].

References

1. Campistron, G., Guijs, R.M., and Geffard, M. (1986). *Brain Res.*, **365**, 179–84.
2. Fonnum, F., Soreide, A., Krale, I. *et al.* (1981). In *Glutamate as a neurotransmitter* (ed. G. Di Chiara and G.L. Gessa) p. 29. Raven Press, New York.
3. Kumoi, K., Saito, N., and Tanaka, C. (1987). *Brain Res.*, **416**, 22–33.
4. Ottersen, O.P. and Storm-Mathisen, J.S. (1984). In *Handbook of chemical neuroanatomy* (ed. A. Björklund, *et al.*), pp. 141–224. Elsevier, Amsterdam.
5. Ottersen, O.P. and Storm-Mathisen, J.S. (1985). *Neuroscience*, **16**, 589–606.
6. Kumoi, K., Saito, N., Kuno, T., and Tanaka, C. (1988). *Brain Res.*, **439**, 302–10.
7. Tohyama, M. (1990). *Brain Science*, **1**, 55–64 (in Japanese).
8. Walas, I. (1980). In *Etats deficitaires cerebraux liés à l'age* (ed. R. Tissot), p. 47. Librairie de l'Université, Geneva.
9. Yoshida, M., Teramura, M., Sakai, M. *et al.* (1987). *Brain Res.*, **410**, 169–73.

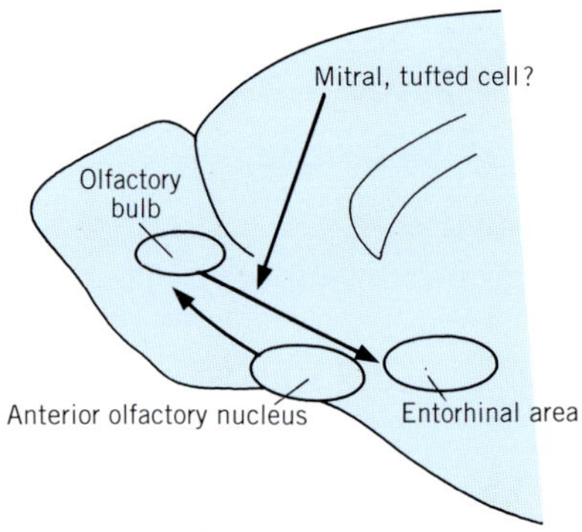

Fig. 8.1 Asparaginergic innervation in the olfactory bulb.

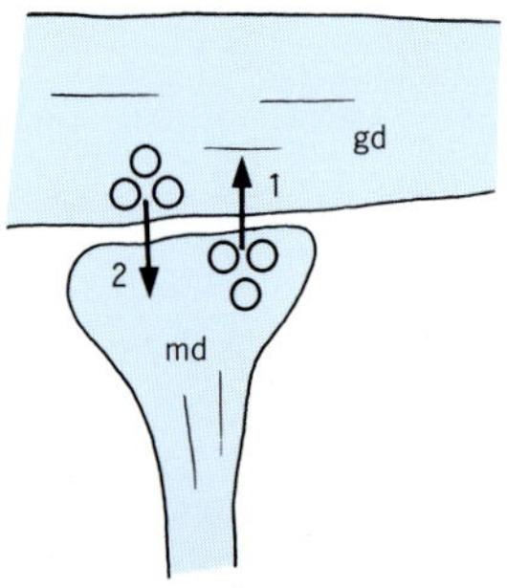

Fig. 8.2 Inhibition effect of Asp on the mitral cells. Asp released from the dendrites of the mitral cells (md) has an excitatory effect on the granular cells (1). This stimulation causes the release of GABA from the dendrites of the granular cells (gd) which inhibits the mitral cell activity (2).

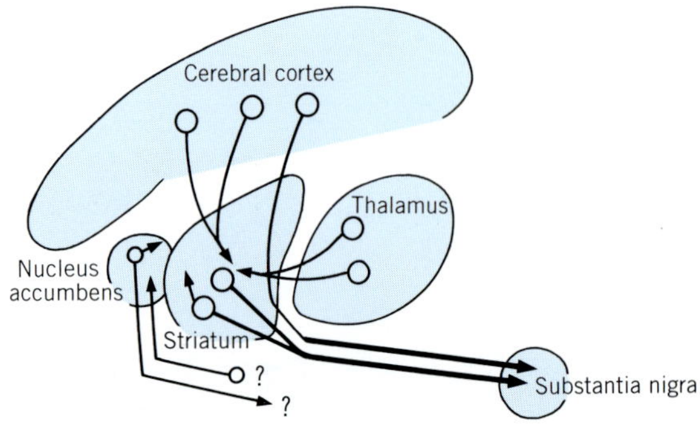

Fig. 8.3 Asparaginergic innervation in the striatum and nucleus accumbens.

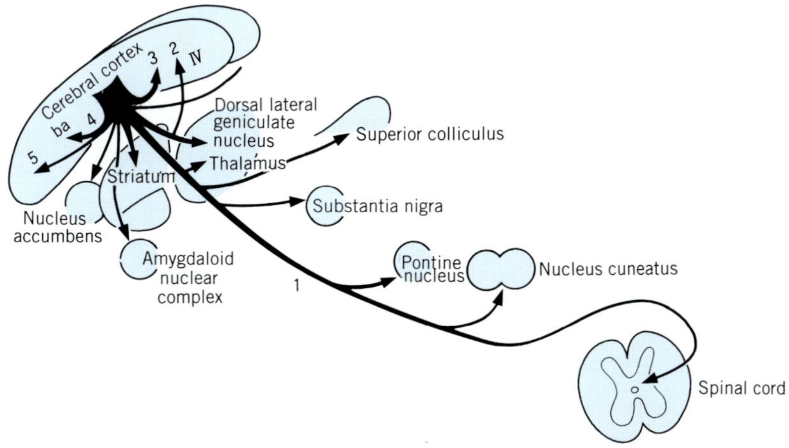

Fig. 8.4 Cortical asparaginergic neuron system: (1) Asp-containing pyramidal tract; (2) Asparaginergic recurrent axons from the visual cortex to the lateral geniculate body, terminating in layer IV of the visual cortex; (3) association fibres containing Asp; (4) association or commissurial fibres terminating in the barrel structure of the cerebral cortex; (5) Asp-containing crossing fibres.

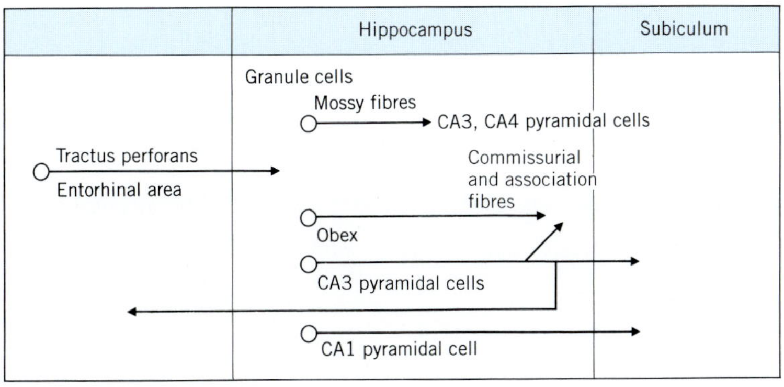

Fig. 8.5 Asparaginergic innervation of the hippocampus.

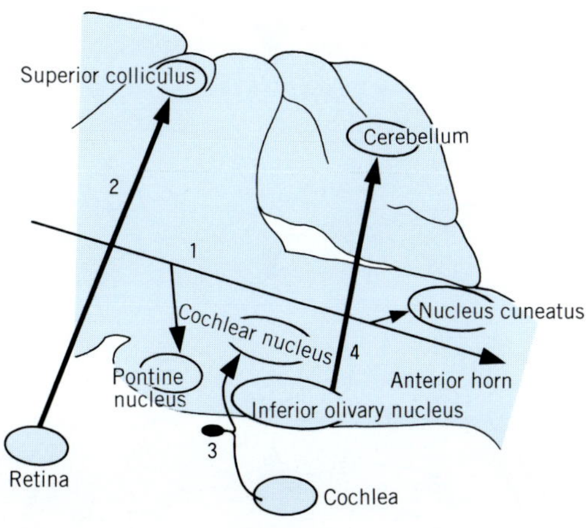

Fig. 8.6 Asparaginergic innervation of the lower brain stem: (1) Asp-containing pyramidal tract; (2) Asp-containing fibres from retina to superior colliculus; (3) Asp-containing cochlear nerves; (4) Asp-containing climbing fibres.

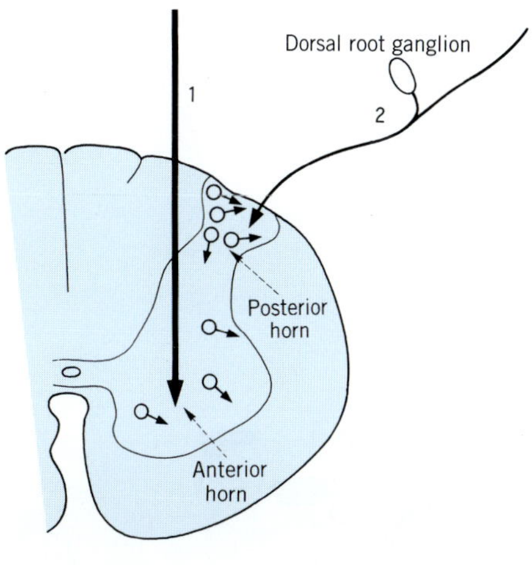

Fig. 8.7 Asparaginergic innervation of the spinal cord: (1) Asp-containing pyramidal tract fibres projecting to the anterior horn; (2) Asp-containing primary sensory fibres projecting to the posterior horn. ○→ Asp-containing local circuit neurons.

Aspartate (immunocytochemistry)

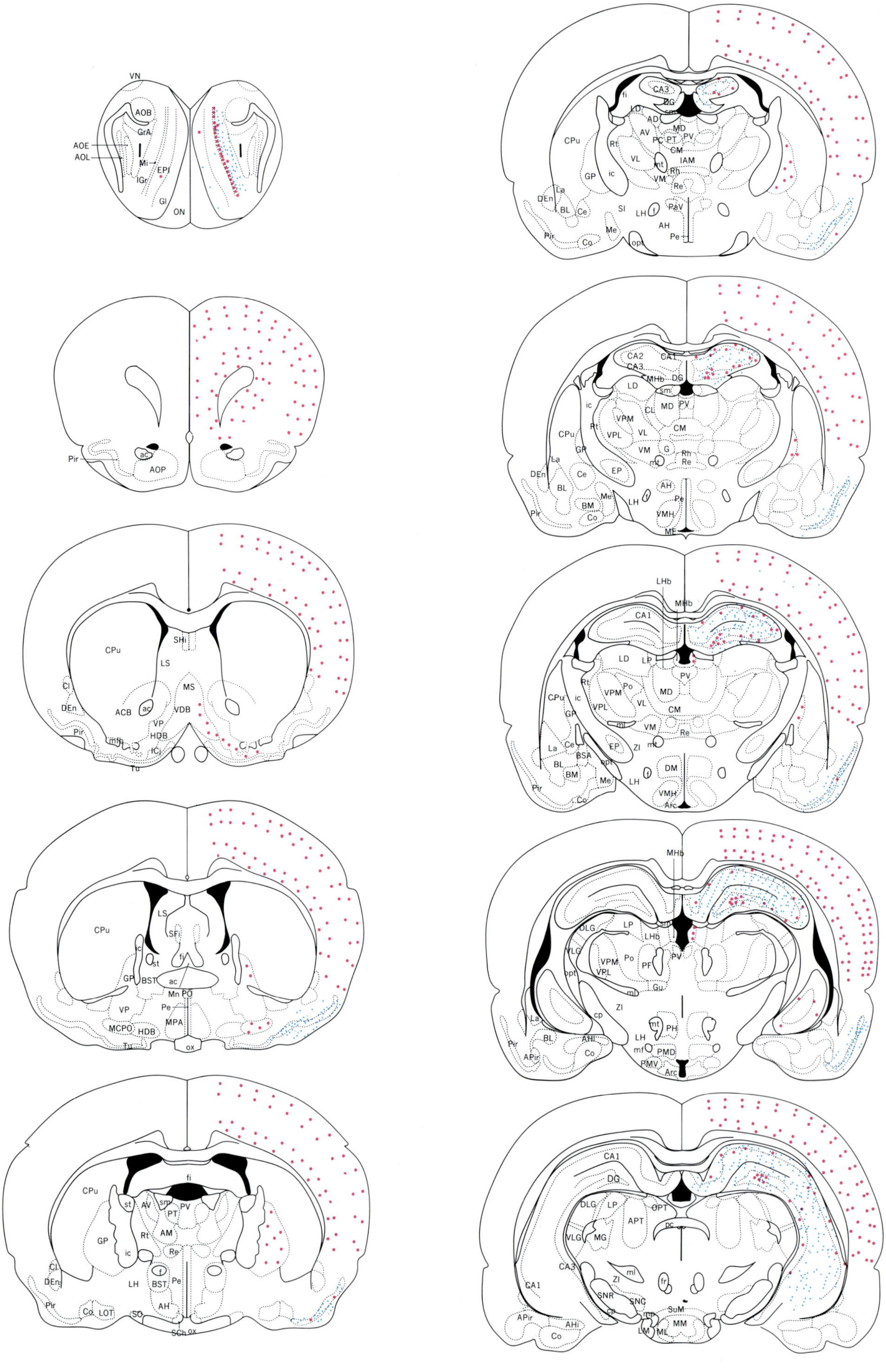

Aspartate (immunocytochemistry)

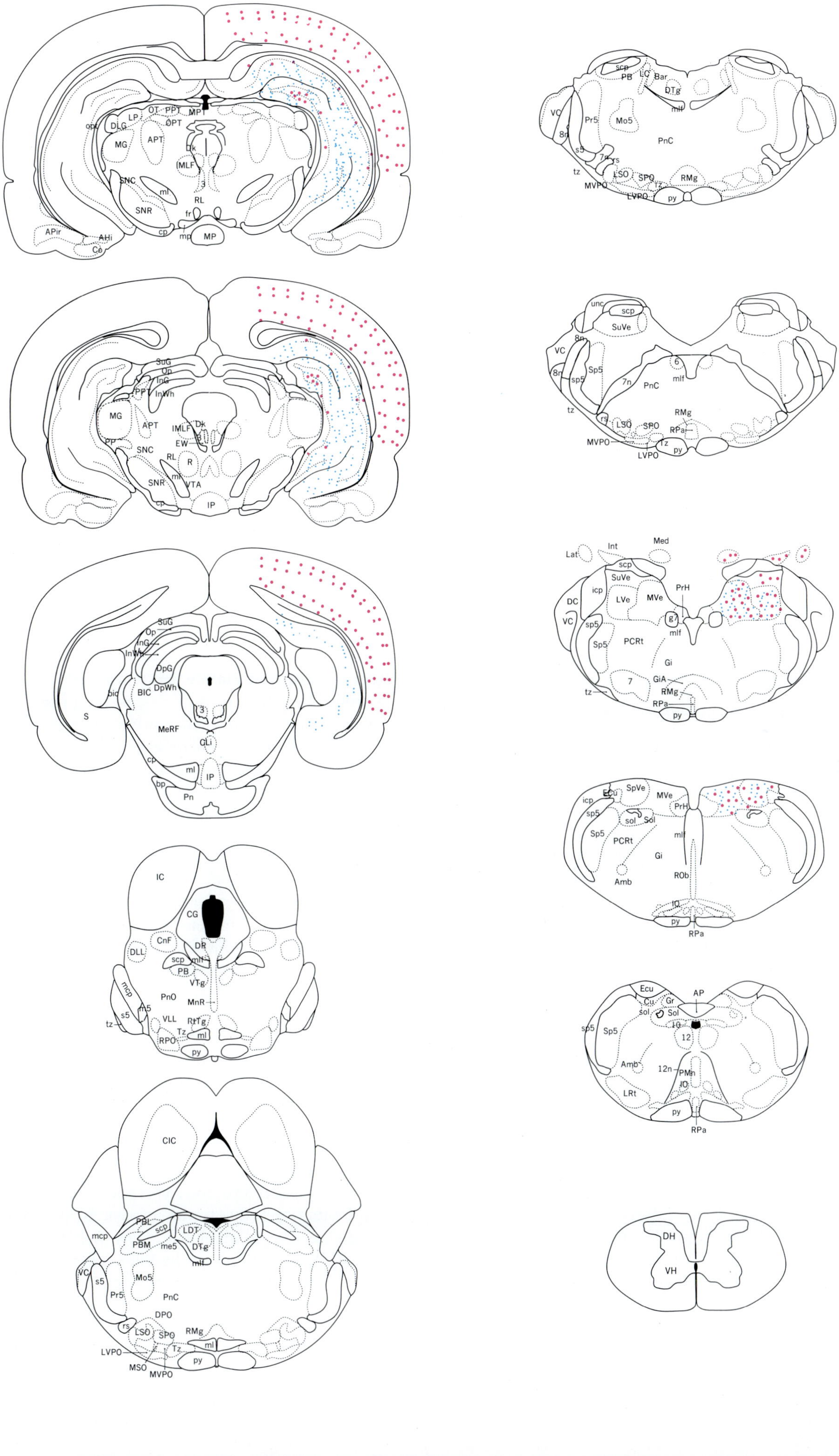

9 The taurine system

Taurine

Neurotransmitter functions

For a long time, taurine has been suggested as a potential inhibitory neurotransmitter. The main reasons for such a belief are as follows:

- wide and uneven distribution in the brain, especially in cerebellum and olfactory bulb[33]
- presence of taurine in neurons, nerve terminals and synaptic vesicles[13,49]
- on electric stimulation or K^+ stimulation, taurine is released and it is dependent on extracellular Ca^{2+} [2,3,8,9,12,20–33] (however, there is a report inferring that taurine release is not dependent on Ca^{2+})
- suggested existence of specific receptor for taurine at the postsynaptic membrane[31–33,35,55]
- inhibitory actions of taurine on the motor neuron in the anterior horn of spinal cord,[1,11,14,42] some neurons in the brain stem,[17] hippocampus,[59] cerebral cortex,[27] cerebellum,[15,36,39,44] and thalamus neurons[10]
- suggested existence of uptake function of taurine in synapse[29,54]
- participation in Ca^{2+} transport of taurine[13,31].

Function in brain development

Taurine levels in the brain are high during the embryonic period, and decline to reach the adult level at the time of weaning[11,16,49,58]. Until the perinatal stage and at birth synaptogenesis is still not established, and it is inferred that taurine, which exists in high concentration during this period and whose concentration subsides later during the growth phase, may have some influence on brain development.

Since the transient high concentration of taurine during the embryonic period is observed in brain as well as in liver, it may also be thought that the transient rise in taurine levels in brain could be associated with transmigration from mother's milk[58]. However, in retina, the taurine levels increased with growth[5,36]. This finding contradicts the proposal that the alteration of taurine content in the nervous system is influenced by the mother's milk.

Localization

Glial cells

Many reports have shown that taurine is localized in the neurons. However, some studies have presented evidence that taurine is also found in the glial cells. Stimulation of β-adrenergic receptors in cultured astroglial cells resulted in cAMP-dependent release of taurine[56]. Furthermore, labelled taurine was taken up by glia[1,4]. Immunohistochemical studies also differ among the different research groups[5,37,46,42]. While some report that immunoreactivity is mostly localized in the neurons and a few cases in glial cells[46], others have demonstrated that immunopositive structures have been observed in many glial cells, particularly in the oligodendrocytes[5,62]. It is still not clear whether taurine is produced in glial cells or not.

Neurons

The types of neurons containing taurine are shown in Table 9.1, and the taurine system postulated in spinal cord is shown in Figs 9.1 and 9.2.

Most of the data presented here is based on immunohistochemical analyses; however, since the antiserum used may not be specific, there exists a reasonable doubt whether the reported localizations of taurine are accurate enough to make a sound inference.

Olfactory bulb

The olfactory bulb is a site which contains taurine in abundance. Immunohistochemically, some parts of the periglomerular cells and granular cells are labelled by the antiserum.[62]

Cerebral cortex

Reports exist which imply that taurine inhibits the cortical neurons[27] and that K^+ [2,3,8,9,20,21], in the presence of Ca^{2+}, evokes release of labelled taurine from cerebral cortical slices. In addition, stimulation of the brain stem reticular formation resulted in the release of taurine into the cerebral cortex[8]. This suggests the presence of an ascending taurinergic system from the brain stem reticular formation.

Hippocampus

Taurine causes hyperpolarization of pyramidal cells[59]. An injection of GABA increases extracellular taurine content in hippocampus[30], which suggests the existence of a link between the GABA system and taurine system in the hippocampus. Uneven distribution of taurine within the hippocampus has been revealed by immunohistochemistry.[46] Pyramidal cells in CA1–CA3, many cells in the hilus region, and a small number of granular cells show immunoreactivity for taurine. These findings reveal that taurine is abundantly present in the efferent projections (pyramidal cells) and the commissural system of hippocampus (pyramidal cells in the hilus and CA2 region). But in mossy fibres (granular cell system), taurine concentration is less. Therefore, the hyperpolarization of pyramidal cells by taurine is due to the intrinsic hippocampus recurrent ramus of pyramidal cell axons (Schaffer fibres) or to the commissural fibres of the hilus region. Also, the existence of taurine of extrinsic origin cannot be denied. Contrary to these results, one report presents data indicating that dentate gyrus in the hippocampus is immunostained by the taurine antiserum[62]. Based on this report, the taurine system in the hippocampus may be located in mossy

Table 9.1. Central taurinergic system

Brain regions	Cell population containing taurine	Possibility of the presence of the extrinsic origin
Olfactory bulb	Periglomerular cells	
Striatum	Large and medium aspiny neurons (non-projecting type) There is a possibility that projecting neurons also contain taurine	+
Nucleus accumbens		
Olfactory tubercule	Local circuit neurons	+
Cerebral cortex	Pyramidal cells of CA_{1-3}	
Hippocampus	Commissural neurons of Hilus A few granular cells	
Septa area	Lateral septal area	+
Thalamus	Thalamic reticular nucleus	+
Cerebellum	Purkinje cells Stellate cells	
Spinal cord	Local circuit neurons	
Retina	Photoreceptors	

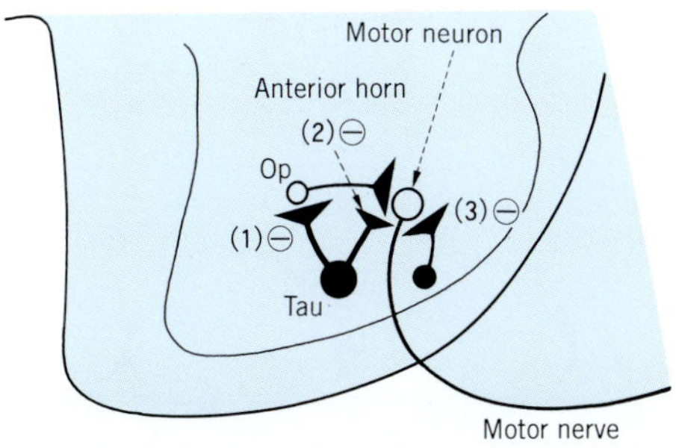

Fig. 9.1 Possible mechanism of the inhibitory effect of taurine- (Tau-) containing local circuit neurons on the anterior horn neurons: (1) Tau-containing nerve ending terminates axosomatically on the opiate peptide (Op)-containing neurons; (2) Tau-containing nerve ending terminates axoaxonically on the Op-containing neurons; (3) Tau-containing nerve terminal inhibits the activity of the anterior horn neurons directly.

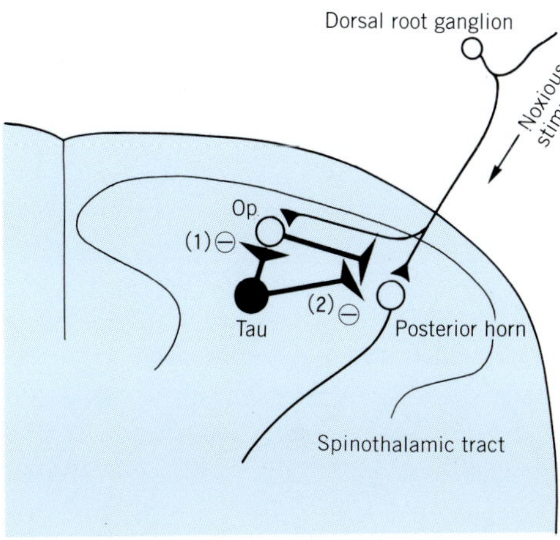

Fig. 9.2 Possible interaction between taurinergic local circuit neurons and those containing opiate peptide (Op) in the dorsal horn of the spinal cord. Tau-containing nerve terminals inhibit Op-containing neurons axosomatically (1) or axoaxonically (2).

fibres and hyperpolarization of pyramidal cells in turn may depend on the action of the taurine located in the mossy fibres.

Diencephalon

Immunohistochemically labelled cells exist in the reticular thalamic nucleus[62]. In the hypothalamus, uptake of taurine and binding sites of taurine have been reported[34].

Basal ganglia

Taurine is abundant in the basal ganglia. When corpus striatum is destroyed by kainic acid, its taurine content decreases by 20–30 per cent[41,48]. This suggests that a portion of the taurine in corpus striatum is supplied intrinsically, and the rest is extrinsic. There is a lack of unanimity in the immunohistochemical data reported from different research groups. Some indicate that only medium-sized cells are immunostained[5,60], while others show that both the medium-sized and large cells are also immunostained[62]. In addition to the corpus striatum, neurons of the nucleus accumbens and olfactory tubercle are also labelled by the taurine antiserum[62].

Cerebellum

In the brain, taurine is most abundant in the cerebellum. The action potential of Purkinje cells is inhibited by taurine[15,39,44,45], and this effect is stronger when taurine is injected into the dendrites, but inhibited by GABA antagonist[46]. These results show that taurine acts on the dendrites of Purkinje cells in coordination with GABA. Depletion of stellate cells by X-irradiation resulted in a decrease in the content of taurine[38]. Thus, it has been inferred that taurine is localized in the stellate cells and acts on the dendrites of Purkinje cells. Since then, immunohistochemical findings using antibodies to taurine-synthesizing enzyme have shown that, in addition to stellate cells and basket cells, Purkinje cells were also labelled[6,7]. There are differing results on the uptake of the labelled taurine. Some reports indicate that it was taken up into Purkinje cells as well as glial cells[3,6,7,44], whereas others found that only Purkinje cells were labelled[37]. Thus, at present, the localization of taurine in the cerebellum has not been settled without doubt.

Spinal cord

Taurine acts inhibitorily on the neurons of anterior horn[1,11]. Injection of opioid peptides causes temporary akinesia, the duration of which is shortened by taurine[19]. This suggests the possibility that axons of taurine-containing cells terminate in the anterior horn neurons and act inhibitorily to interneurons that contain the opioid peptides, which have an inhibitory action. If so, taurine should function excitatorily to motor neurons. There is a possibility that taurine neurons may reach directly to motor neurons (Fig. 9.1; 3) and also terminate at opioid peptide-containing neurons (Fig. 9.1; 1, 2). On the other hand, opioid peptide production increases in the interneurons of dorsal horn of spinal cord in the nociceptive model animals[43]. Since the effect of temporary akinesia caused by opioid peptide is reversed by taurine[19], it can be inferred that the taurine-containing neurons in the dorsal horn of spinal cord also act inhibitorily to the opioid peptide-containing neuron system (Fig. 9.2).

Taurine (immunocytochemistry)

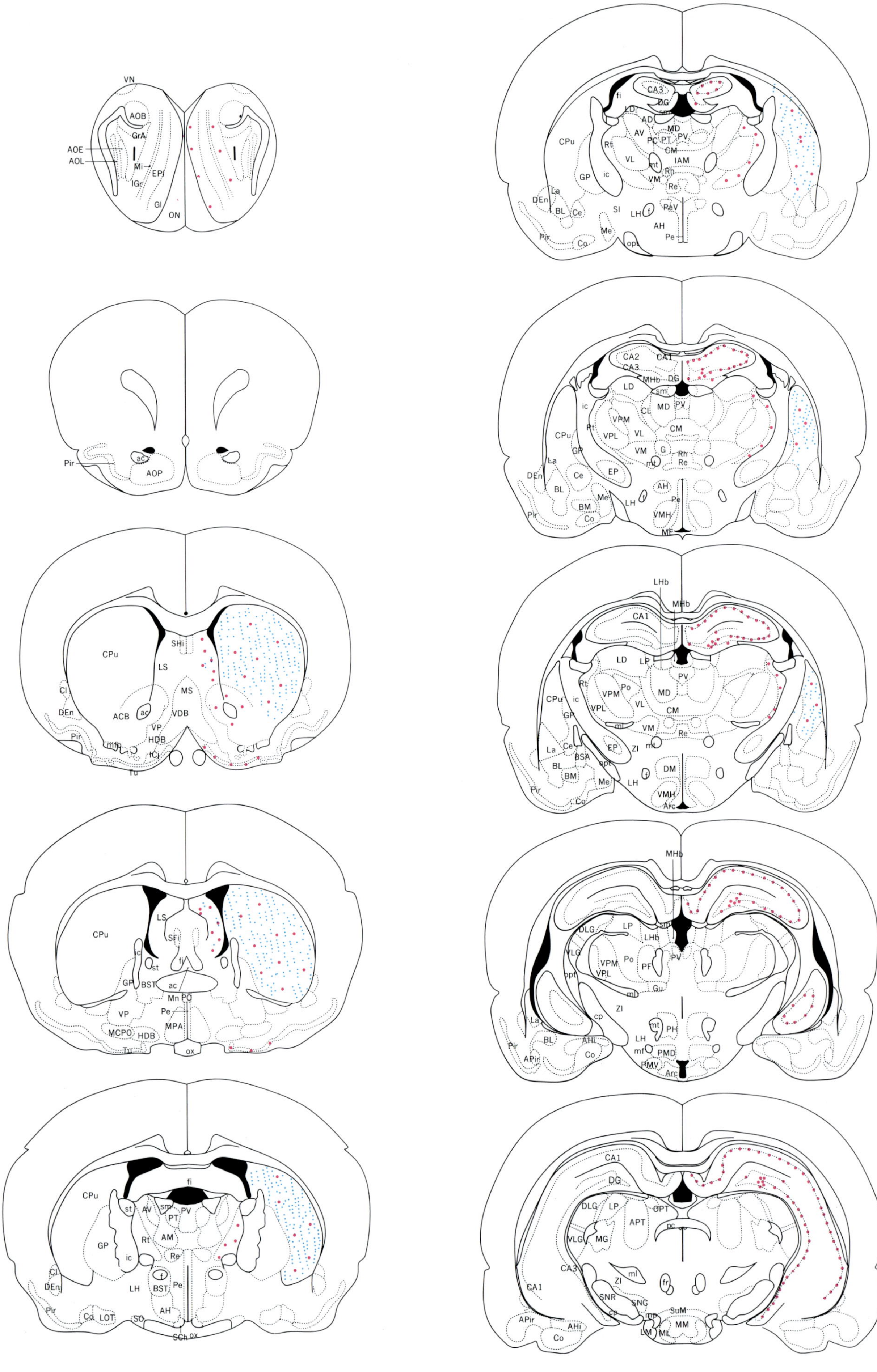

Taurine (immunocytochemistry)

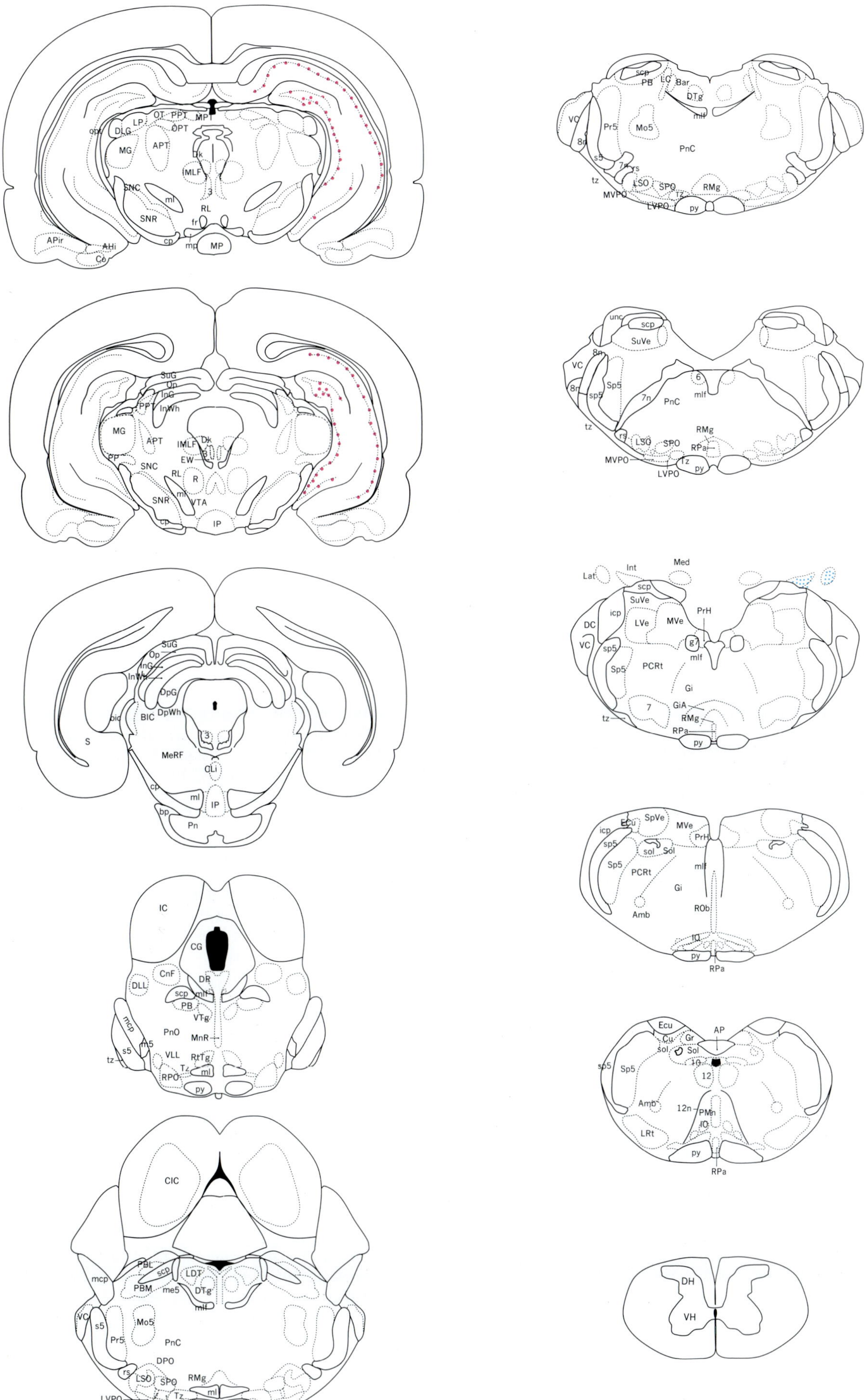

Retina

The taurine content in the retina is exceedingly high[32,36,45,47]. The maximum concentration is seen in the photoreceptor cells of the retina[18,25,45,47,52,57,61]. K^+ stimulation and photic stimulation release this stored taurine[24,26,35,40,42,50,51]. It has been shown that the uptake of labelled taurine occurs in photoreceptor cells as well as in amacrine cells, pigmental epithelial cells, and glial cells[14,26,28,33,40,45,51,52,61]. Photoreceptor cells also degenerate during experimental taurine deficiency in animals[18,53]. These facts suggest that taurine plays an important role, at least in the function of photoreceptors.

References

1. Barker, J.L., Nicoli, R.A., and Padjen, A. (1975). *J. Physiol.*, **245**, 521–48.
2. Baughan, R.W. and Gilbert, C.D. (1980). *Nature (Lond.)*, **287**, 848–9.
3. Bernardi, N., Assumpcao, J.A., Dacke, C.G., and Davidson, N. (1984). *Pflugers Arch.*, **401**, 193–7.
4. Borg. J., Ramaharobandro, N., Mark, J., and Mandel, P. (1980). *J. Neurochem.*, **34**, 1113–22.
5. Campistron, G., Geffard, M., and Bujis, R.M. (1986). *J. Neurochem.*, **46**, 862–8.
6. Chan-Palay, V., Lin, C.-T., Palay, S. *et al.* (1982). *Proc. Natl Acad. Sci. USA*, **79**, 1695–9.
7. Chan-Palay, V., Palay, S.L., and Wu, J.-Y. (1982). *Proc. Natl Acad. Sci. USA*, **79**, 4221–5.
8. Clark, R.M. and Collins, G.G.S. (1976). *J. Physiol. (Lond.)*, **262**, 383–400.
9. Collins, G.G.S. and Topiwala, S.H. (1974). *Br. J. Pharmacol.*, **50**, 451–2.
10. Curtis, D.R. and Tebecis, A.K. (1972). *Exp. Brain Res.*, **16**, 210–18.
11. Curtis, D.R. and Watkins, J.C. (1963). *Biochem. Pharmacol.*, **12**, 1443–4.
12. Davidson, N. (1979). *J. Physiol. (Paris)*, **75**, 673–6.
13. De Belleroche, J.S. and Bradfold, H.F. (1973). *J. Neurochem.*, **21**, 441–51.
14. Ehinger, B. (1973). *Brain Res.*, **60**, 512–16.
15. Frederickson, R.C., Neuss, M., Morzorati, S.L., and McBride, W.J. (1978). *Brain Res.*, **145**, 117–26.
16. Gaull, C.E. and Rassin, D.K. (1979). In *Neuronal growth and differentiation* (eds. E. Meisami, and M.A.B. Braizer), pp. 461–77. Raven Press, New York.
17. Haas, H.L. and Hosli, L. (1973). *Brain Res.*, **52**, 399–402.
18. Hayes, K.C., Carey, R.E., and Schmidt, S.V. *et al.* (1973). *Science*, **188**, 949–51.
19. Izumi, K., Munekata, E., Yamamoto, H. *et al.* (1980). *Peptides*, **1**, 139–46.
20. Jasper, H.H. and Koyama, I. (1969). *Can J. Physiol. Pharmacol.*, **47**, 889–905.
21. Kacmarek, L.K. and Adey, W.R. (1974). *Brain Res.*, **76**, 83–94.
22. Kaczmarek, L.K., and Davison, A.N. (1974). *J. Neurochem.*, **19**, 2355–62.
23. Katz, R.L., Case, T.N., and Kopin, I.J. (1969). *J. Neurochem.*, **16**, 961–7.
24. Kennedy, A.J. and Neal, M.J. (1978). *Exp. Eye Res.*, **26**, 71–5.
25. Kennedy, A.J. and Voaden, M.J. (1974). *Biochem. Soc. Trans.*, **2**, 1256–8.
26. Kennedy, A.J. and Voaden, M.J. (1976). *J. Neurochem.*, **27**, 131–7.
27. Krnjevic, K. and Puil, E. (1976). In *Taurine* (eds. R.J. Huxtable and A. Barbeau), pp. 201–6. Raven Press, New York.
28. Lake, N., Marshall, J., and Voaden, M.J. (1978). *Exp. Eye Res.*, **27**, 713–18.
29. Lahdesmaki, P. and Oja, S.S. (1973). *J. Neurochem.*, **20**, 1411–17.
30. Lerma, J., Herranz, A.S., Herreras, O. *et al.* (1985). *J. Neurochem.*, **44**, 983–6.
31. Lipton, J.M. (1982). In *Taurine* (ed. R.J. Huxtable and H. Pasantes-Morales), p. 373. Plenum Press, New York.
32. Lombardini, J.B. (1976). In *Taurine* (ed. R.J. Huxtable and A. Barbeau), pp. 311–26. Raven Press, New York.
33. Lombardini, J.B. (1983). In *Sulfur amino acids: biochemical and clonical aspects* (ed. K. Kuriyama *et al.*) pp. 251–62. Alan R. Liss, New York.
34. Lopez-Colome, A.M. (1982). In *Taurine in nutrition and neurology* (ed. R.J. Huxtable and H. Pasantes-Morales), pp. 293–310. Plenum Press, New York.
35. Lopez-Colome, A.M., Salceda, R., and Pasantes-Morales, H. (1978). *Neurochem. Res.*, **3**, 431–41.
36. Macaione, S., Ruggieri, P., De Luca, F., and Tucci, G. (1974). *J. Neurochem.*, **34**, 1047–52.
37. Madsen, S., Ottersen, O.P., and Storm-Mathisen, J. (1985). *Neurosci. Lett.*, **60**, 255–60.
38. McBride, W.J., Nadi, N.S., Altman, J., and Aprison, M.H. (1976). *Neurochem. Res.*, **1**, 141–52.
39. McBride, W.J. and Frederickson, R.C.A. (1978). In *Taurine and neurological disorders* (ed. A. Barbeau and R.J. Huxtable), pp. 415–51. Raven Press, New York.
40. Neal, J., Peacock, D.G., and White, R.D. (1973). *Br. J. Pharmacol.*, **47**, 656–7.
41. Nicklas, W.J., Duvosin, R.C., and Berl, S. (1979). *Brain Res.*, **241**, 197–206.
42. Nicoll, R.A., Padjen, A., and Barker, J.K. (1976). *Neuropharmacology,* **15**, 45–53.
43. Noguchi, K., Morita, Y., Kiyama, H. *et al.* (1989). *Mol. Brain Res.*, **5**, 227–34.
44. Okamoto, K. and Sakai, Y. (1980). *Br. J. Pharmacol.*, **69**, 407–13.
45. Orr, H.T., Cohen, A.I., and Carter, J.A. (1976). *Exp. Eye Res.*, **23**, 377–84.
46. Ottersen, O.P., Madsen, S., Meldrum, B.S., and Storm Mathisen, J. (1985). *Exp. Brain Res.*, **59**, 457–62.
47. Pasantes-Morales, H., Klethi, J., Urban, P.F., and Mandel, P. (1974). *Exp. Brain Res.*, **19**, 131–41.
48. Perry, T.L. (1978). In *Taurine and neurological disorders* (ed. A. Barbeau and R.J. Huxtable), pp. 441–51. Raven Press, New York.
49. Rassin, D.K., Sturman, J.A., and Gaull, G.E. (1981). *J. Neurochem.*, **37**, 740–8.
50. Salceda, R., Lopez-Colome, A.M., and Pasantes-Morales, H. (1977). *Brain Res.*, **135**, 186–91.
51. Schmidt, S.Y. (1978). *Exp. Eye Res.*, **26**, 529–35.
52. Schmidt, S.Y. and Berson, E.L. (1976). *Exp. Eye Res.*, **27**, 191–8.
53. Schmidt, S.Y., Berson, E.L., and Hayes, K.C. (1976). *Invest. Ophthalmol.*, **15**, 47–52.
54. Schmidt, R., Sieghart, W., and Karobath, M. (1976). *Brain Res.*, **116**, 158–64.
55. Segawa, T., Inoue, A., and Ochi, T. (1982). *Adv. Exp. Med. Biol.*, **139**, 31–324.
56. Shain, W., Connor, J.A., Madelian, V., and Martin, D.L. (1989). *J. Neurosci.*, **97**, 2306–12.
57. Starr, M.S. and Voaden, M.J. (1972). *Vision Res.*, **12**, 1261–9.
58. Sturman, J.A. and Gaull, G.E. (1975). *J. Neurochem.*, **25**, 831–6.
59. Taber, K.H. and Thalmann, R.H. (1983). *Soc. Neurosci. Abstr.*, **9**, 414.
60. Tomida, Y. and Kimura, H. (1987). *Acta Histochem. Cytochem.*, **20**, 31–40.
61. Voaden, M.J., Lake, N., Marshall, J., and Morijaria, B. (1977). *Exp. Eye Res.*, **25**, 249–57.
62. Yoshida, M., Karasawa, N., Ito, M. *et al.* (1986). *Neurosci. Res.*, **3**, 356–63.

Adenosine

10 The adenosine system

Adenosine

Physiological function

Adenosine exerts various physiological actions. These include bronchial constriction, platelet aggregation, and dilatation of coronary and other blood vessels. These actions are mediated via specified receptors. Xanthine derivatives such as theophylline, theobromine, and caffeine bind to the adenosine receptors. These xanthine derivatives are contained in popular drinks such as coffee, brown tea, and Coca-Cola, and they are also used widely as drugs (diuretics and smooth muscle relaxants).

Adenosine shows inhibitory effects on neurons. When adenosine or ATP is injected intraventricularly into animals, it produces calming behaviour and an anticonvulsive effect. In humans, a hereditary disease condition exists which is caused by adenosine deaminase deficiency. In these patients, adenosine is not degraded easily, resulting in serious immunodeficiency, neurosymptoms of athetosis, irritability, spasticity, nystagmus, etc. This disease shows that adenosine has some vital physiological effects on neurons. Adenosine, when applied iontophoretically, inhibits neuroactivities in Purkinje cells of cerebellum, hippocampus, corpus striatum, thalamus, superior colliculus, olfactory bulb, etc[5]. In intracellular recordings, inhibition by adenosine results in hyperpolarization of membrane, but without alteration in membrane resistance. Thus, the property of generating an action potential is not changed, but the magnitude of the action potential is decreased. This shows that the actions of adenosine depend on the inhibition of release of excitatory transmitters. The fact that adenosine does not inhibit the firing of cortical neurons that is stimulated by acetylcholine and glutamate also shows that the site of its action is presynaptic.

Adenosine is now widely accepted as the major inhibitory neuromodulator in the central nervous system besides GABA. It is also thought to be an endogenous neuroprotective metabolite. Its modulation of glutamate release may underlie this neuroprotection (for a review, see Deckert and Gleiter, *J. Neural Transmission*, Suppl. **43**, 23–31, 1994).

Biosynthesis

In slices of cerebral cortex from guinea pig, release of adenosine and its nucleotide due to nerve stimulation has been confirmed. High K^+ stimulation also results in the release of ^{14}C-adenosine derivatives from the cerebral cortex slice and synaptosome. The released components are made up of adenosine (50 per cent), adenine nucleotide (5–10 per cent), with the balance being inosine and hypoxanthine[2]. Since the release of these substances by high K^+ stimulation is not influenced by tetrodotoxin, a Na^+ channel blocker, it seems to depend on direct depolarizing action of K^+, independent of Na^+ flow. The released ATP is hydrolysed rapidly extracellularly, and then adenosine, inosine, and hypoxanthine are produced. If purinergic neurons exist, it is still unclear whether ATP or adenosine is released from those neurons.

How, then, is adenosine of the nervous system synthesized? The following steps are considered as a pathway for such adenosine synthesis[5].

1. dephosphorylation of 5′-AMP that originates from ATP, cyclic AMP
2. dephosphorylation of 3′-AMP that is produced by the degradation of RNA
3. from *S*-adenosyl-homocystein (SAH) that is produced by methylation of *S*-adenosyl homocysteine (which turns out to be an important parent compound in polyamine synthesis)
4. from adenine and ribose-1-phosphate.

For step 1, 5′-nucleotidase and other enzymes are involved. 5′-Nucleotidase is an ecto-enzyme that exists at the membrane surface of glia. Steps 2 and 3 are also critical in adenosine synthesis, but it is hard to see step 4 being an important pathway because the amount of substrate adenine is limited, and the PNPase activity is also low. There is no concrete conclusion as to whether adenosine is released or produced from released ATP. Inactivation of adenosine occurs extracellularly by two processes:

- re-uptake (or transport) into the intracellular compartment, and
- in the intracellular compartment, a degradation into inactive inosine by adenosine deaminase (ADAse).

A high affinity uptake system has been shown to exist (K_m 19 mM) in slices of guinea pig cerebrum[2]. Two kinds of uptake system (fast and slow) have been described. The fast uptake system is inhibited by dipyridamole. The brain localization of adenosine transport has been investigated with the use of nitrobenzylthioinosine (NBI), which binds to adenosine uptake sites specifically and inhibits the uptake of adenosine. The first step in the degradation of adenosine is its transformation into inactive inosine by ADAse, a cytoplasmic enzyme with wide distribution in the brain. 5′-AMP deaminase is an ecto-enzyme that degrades the extracellular AMP (Fig. 10.1).

On the presynaptic action of adenosine, it has been shown that in the centre, it inhibits the release of acetylcholine in the cerebal cortex of guinea pig. This effect is blocked by theophylline. Theophylline and caffeine increase the release of acetylcholine. Adenosine also inhibits the release of noradrenaline, acetylcholine, dopamine, and serotonin in the cerebral cortex slice sample. This is not a strong inhibition, and 15–30 per cent of the release is inhibited by 10^{-5} M adenosine. Adenosine weakly inhibits the release of GABA. Such presynaptic actions have been considered to occur due to the decrease of Ca^{2+} permeability of membrane at the nerve ending. However, it is

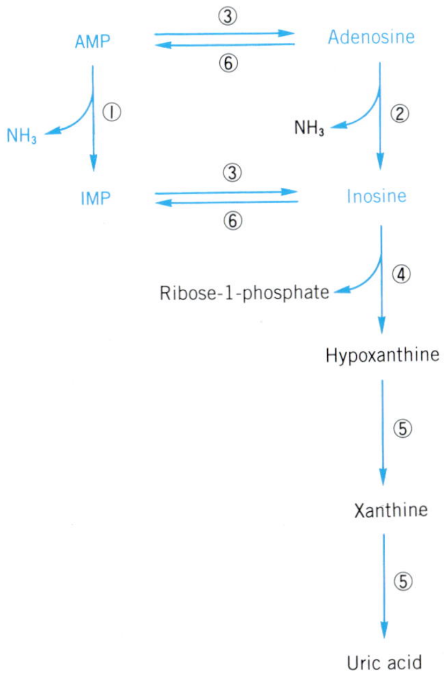

① AMP deaminase	④ nucleosidase
② adenosine deaminase	⑤ xanthine oxidase
③ 5′-nucleosidase	⑥ adenosine kinase

AMP : adenosine-5′-monophosphate
IMP : inosine-5′-monophosphate
R-1-P : ribose-1-phosphate

Fig. 10.1 Catabolism of adenosine and adenine nucleotides: (1) AMP deaminase; (2) adenosine deaminase; (3) 5′-nucleosidase; (4) nucleosidase; (5) xanthine oxidase; (6) adenosine kinase (from ref. 5).

also reported that mobilization of Ca^{2+} from the Ca^{2+} pool in the nerve ending is inhibited by adenosine.

In the central nervous system, adenosine (or ATP) is released from nerve endings and binds to the A1 receptor in the presynaptic membrane and then inhibits adenylate cyclase activity, thereby decreasing the intracellular cyclic AMP. This in turn inhibits the Ca^{2+} influx or shift from the Ca^{2+} pool, which leads to the inhibition of release of other neurotransmitters such as acetylcholine. Meanwhile, the nerve endings have been thought to reabsorb the extracellular surplus adenosine and degrade it to inactive inosine by intracellular ADAse.

Distribution and projection tract

It is still not known for sure which neuron groups in the brain use adenosine as a neurotransmitter. Morphological methods which have been used to clarify the adenosinergic system include:

- observation of the distribution of receptors or uptake sites by autoradiography using isotope-labelled ligands
- immunohistochemical observation of the cell bodies or nerve endings that contain adenosine, using an antibody to adenosine
- observation of the distribution of cell bodies or nerve endings that contains an high concentration of an enzyme, which is involved in the synthesis or degradation of adenosine.

The main synthetic pathway of adenosine is still not clear, since adenosine is also an intermediate metabolite. Therefore, not only adenosinergic neurons but also other neurons may contain high concentration of adenosine. Adenosinergic neurons are expected to contain ADAse, which degrades adenosine, in high concentration. Therefore, distribution of ADAse in the brain has been investigated immunohistochemically, using an antibody to ADAse[3].

Neuron groups that contain ADAse in highest concentration include the tuberal mammillary nucleus, caudal mammillary nucleus, and posterior caudal mammillary nucleus[3,6]. These neuron groups are called the posterior hypothalamic magnocellular nuclei (PHMN). Furthermore, PHMN also contain histidine decarboxylase and glutamic acid decarboxylase, which are the markers for histamine and GABA[6]. ADAse-positive neurons in PHMN project to wide forebrain regions including hypothalamus and cerebral cortex; and in brain stem, projections to mesencephalic trigeminal nucleus and substantia nigra have been demonstrated. In addition, ADAse-positive cells are observed in corpus striatum, lateral septal nucleus, superior colliculus, Purkinje cells, and pontine reticular formation. There are projections from corpus striatum to substantia nigra[10] and from lateral septal nucleus to medial habenular nucleus via stria medullaris[9]. ADAse-positive neurons that distribute in the surface layer of the superior colliculus project to the lateral posterior nucleus of the thalamus[4].

Akasu *et al.*[1] reported that a slow hyperpolarizing synaptic potential in pelvic neurones of vesical parasympathetic ganglia, produced by the stimulation of preganglionic nerves, is also elicited by application of adenosine, indicating that preganglionic neurons innervating vesical ganglia are adenosinergic. The slow hyperpolarizing synaptic potential is inhibited by caffeine and ADAse, but is enhanced by dipyridamole, an uptake inhibitor of adenosine. It is conceivable that adenosine functions as a chemical mediator in this system. The preganglionic neurons of parasympathetic nervous system located in the brain stem also show immunoreactivities to ADAse as well those in spinal parasymphathetic nuclei[7], which indicates that ADAse is a useful marker for the adenosinergic system.

References

1. Akasu, T., Shinnick-Gallagher, P., and Gallagher, J.P. (1984). *Nature*, **311**, 62–5.
2. Kuroda, Y. and MacIlwan, H. (1973). *J. Neurochem.*, **21**, 889–900.
3. Nagy, J.I., LaBella, L.A., Buss, M., and Daddona, P.E. (1984). *Science*, **224**, 166–8.
4. Miguell-Hidalgo, J.J., Senba, E., Matsutani, S. *et al.* (1989). *Brain Res.*, **476**, 189–93.
5. Phillis, J.W. and Wu, P.H. (1981). *Progr. Neurobiol.*, **16**, 187–239.
6. Senba, E., Daddona, P.E., Watanabe, T. *et al.* (1985). *J. Neurosci.*, **12**, 3393–402.
7. Senba, E., Daddona, P.E., and Nagy, J.I. (1987). *Neuroscience*, **20**, 487–502.
8. Snyder, S.H. (1985). *Ann. Rev. Neurosci.*, **8**, 103–24.
9. Staines, W.A., Yamamoto, T., Dewar, K. *et al.* (1988). *Brain Res.*, **455**, 72–87.
10. Yamamoto, T., Staines, W.A., Dewar, K. *et al.* (1988). *Brain Res.*, **474**, 112–24.

Adenosine deaminase (immunocytochemistry)

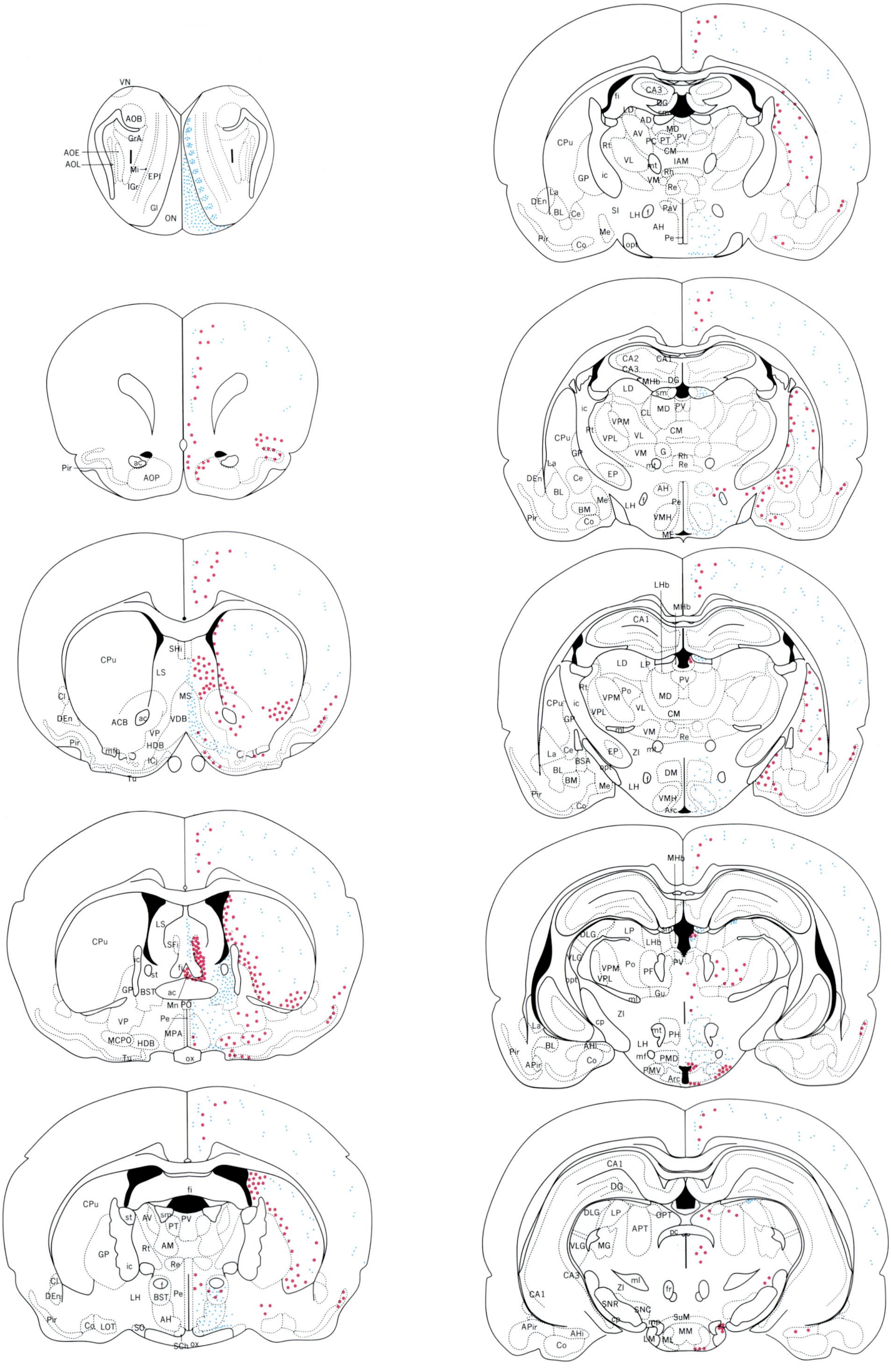

Adenosine deaminase (immunocytochemistry)

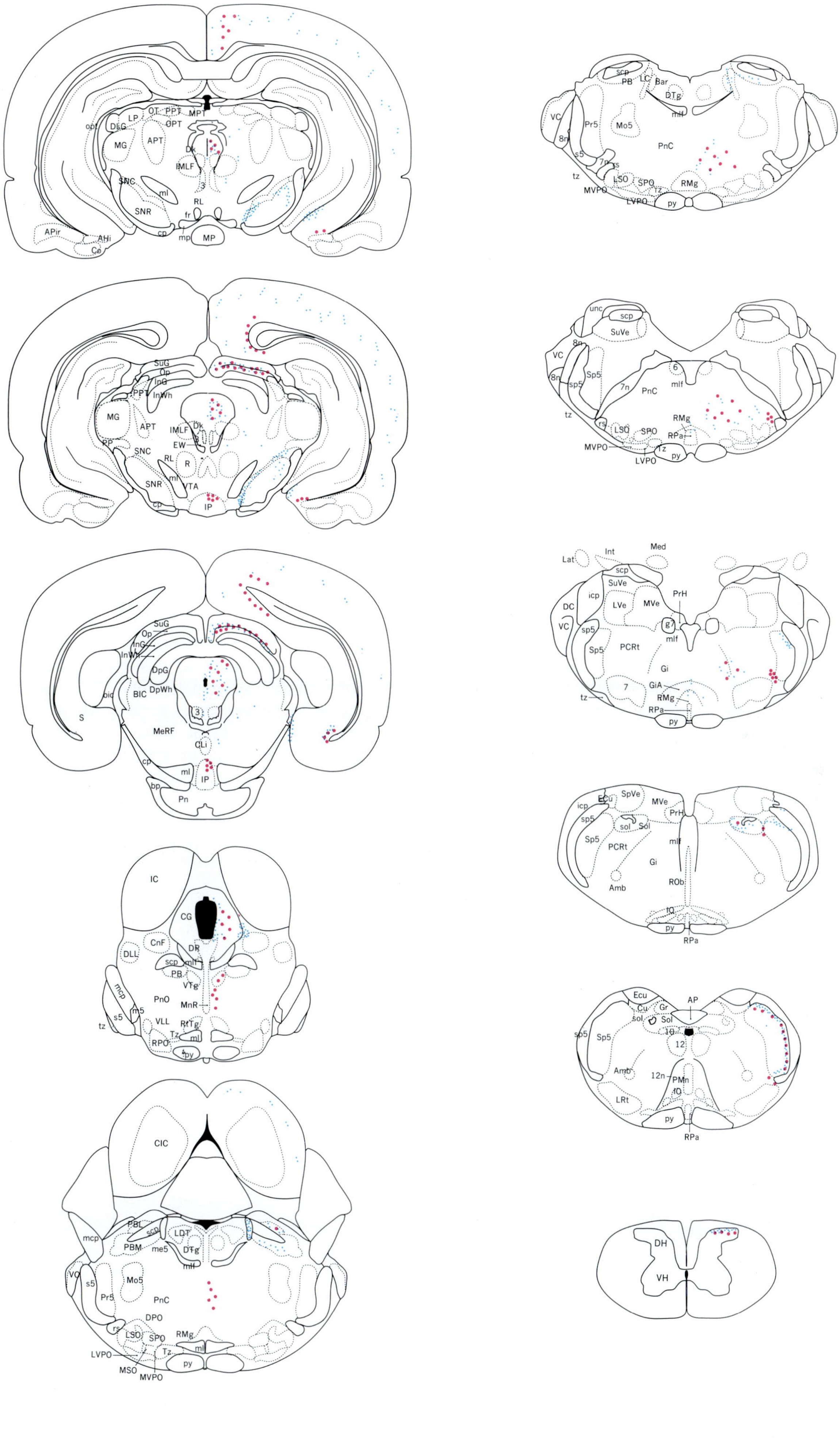

Adenosine receptor

Subtypes

Adenosine receptors in the central nervous system have been classified into A_1 and A_2 by van Calker *et al.*[15]. The A_1 receptor is a high affinity site (affinity constant of adenosine: 10–100 nM), and it inhibits adenylate cyclase and decreases cyclic AMP. Conversely, the A_2 receptor is a low affinity site (affinity constant of adenosine: 2–20 μM), which activates adenylate cyclase and increases the cyclic AMP. These receptors show the following different pharmacological affinities for various analogues of adenosine (Fig. 10.2).

A_1: N^6-cyclohexyladenosine (CHA), N^6-phenylisopropyladenosine (PIA) > 5′-*N*-ethylcarboxamidoadenosine (NECA)
A_2: NECA > PIA

The adenosine receptor protein from rat brain has been purified[11] and recently the structures of cDNA of the A_1 and A_2 receptors have also been clarified using molecular biological methods. Clones RDC7 and RDC8[6] obtained from homology screening with a cDNA sequence of G-protein coupled receptor of β-adrenergic receptor and muscarinic ACh receptor were identical with two types of adenosine receptors, A_1 and A_{2A}, respectively[9]. Even more recently, two other receptor types, corresponding to A_{2B} subtype[13] and to a novel subtype called A_3[3,16]. The structure of the A_1 receptor is shown in Fig. 10.3. Thus adenosine acts on a family of G-protein coupled receptors called adenosine receptors or P_1 purinoceptors. Four members of this family have been cloned: A_1, A_{2A}, A_{2B}, and A_3. Adenosine regulates the release of several neurotransmitters acting at A_1 receptors that are widely distributed in the central nervous system. Since adenosine acting at A_{2A} receptors modulate dopaminergic transmission, A_{2A} receptor mRNA is expressed in the dopamine rich regions of the brain, such as the striatum and olfactory tuberculum. Although the level of A_{2B} receptor mRNA is low in the brain, this receptor type can be detected[13]. In astrocytes the most abundant receptor appears to be the A_{2B} receptor. Activation of A_{2B} receptor usually requires high concentration of adenosine. It has been proposed that A_3 plays a role in the pathophysiology of cerebral ischemia.

The purine nucleotide ATP acts on receptors called P_2 purinoceptors, which consist of at least two main families: G protein coupled receptors (P_{2Y} and P_{2U}) and ion channels (P_{2X}). The first cloned P_2 purinoceptor came from a neuroblastoma cell line[8]. The ATP receptors are located on both neurons and glial cells, suggesting that ATP may be a mediator of neuron–glial interactions. P_{2X} receptors are cation-selective ion channels and P_{2X} mRNA is expressed by apoptotic thymocytes and is abundant in the pituitary gland, suggesting roles for ATP-gated ion channels in endocrine and immune function[14].

Distribution

When the distribution of the A_1 receptor was observed by means of autoradiography using ^{3}H-labelled CHA, a high density of labelling was seen in stratum oriens, pyramidale, and radiatum of hippocampus, the molecular layer of cerebellum, and the surface layer of the superior colliculus[2,6]. But, after blocking the high affinity receptor (A_1) by PIA, and when the distribution of A_2 receptor was examined using ^{3}H-NECA, high concentrations were observed in the olfactory tubercle, nucleus accumbens, corpus striatum, lateral amygdaloid nucleus, etc. High concentrations of both A_1 and A_2 receptors were seen in cerebral cortex, anteromedial, dorsomedial, and posterolateral nuclei of thalamus, posterior thalamic nuclear group, and medial geniculate body. However, adenosine receptors in the hippocampus were mostly A_1 receptors.

In recent investigations, NECA has been proved to bind specifically to ribose uroamide binding sites, in addition to adenosine. Using [2-p-(2-carboxyethyl) phenethylamino]-5′-*N*-carboxamidoadenosine (CGS 21680), an agonist that binds specifically to A_2 receptor, it has become possible to observe the distribution of A_2 recep-

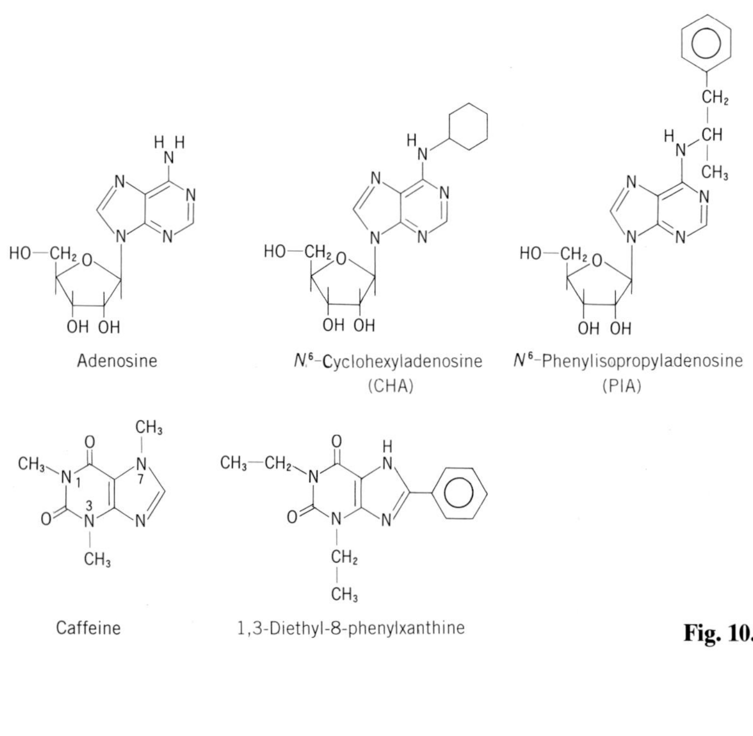

Fig. 10.2 Adenosine and xanthine derivatives (from ref. 8).

tors, without blocking A_1 receptors[5]. In the binding experiment, CGS 21680 has a high affinity to A_2 receptor (K_d = 15 nM) so that it needs 180-fold higher concentration to bind to the A_1 receptor (K_i = 2600 nM).
The efficiency of binding pattern is:

$$\text{CGS 21680} = \text{NECA} > \text{2CADO} > \text{R-PIA} > \text{CHA} > \text{S-PIA}.$$

Depending on the autoradiographic studies that had used CGS 21680 binding sites have been revealed in corpus striatum, nucleus accumbens septi, and olfactory tubercle.

By Northern blot analysis, high levels of A_1 receptor mRNA have been identified in cerebral cortex, cerebellum, and hippocampus. Intermediate levels have been found in the olfactory bulb, corpus striatum, and mesencephalon[10]. *In situ* hybridization histochemistry has shown additional locations of high receptors concentration. These include thalamus, medial geniculate body, and ventral tegmental area. In the cerebellum, the expression of A_1 receptor mRNA was concentrated in the granular cell layer[10]. In Weaver mice, characterized by 80 per cent deficit in granule cells, ^{3}H-CHA binding was less than 80 per cent[3]. In the surface layer of superior colliculus, a high concentration of ^{3}H-CHA binding was observed, but A_1 receptor mRNA was not observed. Since in rats enucleated to abolish the labelling in the superior colliculus opposite to the enucleation, ^{3}H-CHA binding in the contralateral superior colliculus was not observed, A_1 receptor is considered to be produced in the retinal ganglion cells and expressed in the superficial layer of superior colliculus[4]. High levels of A_2 receptor mRNA were identified by *in situ* hybridization in corpus striatum, nucleus accumbens, and olfactory tubercle, similar to the distribution of the A_2 receptor clarified by CGS 21680 binding[12].

Re-uptake mechanism

Although adenosine has been proved to be an inhibitory neuromodulator in the central nervous system, demonstration that adenosine released into the synaptic cleft is inactivated has become a topic of urgent relevance. Among the points of interest, one is the re-uptake mechanism and the other is degradation by ADAse. Theoretically, the uptake sites should overlap with the distribution of A_1 and A_2 receptors. The distribution of adenosine uptake site has been demonstrated by autoradiography using ^{3}H-nitrobenzylthioinosine (NBI), a strong inhibitor of adenosine uptake[1]. In comparison with the distribution of the adenosine receptor, the regions which had higher concentration of adenosine re-uptake sites were corpus striatum, nucleus accumbens, dorsal septal nucleus, piriform lobe, substantia nigra, superficial layer of superior colliculus, trigeminal nucleus caudalis, and substantia gelatinosa of the spinal cord dorsal horn. Thalamus also contained signals for both adenosine re-uptake sites and adenosine receptors, though the distributions were slightly different. In comparison, high levels of A_1 receptors are observed in anterior and median nuclei and gelatinous nucleus, while re-uptake sites are located in the paraventricular nucleus, rhomboid nucleus, and reuniens nucleus. In cerebellum and hippocampus, while the A_1 receptors are observed in high concentrations, re-uptake sites are detected only in low concentrations. In these regions, the activity of adenosinergic neurons is considered to be not so great. In the amygdaloid complex, the receptors are concentrated in the central nucleus, while the re-uptake sites are distributed in cortical and medial nuclei. In the nucleus of the solitary tract, central grey matter, and hypothalamus, high concentrations of re-uptake sites are observed, but both A_1 and A_2 receptors are in low concentration. In these areas, other adenosine receptors may exist or NBI may participate in the uptake of nucleoside other than adenosine. As established in the noradrenaline series and dopamine systems, identification of adenosine transporters and their distribution in the brain will provide more insights into adenosinergic neurotransmission.

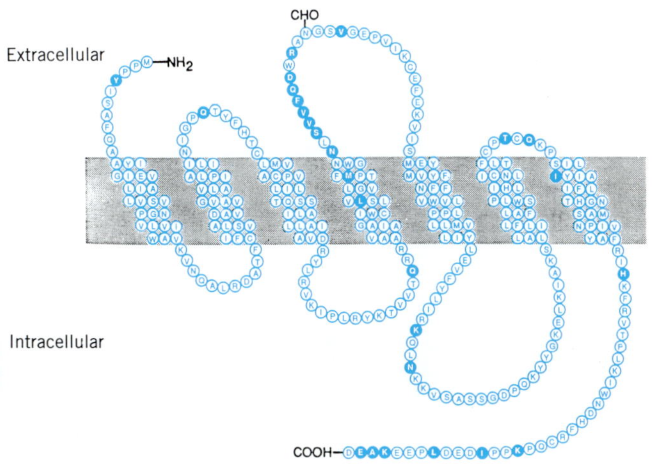

Fig. 10.3 Structure of the adenosine A_1 receptor (rat). Solid circles, amino acids that differ between the rat and dog clones; CHO, potential *N*-linked glycosylation site (from ref. 9).

References

1. Bisserbe, J.C., Patel, J., and Marangos, P.J. (1985). *J. Neurosci.*, **5**, 544–50.
2. Fastbom, J., Pazos, A., and Palacios, J.M. (1987). *Neuroscience*, **22**, 813–26.
3. Fredholm, B.B. (1995). *Pharmacol. Toxicol.*, **76**, 228–39.
4. Goodman, R.R., Kuhar, M.J., Hester, L., and Snyder, S.H. (1983). *Science*, **220**, 967–9.
5. Jarvis, M.F. and Williams, M. (1989). *Eur. J. Pharmacol.*, **168**, 243–6.
6. Lee, K.S. and Peddington, M. (1986). *Neuroscience*, **19**, 535–49.
7. Libert, F., Parmentier, M., Lefort, A. *et al.* (1989). *Science*, **244**, 569–72.
8. Lustig, K.D., Shiau, A.K., Brake, A.J. *et al.* (1993). *Proc. Natl. Acad. Sci. USA*, **90**, 5113–17.
9. Maenhaut, C., Van Sande, J., Libert, F. *et al.* (1990). *Biochem. Biophys. Res. Com.*, **173**, 1169–78.
10. Mahan, L.C., McVittie, L.D., Smyk-Randall, E.M. *et al.* (1991). *Mol. Pharmacol.*, **40**, 1–7.
11. Nakata, H. (1990). *J. Biol. Chem.*, **264**, 671–7.
12. Schiffmann, S.N., Libert, F., Vassart, G. *et al.* (1990). *Brain Res.*, **519**, 333–7.
13. Stehle, J.H., Rivkees, S.A., Lee, J.J. *et al.* (1992). *Mol. Endocrinol.*, **6**, 384–93.
14. Surprenant, A., Buell, G., and North, R.A. (1995). *TINS*, **18**, 224–9.
15. Van Calker, D., Muller, M., and Hamprecht, B. (1979). *J. Neurochem.*, **33**, 999–1005.
16. Zhou, Q.Y., Li, C., Olah, M.E. *et al.* (1992). *Proc. Natl. Acad. Sci. USA*, **89**, 7432–6.

Adenosine receptor (binding sites)

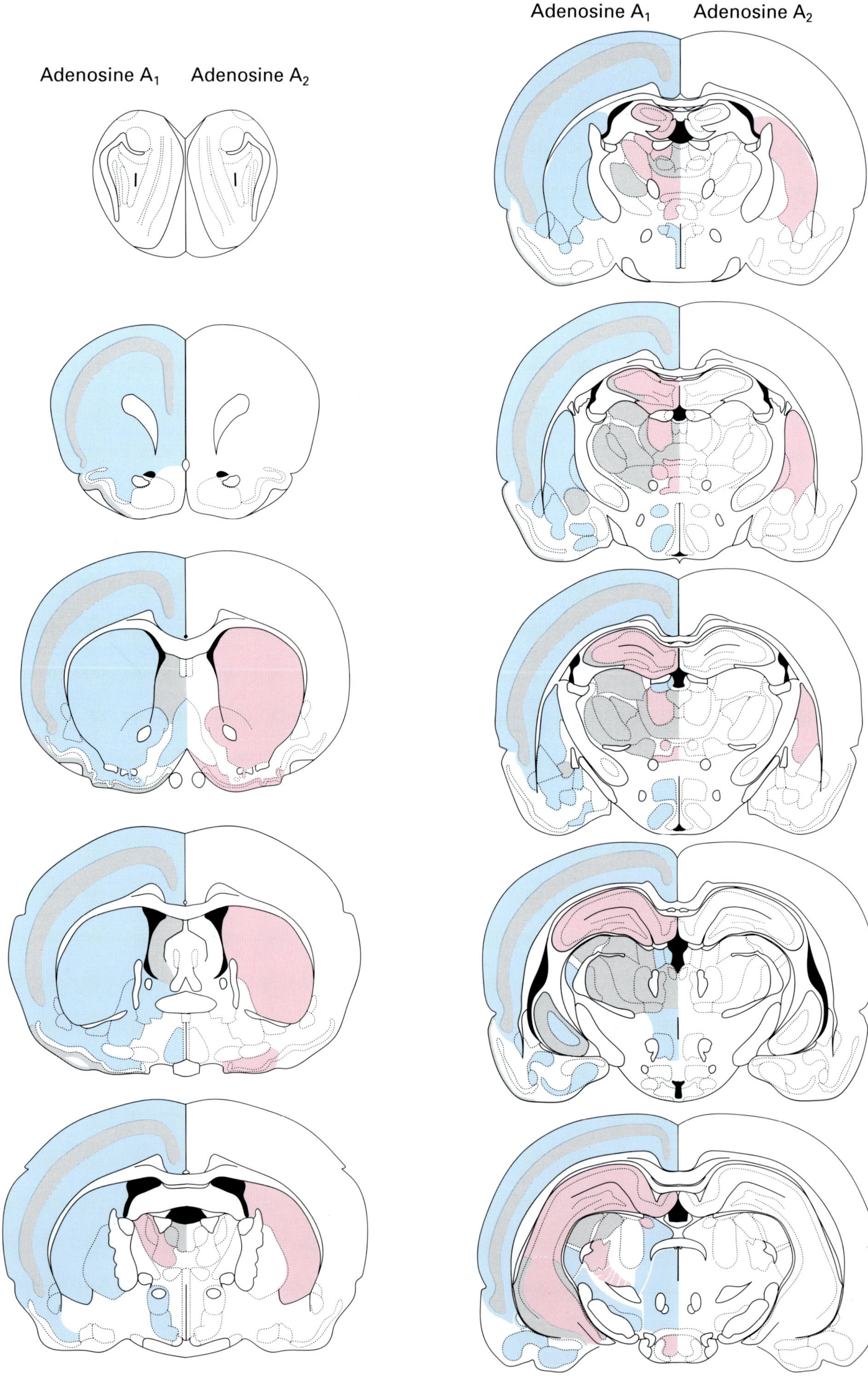

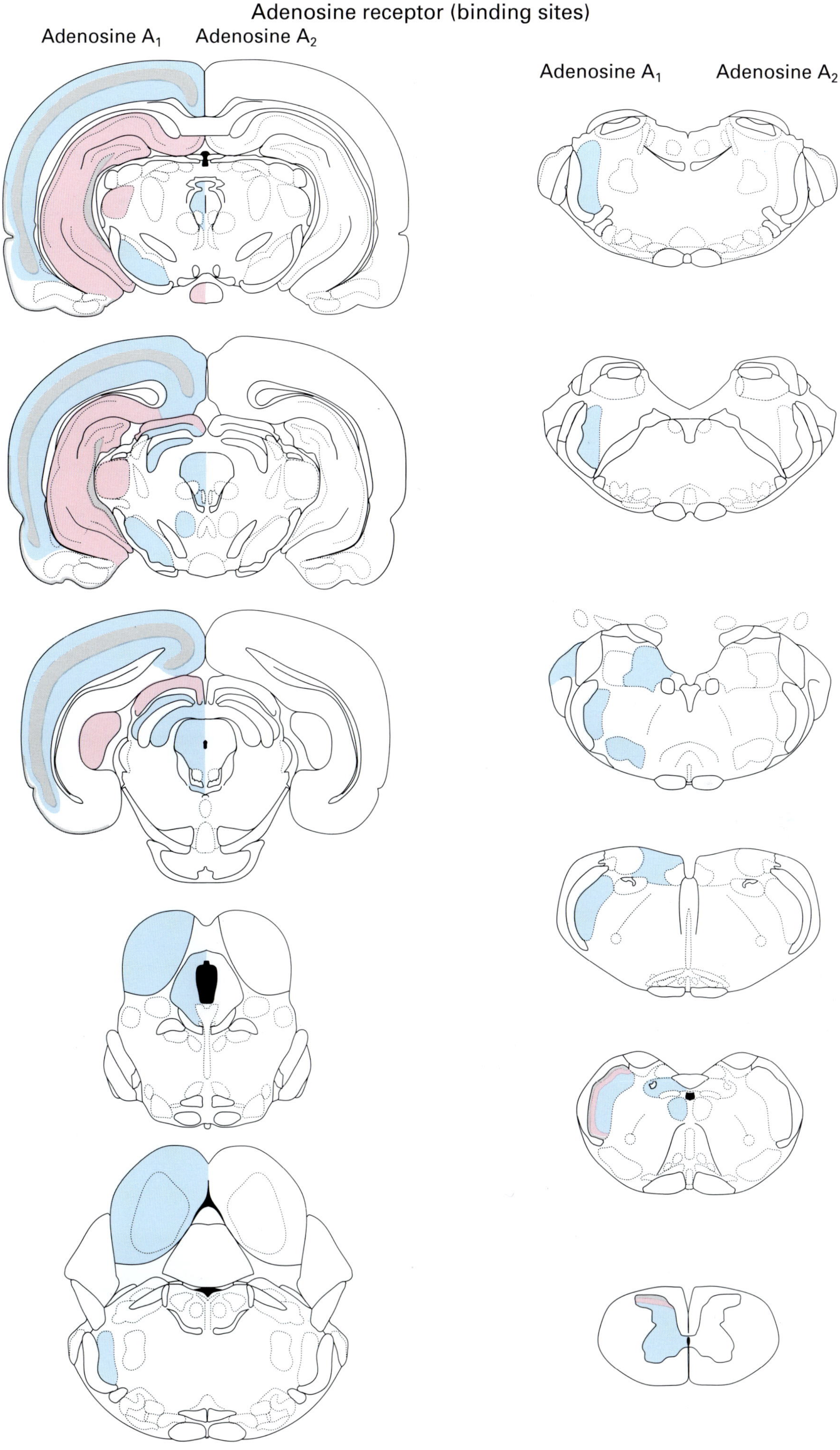
Adenosine receptor (binding sites)
Adenosine A_1
Adenosine A_2
Adenosine A_1
Adenosine A_2

Growth factors

11 The nerve growth factor system

Nerve growth factor

The nerve growth factor family

Nerve growth factor (NGF) is a natural polypeptide, isolated from the submaxillary gland of mouse. Its β subunit is made up of 118 amino acids and exists as a dimer.

The primary structure of the neurotrophic factor extracted from pig brain (brain derived neurotrophic factor, BDNF) has also been determined and this BDNF turned out to be highly homologous to NGF[12]. Since then, based especially on the homologous regions of gene arrangement between NGF and BDNF, a third factor (neurotrophin-3, NT-3) was identified by the PCR method[6].

Recently, genes for NT-4[3] and NT-5[1] have been identified by a similar strategy. It is now believed that NGF and related molecules form a family with at least five members.

Distribution of nerve growth factor

Due to its established history and the existence of a specific antibody, NGF has been quantitated in most regions of the brain and its distribution has also been described. Various methods have been employed for biochemical quantification of NGF and reliable data have been accumulated. In the hippocampus, only extremely low concentrations of NGF, of the order of several ng/g wet tissue, have been detected. So far, therefore, particularly positive results regarding the distribution of NGF by immunohistochemical methods have not been attained. The distribution of NGF mRNA levels in the tissues has also been studied using *in situ* hybridization. However, owing to the problems of the specificity of hybridization and other related questions, it is difficult to reach a meaningful conclusion. For these reasons, the localization of distribution of NGF has not been described in this atlas.

Some reports exist on the distribution of mRNA of BDNF[5] and NT-3 by *in situ* hybridization. However, since the localization of distribution has not been investigated in detail in these reports, they have not been included here. However, it has been confirmed that among the NGF family members, BDNF is widely distributed in comparison to NGF and NT-3.

Nerve growth factor receptor

Subtypes and structure

Pharmacological experiments show that the NGF receptor system consists of two types. One is a low affinity receptor ($K_d = 10^9$M) and the second one is a high affinity receptor ($K_d = 10^{11}$M). Various experiments suggest that it is the high affinity receptor (H-NGFR) which directly mediates the physiological action of NGF. The low affinity receptor was first identified at the molecular level. Specific monoclonal antibodies were produced to rat L-NGFR and human L-NGFR (192 IgG to rat, MC 20.4 to human) and by using these as probes, tissue distribution[14,15] and gene cloning[16] have also been reported. The L-NGFR is a polypeptide that penetrates the membrane once, judging from the deduced amino acid sequence, and consists of 426 (rat) amino acids. The intracellular domain consists of 155 amino acids, but lacks the structure to mediate known signal transduction (for instance kinase domain and the ATP binding region, etc.). The molecular weight of L-NGFR, calculated from cDNA, is around 75 kDa, and this corresponds to the biochemical data reported previously.

Though molecular details regarding L-NGFR have accumulated, not much is known about H-NGFR. First steps toward solving this puzzle began in 1991. *trk*, a proto-oncogene, which had not been linked to studies on NGF till then, was demonstrated to be a constitutive element of H-NGFR[4]. The trk protein, with a molecular weight of 140 kDa, penetrates the membrane once and has a tyrosine kinase domain in its intracellular portion.

In an analysis of cells with forcibly expressed *trk* gene, two sets of results were obtained. One is, in this case of only *trk* expression, the affinity for NGF has a K_d around 10^{-9}M, which is almost similar to the value for L-NGFR. When L-NGFR and *trk* were expressed simultaneously, the K_d value is around 10^{11}M and this agrees with the value of H-NGFR that has been reported to date. On the other hand, a report[8] exists that even during only *trk* expression, the K_d value was around 10^{11}M. In this case, *trk* possibly acts as H-NGFR single-handedly. At present, it has not been resolved which model is valid, but some verifying experiments have been reported recently. In one of these experiments, a modification to the NGF molecule was introduced artificially. Substitution of Lys-32, Lys-34, and Lys-95 of NGF with alanine resulted in loss of binding ability to L-NGFR, but did not affect the binding with *trk*[7]. When this NGF variant molecule was bound to cells with NGF reactivity (PC 12 cells), the physiological activity of NGF was established. This shows that the physiological action of NGF is not dependent on the binding with L-NGFR. Thus the physiological relevance of L-NGFR needs reappraisal. The above-mentioned experiment suggests that trk alone is sufficient to function as a signal-transducing receptor for NGF and that the previous model of both L-NGFR and *trk* combining to form H-NGFR is now in doubt. The same thing has been proved in an experiment which used MAH cells, an immortalized sympathoadrenal progenitor cell line which does not express L-NGFR, but does express trk protein[2].

The experimental results shows that differentiation to sympathetic ganglion precursor tissue occurs when the NGF produced acts only on trk protein. A scan of recent experimental results also suggest that the postulate that trk protein by itself mediates the physiological action of NGF seems more acceptable, but detailed experiments are needed to confirm this. It is now clear that *trk* forms a gene family, and trkB[9] with homology to *trk* and *trk*C[11] are members of this family. As *trk* mediates the physio-

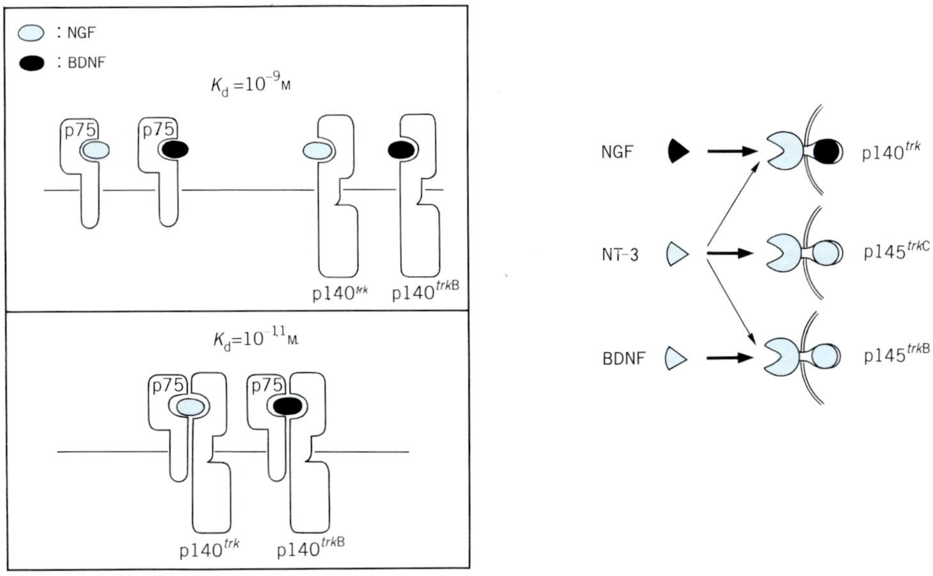

Fig. 11.1 Schematic representation of putative NGF receptor system. p75 denotes the low affinity NGF receptor and p140 indicates *trk* family protein. NGF preferentially binds to p140^{trk}, while BDNF binds to p145^{trkB}. NT-3 can associate with p140^{trk} and p145^{trkB} in addition to p145^{trkC} (from Bothwell, M. (1991). *Cell*, **65**, 915–18).

logical activity of NGF, it has been thought that *trk*B and *trk*C also function as receptors of BDNF and NT-3 respectively. But such one-to-one linking in the frame of NGF-*trk*, BDNF-*trk*B and NT-3-*trk*C may not be feasible because NT-3 possesses the binding ability with *trk* and *trk*B as well. So, at present, it seems that the situation is more complex than our current understanding (Fig. 11.1).

Distribution

Though remarkable progress has been made recently on the existence of L-NGFR, its significance remains unclear. However, it is important to learn the distribution of L-NGFR that has been studied in detail so far; so comparative examination of the distribution of the *trk* family deserves attention. In this atlas, detailed distribution of localization of L-NGFR, as recognized by the specific monoclonal antibody, is presented[14,15]. This map reveals that the main localization area is located at the basal forebrain region containing the cholinergic neurons. Expression of L-NGFR mRNA has also been examined[10,17] and this is also shown. Though the expression of both protein and mRNA are found in a limited area, would the expression of the *trk* family correlate with this? At present, details on the localization of distribution of the *trk* family during the maturation of the nervous system are still unclear, and this gap urgently needs to be filled. Since L-NGFR alone is unable to mediate the physiological activity of NGF, it needs the existence of the *trk* family. We are now examining the expression of the *trk*B gene, and we have found its expression in a wider area, different from that of L-NGFR. In order to understand correctly the function of the NGF family in the central nervous system, further progress on the details of the *trk* family is needed.

References

1. Berkemeier, L.R., Winslow, J.W., Kaplan, D.R. *et al.* (1991). *Neuron*, **7**, 857–66.
2. Birren, S.J., Verdi, J.M., and Anderson, D.J. (1992). *Science*, **257**, 395–7.
3. Hallböök, F., Ibáñez, C.F., and Persson, H. (1991). *Neuron*, **6**, 845–58.
4. Hampstead, B.L., Martin-Zanka, D., Kaplan, D.R. *et al.* (1991). *Nature*, **350**, 678–83.
5. Hofer, M., Pagliusi, S.R., Hohn, A. *et al.* (1990). *EMBO J.*, **9**, 2459–64.
6. Hohn, A., Leibrock, J., Bailey, K., and Barde, Y.A. (1990). *Nature*, **344**, 339–41.
7. Ibáñez, C.F., Ebendal, T., Barbany, G. *et al.* (1992). *Cell*, **69**, 329–41.
8. Klein, R., Jing, S., Nanduri, V. *et al.* (1991). *Cell*, **65**, 189–97.
9. Klein, R., Nanduri, V., Jing, S., Lamballe, F. *et al.* (1991). *Cell*, **66**, 395–404.
10. Koh, S., Oyler, G.A., and Higgins, G.A. (1989). *Exp. Neurol.*, **106**, 209–21.
11. Lamballe, F., Klein, R., and Barbacid, M. (1991). *Cell*, **66**, 967–79.
12. Leibrock, J., Lottspeich, F., Hohn, A. *et al.* (1989). *Nature*, **341**, 149–52.
13. Phillips, H.S., Hains, J.M., Laramee, G.R. *et al.* (1990). *Science*, **250**, 290–2.
14. Pioro, E.P. and Cuello, A.C. (1990). *Neuroscience*, **34**, 57–87.
15. Pioro, E.P. and Cuello, A.C. (1990). *Neuroscience*, **34**, 89–110.
16. Radeke, M.J., Misko, T.P., and Hsu, C. (1987). *Nature*, **325**, 593–7.
17. von Bartheld, C.S., Heuer, J.G., and Bothwell, M. (1991). *J. Comp. Neurol.*, **310**, 103–29.

NGF receptor (immunocytochemistry)

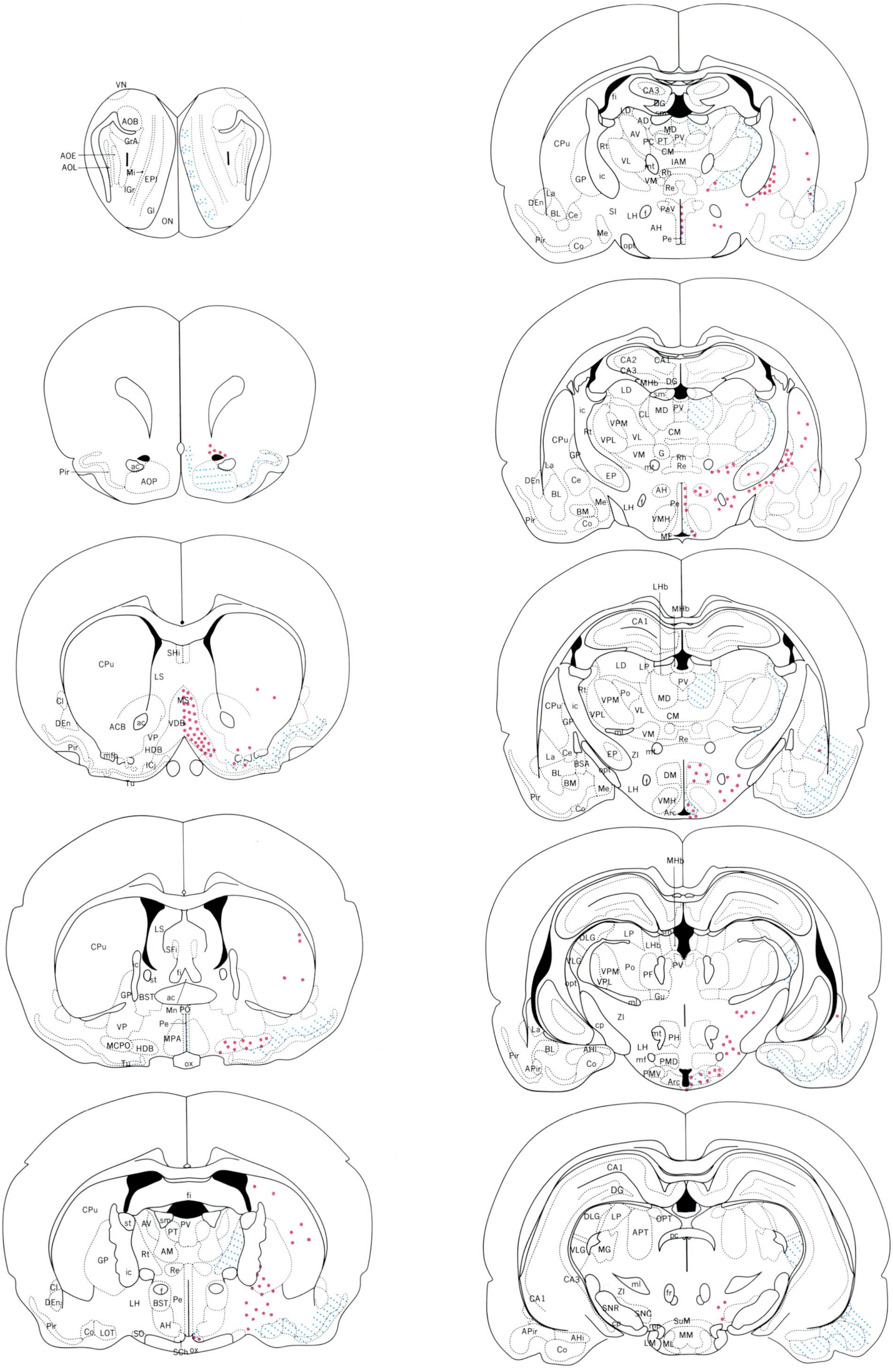

NGF receptor (immunocytochemistry)

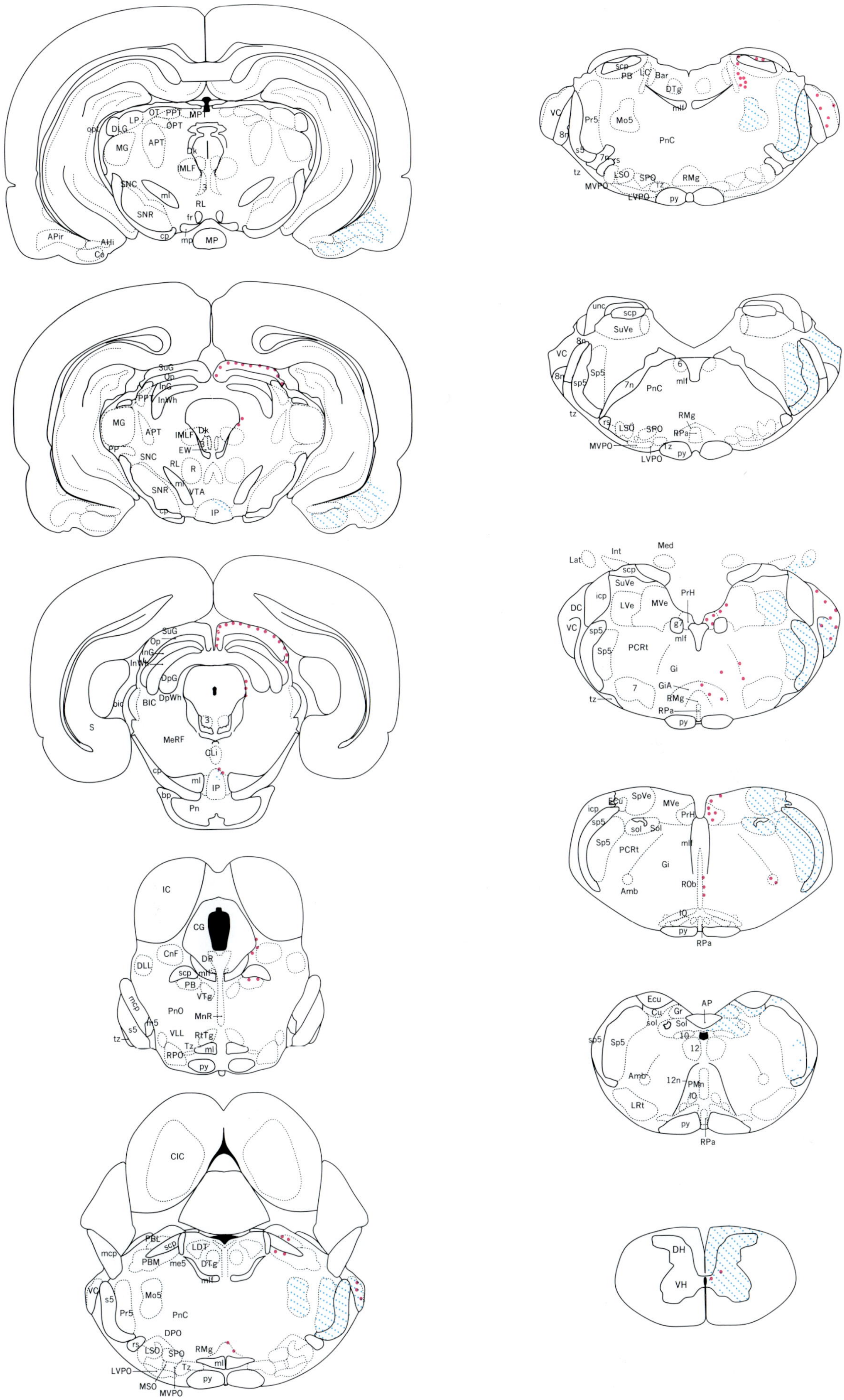

12 The fibroblast growth factor system

Fibroblast growth factor

Structure

Fibroblast growth factor (FGF), as its name suggests, was originally identified as a peptide growth factor, extracted from cattle hypophysis, with mitogenic activity for fibroblasts. FGF was considered to be of two types; basic FGF (bFGF) and acidic FGF (aFGF). However, structurally similar factors have been identified in recent years and it is clear that they form a family consisting, at present, of seven types (Table 12.1).

bFGF is a polypeptide consisting of 146 amino acid residues, and aFGF consists of 140 amino acid residues. There is approximately 55 per cent identity between these two types. The amino acid sequence of bFGF is very much conserved among different species, and only two of the 146 amino acid residues vary between human and bovine bFGF. However, aFGF exhibits more difference in its amino acid residues among species. Eleven of the 140 amino acid residues vary between human and bovine. Gene cloning has been carried out in both bFGF and aFGF, and it has become clear that neither type has a typical signal sequence. This suggests that neither aFGF nor bFGF has a mechanism for extracellular secretion. At present, the secretory mechanism of FGF has not been much clarified[2].

Physiological function

After its discovery, FGF was implicated in various physiological functions. At present, it is not an exaggeration to state that it is difficult to find tissues that do not react to FGF. Table 12.2 shows the type of cells that bFGF and aFGF act upon. In these cells, the main physiological function of FGF is to facilitate mitotic propagation and cell differentiation. These data show that FGF acts on various types of cells. The main functions of FGF at the tissue level, *in vivo*, are mesodermal tissue induction in fetus development, induction of limb or lens regeneration in lower animals, induction of wound healing, vascularization of blood vessels (angiogenesis), neurotrophic action, etc. Among these, the function of FGF in vascularization of blood vessels has been largely investigated since its discovery. Recently, attention has also been paid to functions such as mesodermal tissue induction and neurotrophic action.

An opinion has been frequently expressed in the past four or five years that FGF could be the true form of the 'organizer', long known in the field of embryology. But this opinion does not have a sound basis. However, through the induction of homeobox gene expression, it seems clear that FGF plays an important role in morphogenesis. Since 1986, studies have been in progress in cultured central nervous system cells on the neurotrophic role of FGF. It has been shown that the contents of both bFGF and aFGF are high in the central nervous system and this suggests that FGF possibly functions as a neurotrophic factor, in combination with NGF and other factors.

The fibroblast growth factor family

It has become clear that FGF forms a family of growth factors consisting of seven members. Apart from aFGF and bFGF, the other members of the family are *int-2*[6], isolated as the oncogene of mouse mammary cancer; *HST1*[8], an oncogene of human stomach cancer; FGF-5[9], an oncogene of human bladder tumour; *hst-2*/FGF-6[4], homologous to *hst-2*, and KGF[1], keratinocyte growth factor derived from fibroblast lines of lung of human fetus. Of these seven members, only aFGF and bFGF are lacking in signal sequence and the other five members are secretory proteins. A highly homologous region among

Table 12.1. The FGF family (adapted from Sakamoto, H. (1992) *Exp. Med.*, **10**, 19, in Japanese)

	Original name	Commonly used abbreviations	Position on human chromosome
FGF–1	Acidic fibroblast growth factor Endothelial cell growth factor Heparin binding growth factor-1	Acidic FGF/aFGF ECGF HBGF-1	5
FGF–2	Basic fibroblast growth factor Heparin binding growth factor-2	Basic FGF/bFGF HBGF-2	4q25
FGF–3	*int-2*	*int-2*	11q13
FGF–4	Transforming gene from human stomach-1 Heparin-binding secretary transforming factor-1 Kaposi's sarcoma FGF	*hst-1* HSTF-1 ks FGF/K-FGF	11q13
FGF–5	Fibroblast growth factor-5 Fibroblast growth factor-6	FGF-5 FGF-6	4q21 12p13
FGF–6	*hst-1* related gene	*hst-2*	
FGF–7	Keratinocyte growth factory	KGF	FGF-7

these members, consisting of around 120 amino acid residues, occupies the core area of each protein (Fig. 12.1). The physiological actions of these members are still unclear, though all members, excluding KGF, possess mitogenic actions to fibroblast. KGF is a specific growth stimulator of epithelial cells[1].

Distribution

Until now, the nature and distribution of aFGF and bFGF in tissues have been examined by immunohistochemical techniques using the antibody. Since the amino acid sequences of both FGFs have a high degree of homology, it is difficult to distinguish them by immunostaining. To solve this problem, an attempt has been made using antibody against peptides such as synthetic peptides formed from the regions which do not show homology. The immunopositive nature of bFGF has been clarified by this technique[3,5]. bFGF differs from aFGF in the 24 amino acid residues at the N-terminal end. Antibody produced to this region is bFGF specific and has been confirmed not to cross-react with aFGF[7].

Among the other members, the existence of aFGF, *int-2*, *hst-1*, FGF-5, FGF-6, and KGF during the maturation of central nervous system has been confirmed, but their localization of mRNA expression in tissues or protein by immunohistochemical technique has not so far been examined. It is clear that, quantitatively, bFGF and aFGF have a high expression profile, and other members of the family show a low profile.

Table 12.2. Cell types for whcih bFGF or aFGF is mitogenic or affects differentiation (adapted from ref. 2)

	bFGF	aFGF
Normal diploid cells		
Glial or astroglial cells	+ (D)	+
Oligodendrocytes	+ (D)	+
Trabecular meshwork cells	+	?
Endothelial cells from capillary, large vessels, and endocardium		
Corneal endothelial cells	+ (D)	+ (D)
Fibroblasts	+	+
Myoblasts	+ (D)	+ (D)
Vascular smooth muscle	+	+
Chondrocytes	+ (D)	+ (D)
Osteoblasts	+	+
Blastoma cells	+	?
Adrenal cortex cells	+	+
Glanulosa cells	+	+
Prostatic endothelial cells	+	+
Mesothelial cells	+	+
Neuronal cells	+	?
Established cell lines		
Rat fibroblast-1	+	+
BALB/c3T3	+	+
Swiss 3T3	+	+
BHK-21	+	+
A-204 rhabdomyosarcoma	+	?
PC-12	(D)	(D)

+, positive effects of cell proliferation; (D), differentiation; ?, effect on cell proliferation not determined.

References

1 Finch, P.W., Rubin, J.S., Miki, T. *et al.* (1989). *Science*, **245**, 752–5.
2. Gospodarowicz, D., Ferrara, N., Schweigerer, L., and Neufeld, G. (1987). *Endocrine Rev.*, **8**, 95–114.
3. Iwata, H., Matsuyama, A., Okumura, N. *et al.* (1991). *Brain Res.*, **550**, 329–32.
4. Marics, I., Adelaide, J. Raybaud, F. *et al.* (1989). *Oncogene*, **4**, 335–40.
5. Matsuyama, A., Iwata, H., Okumura, N. *et al.* (1992). *Brain Res.*, **587**, 49–65.
6. Moore, R., Casey, G., Brookes, S. *et al.* (1986). *EMBO J.*, **5** 919–24.
7. Okumura, N., Takimoto, K., Okada, M., and Nakagawa, H. (1989). *J. Biochem.*, **106**, 904–9.
8. Taira, M., Yoshida, T., Miyagawa, K. *et al.* (1987). *Proc. Natl Acad. Sci. USA*, **84**, 2980–4.
9. Zahn, X., Bates, B., Hu, X., and Goldfalb, M. (1987). *Mol. Cell. Biol.*, **8**, 3487–95.

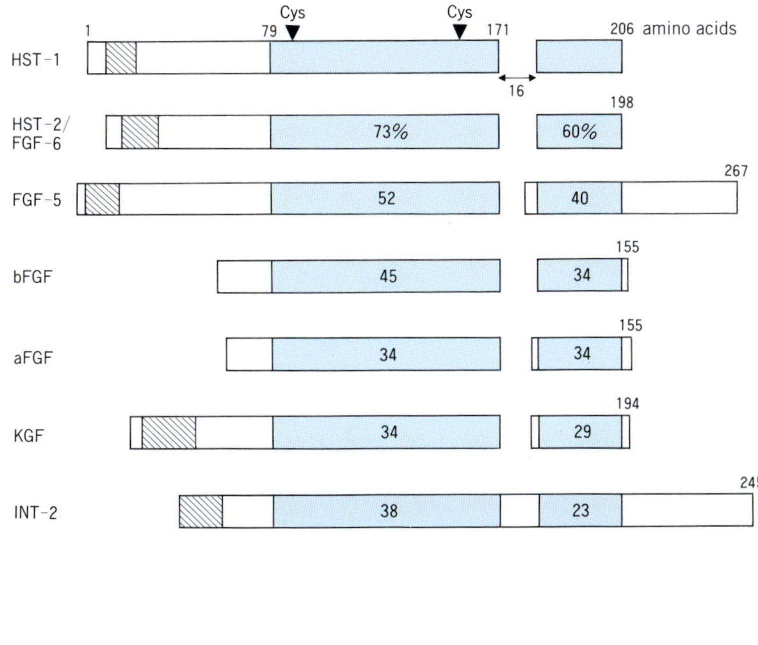

Fig. 12.1 Schematic representation of seven members of the FGF family. ▧ Hydrophobic amino acid stretch which can serve as signal peptide; □, homologous region with HST-1 and percentage homology; Cys, the positions of the cysteine residues which are conserved among seven members (adapted from Sakamoto, H. (1992). *Exp. Med.* (in Japanese), **10**, 19).

FGF (immunocytochemistry)

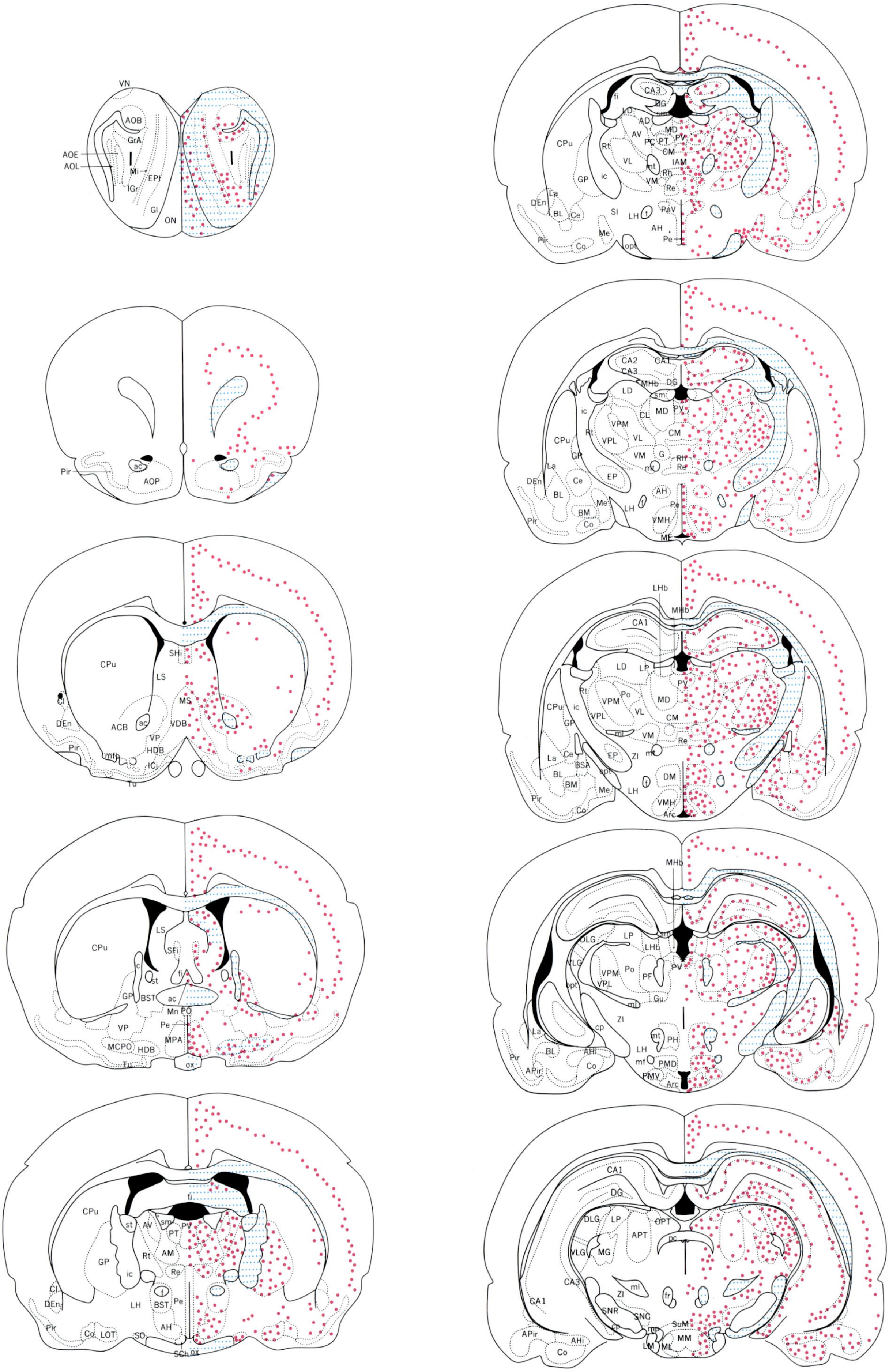

FGF (immunocytochemistry)

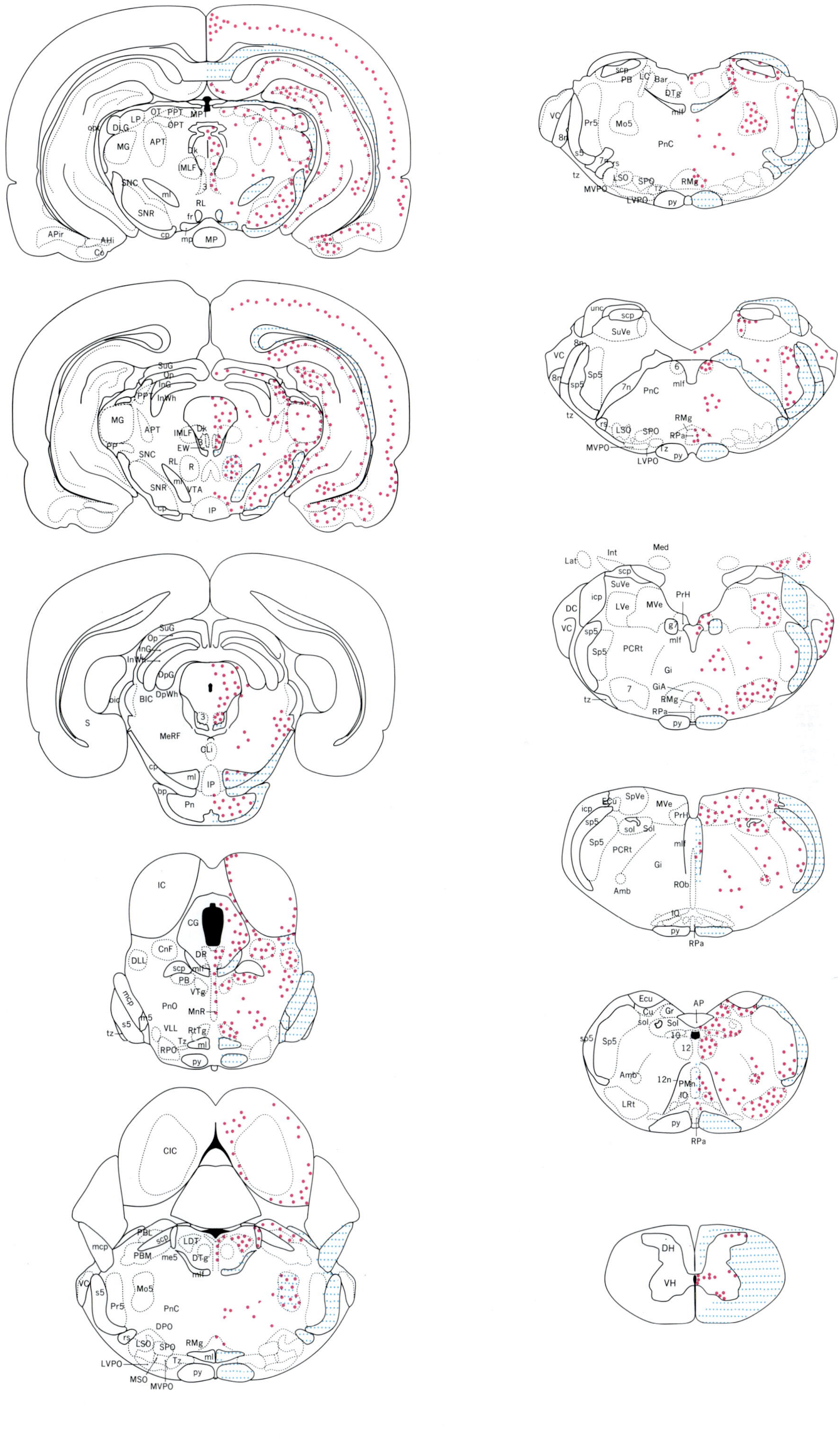

Fibroblast growth factor receptor

Structure

As described above, FGF (a heparin-binding growth factor) performs various functions in many types of cells. The existence of a high-affinity receptor at the surface of the reacting cells has been observed. However, knowledge about this receptor at the molecular level has remained elusive for a long time since the discovery of FGF. In 1989, the chicken FGF receptor was first cloned by Lee *et al.*[6] and analysis of FGF receptor at the gene level then rapidly advanced. Figure 12.2 shows the structure of the FGF receptor cloned by Williams *et al.* It is a transmembrane protein with three extracellular immunoglobin-like domains and an intracellular tyrosine kinase domain. Fundamentally its structure is similar to that of EGF receptor or PDGF receptor. Also, the tyrosine kinase domain is divided into two parts, with a centrally located intervening sequence consisting of 14 amino acid residues.

In the analysis of the gene sequence of this chicken FGF receptor, an interesting fact became evident. The human cancer *fms*-like gene (*flg*) was found to possess almost a similar structure to the chick FGF receptor[10]. Since then, the gene of mouse FGF receptor has been successively isolated[9] and these were designated generically as *flg* type. Apart from this, the *bek* type of FGF receptor, the other subtype which is highly homologous to *flg* type, was also identified[2]. It became clear that the FGF receptors form a gene family (see next section, on subtypes). Thus, the structure of high-affinity receptors bearing signal transmission of FGF, which had remained unclear for a long time, is now understood. It has also been shown to have a signal transmission mechanism mediated by tyrosine kinase, similar to other growth factors like EGF. These rapid advances will surely lead to resolution of the action mechanism of FGF.

Subtypes

As mentioned above, by 1990 it had become evident that the FGF receptors (FGF-R) form a gene family. The members of this family which have been identified so far, and their structural homologies, are presented in Table 12.3. Apart from the already mentioned *flg* and *bek* types, FGF-R3[5] and FGF-R4[8] are the newly described subtypes. Excluding these four main types, the existence of some variants due to mRNA splicing were also reported. For instance, among the *flg* types, one that deletes the terminal immunoglobulin-like domain[9], one that possesses only an extracellular domain and without a structure to penetrate the membrane[4], one with a replaced amino acid at the membrane penetrating area[3], and others have been reported. It appears that variants in the FGF receptor family will tend to become more complex. The problem is associated with the fact that FGF ligands also form a family of seven types of members; i.e. there exists a possibility of interaction between the ligand and receptor. The fact remains that a one-to-one interaction between FGF and a member of the receptor family is absent and redundancy is present in quite a few systems studied recently[2,7,10].

Distribution

So far, the action of FGF on the nervous system has been examined using primary culture of neurons. Walicke *et al.*[11] first reported that bFGF increased the neuronal survival and axonal extension in a highly purified population of fetal rat hippocampal neurons. Since this report, the action of FGF as a neurotrophic factor has been given much attention. Walicke and her colleagues sliced the fetal brain into smaller sections and used the cells cultured from these sections for their experiments. The cultured cells of most regions reacted positively to bFGF, regardless of its efficacy[12]. This suggested that FGF could act on cell groups covering a wide area of the central nervous system. However, this culture system has a limit in identifying a target cell in the central nervous system with extreme complexity. Thus, a more detailed reference at the tissue level was needed. We had learned from the study of Lee *et al.* about the cloning of chicken FGF receptor mentioned earlier, and about the *flg* gene that is believed to be the human FGF receptor. Following these steps, we first isolated a partial clone of *flg* from human endothelial cell mRNA using the PCR method, and studied the expression of the rat FGF-receptor mRNA in the matured central nervous system using *in situ* hybridization[13]. The results obtained are presented in this atlas.

From the revealed distribution, one can infer that the FGF receptor mRNA is mainly expressed in the neurons, and not in the glial cells of the normal central nervous system. Though the expression of FGF receptor mRNA was seen widely in the central nervous system, it was intensively expressed in some specific neurons, without equal expression in all neurons. Due to varying experimental conditions, it is difficult to compare the results of ours with those of past experiments in cultured cells. For example, whereas we used a mature nervous system, the cell culture experiments were conducted in fetal brain.

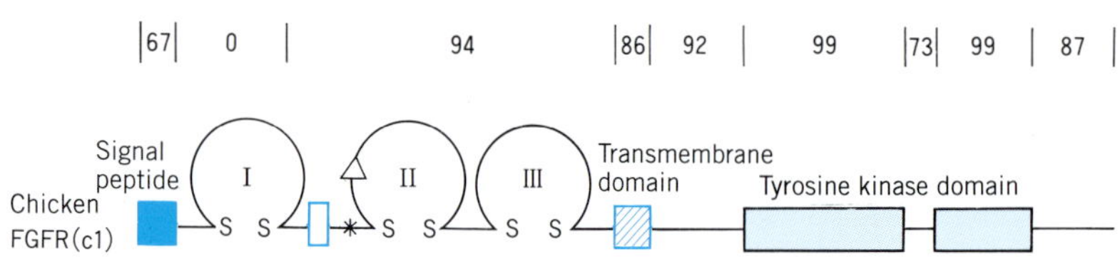

Fig. 12.2 Schematic representation of chicken FGF receptor (*flg* type): in the extracellular region there are three immunoglobulin-like domains (I, II, III) (adapted from ref. 4).

Table 12.3. Homologies between subtypes of the FGF receptor family

	FgFR4 vs. *flg*	**FGFR4 vs. *bek***	**FGFR4 vs. FGFR3**	***flg* vs. *bek***	***flg* vs. FGFR3**	***bek* vs. FGFR3**	
Domains							
I	22	21	26	38	21	27	136
II	63	70	78	81	66	72	235
III	74	72	74	79	82	81	342
TM }	32	31	34	66	41	45	
JM }							455
TK	75	78	78	88	83	87	561
KI	14	29	36	50	43	50	575
TK	86	84	86	92	91	92	734
CT	38	48	46	60	44	56	
overall identity	56	57	60	71	61	65	

I, outer immunoglobuin-like domain; II, middle immunoglobulin-like domain; III, inner immunoglobulin-like domain; TM, transmembrane domain; JM, juxtamembrane region; TK, tyrosine kinase domain; KI, kinase insertion; CT, C-terminal region (adapted from ref. 8)

The expression of FGF receptor mRNA was strongly observed in the hippocampus where FGF showed strong effects in culture. Our experiment also showed an intense signal in the cells such as the cholinergic neurons of the laterodorsal tegmental nucleus, which is too small a region for culture experiments. The physiological action of FGF on these cell groups needs to be addressed in future research.

The FGF receptors form a family, and the variety and differences of expression styles shown by each subtype at the tissue level is an interesting phenomenon. Since the distribution of FGF receptor mRNA mentioned above is thought to be mainly of the *flg* type, investigations are needed on the expressions of FGF-R3 and FGF-R4 of *bek* type. To investigate each subtype individually, we are making use of an oligonucleotide probe which is different from the cDNA probe described above, and are studying the expressions of *flg* and *bek* in the mature nervous system. Compared to the *flg* data, the expression of *bek* type was observed mostly in oligodendroglia rather than neurons[1]. This suggests that different cell types express FGF receptors of different subtypes. One can wonder whether this pattern is a solution to the diversity of the FGF system.

References

1. Asai, T., Wanaka, A., Kato, H. *et al.* (1993). *Mol. Brain Res.*, **17**, 174–8.
2. Dionne, C.A., Crumley, G., Bellot, F. *et al.* (1990). *EMBO J.*, **9**, 2685–92.
3. Hou, J., Kan, M., McKeehan, K. *et al.* (1991). *Science*, **251**, 665–8.
4. Johnson, D.E., Lee, P.L., Lu, J., and Williams, L.T. (1990). *Mol. Cell. Biol.*, **10**, 4728–39.
5. Keegan, K., Johnson, D., Williams, L.T., and Hayman, M. (1991). *Proc. Natl Acad. Sci. USA*, **88**, 1095–9.
6. Lee, P.L., Johnson, D.E., Cousens, L.S. *et al.* (1989). *Science*, **245**, 57–60.
7. Mansukhani, A., Moscatelli, D., Tallarico, D. *et al.* (1990). *Proc. Natl Acad. Sci. USA*, **87**, 4378–82.
8. Partanen, J., Makela, T., Eerola, E. *et al.* (1991). *EMBO J.*, **10**, 1347–54.
9. Reid, H.H., Wilks, A.F., and Bernard, O. (1990). *Proc. Natl Acad. Sci. USA*, **87**, 1596–600.
10. Ruta, R., Howk, R., Ricca, G. *et al.* (1988). *Oncogene*, **3**, 9–15.
11. Walicke, P., Cowan, W.M., Ueno, N. *et al.* (1986). *Proc. Natl Acad. Sci. USA*, **83**, 3012–16.
12. Walicke, P.A. (1988). *J. Neurosci.*, **8**, 2618–27.
13. Wanaka, A., Jonson, E.M., Jr, and Milbrandt, J. (1990). *Neuron*, **5**, 267–81.

FGF receptor (ISH)

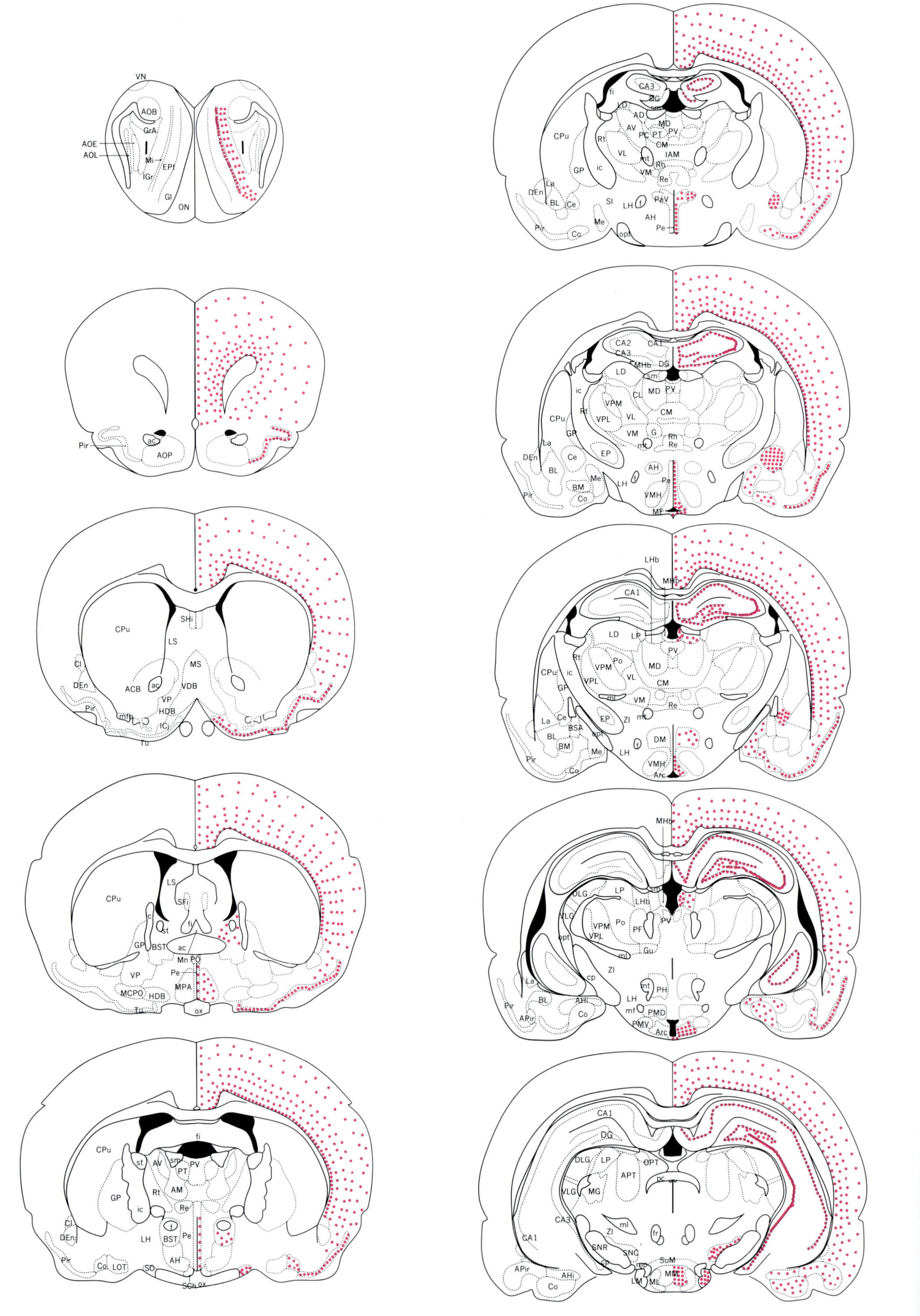

FGF receptor (ISH)

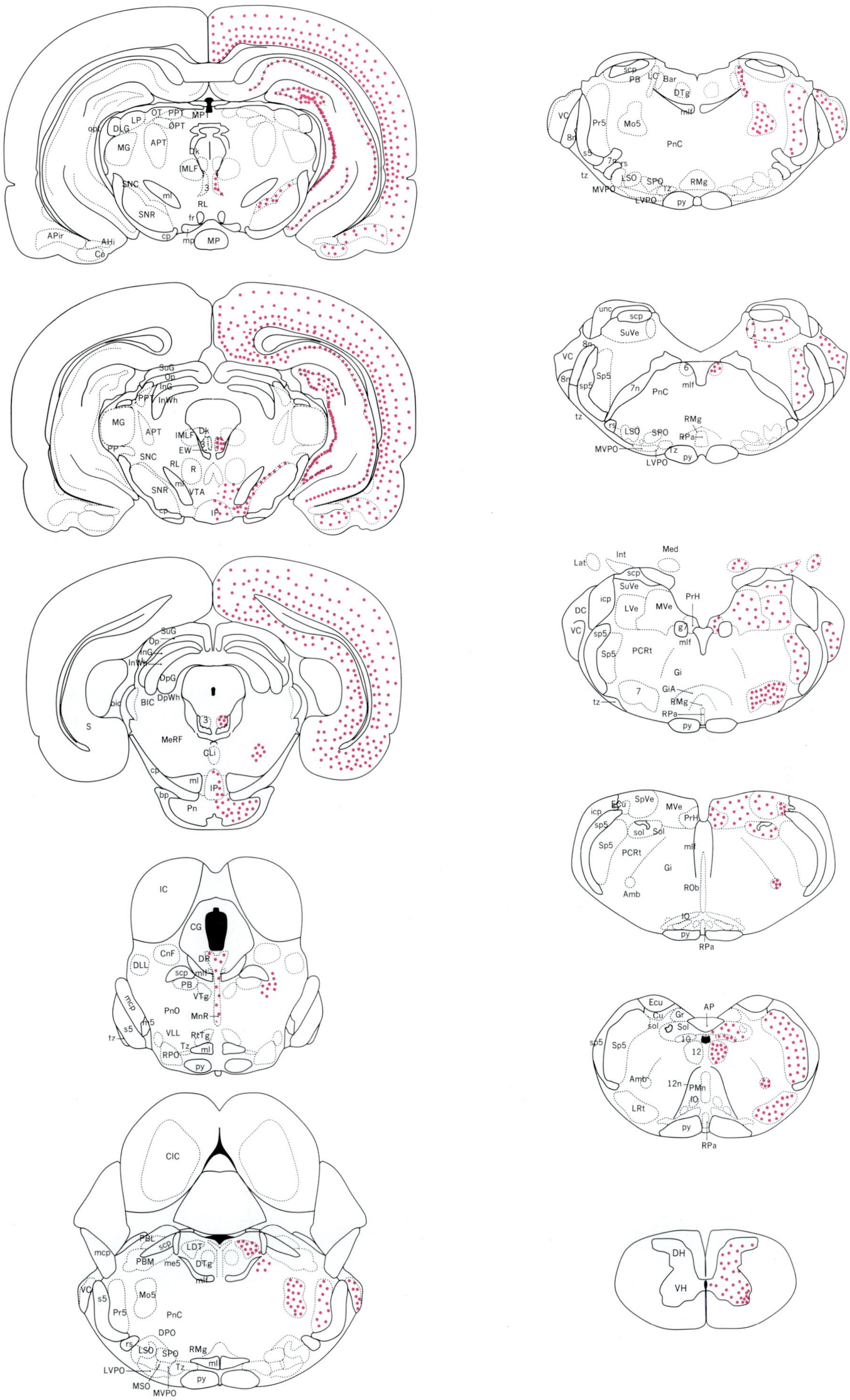

13 The pleiotrophin system

Pleiotrophin

Structure

Pleiotrophin (PTN) is a polypeptide, isolated from the bovine uterus using heparin affinity chromatography. It has a molecular weight of 17 000 and possesses the effect of facilitating the propagation of fibroblasts[4]. The amino acid sequence of its N-terminal end has been determined and it became evident that PTN is different from the FGF and other neurotrophic factors reported so far, and that it shows 100 per cent homology to p18 isolated from perinatal rat brain[6]. The research group at the University of Helsinki that purified p18 and the research group at the Washington University, St. Louis, that purified the PTN succeeded in cloning the gene of this polypeptide in 1990[3,5]. The amino acid sequence determined from the cDNA has some special features. First, 32 residues at the N-terminal end of the total 168 amino acid residues showed hydrophobicity. This was considered as the signal peptide, and it became clear that PTN is a secretory protein (Fig. 13.1). Secondly, this sequence is not homologous to those of the already identified members of the growth factor family (NGF, FGF, EGF, and others). Also, one special feature of this protein is that after secretion, it is rich in lysine residues at the N- and C-terminal ends.

Distribution and physiological function

PTN/p18, though initially isolated as a mitogen of fibroblasts, was found to be highly expressed in brain and its expression increased with time during the developmental process[3,5,6] (Fig. 13.2). This suggests that PTN is involved in the development and maintenance of neurons, and in fact, purified PTN induces axonal elongation to cultured neurons of rat fetal brain origin[3,6].

We have investigated the expression of PTN mRNA at tissue level using *in situ* hybridization, as a step towards identifying the functions of PTN in the nervous system[7]. In the rat central nervous system, during the early stage of development (embryo 10–11 days), its expression was observed in the undifferentiated neuroepithelium. When development progresses (embryo 17–20 days), PTN mRNA was strongly expressed at the external layer consisting of mainly undifferentiated precursor cells. After birth, during the first two weeks of the perinatal period, the expression intensity of PTN mRNA became stronger and its signal was detected in a wide range of cell groups. Corresponding results obtained by the northern blot method are shown in Fig. 13.2. During maturation, the gene expression of PTN decreased to the extent that in the mature central nervous system, its expression was limited to certain population of neurons. PTN mRNA expression in the mature nervous system is shown in this atlas.

Based on the investigations on PTN mRNA expression in the mature nervous system, the physiological actions of PTN can be summarized as follows:

- During early stages of development in the nervous system, until midgestation, PTN mRNA is expressed mainly in undifferentiated precursor cells. It is believed that PTN is produced and secreted in these cells and autocrine tissues, to accelerate differentiation and growth of precursor cells.
- During the first two weeks of the perinatal period, neurons extend to axons and form neural circuits. Thus, an increase and strengthening of expression of PTN mRNA occurring simultaneously suggests some identity of PTN with axonal elongation and synapse formation. This inference is supported by the fact that PTN accelerates neurite elongation in cultured neurons.
- In the mature nervous system, which differs from the early stages after birth, PTN is mainly produced in specific neurons and is considered to act on more limited cell groups.

It becomes evident that the identification of receptors of PTN is indispensable to elucidate the physiological actions of PTN.

```
                          20           ↓   40
Bovine MQTPQYLQQRRKFAAAFLAFIFILAAVDTAEAGKKEKPEKKVKKSDCGEWQWSVCV
Human  --AQ--Q-------------------------------------------------
Rat    -SSQ--Q------------L------------------------------------
                                                      *       *

          60                  80                  100
Bovine PTSGDCGLGTREGTRTGAECKQTMKTQRCKIPCNWKKQFGAECKYQFQAWGECDLN
Human  --------------------------------------------------------
Rat    --------------------------------------------------------
            *             *        *   *         *         *

             120                 140                 160
Bovine TALKTRTGSLKRALHNADCQKTVTISKPCGKLTKSKPQAESKKKKEGKKQEKMLD
Human  -----------------E----------------P--------------------
Rat    ----------------------------------P--------------------
                         *         *
```

Fig. 13.1 The primary structures of bovine, human, and rat pleiotrophin. Signal peptide sequence spans from N-terminus to the position indicated by arrow. For the human and rat pleiotrophin, only the amino acid residues different from bovine pleiotrophin are indicated (adapted from ref. 3).

Subtypes

With reference to the amino acid sequence of PTN, a factor exhibiting 50 per cent homology has been identified. This purported intracellular receptor for retinoic acid, which induces differentiation of embryonal carcinoma cells, was observed to be highly expressed in mid gestation mouse embryo and adult kidney and was named MK[1,2]. MK also possesses high affinity to heparin and has similar physicochemical properties to PTN. However, the expression of MK mRNA at tissue level is different from PTN in that it is relatively milder in the nervous system. Since the expression of MK mRNA is strong during the early to middle embryonic stages, and the expression later becomes weakened, one can infer that its role in development and maturation process is different from that of PTN[1]. At present, it is also unclear whether MK has an axonal elongation effect (like that of PTN) on neurons and a mitogenic effect on fibroblasts or not. Further progress is needed in this area. Since the homology of amino acid sequence and the position of cysteine residues in the sequence (which contributes to the disulfide S–S bond) is similar for both PTN and MK, it is highly possible that these two form a family of growth factors (Fig. 13.3). Alternatively one can also consider the possibility of PTN constituting a gene family with some members, in addition to MK, analogous to FGF and its members.

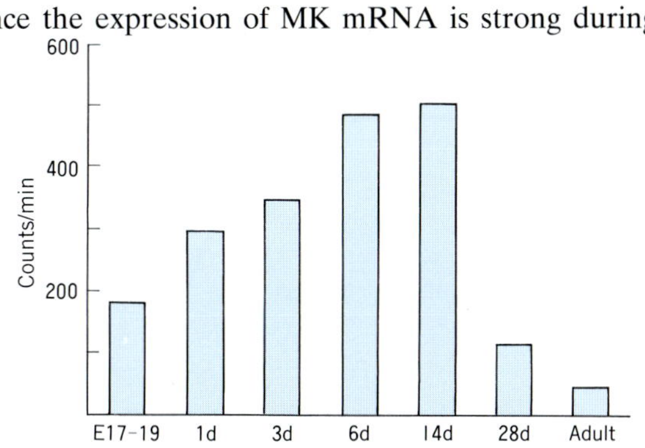

Fig. 13.2 Developmental expression of pleiotrophin mRNA in the rat central nervous system. This graph shows the result of the semiquantitative analyses of northern blotting (adapted from ref. 5).

References

1. Kadomatsu, K., Huang, R.P., Suganuma, T. *et al.* (1990). *J. Cell Biol.*, **110**, 607–16.
2. Kadomatsu, K., Huang, R.P., Suganuma, T. *et al.* (1988). *Biochem. Biophys. Res. Commun.*, **151**, 1312–18.
3. Li, Y.-S., Milner, P.G., Chauhan, K. *et al.* (1990). *Science*, **250**, 1690–4.
4. Milner, P.G., Li, Y.-S., Hoffman, R.M. *et al.* (1989). *Biochem. Biophys. Res. Commun.*, **169**, 1096–103.
5. Merenmies, J. and Rauvala, H. (1990). *J. Biol. Chem.*, **265**, 16721–4.
6. Rauvala, H. (1989). *EMBO J.*, **8**, 2933–41.
7. Wanaka, A., Caroll, S.L., and Milbrandt, J. (1993). *Dev. Brain. Res.*, **72**, 133–44.

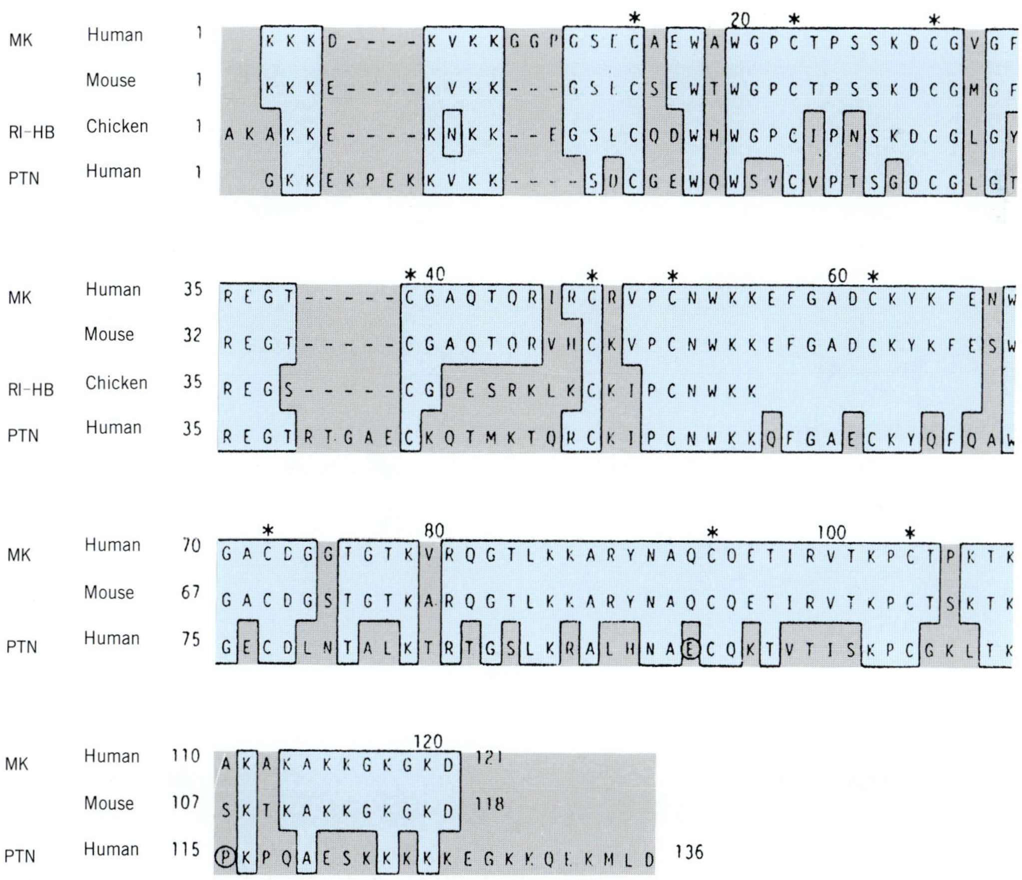

Fig. 13.3 Comparison and alignment of human and mouse MKs and human PTN. * indicates the cysteine residues conserved among these peptides. (From Tsutsui, J. *et al.* (1991). *Biochem. Biophys Res. Commun.*, **176**, 792–7.)

Pleiotrophin (ISH)

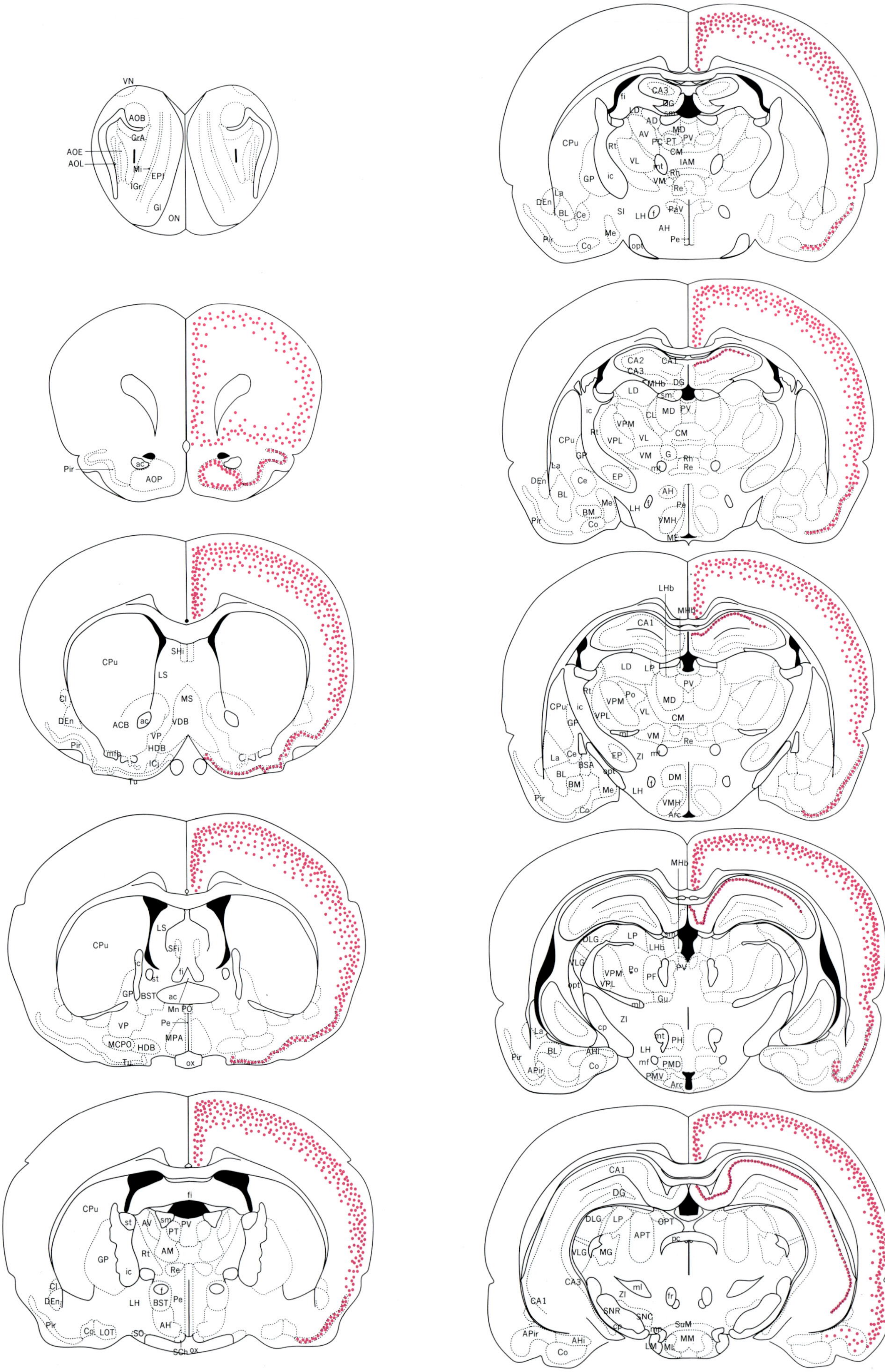

Pleiotrophin (ISH)

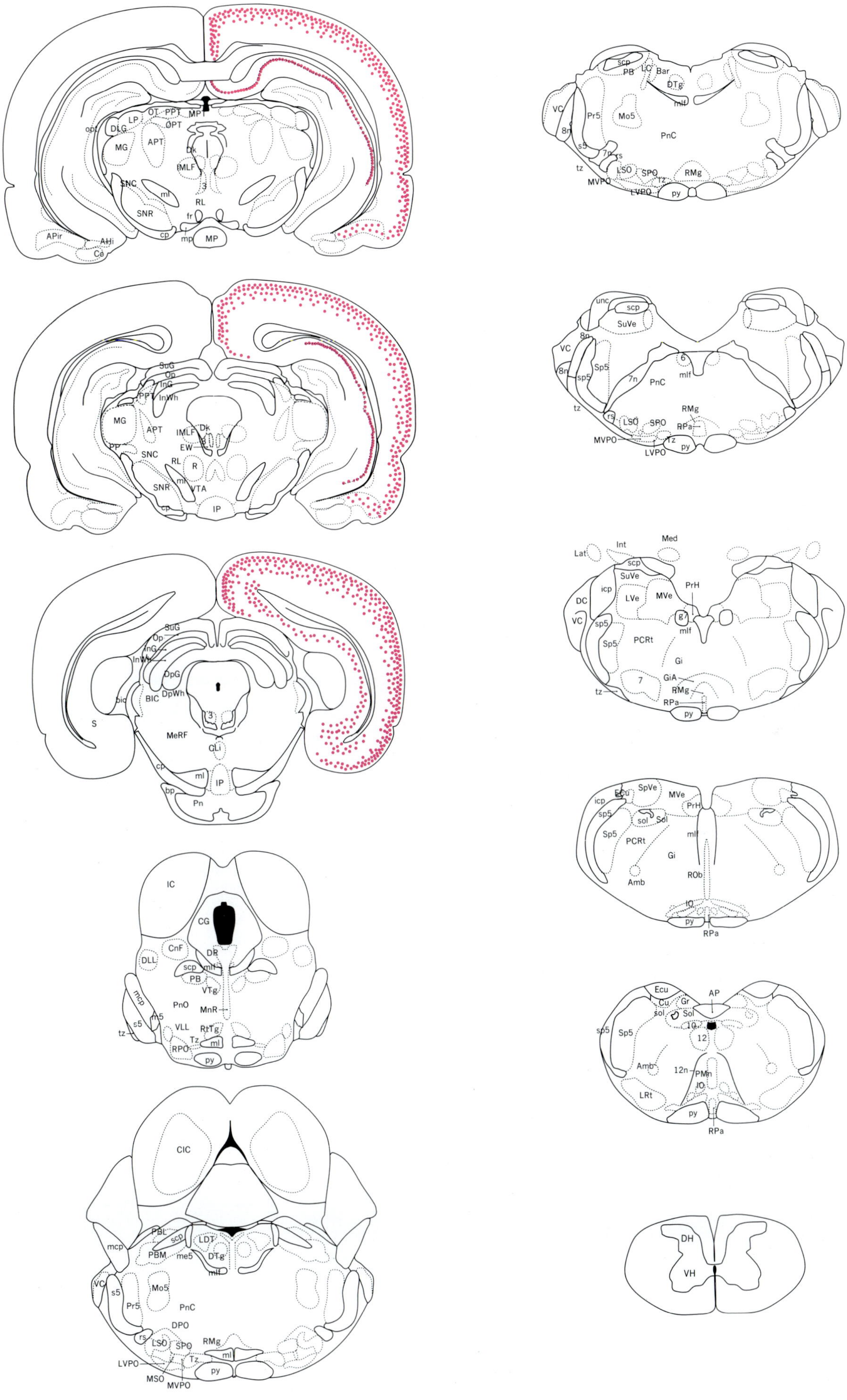

Peptides

14 The tachykinin system

Tachykinin

Classification

Tachykinins (Fig. 14.1) are polypeptides which are widely distributed in vertebrates and invertebrates. They are characterized by the common amino acid sequence Phe-X-Gly-Leu-Met-NH_2 at the C-terminal end. They possess multiple physiological actions, including vasodilation, contraction of smooth muscles of intestine and urinary bladder, saliva secretion, and induction of low blood pressure in mammals. It has also been found that each specific physiological action depends on the amino acid sequence at the C-terminal end.

The first tachykinin, substance P (SP), was discovered by von Euler and Gaddum in 1931 as an active component in the brain and small intestine of horses. Since then, polypeptides which have similar physiological actions to SP have been isolated from amphibia and molluscs. After the determination of the amino acid sequence of SP by Leeman *et al.* in 1971, the polypeptide group of compounds that possess similar physiological actions to that of SP were given the generic name of tachykinins. Furthermore, in 1983, Kanazawa *et al.*[4] and Matsuo *et al.*[2] reported two types of tachykinins from pig spinal cord concurrently. They designated different names for the same compounds; neurokinin α and neuokinin β were used by the former group, while neuromedin L and neuromedin K were chosen by the latter group. Subsequently, at the Satellite Symposium on Substance P and Neurokinins, held at the 30th International Congress of the International Union of Physiological Sciences (IUPS, Montreal, 1986), the names neurokinin A and neurokinin B were proposed and adopted (Table 14.1).

At present, at least three types of tachykinins—SP, neurokinin A (NKA) and neurokinin B (NKB)—are known to exist in the nervous system and peripheral tissues of mammals. Another school of thought believes in the existence of five types, which include neuropeptide K and neuropeptide γ. However, whether these two peptides actually exist is a matter of dispute. The function of mammalian tachykinins at the gene level has also been investigated in detail[3,6,9], and is presented in Fig. 14.2. Preprotachykinin A (PPT-A) gene encodes both SP (exon 3) and NKA (exon 6) as shown in Fig. 14.2. Three types of mRNA, namely the αPPT-A mRNA containing only SP, βPPT-A and γPPT-A mRNA containing both SP and NKA are expressed by alternative splicing of RNA precursor following transcription. The difference between βPPT-A and γPPT-A depends on the selection containing exon 4 coding spacer peptide or not. The splicing of PPT-A precursor mRNA is tissue specific, with γPPT-A mRNA constituting more than 80 per cent, βPPT-A mRNA forming almost 20 per cent and αPPT-A mRNA a smaller percentage in the central nervous system. In the periphery of the enteric tube, thyroid gland, and other tissues, βPPT-A mRNA is produced in larger amounts. On the other hand, NKB differs from SP and NKA in being coded at the third exon upstream from the 3′ terminal on the PPT-B gene. Though the expression site and functions of the PPT-B gene are different from those of the PPT-A gene, the gene structures of both are very similar and both could have evolved from a common ancestor. Furthermore, one part of the 5′ terminal exon of the PPT-B gene and the 3′ exon of myosin-single-stranded-like protein gene are overlapping in the same direction; this example of overlapping genes in the mammalian system is highly interesting.

Distribution

Since the early 1970s, from the investigations of Otsuka *et al.* on the bovine dorsal roots of spinal nerves, it has been proposed that tachykinin SP functions as a neurotransmitter or as a neuromodulator of primary sensory neurons in pain transmission[5,8].

Investigations on the distribution of SP in brain have been carried out in detail by means of immunohistochemistry or radioimmunoassay. It has been clarified that SP-positive cells and nerve endings are distributed widely in the brain. The additional existence of many projections has also been presented[1,7].

However, since the structures of tachykinins are remarkably similar, it is difficult to infer that the anti-SP antibody specifically recognizes only SP. It is possible that NKA or NKB would also be recognized concurrently. Also, by *in situ* hybridization using a specific synthetic oligonucleotide probe to SP, three types of mRNA produced by the alternative splicing of RNA precursor following transcription from PPT-A gene could not be clearly recognized. This suggests the necessity of investigating the existence of three possibilities in the neutral circuit of a cell: only SP, only NKA, or coexistence of SP and NKA.

Since antibodies that specifically recognize SP, NKA, or NKB for use in immunohistochemistry were unavailable, this atlas presents the results of localization of SP-positive cells and fibres in the brain, as clarified by immunohistochemistry using the anti-SP antibody[1,7,10]. The results of the localization of PPT-A and PPT-B mRNA were obtained by *in situ* hybridization[11].

Positive cells are spread over wide areas of the brain such as the septal area, corpus striatum, hypothalamus, amygdala, interpeduncular nucleus, colliculus superior, colliculus inferior, central grey matter, dorsal and lateral tegmental area, raphe nuclei group, and especially the ventral medial hypothalamic nucleus, which showed SP positively with very high density. In addition, positive fibres were also observed in widely spread areas of brain such as the corpus striatum, hypothalamus, lower brain stem, and others. SP-positive fibres with very high density were also seen in the substantia nigra and the dorsal horn.

The distribution of SP immunopositive cells corresponds to the distribution of SP and NKB mRNA revealed by *in situ* hybridization. A considerable expression of mRNA was recognized in the olfactory bulb,

Substance P	:	H-Arg-Pro-Lys-Pro-Gln-Gln-PHE-Phe-GLY-LEU-MET-NH_2
Substance K/neurokinin A	:	H-His-Lys-Thr-Asp-Ser-PHE-Val-GLY-LEU-MET-NH_2
Neuromedin K/neurokinin B	:	H-Asp-Met-His-Asp-Phe-PHE-Val-GLY-LEU-MET-NH_2

Fig. 14.1 Amino acid sequence of mammalian tachykinins.

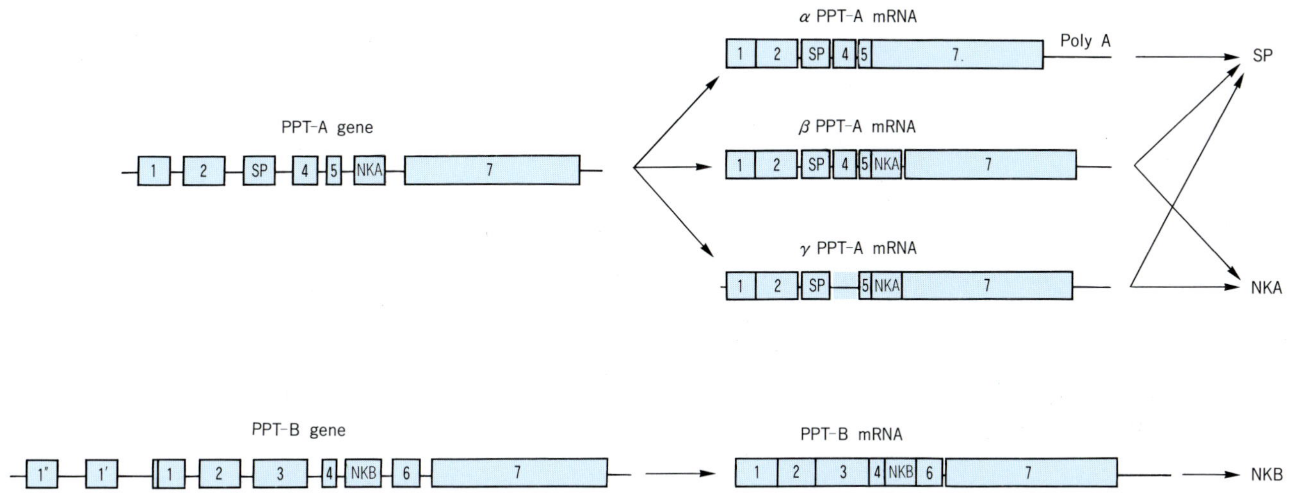

Fig. 14.2 Processing of mammalian tachykinin family genes, transcripts, and peptides.

Table 14.1. Mammalian tachykinin family and its receptor family

Substance P (SP)	NK1 (substance P receptor)
Neurokinin A (NKA) =neurokinin α =substance K =neuromedin L =substance L	NK2 (substance K receptor)
Neurokinin B (NKB) =neuromedin K =neurokinin β	NK3 (neuromedin K receptor)

thalamus, and cerebral cortex, where in the past it was believed that immunopositive cells do not exist or are very rare[11]. In this atlas, the expression of SP mRNA and NKB mRNA is presented as tachykinin mRNA in the same brain areas. However, in some instances the expression of only SP or only NKB is revealed in specific areas. Examples of SP expression only include the zona incerta, dorsal areas of nucleus paraventricularis, dorsomedial hypothalamic nucleus, premammillary nucleus, superior mammillary nucleus, ventral tegmental area, deep grey layer of superior colliculus, inferior colliculus, locus coeruleus, raphe nuclei group excluding raphae obscurus, reticular formation of medulla oblongata, and solitary tract nucleus. Examples of NKB expression only are very rare. They include the pirifom cortex, rostral area of hippocampus, intercalated region of amygdala, bed nucleus of the accessory tract, and lateral mammillary nucleus.

References

1. Buck, S.H., Helke, C.J., Burcher, E. *et al.* (1986). *Peptides*, **7**, 1109–20.
2. Kangawa, K., Minamino, N., Fukuda, A., and Matsuo, H. (1983). *Biochem. Biophys. Res. Commun.*, **29**, 533–40.
3. Kawaguchi, M., Hoshimaru, M., Nawa, H., and Nakanishi, S. (1986). *Biochem. Biophys. Res. Commun.*, **139**, 1040–6.
4. Kimura, S., Goto, K., Ogawa, T. *et al.* (1984). *Neurosci. Res.*, **2**, 97–104.
5. Konishi, S. and Otsuka, M. (1974). *Nature*, **252**, 734–5.
6. Kotani, H., Hoshimaru, M., Nawa, H., and Nakanishi, S. (1986). *Proc. Natl Acad. Sci. USA*, **86**, 7074–8.
7. Ljungdahl, A., Hokfelt, T., and Nilsson, G. (1978). *Neuroscience*, **3**, 861–943.
8. Maggio, J.E. (1988). *Annu. Rev. Neurosci*, **11**, 13–28.
9. Nawa, H., Hirose, T., Takashima, H. *et al.* (1983). *Nature*, **306**, 32–6.
10. Nakanishi, S. (1991). *Annu. Rev. Neurosci.*, **14**, 123–36
11. Warden, M.K. and Young, W.S. III (1988). *J. Comp. Neurol.*, **272**, 90–113.

Tachykinin (immunocytochemistry)

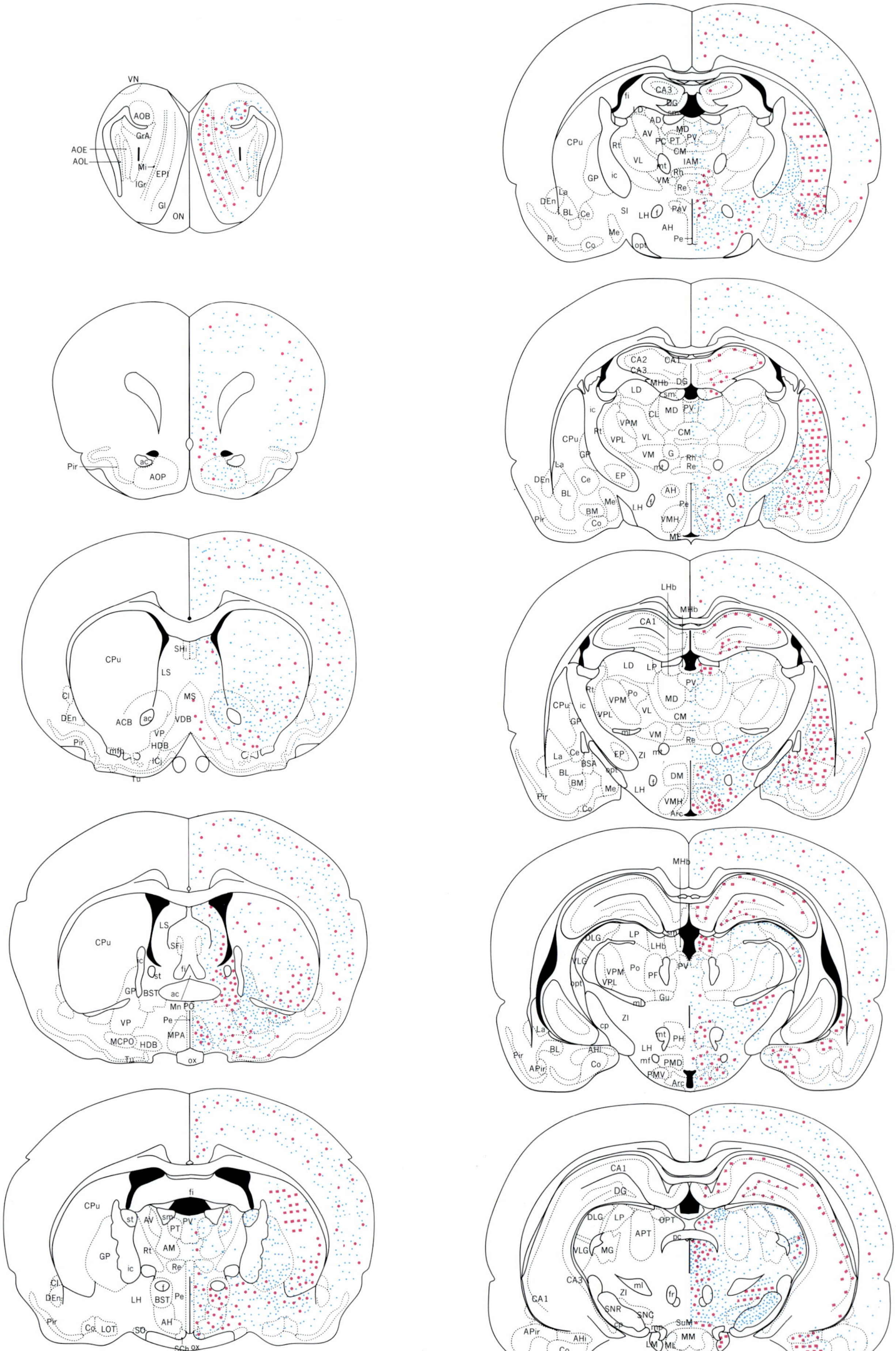

Tachykinin (immunocytochemistry)

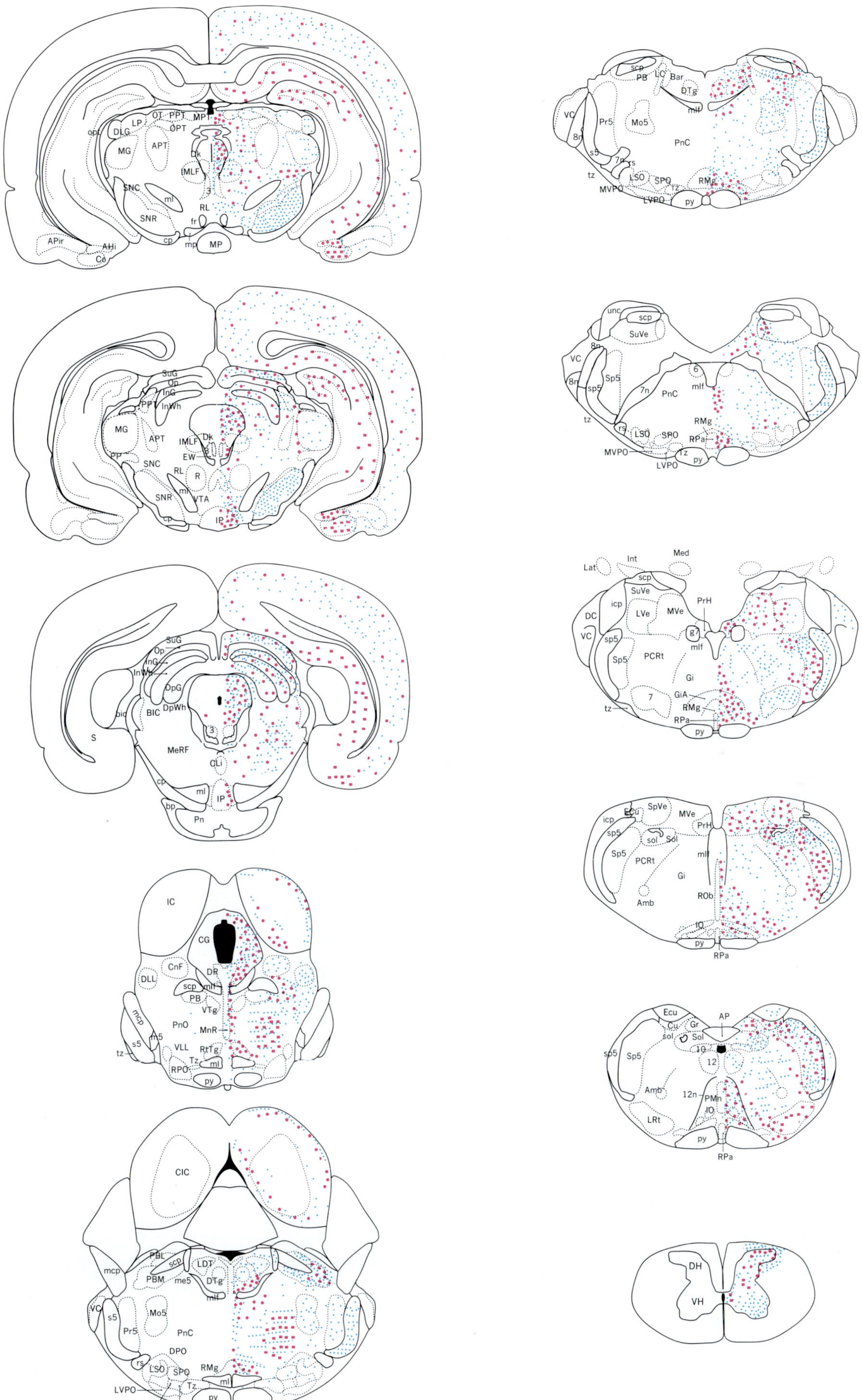

Tachykinin receptor

Subtypes

When considering the transduction cascade of information, its physiological response can be considered to depend on the types of receptors expressing on cells. At present, the tachykinin receptor is known to exist in the nervous system as well as many peripheral tissues, and is known to evoke various physiological actions[5].

The tachykinin receptors were pharmacologically considered as multiple receptor subtypes of SP. This belief was based on potency differences among the tachykinins in bioassays using the constriction of ileum segment. However, recent pharmacological evidence suggests the existence of three types of receptors, SP-P, SP-E, and SP-K made up of SP, kassinin and physalaemin of amphibian tachykinin, and eledoisin. Following the discovery of new mammalian tachykinin in 1984, it also became evident that the endogenous ligand of tachykinin receptors SP-P, SP-E, and SP-K corresponds to SP, NKA, and NKB respectively.

Three distinct types of mammalian tachykinin receptors were cloned. These contain seven transmembrane domains like β-adrenalin receptor, serotonin 1C, and β receptors. These are suggested to belong to the G-protein coupled type of receptors (Fig. 14.3) whose signals were transferred via IP_3/DG[7–9]. The relationships of these receptors and their endogenous ligands have been established from the pharmacological evidence as follows: substance P receptor (SP-R, NK1) to SP, neurokinin A receptor (NK2, SK-R) to NKA, neurokinin B receptor (NK3, NK4) to NKB (Table 14.1).

Distribution

The localization of tachykinin receptors has been demonstrated by receptor-binding autoradiography using an isotope-labelled ligand[1–3,6]. In recent years, knowledge of the antagonists of tachykinin has advanced and their pharmacological analysis is anticipated. Concurrently, it is also possible to analyse the localization of receptor mRNA-producing cells with very high sensitivity by *in situ* hybridization. The data on the localization of tachykinin receptors shown in this atlas has been derived from receptor binding-autoradiography and *in situ* hybridization[4,5].

The localization of binding sites for the tachykinin receptor is distributed widely in brain, including cerebral cortex, and the binding sites of eledoisin, kassinin, and NKB, except SP, are common in almost all areas of the brain.

From these results, some interesting patterns have been elicited. Substantia nigra is used here as an example. From the results of immunohistochemistry using anti-SP antibody and radioimmunoassay, it has been inferred that very high density of immunopositive SP fibres and nerve endings exist. However, the binding sites of SP and SP receptor mRNA producing cells are not identified yet. But binding sites for tachykinins except SP are barely observed only in substantia nigra pars compacta. This so-called mismatch occurs between the ligand and receptors.

Differences in the distribution of binding sites of tachykinins are provided as:

- areas of binding sites for SP only: examples included are corpus striatum, nucleus accumbens, arcuate nucleus, thalamus, dorsal parabrachial nucleus, trigeminal motor

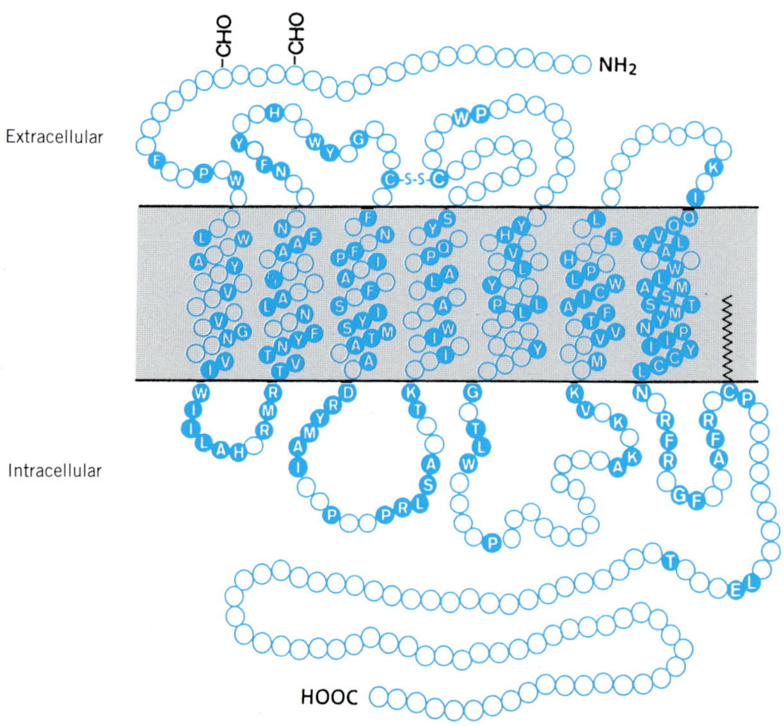

Fig. 14.3 Substance P receptor. Conserved amino acids of the three tachykinin receptors on a transmembrane model of the substance P receptor. Black circles represent amino acids identical among the three tachykinin receptors (from ref. 10).

nucleus, pontine nucleus, ventral cochlear nucleus, reticular formation of medulla oblongata, and others; and
- areas of binding sites of tachykinins other than SP: examples included are suprachiasmatic nucleus, dorsal and lateral tegmental area, substantia nigra pars compacta, and others. In the brain stem, tachykinin excluding SP is observed only in the sensory area, as opposed to the binding site of SP, which is observed in motor and sensory areas. Among the tachykinins excluding SP, there are also slight differences in the location of binding sites in the solitary tract nucleus; eledoisin and kassinin show binding, but NKB does not.

The results obtained by autoradiography of SP and by *in situ* hybridization of SP receptor mRNA show differences in the localization of receptors in some areas of the brain. This may be due to the fact that in *in situ* hybridization, the distribution of synthesizing cell bodies is detected and the binding sites show the location of the receptor itself. In neurons with long processes, the location of receptor protein is far from its synthesizing cell bodies. In some regions of the brain, the localization of SP immunopositive cells and SP receptor mRNA positive cells looks similar, which suggests the possible co-localization of SP and SP receptors in these regions.

Receptor-binding autoradiography also has some demerits in that non-specific binding can be picked up. Also it is difficult in practice to identify the cells which synthesize the receptors. In *in situ* hybridization, it is also not easy to identify the intracellular localization of the receptor. Thus, it is more appropriate to use a combination of methods which utilize receptor specific antibody.

Regarding the SP operating system in the corpus striatum–substantia nigra tract:

- SP-containing cells in striatum project their axons to substantia nigra as well as to the large-sized neurons in striatum which does not express SP;
- the release of acetylcholine was observed when striatum was stimulated by SP[9];
- Gerfen[4] demonstrated the coexistence of SP-R and acetylcholine in the large cells of the striatum by *in situ* hybridization.

From these facts, it can be inferred that SP participates in the release of acetylcholine, as acetylcholine-containing cells express SP-R, and this SP is supplied from the SP

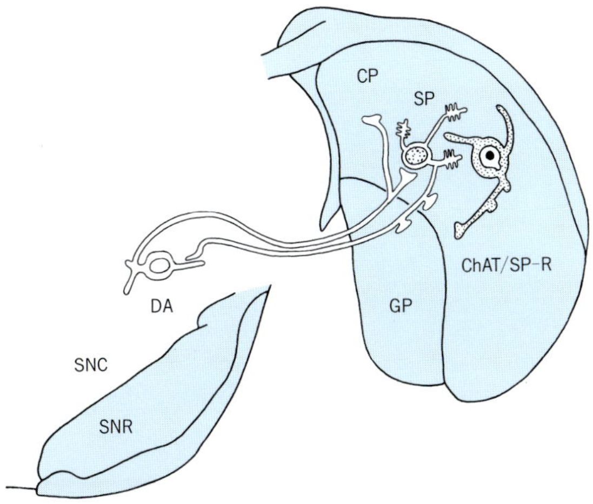

Fig. 14.4 Strionigral SP projection and the lack of SP-R in the substantia nigra. CP, caudate putamen; GP, globus pallidus, SNR, substantia nigra pars reticulata; SNC, substantia nigra pars compacta; DA, dopamine; ChAT, choline acetyl transferase; SP-R, substance P receptor (NK1).

cells present in the corpus striatum. A striking point is that neurons in substantia nigra do not have SP receptor although the abundant SP projection from striatum was observed. Figure 14.4 shows a typical design of the SP operating system in substantia nigra–corpus striatum.

References

1 Buck, S.H., Helke, C.J., Burcher, E. *et al.* (1986). *Peptides*, **7**, 1109–20.
2. Dam, T.V., Escher, E., and Quirion, R. (1988). *Brain Res.*, **483**, 372–6.
3. Dam, T.V., Martinelli, B., and Quirion, R. (1990). *Brain Res.*, **531**, 333–7.
4. Gerfen, C.R. (1991). *Brain Res.*, **556**, 165–70.
5. Maeno, H., Kiyama, H., and Tohyama, M. (1993). *Mol. Brain Res.*, **18**, 43–58.
6. Maggio, J.E. (1988). *Annu. Rev. Neurosci.*, **11**, 13–28.
7. Mantyh, P.W., Gates, T., Mantyh, C.R., and Maggio, J.E. (1989). *Neurosci.*, **9**, 258–79.
8. Masu, Y., Nakayama, K., Tamaki, H. *et al.* (1987). *Nature*, **329**, 836–8.
9. Petitet, F., Glowinsky, J., and Beaujouan, J.C. (1991). *Eur. J. Pharmacol.*, **192**, 203–4.
10. Nakanishi, S. (1991). *Ann. Rev. Neurosci.*, **14**, 123–36.

Tachykinin receptor (binding sites)

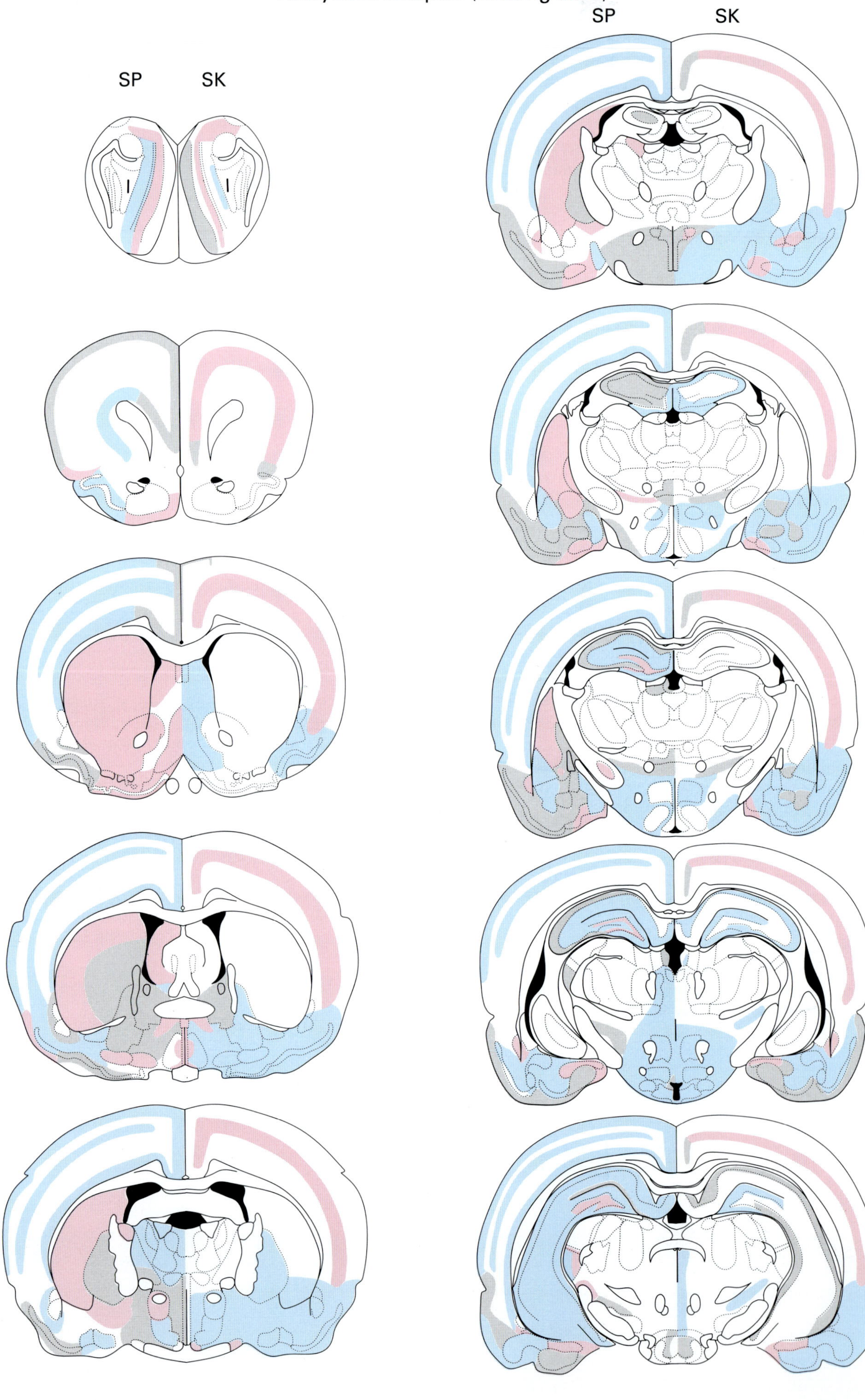

Tachykinin receptor (binding sites)

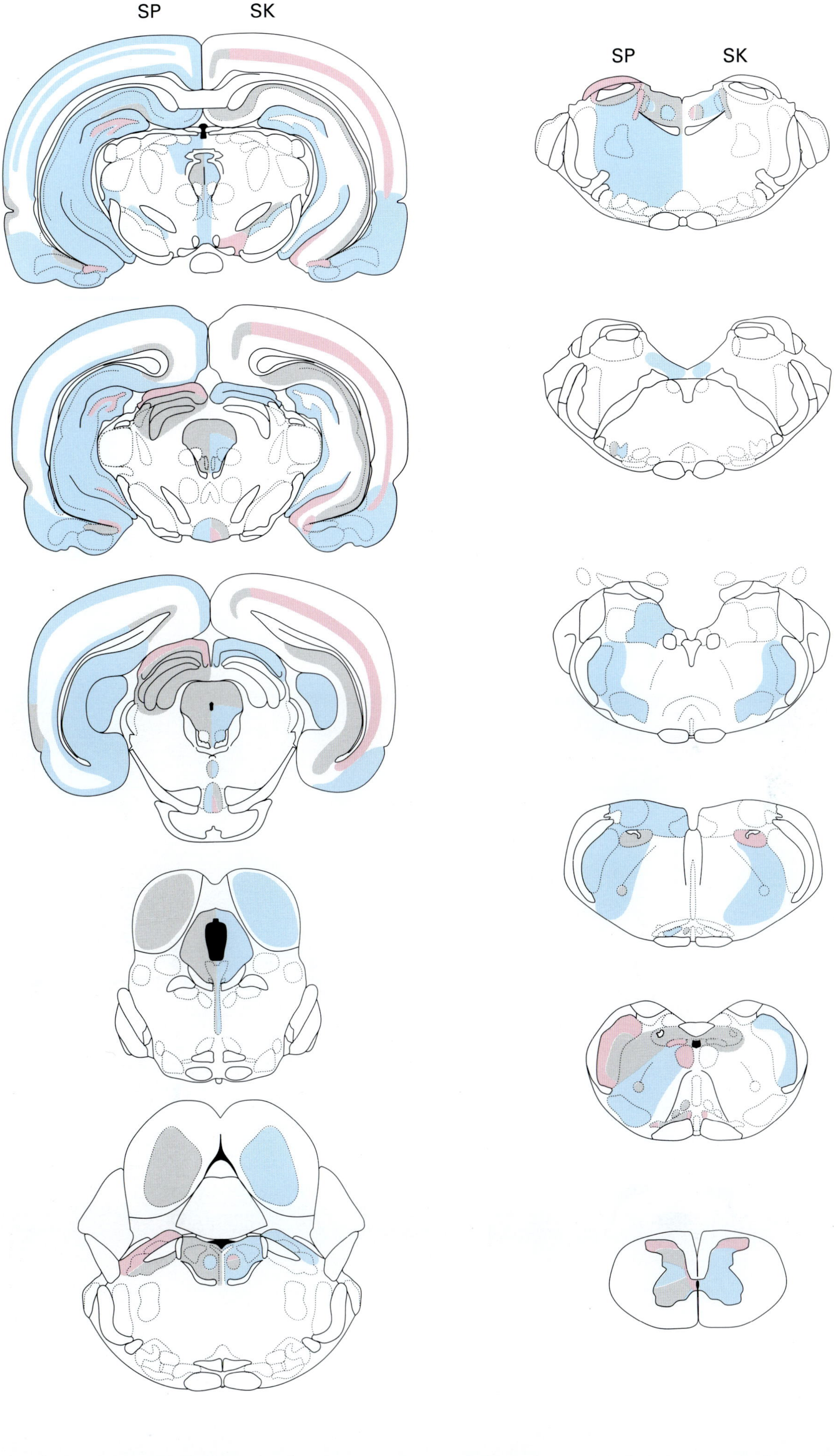

Tachykinin receptor (binding sites)

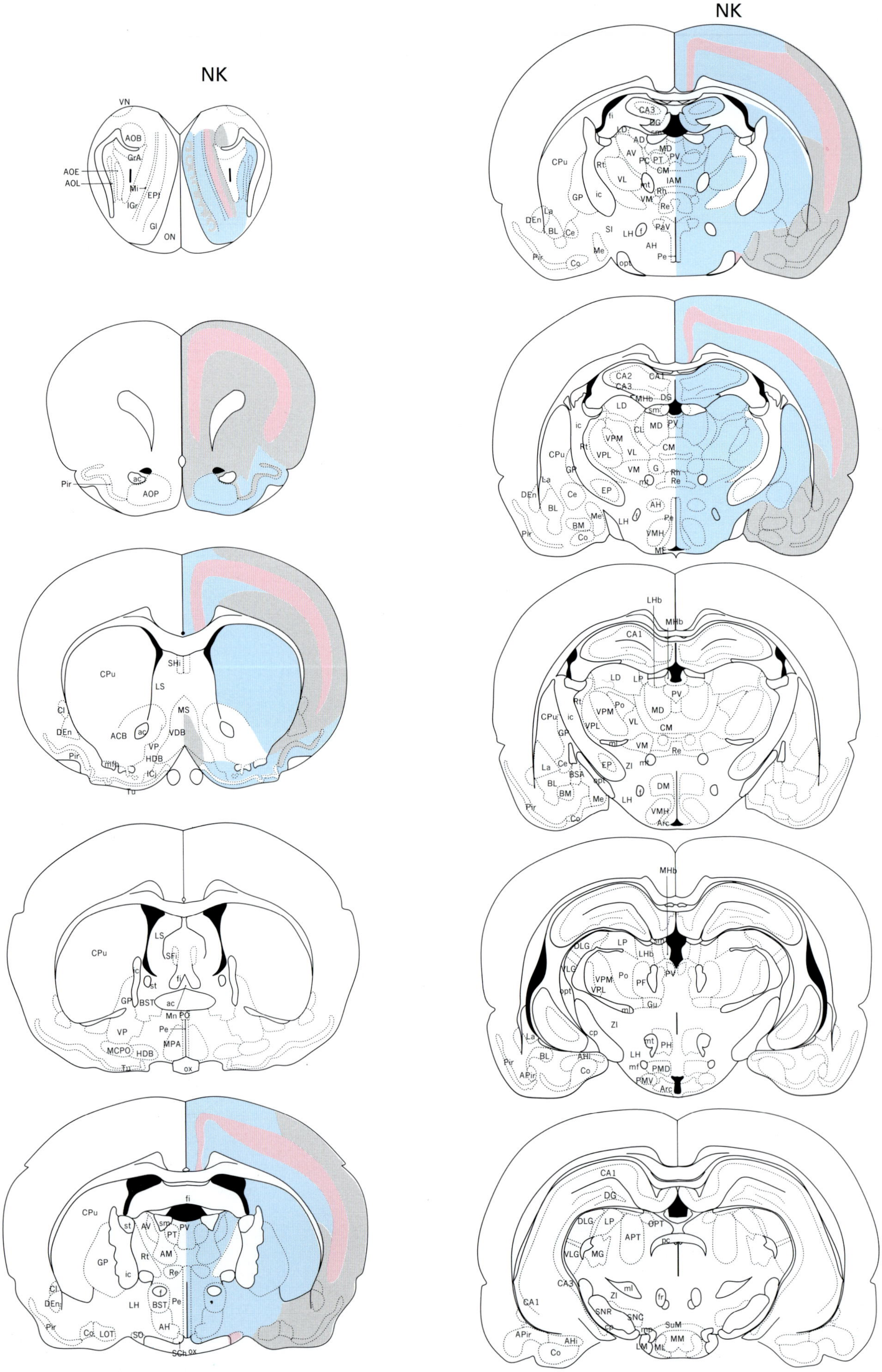

Tachykinin receptor (binding sites)
NK

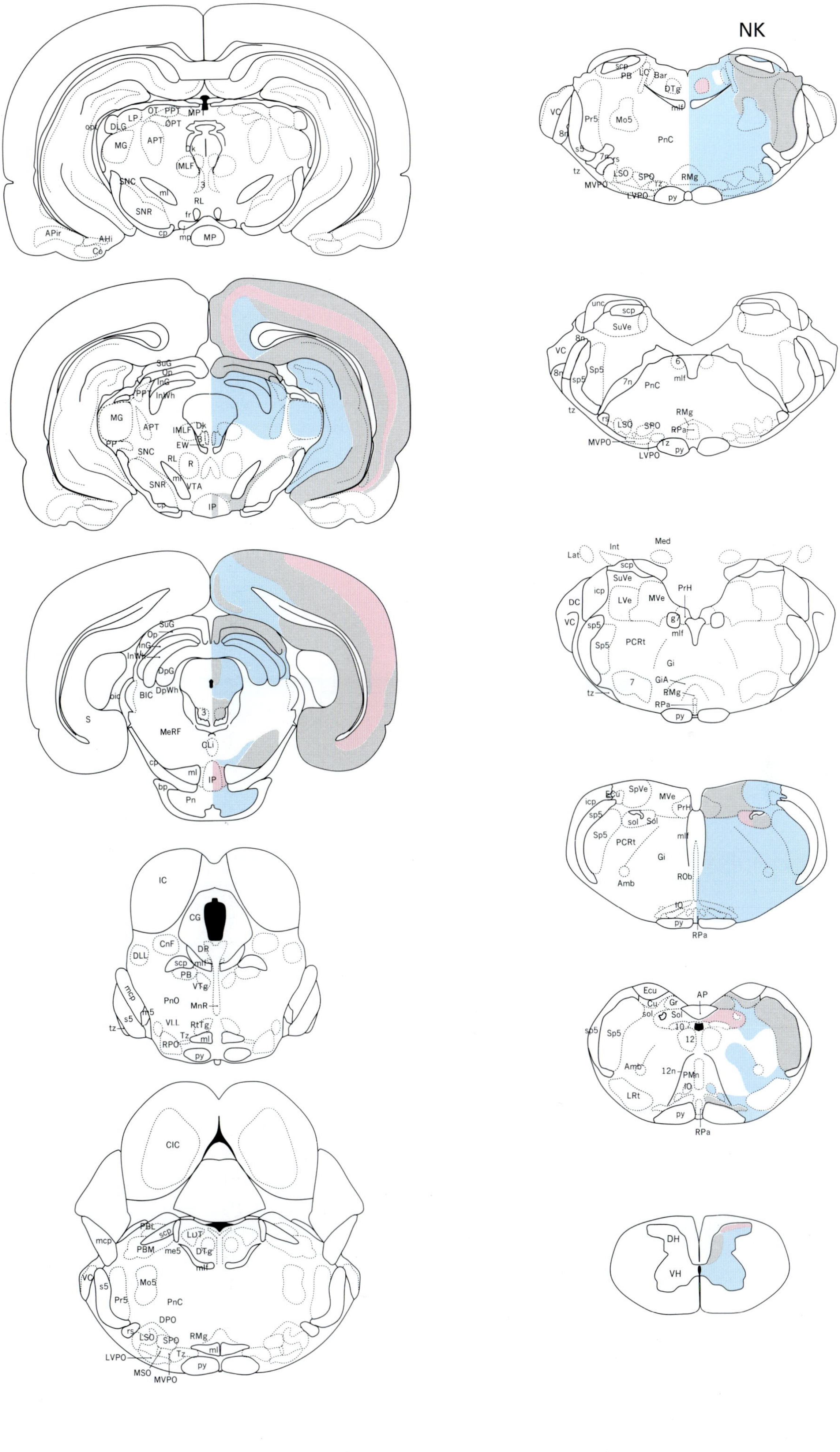

15 The calcitonin gene related peptide system

Calcitonin gene related peptide

Origin

Calcitonin gene related peptide (CGRP) consists of 37 amino acid residues, produced through an alternate splicing of the calcitonin gene[5]. Whereas CGRP is predominantly expressed in the nervous system, calcitonin is mainly present in thyroid gland C cells, which also produce CGRP (Fig. 15.1). Recently a new CGRP gene has been characterized which produces βCGRP; the previously identified form of CGRP is called αCGRP. The amino acid compositions of α and βCGRP are largely analogous. In the rat and human types, the α and β forms of CGRP differ in one and three amino acid residues respectively (Fig. 15.2). However, the densities of binding sites in the brain vary between αCGRP and βCGRP. At sensory neurons, αCGRP binds six times more than βCGRP. In the small intestine, the binding is exclusively of βCGRP. However, in the lower brain stem, the distributions of α- and βCGRP-containing neurons are very similar and it is inferred that the identical neuron produces both forms in this region.

Distribution and projection tract

The distribution of CGRP in the central and peripheral nervous systems has been investigated in detail. The regional distributions of CGRP binding sites in the central nervous system of rats and humans has also been clarified, as shown in Fig. 15.3[1–3].

Physiological function

Since CGRP exists widely and abundantly in the central and peripheral nervous systems, it is believed to be involved in various functions. Some identified physiological actions of CGRP include vasodilation, inhibition of athrocytosis, and inhibition of acidic gastric juice secretion.

CGRP coexists with acetylcholine (ACh) in the motor neuron. Thus CGRP enhances muscle contraction by ACh via cAMP. Also the CGRP increases the striated muscle origin nicotine (n) ACh receptor[4]. It is therefore inferred that the actions of CGRP on the striated muscle are of two kinds. First, CGRP binds to the receptor and thus directly brings about a change in the intracellular

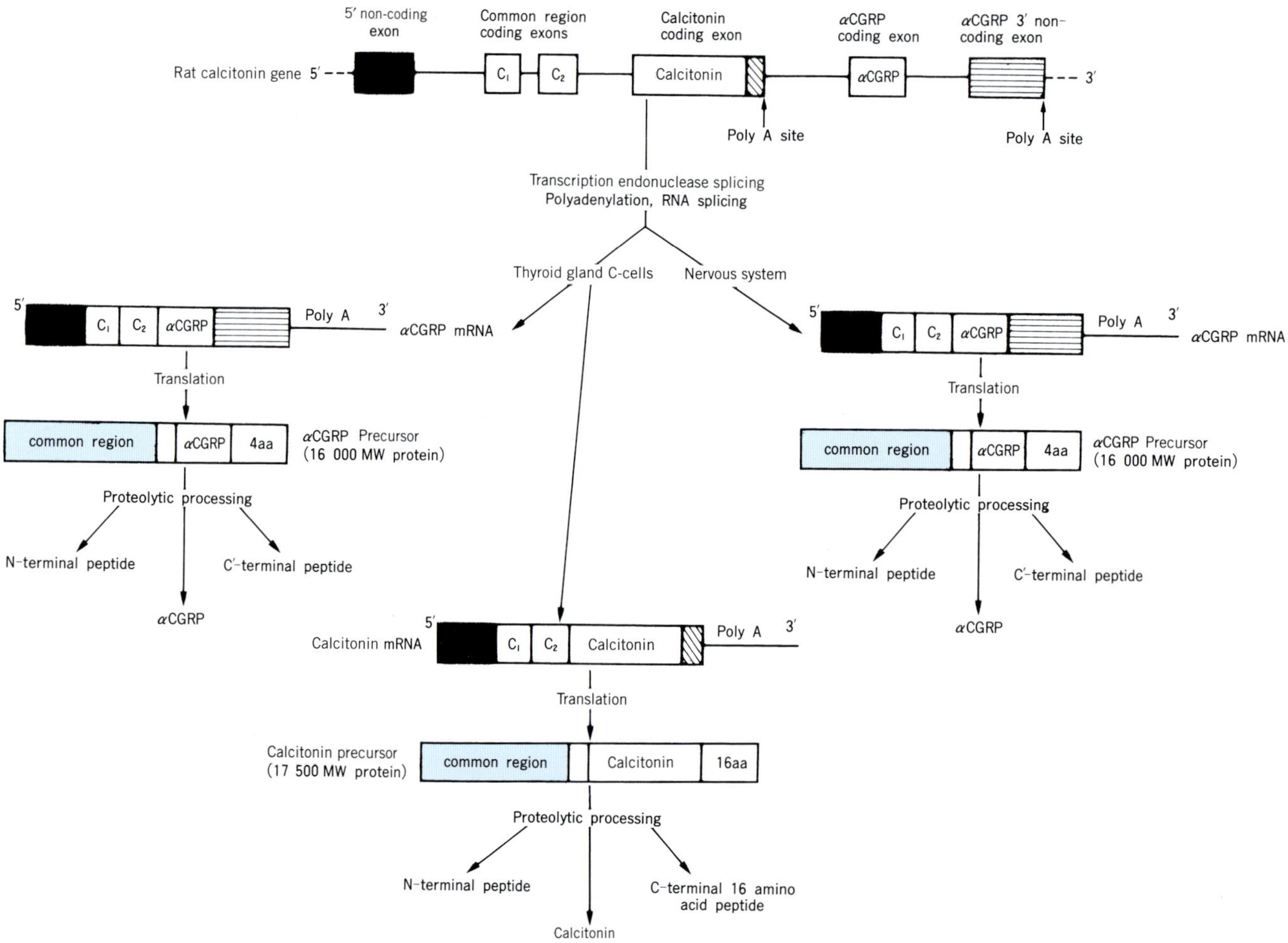

Fig. 15.1 Production of CGRP and calcitonin from the calcitonin gene (from ref. 2).

Fig. 15.2 Amino acid sequences of rat and human α- and β-CGRP. Underlines show differences in the otherwise homologous sequences (from ref. 2).

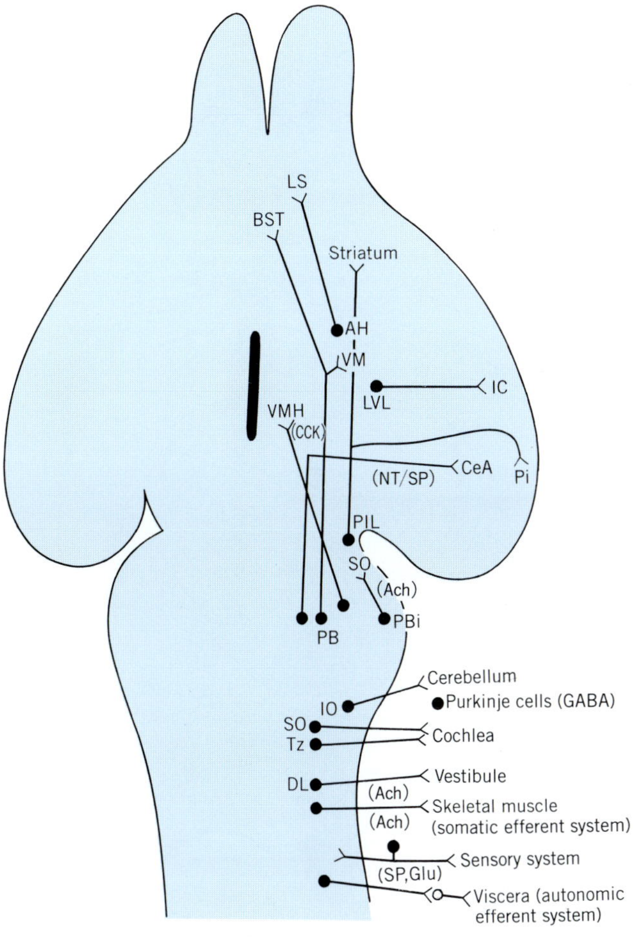

Fig. 15.3 CGRP-containing neuronal pathways in the brain. ACh, acetylcholine; AH, anterior hypothalamic area; BST, bed nucleus of the stria terminalis; CeA, central amygdaloid nucleus; DL, dorsolateral part of the medial longitudinal fasciculus; Glu, glutamate; IC, insular cortex; IO, inferior olive; LVL, lateral part of the ventrolateral thalamic nucleus; LS, lateral septal nucleus, NT, neurotensin; PB, parabrachial nucleus; PBi, parabigeminal nucleus; SC, superior colliculus; SO, superior olive; SP, substance P; Tz, nucleus of the trapezoid body; Pi, piriform cortex; PIL, posterior intralaminar thalamic nucleus; VM, ventromedial thalamic nucleus; VMH, ventromedial hypothalamic nucleus.

metabolism (an increase in cAMP). Secondly, CGRP induces an increase in nACh receptor and thereby causes an increase in ACh activity. Furthermore, CGRP inhibits regeneration of the axon of motor neuron and controls the expression of nACH receptor as mentioned previously. Thus, it can be considered that CGRP also functions as a trophic factor. This trophic function of CGRP in the brain has been proposed for a long time. The multiple functions of CGRP can be demonstrated by excision experiments in the nerves. For instance, axotomy of the motor neuron causes a remarkable increase in αCGRP mRNA of the motor neuron, without any alteration in βCGRP mRNA. This demonstrates that α- and βCGRP exhibit different actions in the motor neuron.

CGRP is also present in the sensory neuron. It is localized in small cells and medium to large cells of the dorsal root ganglion. The small cells containing CGRP also harbour substance P (SP). Processes of medium-sized to large CGRP cells are non-myelinated or covered with a thin myelin sheath. Here, the dense CGRP fibrous web is observed in the gelatinous layer of the dorsal column of spinal cord. Thus, the CGRP which is produced in the dorsal root ganglion neuron is considered to be involved in the transmission of nociceptive stimulation to the central area at the posterior horn level. However, the CGRP or SP that is produced in the sensory ganglion is transported not only to the central area, but also to the peripheral regions. There, CGRP and SP are disengaged from the peripheral branch terminal by nociceptive stimulation, and are considered to be concerned with flare response or the weal response of inflammation. The CGRP and SP in smooth muscle (tunica muscularis) of the digestive tract are also disengaged by various stimuli present in the digestive tract and are considered to be involved in the regulation of enterokinesis.

CGRP is distributed widely in the central nervous system and it can thus be called a neuromodulator (Fig. 15.3). Since CGRP is also expressed in immature brain during the presynaptic phase, and is expressed densely in the Purkinje layer of cerebellum, it can be considered to act as a trophic factor as well.

References

1. Inagaki, S., Matsuda, Y., Nakai, Y., and Takagi, H. (1990). *Brain Res.*, **537**, 263–70.
2. Ishida-Yamamoto, A. and Tohyama, M. (1989). *Progr. Neurobiol.*, **33**, 335–86.
3. Kubota, Y., Inagaki, S., Shimada, S. *et al.* (1991). *Neuroscience*, **40**, 423–8.
4. New, H.V. and Mudge, A.W. (1986). *Nature*, **328**, 809–11.
5. Rosenfeld, M. G., Mermod, J. J., Amara, S. G. *et al.* (1983). *Nature*, **304**, 129–35.

Calcitonin gene related peptide (immunocytochemistry)

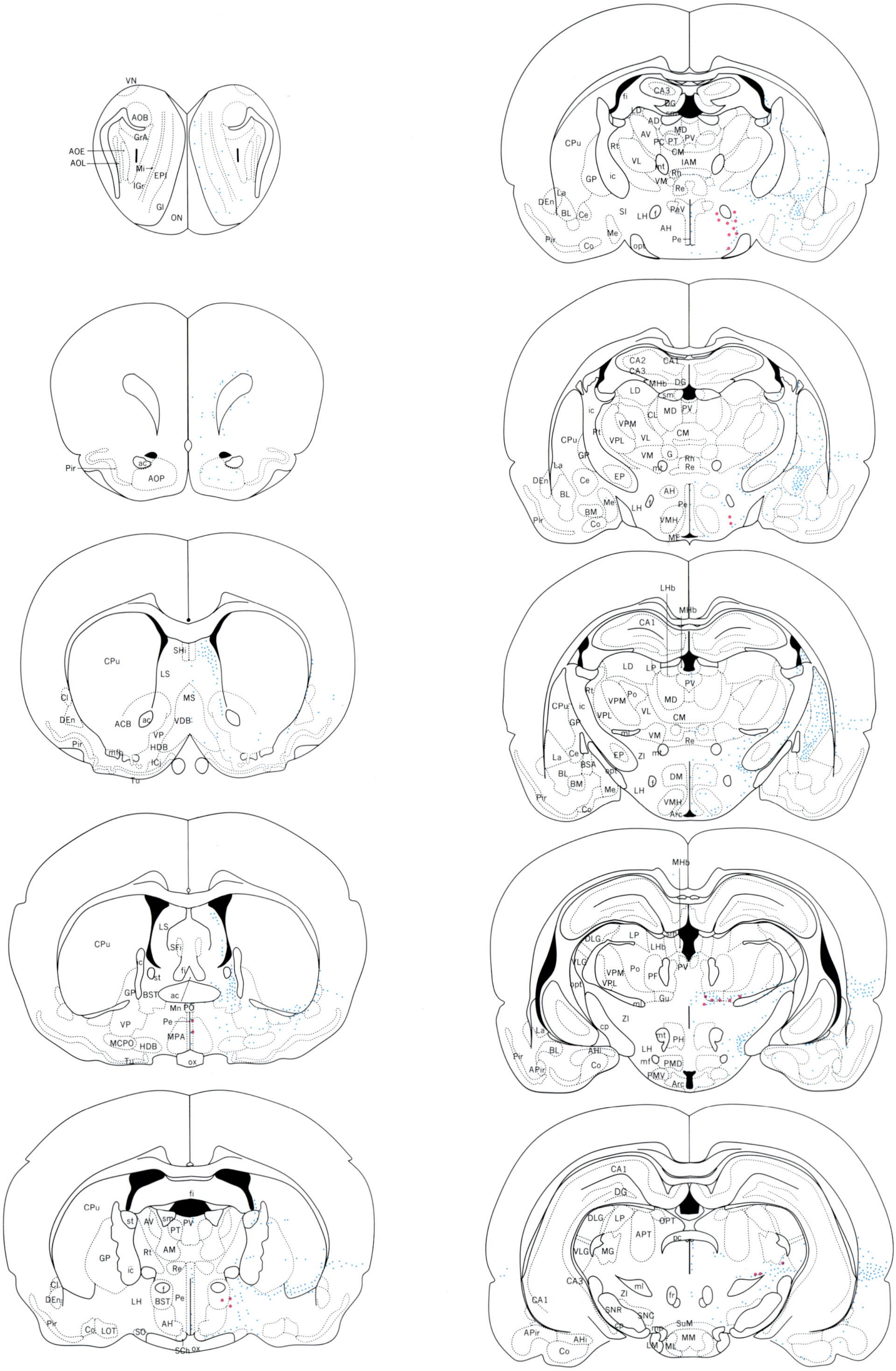

Calcitonin gene related peptide (immunocytochemistry)

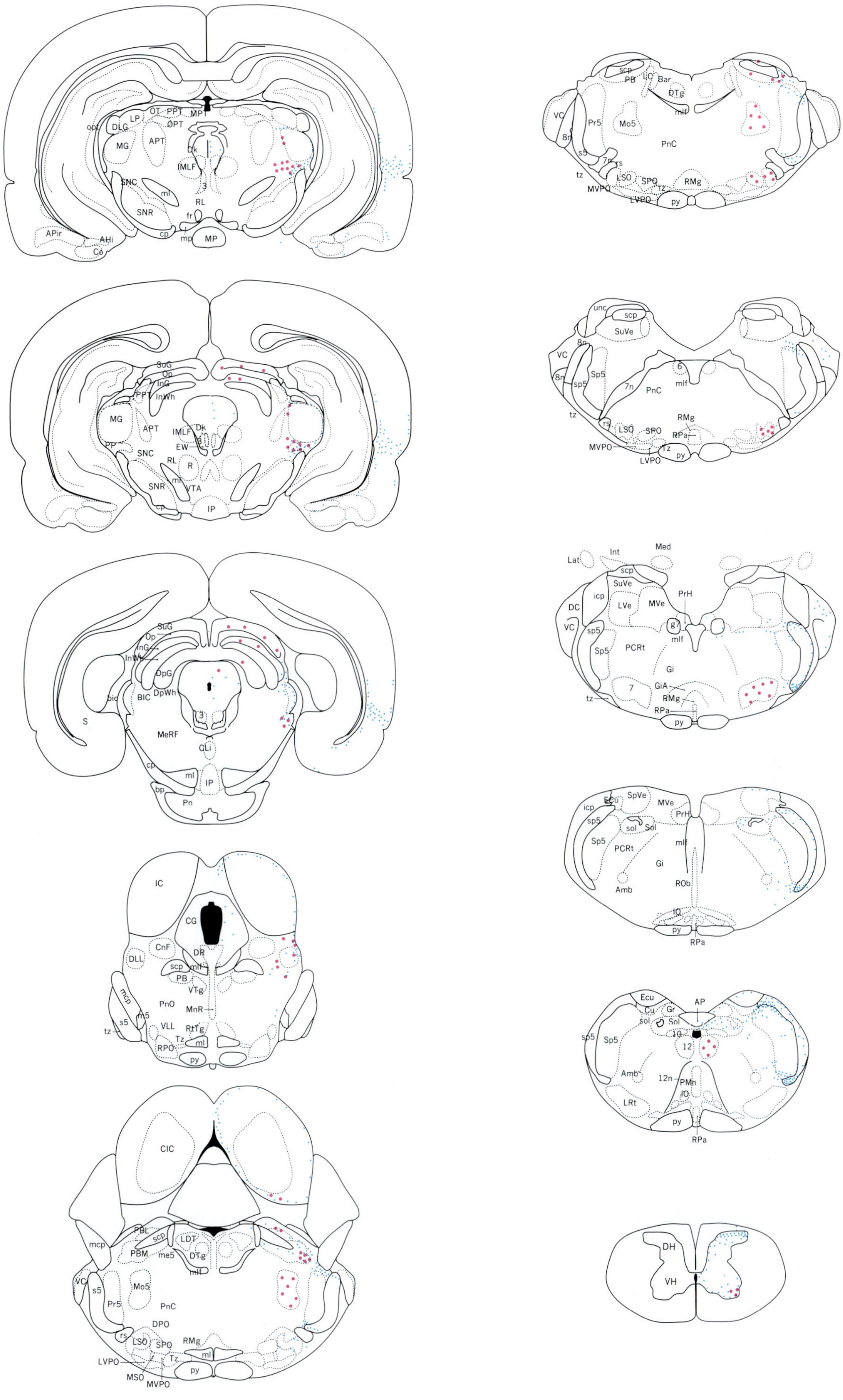

Calcitonin gene related peptide receptor

Distribution of binding site

The distribution of the CGRP binding site has been investigated by making use of [^{125}I]-αCGRP as a ligand[1,2,5,6], and the binding sites were found to be widely distributed in the brain. Exceptionally strong binding was observed with nuclei, amygdaloid complex, pars caudalis of corpus striatum, corpus mamillare, and others. There also exist some regions in which the binding sites do not correspond to the distribution of CGRP.

Subtypes

In the central nervous system, binding sites of CGRP receptor with both high affinity (K_d 0.25 nmol/l, B_{max} 8 fmol/mg membrane protein) and low affinity (K_d 80 nmol/l, B_{max} 120 fmol/mg membrane protein) have been reported[8].

One report[4] suggests that $CGRP_1$ and $CGRP_2$ receptors can be distinguished by a pharmacological criterion, in that $CGRP_1$ receptor does not possesses an affinity to [Cys(ACM)^{2,7}]α-hCGRP. $CGRP_1$ receptor also inhibits the activity of the C-terminal end of CGRP. Thus $CGRP_{8-37}$ turns out to be an inhibitor of $CGRP_1$ receptor, which exhibits 20–50 times affinity to $CGRP_{12-37}$. It has also been reported that athrocytosis of CGRP mediates this receptor.

Conversely, $CGRP_2$ receptor is a strong agonist to [Cys(ACM)^{2,7}] α-CGRP and possesses strong resistance to activity inhibition by $hCGRP_{8-37}$.

Table 15.1. Pharmacological characteristics of the subtypes of CGRP receptor (CGRPq and CGRP2) (from ref. 4)

	$CGRP_1$	$CGRP_2$
[Cys(ACM)^{2,7}] α-hCGRP	Inactive	Agonist; high affinity
$hCGRP_{12-37}$	Antagonist	Inactive
$jCGRP_{8-37}$	Potent antagonist	Weak antagonist

Medication by intracellular metabolism

If striated muscle is treated with αCGRP, the amount of cAMP production is increased (Table 15.2)[7]. The binding of αCGRP to its receptor results in the activation of adenylate cyclase and causes an increase in the production of cAMP in the muscle. This results in triggering the contraction of striated muscle. Even in the SK-N-MC cell line of human neuroblastoma, this elevation in the production of cAMP by α,β $CGRP_2$ has been reported.

Structure

CGRP binding fractions were concentrated with an affinity column of chemically synthesized CGRP, and purified as a protein of apparent molecular weight 66 000[8]. Following the isolation of CGRP receptors, monoclonal antibodies were raised against the receptor protein, and were used to clone the receptor gene in an expression cloning system. The CGRP receptors is N-glycosylated, and the Mr of deglycosylated receptor protein is 45 000. The fully characterized intact CGRP receptor (protein purification, SDS/PAGE followed by silver staining, cross-linking of [^{125}I]CGRP to its receptors with bifunctional reagents, and by Western blot analysis with CGRP-specific monoconal antibodies) has shown that the Mr of the CGRP receptor is as large as 66 000, consisting of a solitary subunit. GTP causes a marked increase in the affinity of CGRP for its receptor, suggesting the formation of G-proteins involved in CGRP binding to its receptors and signaling of its effector system. Different responses elicited by CGRP administered by peripheral and intracerebroventricular routes may be mediated through different receptors, although there are no data to support this hypothesis. However, there is probably more than one class of receptor for CGRP with different selectivity to α and β CGRP. The canine orphan receptor gene RDC-1 was identified as

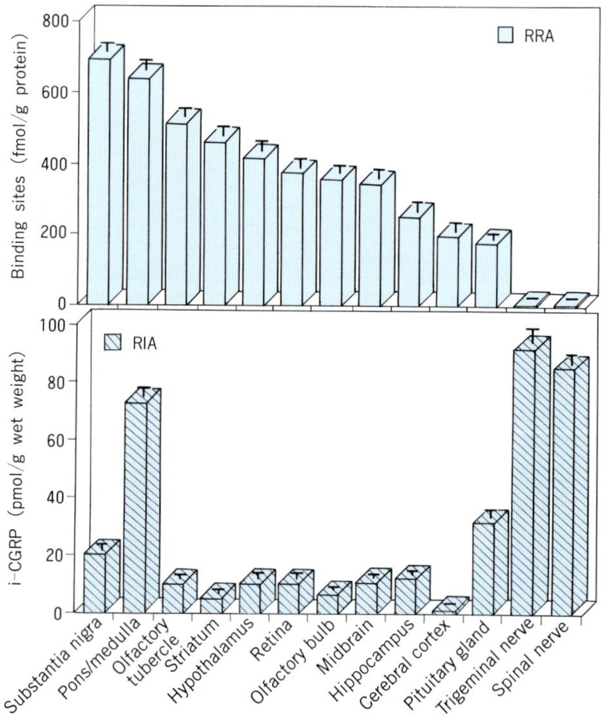

Fig. 15.4 CGRP binding sites and CGRP-like immunoreactivity in the brain. RRA, radioreceptor assay; RI, radioimmunoassay (from ref. 8).

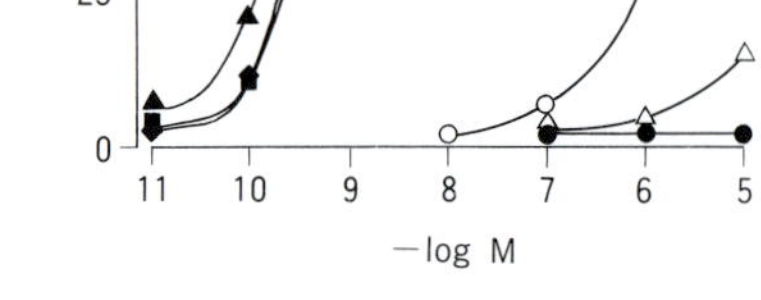

Fig. 15.5 Stimulation of cAMP in SK-N-MC cells. Presence of hCGRP-II (▲), hCGRP-I (■), [C^{γ}-(4-azido-anilino)Asp^3]hCGRP-I (◆), $hCGRP-I_{8-37}$ (●), DAPamide (○) and sCT (△) (from ref. 3)

Table 15.2. Effect of CGRP (0.1 μM) and phosphodiesterase inhibitors on cAMP levels in mouse diaphragm (from ref. 7)

Treatment	Cyclic AMP (fmol/mg tissue)			*n*
	CGRP (−)	CGRP (+)	(+)/(−)	
Control	312 ± 42	571 ± 89	1.83	5
Theophylline (1.0 mM)	608 ± 98[a]	1409 ± 251[a]	2.31	5
Papaverine (30 μM)	769 ± 135[a]	2829 ± 452[a]	3.67	5
Ro 20–1724 (0.3 mM)	2051 ± 651[a]	5461 ± 1455[a]	2.66	5

P , 0.01

the CGRP-1 receptor gene in 1989 (based on PCR). This RDC-1 gene encodes a protein that confers sensitive and specific responsiveness to CGRP in transfected cells (ED_{50} is around 1×10^{-9} M). The second receptor was cloned from human lung cDNA library by using a partial sequence of clone encoding incomplete 800-bp cDNA as a probe. Although the activation of cAMP was demonstrated in HEK 239 cells stably expressing this cDNA, it has not yet been proved that this is a CGRP receptor.

References

1. Henke, H., Sigrist, S., Lang, W. *et al.* (1987). *Brain Res.*, **410**, 404–8.
2. Inagaki, S., Kito, S., Kubota, Y. *et al.* (1986). *Brain Res.*, **374**, 287–98.
3. Muff, R. (1992). In *Calcitonin gene-related peptide* (ed. Y. Tache, *et al.*), pp. 106–18. New York Academy of Sciences.
4. Quirion, R. (1992). In *Calcitonin gene-related peptide* (ed. Y. Tache, *et al.*), pp. 88–105. New York Academy of Sciences.
5. Seifert, H., Chesnut, J., DeSouza, E. *et al.* (1985). *Brain Res.*, **346**, 195–8.
6. Skofitsch, G. and Jacobowitz, D.M. (1986). *Peptides*, **4**, 975–86.
7. Takami, K., Hashimoto, K., Uchida, S. *et al.* (1986). *Japan J. Pharmacol.*, **42**, 345–50.
8. Wimalawansa, S.J. (1996). *Endocrine Reviews*, **17**, 533–85.

Calcitonin gene related peptide receptor (binding sites)

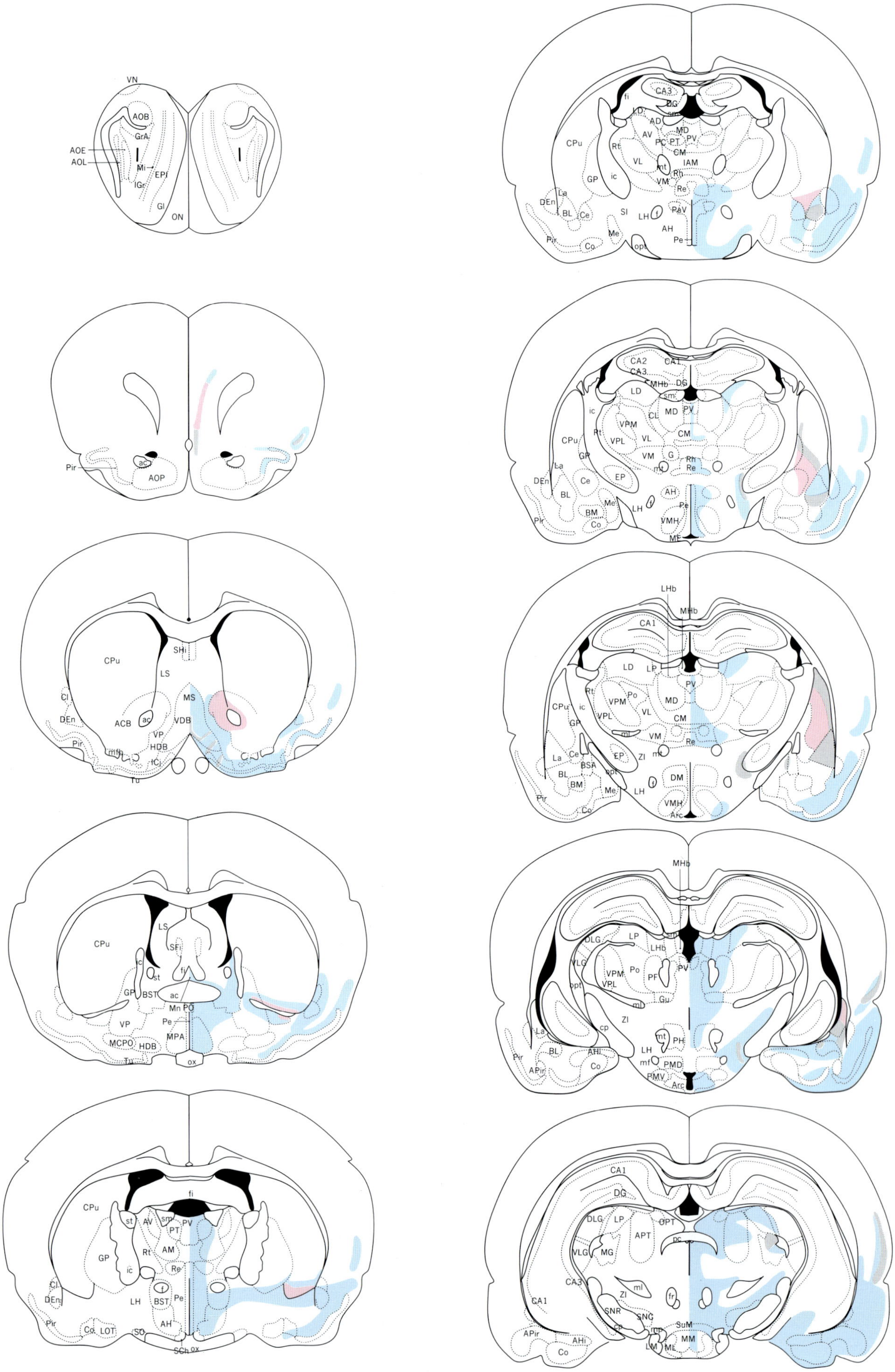

Calcitonin gene related peptide receptor (binding sites)

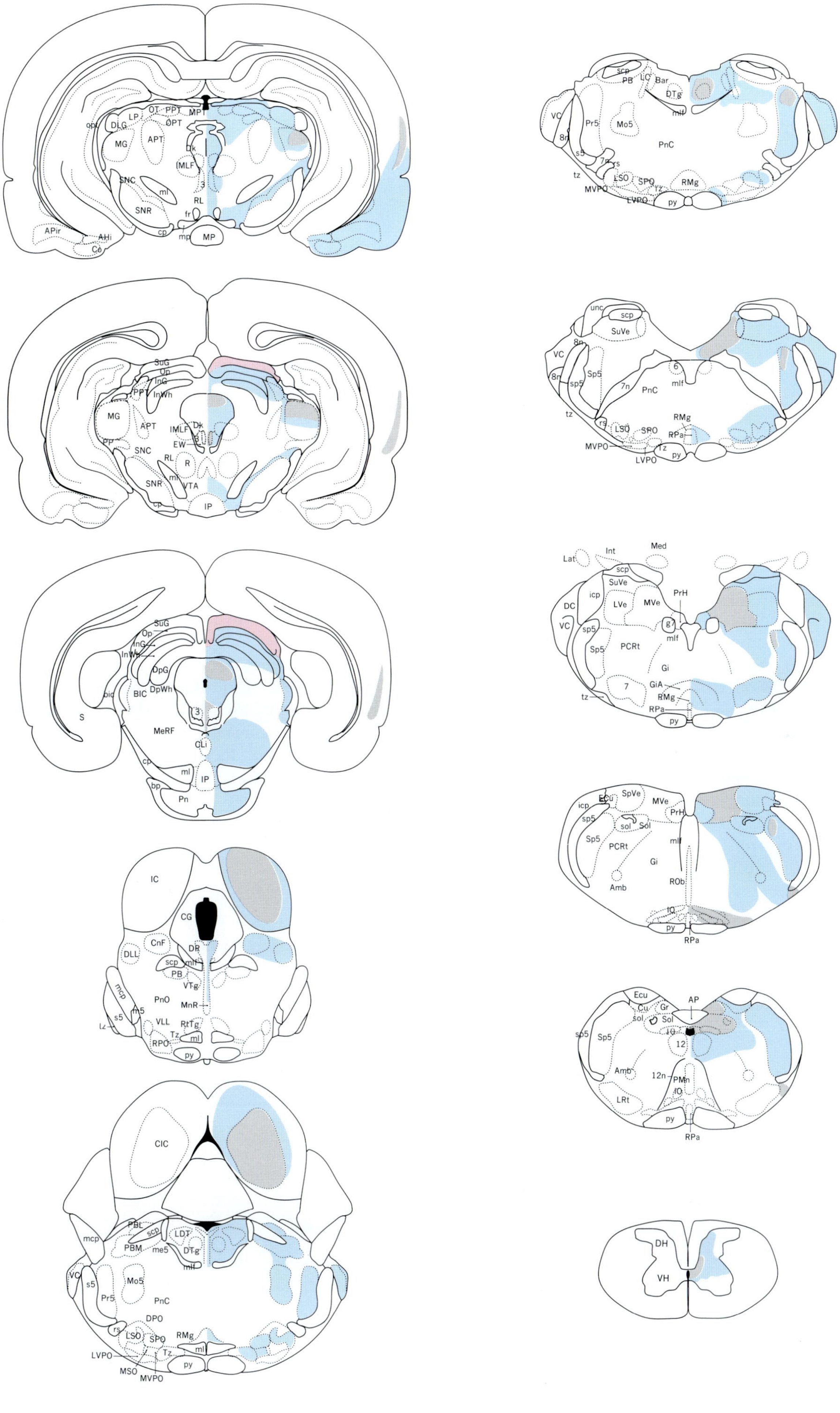

16 The cholecystokinin system

Cholecystokinin

Cholecystokinin (CCK) was originally isolated as a gastrointestinal peptide hormone that stimulates contraction of gallbladder and releases pancreatic enzymes. Subsequent studies have shown that CCK is also widely distributed in the central nervous system. Compared to other peptide hormones, CCK is abundantly present in the cerebral cortex and its neural functions are therefore of much interest.

Structure

The CCK of rat, cleaved at the N- and C-terminal ends from prepro-CCK, consists of 115 amino acid residues, as shown in Fig. 16.1[8]. When the Gly residue at the C-terminal end is excised from the pro-CCK that is formed, and the adjacent Phe residue is α-carboxyamidated, CCK becomes physiologically functional. The presence of a four amino acid sequence at the C-terminal, Trp-Met-Asp-Phe-NH_2, is important for its physiological activity. The sulfonation of a Tyr residue from the C-terminal end is also an essential criterion for CCK's physiological activity or receptor binding. As the cleavage progresses from the N-terminal end, the loss of amino acids leads to the formation of CCK-58, CCK-33, CCK-8, CCK-4, etc. More than 90 per cent of the CCK present in the central nervous system is reported to be CCK-8. Since the five amino acid residues at the C-terminal end of CCK are the same as in gastrins, cross-reaction of antibody needs to be taken care of during immunohistochemical investigations.

Physiological function

Since the report by Gibbs *et al.*[3] in 1973 on the regression of food intake by CCK, many other physiological actions in which CCK plays a role have been identified.

Satiety effect

Food ingestion is inhibited by intraperitoneal as well as intracerebral infusion of CCK. It is suggested that the peripheral CCK mediates this effect via the vagus nerve, and the central CCK via the projection from CCK-positive neurons located at the lateral regions of lateral parapeduncular nucleus to the ventromedial hypothalamus.

Link to pain

CCK in low concentration provides an opposing effect to the analgesic action caused by opiate, but in high concentrations, CCK itself causes an analgesic action[2]. CCK is believed to act on the central grey matter as well as the dorsal column of the spinal cord. Since *in situ* hybridization results shows higher concentration of CCK mRNA in the thalamus, it is also suggested that CCK may participate in the actions of the thalamocortex.

Link to psychoneurosis

CCK-positive cells of the ventral tegmental area contain dopamine as well[4]. With 6-OHDA treatment, which destroys catecholamine-containing neurons, the CCK-positive terminals of the olfactory tubercle accumbens nucleus, nuclei of stria terminalis, and central amygdaloid nucleus were also found to disappear[9]. This suggests that

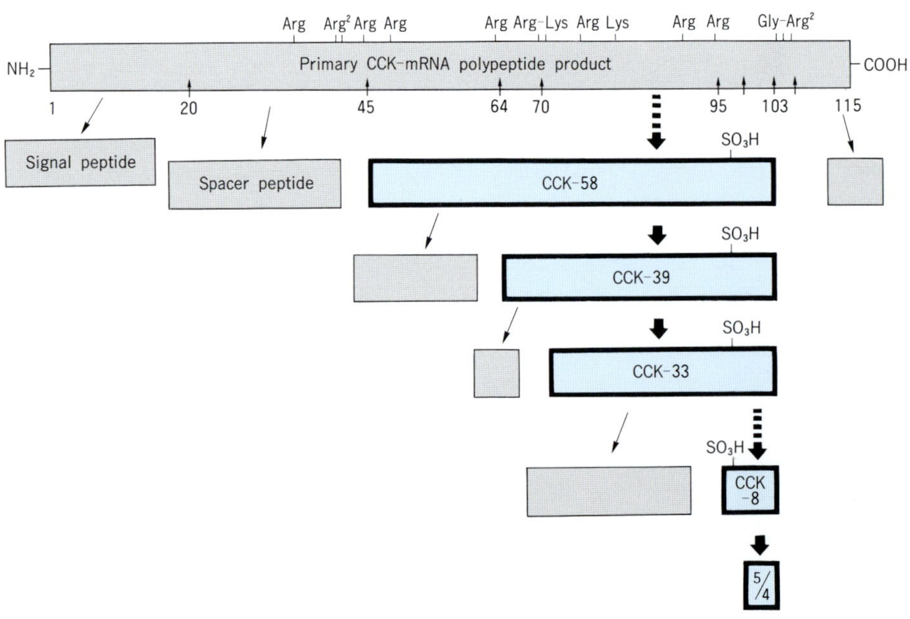

Fig. 16.1 Biosynthesis of CCK.

CCK-positive cells of the ventral tegmental area give rise to axons to those nuclei to release dopamine and CCK. Furthermore, CCK is also known to inhibit the release of dopamine or alter the activity of dopamine receptors[8]. Since CCK is closely linked with dopamine neurons, by extension CCK is also postulated to be involved in mental disease that is clearly attributed to aberration of dopamine neurons. For instance, there is a report stating that CCK receptors in schizophrenic patients are depressed in the frontal lobe cortex and hippocampus.

Distribution

Immunochemical investigations of CCK-8 show that it is widely distributed in the brain, excluding the cerebellum. Higher concentration of CCK-8 has been mapped in regions such as the forebrain, cerebral cortex, hippocampus, amygdaloid complex, central grey matter of midbrain, substantia nigra, and ventral tegmental area[1,6]. CCK distribution is highest in the forebrain, and CCK-positive cells are abundantly distributed together with the nerve fibres in the paraglomerular layer of the olfactory tubercle, surface layer of layers II–III, and deep layer of layers V–IV of the cerebral cortex, from the anterior part of the amygdala to the lateral nucleus group of amygdaloid complex, entorhinal cortex, piriform cortex, subiculum, and py ramidal cell layer of hippocampus CA1–3. In the hypothalamus, the CCK cell bodies are distributed in the preoptic region, paraventricular nucleus, supraoptic hypothalamic nucleus, and dorsomedial hypothalamic nucleus. In the dorsomedial hypothalamic nucleus, mainly nerve fibres are seen. However, *in situ* hybridization analysis did not reveal the expression of CCK mRNA in the preoptic region or dorsomedial hypothalamic nucleus.

A large number of nerve cells that express CCK mRNA are observed at the medial and superior nuclei of the mammillary body, which incorporate the posterior part of the hypothalamus. Among these, a few CCK-immunopositive cells are identified and traced to project into the anterior nuclei of the thalamus. These cells are absent in the mammillary body. Furthermore, though most regions of the thalamic nucleus show mRNA expression for CCK, immunopositive cells are rarely observed except in one region. There is a noticeable difference in the distribution of CCK in the thalamus or hypothalamus and between the prepro-CCK mRNA and CCK immune activity. However, in most of the other regions of the brain, this variability is not visible. Why is a large quantity of CCK mRNA distributed in the thalamus? It could be that CCK in the thalamus is transported at a rapid rate from the cell body to the periphery and is detected by *in situ* hybridization histochemistry, which has a high sensitivity in detecting CCK positive cells[5]. Pro-CCK, which is defined as the product extended beyond Gly at the C-terminus, is concentrated in the cerebral cortex from the embryonic to the juvenile period, but pro-CCK concentration decreases remarkably in adult rats. In comparison, the concentrations of CCK-33 and CCK-8 (carboxyamidated CCKs, where the C-terminal Phe residue is α-carboxyamidated) increase after postpartum day 7. These observations suggest that the expression of transmitter active CCK peptides in the brain is mostly regulated at the post-translational rather than the transcriptional level[7]. Thus, the physiological significance of pro-CCK is further revealed by its cerebral distribution.

References

1. Cho, H.J., Shiotani, Y., Shiosaka, S. *et al.* (1983). *J. Comp. Neurol.*, **218**, 25–41.
2. Faris, P.L. (1983). *Science*, **219**, 310–12.
3. Gibbs, J., Young, R.C., and Smith, G.P. (1973). *J. Comp. Physiol.*, **84**, 488–95.
4. Hokfelt, T., Skiboll, L., Rehfeld, J.F. *et al.* (1980). *Neuroscience*, **5**, 2093–124.
5. Ingram, S.M., Krause, II, R.G., Baldino, F., Jr., and Skeen, L.C. (1989). *J. Comp. Neurol.*, **287**, 260–72.
6. Kiyama, H., Shiosaka, S., Kubota, Y. *et al.* (1983). *Neuroscience*, **10**, 1341–59.
7. Mogensen, N.W., Hilsted, L., Bardram, L., and Rehfeld, J.F. (1990). *Dev. Brain Res.*, **54**, 81–6.
8. Rehfeld, J.F. (1985). *J. Neurochem.*, **44**, 1–10.
9. Voigt, M.M. and Wang, R.Y. (1984). *Brain Res.*, **296**, 189–93.

Cholecystokinin (immunocytochemistry)

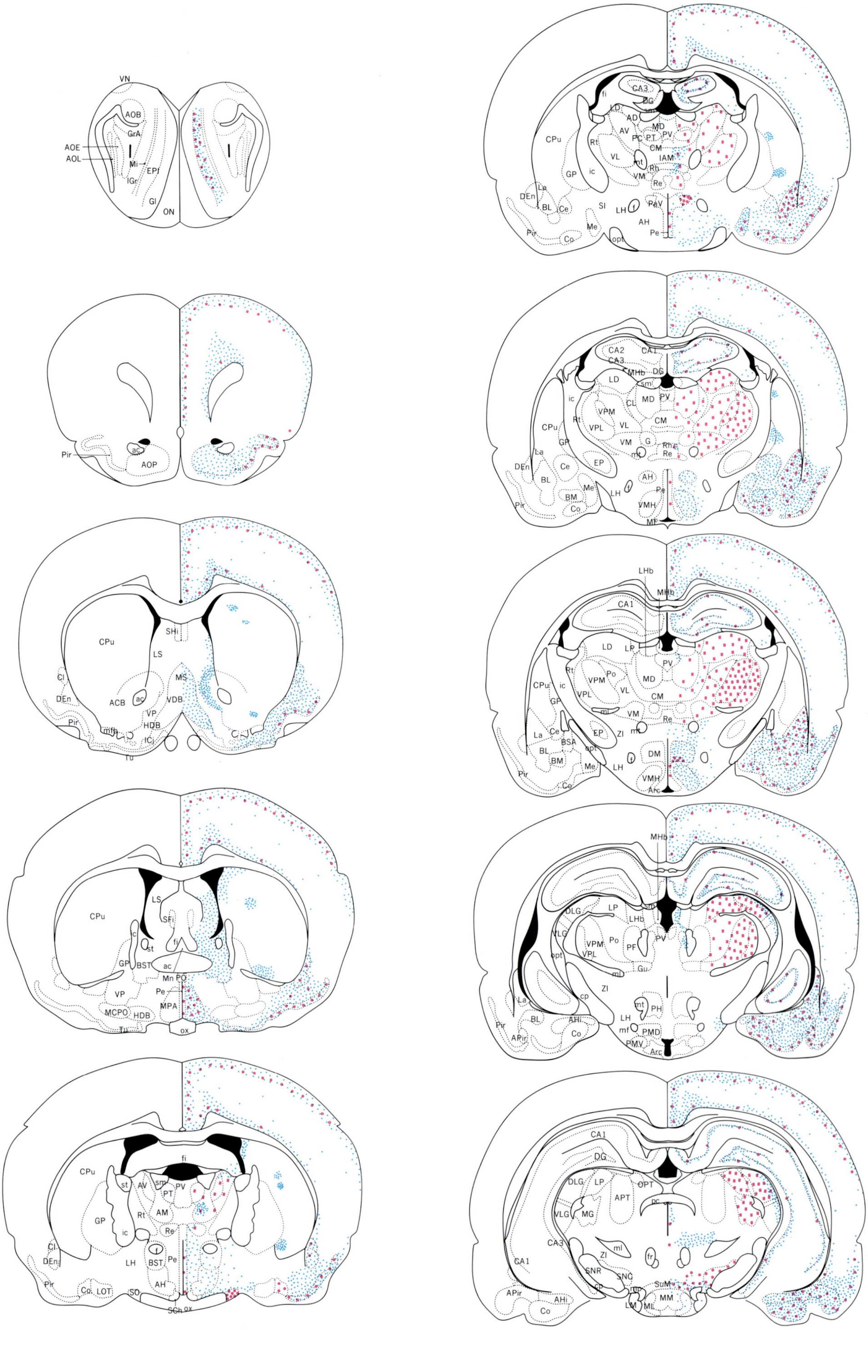

Cholecystokinin (immunocytochemistry)

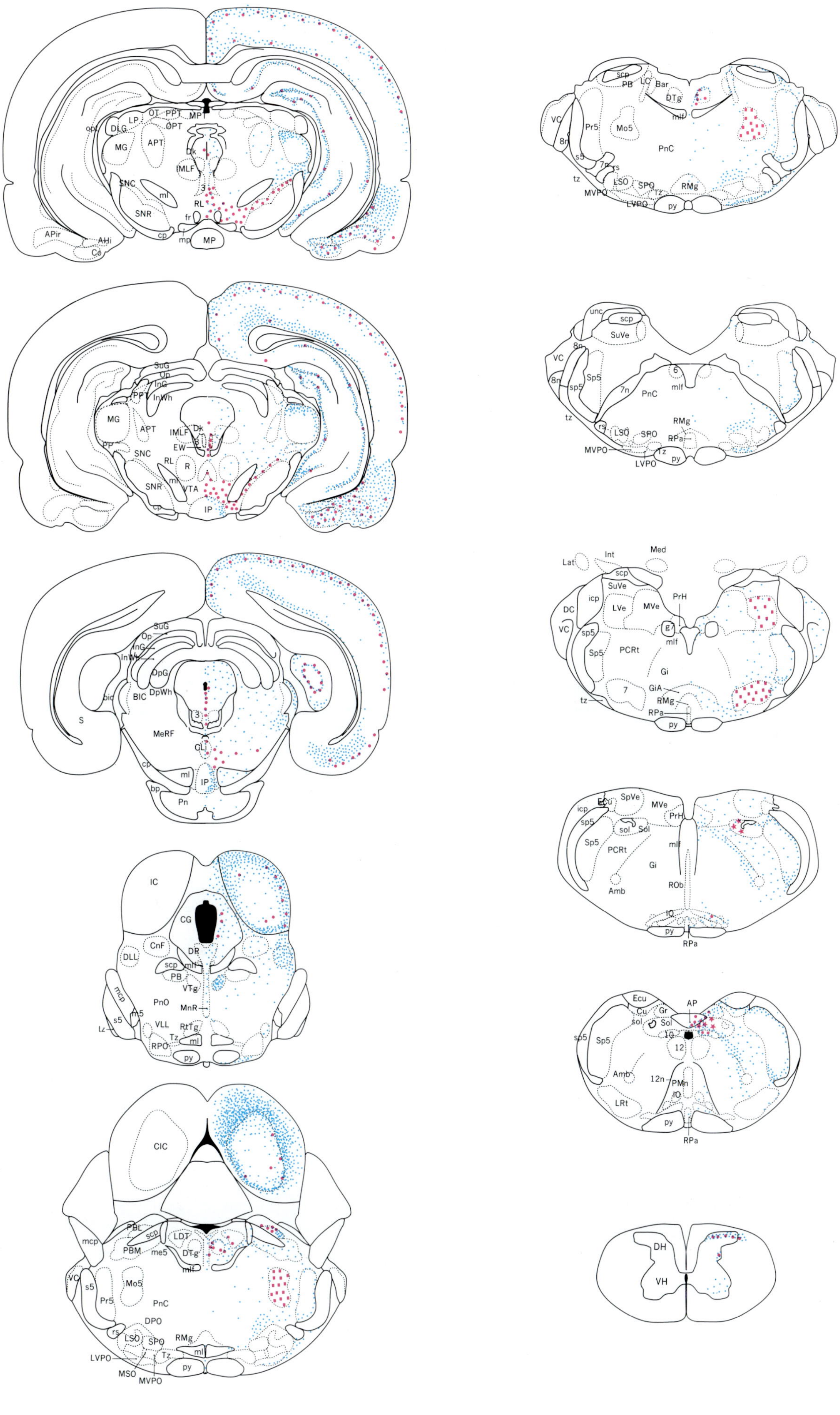

Cholecystokinin receptor

Types of binding sites

The binding sites of CCK vary between the brain and peripheral organs[3,7], especially the acinar region of the pancreas. Studies which have used [^{125}I]-Bolton–Hunter (BH)-CCK-33 or [^{125}I]-BH-CCK 8 have shown that binding sites with both high and low affinities exist in the acinar region of the pancreas. However, brain possesses only the high affinity type of receptors. The receptors of the central nervous tissues are called CCK-B, and those of peripheral tissues are identified as CCK-A. However, the peripheral types of CCK receptors were also observed in

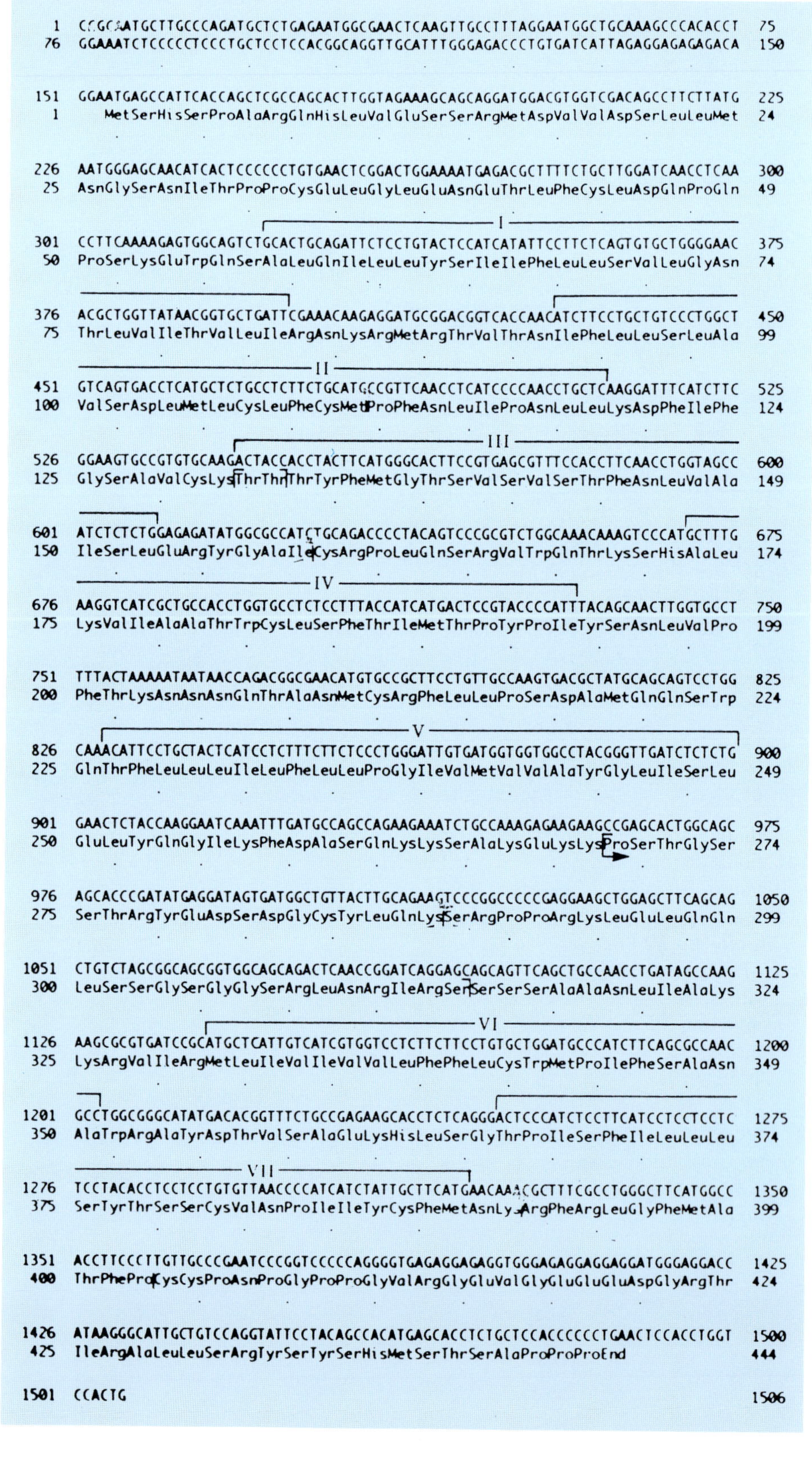

Fig. 16.2 CCK-A receptor nucleotide and amino acid sequences (from ref.5).

```
CCK-B gastrin
     Human      V V R M L L V I V V L F F L C W L P[V]Y S A N T
     Canine     - - - - - - - - - - - - - - - - - -[L]- - - - -
     Rat        - - - - - - - - - L - - - - - - - -[V]- - V - -
     Mastomys   - - - - - - - - - L - - - - - - - -[I]- - - - -

CCK-A
     Rat        - I - - - I - - - - - - - - - - M -[I]F - - - A
```

Fig. 16.3 Comparison of the amino sequences (single letter code) corresponding to transmembrane domain VI of the human, canine, rat, and mastomys CCK-B gastrin and the rat CCK-A receptors, as deduced from the corresponding cDNAs. Dashes indicate identity with the human receptor. The boxed amino acids correspond to human ^{349}Val (from ref. 5).

the spinal cord[1,4], so it becomes questionable whether the differentiation of CCK receptors into central and peripheral types makes sense.

Localization

Autoradiographical studies making use of [^{125}I]-BH-CCK-33, [^{125}I]-BH-CCK-8, and [^{3}H]-L-364, 718 as ligands have identified the distribution of CCK binding sites[1,2] (Fig. 16.2). A single cDNA encoding a single protein is present in both pancreas and gallbladder and can account for both high and low affinity CCK-A receptors found on pancreatic acini. A single cDNA encoding a single CCK-B receptor protein is present both centrally and in the periphery, including the gastrointestinal system. The CCK-B receptor cloned from the brain has been actually characterized as a gastrin receptor.

Physiological function

CCK-A receptors mediate many functions, which include physiological contractions of the gallbladder, pancreatic enzyme secretion, relaxation of the sphincter of Oddi, and potentiation of insulin secretion. CCK-A receptors exist in the pituitary gland, myenteric plexus, postremal area, solitary tract nucleus, interpeduncular nucleus, posterior part of subthalamus, and mesencephalon. In the mesencephalon, the CCK-containing dopaminergic neurons may be involved in the pathogenesis of schizophrenia, Parkinsonism, and drug addition. In structures such as the solitary tract nucleus and the area postrema, the CCK-A receptors are postulated to be implicated in feeding and transmission of viscerosensory information.

CCK-A is involved in the inositol phospholipid metabolism which mediates the G protein and this involvement has been clearly demonstrated in the pancreas. Specifically, G protein activates phospholipase C and as a result phosphatidylinositol is transformed into diacylglycerol (DG) and inositol triphosphate (IP_3). The IP_3 mobilizes Ca^{2+} that activates protein C kinase which leads to chloride channel opening in the plasma membrane. The binding sites of CCK-B exist widely in the cerebral cortex and corpus striatum in the brain, but physiological actions of CCK-B receptors remain unclear. The CCK-B receptor has been implicated in mediating many of the central effects of cholecystokinin, including anxiety, panic attacks, satiety, and analgesia. CCK is also thought to have a trophic effect on the gastro-intestinal tract. CCK-8 and gastrin promoted the cell growth as well as an increase of tyrosine phosphorylation of MAPK and p125FAK, suggesting that the CCK-B/gastrin receptor transmits mitogenic signals by cross-talking with the tyrosine kinase cascades.

Structure

The primary structure of CCK-A has recently been elucidated[5]. Fig. 16.2 shows the nucleotide and amino acid sequences from the CCK-A receptor. The human CCK-A receptor has an open reading frame encoding 429-amino acid protein with seven putative hydrophobic transmembrane domains (I–VII) and CCK-B has that encoding 448-amino acid protein. There is significant homology with members of the β-adrenergic family of G protein-coupled receptors. Gastrin-stimulated phosphatidylinositol hydrolysis and intracellular Ca^{2+} mobilization in COS-7 cells expressing CCK-A receptor suggests signalling through phospholipase C. The CCK-A and CCK-B receptors are 48 per cent homologous and each receptor is highly conserved between species. The human CCK-A receptor gene maps to chromosome 4, and both CCK-A and CCK-B receptors consist of five exons interrupted by four introns.

References

1. Hill, D.R., Shaw, T.M., and Woodruff, G.N. (1988). *Neurosci. Lett.*, **89**, 133–9.
2. Hill, D.R., Campbell, N.J., Shaw, T.M., and Woodruff, G.N. (1987). *J. Neurosci.*, **7**, 2967–76.
3. Innis, R.B. and Snyder, S.H. (1980). *Proc. Natl. Acad. Sci. USA*, **77**, 6917–21.
4. Morgan, T.H., Robinson, P.H., Godrich, M.S., and McHugh, P.R. (1986). *Brain Res.*, **362**, 175–9.
5. Wank, S.A., Pisegna, J.R., and DeWeerth, A. (1994). *Ann. N.Y. Acad. Sci.*, **713**, 49–66.
6. Kopin, A.S., Beinborn, M., Lee, Y.-M. *et al.* (1994). *Ann. N.Y. Acad. Sci.*, **713**, 67–78.
7. Williams, J.A., Bailey, A.C., and Roach, E. (1988). *Am. J. Physiol.*, **254**, G513–G521.

Cholecystokinin receptor (binding sites)

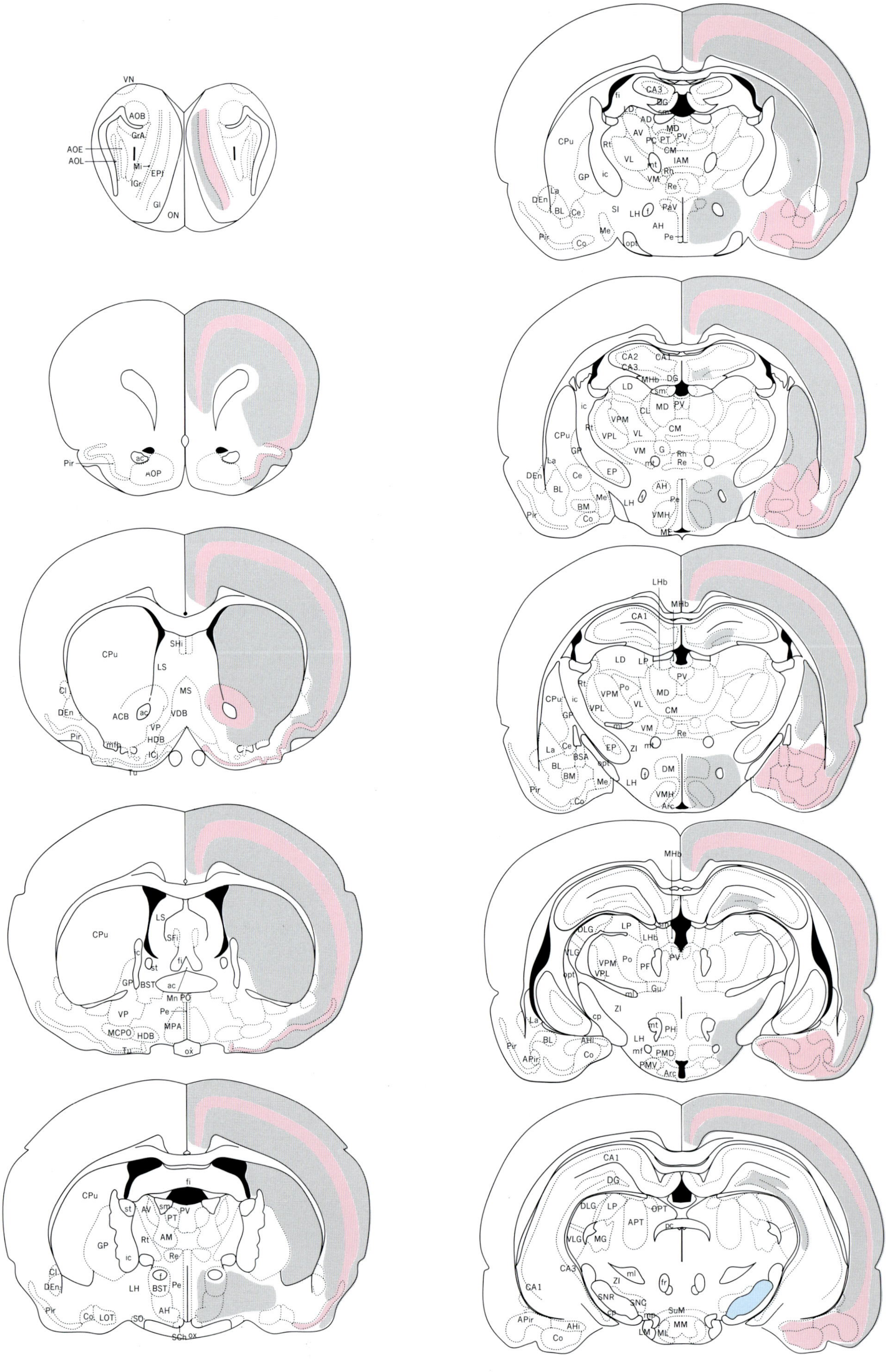

Cholecystokinin receptor (binding sites)

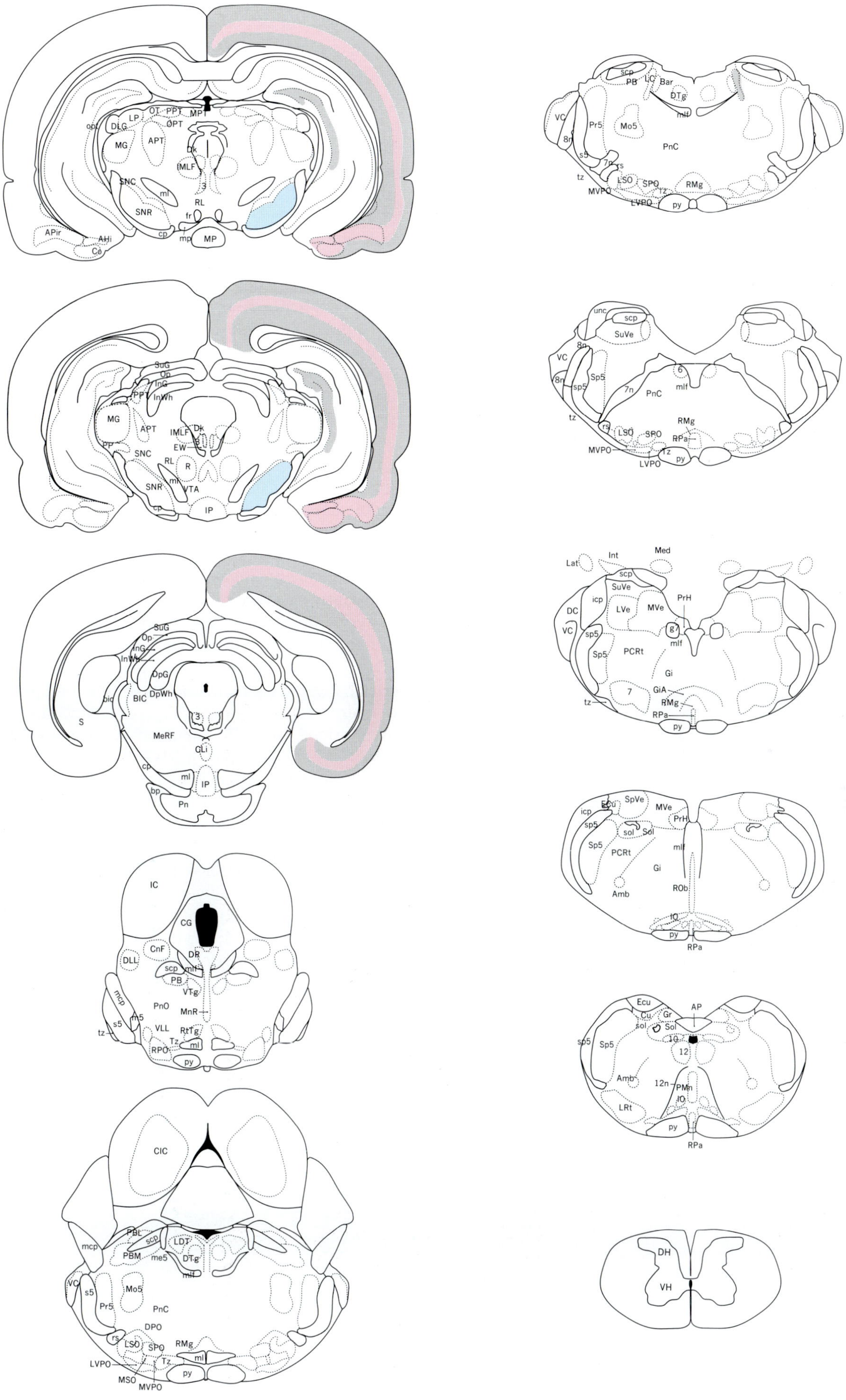

17 The neurotensin system

Neurotensin

Structure and function

Neurotensin is a peptide that was first isolated from bovine hypothalamus by Carraway and Leeman[3] in 1973, and consists of 13 amino acid residues. This peptide is also known to be widely distributed in the digestive tract, and is known as one of the brain–intestine peptides. Neurotensin has been reported to exert pharmacological influence on the dopamine system of the substantia nigra and corpus striatum, and cause antinociception, hypothermia, etc. However, its physiological roles in the nervous system are still unclear[5].

As an analogous peptide to neurotensin, in 1984 Minamino *et al.* extracted an endogenous peptide, neuromedin N, which consists of six amino acid residues, among which four residues of the C-terminal end are common with neurotensin (Fig. 17.1)[11]. Later, neurotensin and neuromedin N were proved to be coded on a single gene by Dobner *et al.*[1,4,6]. The preproneurotensin/neuromedin N gene codes a precursor protein that consists of 170 amino acid residues, among which two peptides resemble neurotensin. From the N-terminal end, the neuromedin-analogous peptide that consists of six amino acid residues, neuromedin N, and neurotensin have been coded (Fig. 17.1). At present, the functions of these peptides are unclear.

Distribution

One of the characteristics of localization of neurotensin in the brain is a large alteration in the amount of expression in the course of ontogeny. With respect to the ontogeny of neurotensin in the brain, three types have been classified by the pattern of temporal alteration[9]. Type 1 appears very abundantly from embryonal anaphase, but is rarely detected 1 or 2 weeks after birth. Type 2 expresses neurotensin abundantly for several weeks following birth, but after that the expression amount is depressed and then can be detected only at a very low level in adults. Type 3 gains gradually after birth and is then retained. Types 1 and 2 are observed abundantly in the forebrain. Those belonging to type 1 include mitral cells of the olfactory bulb cells of the layers II, III, and VI of the cerebral cortex, cingulate gyrus, and cells located in the deep layer of the rostral part of the nucleus of the spinal tract of the trigeminal nerve. Many cells of the forebrain area belong to type 2, and these include the pyramidal cells of Ammon's horn CA1 of hippocampus, subiculum, and caudate putamen. Fibres projected from neurons belonging to types 1 and 2 also show abundant immunoreactivity transiently. For example, in the areas of the mammillary body that receive projection from the subiculum, the anterior ventral thalamic nucleus that receives projection from the cingulate cortex, the lateral area of substantia nigra that receives projection from the caudal area of corpus striatum, and the lateral olfactory tract that is a bundle of output fibres of mitral cells of the olfactory bulb, extremely abundant neurotensin positive fibres are transiently observed and become difficult to identify after 3 weeks postnatally. Those that belong to type 3 are present in Broca's diagonal band, the medial preoptic area, piriform lobe, amygdaloid complex, dorsal raphe nuclei, ventral tegmental area, etc. The ontogeny pattern of neurotensin expression that is classified into three types as described above by immunochemistry is similar to the results obtained from mRNA expression by *in situ* hybridization[9,12]. Apart from this, there are salient points in the distribution of neurotensin that need attention. Neurotensin mRNA is positive in the hypoglossal nucleus, but the peptide has not been identified by immunohistochemistry. The expression of neurotensin mRNA in the hypoglossal nucleus shows an expression of type 2 (indicated in the map)[8]. Mismatch, such as this type of peptide and mRNA to code it, has been reported to exist in some areas even among other peptides, but the reason for this is unclear: whether the mRNA would not have been

Neurotensin	pGlu-Leu-Tyr-Glu-Asn-Lys-Pro-Arg-Arg-Pro-Tyr-Ile-Leu
Neuromedin N	Lys-Ile-Pro-Tyr-Ile-Leu
Neuromedin N-like peptide	Lys-Phe-Pro-Thr-Ala-Leu

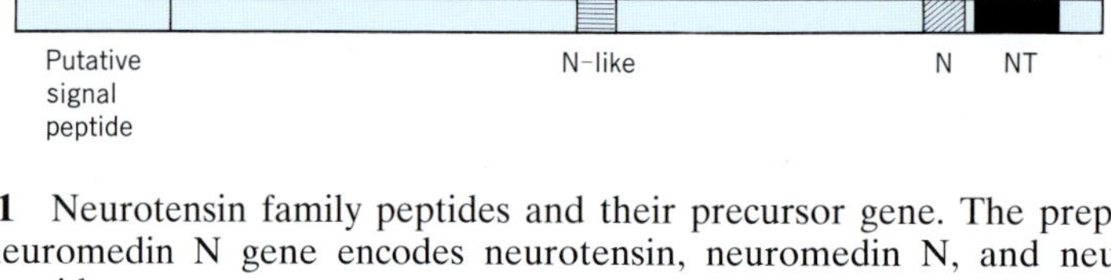

Fig. 17.1 Neurotensin family peptides and their precursor gene. The preproneurotensin/neuromedin N gene encodes neurotensin, neuromedin N, and neuromedin N-like peptide.

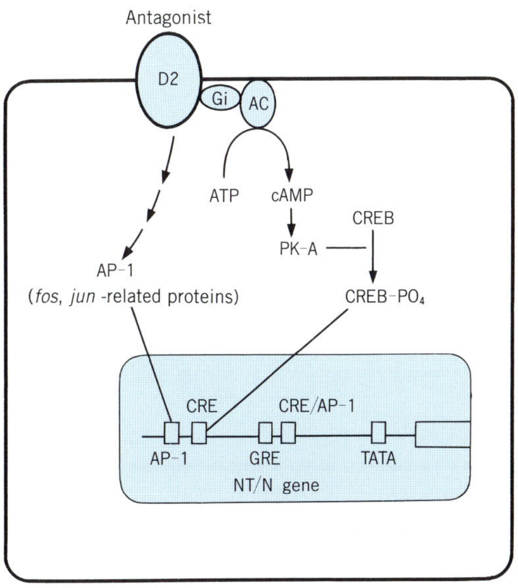

Fig. 17.2 Proposed theoretical model for dopamine D2 receptor mediated regulation of NT/N gene transcription. Blockage of D2 receptors has been shown to increase intracellular levels of cAMP and also cause transient activation of c-*fos* related genes. The increase of cAMP cause an activation of PKA leading to activation of NT/N gene transcription via CREB. The fos family proteins also activate NT/N gene via AP-1. Gi, inhibitory G protein; AC, adenylate cyclase; CRE, cAMP responsive element; CREB, cAMP responsive element binding protein; GRE, glucocorticoid responsive element (from ref. 10).

translated into protein, or whether it is a problem of processing of precursor protein. Furthermore, in immunohistochemistry, colchicine was used to identify the peptide-containing cell bodies. This was done to prevent an axonal flow by colchicine and to make the produced peptide accumulate in cell bodies and then to make detection by the antibody easier. The neurotensin-positive cells are observed very abundantly in the paraventricular nucleus of the hypothalamus in the colchicine-treated brain, but the mRNA could be hardly detected in the normal brain by the *in situ* hybridization method. However, since neurotensin mRNA-positive cells also became abundantly visible on colchicine treatment, it becomes apparent that colchicine itself may accelerate the production of neurotensin in the part of neurons where colchicine is expected to participate in the transcription regulation of the neurotensin gene directly or indirectly[7]. These neurons do not produce neurotensin actively under physiological conditions.

Regulation of transcription expression

The gene for neurotensin has been cloned by Kislauskis *et al.*[6]. In the -200 to -50 bases of the 5′ flanking region of neurotensin gene are located AP-1, CRE (cAMP response element), GRE (glucocorticoid response element), etc. and these were considered to regulate the transcription of this gene. Practically, when shortening this area and observing the expression amount of protein that was coded downstream (a deletion assay), the expression amount decreased severely with the loss of this area. In experiments using the cell line PC12, the transcription regulation of neurotensin in PC12 cells responds to various types of external stimulation, and its expression amount turns out to be several hundred times greater through the combined stimuli of NGF, dexamethasone, forskolin, and lithium[6]. Stimulation that excluded any one of these compounds resulted in considerable decrease in expression. Meanwhile, it has been shown that injection of haloperidol *in vivo* results in neurotensin mRNA expression in the lateral area of corpus striatum within an hour, and this reaches a maximum in around 7 hours[2,10]. This is considered to participate in the rise of intracellular cAMP level by blocking the D2 receptor that exists in the corpus striatum and by a temporary rise of c-*fos* expression (Fig. 17.2).

References

1. Alexander, M.J., Millar, M.A., Dorsa, D.M. *et al.* (1989). *Proc. Natl Acad. Sci. USA*, **86**, 5202–6.
2. Augood, S.J., Kiyama, H., Faull, R.L., and Emson, P.C. (1991). *Mol. Brain Res.*, **9**, 341–6.
3. Carraway, R. and Leeman, S.E. (1976). *J. Biol. Chem.*, **251**, 7045–52.
4. Dobner, P.R., Barber, D.L., Villa-Kamaroff, L., and Mc-Kierman, C. (1987). *Proc. Natl Acad. Sci. USA*, **84**, 3516–20.
5. Emson, P.C. (1986). In *Handbook of chemical neuroanatomy*, Vol. 4 (ed. A. Björklund and T. Hökfelt). pp. 355–405. Elsevier, Amsterdam.
6. Kislauskis, E., Bullock, B., McNeil, S., and Dobner, P.R. (1988). *J. Biol. Chem.*, **263**, 4963–8.
7. Kiyama, H. and Emson, P.C. (1991). *Mol. Brain Res.*, **9**, 353–8.
8. Kiyama, H., Emson, P.C., and Tohyama, M. (1990). *Dev. Brain Res.*, **58**, 293–6.
9. Kiyama, H., Sato, M., and Emson, P.C. (1992). In *Handbook of chemical neuroanatomy*, Vol. 10, pp. 399–431. Elsevier, Amsterdam.
10. Merchant, K.M., Dobner, P.R., and Dorsa, D.M. (1992). *J. Neurosci.*, **12**, 652–63.
11. Minamino, N., Kangawa, K., and Matsuo, H. (1984). *Biochem. Biophys. Res. Commun.*, **122**, 542–9.
12. Sato, M., Lee, Y., Shiosaka, S. *et al.* (1990). *Dev. Brain Res.*, **54**, 249–55.

Neurotensin (immunocytochemistry)

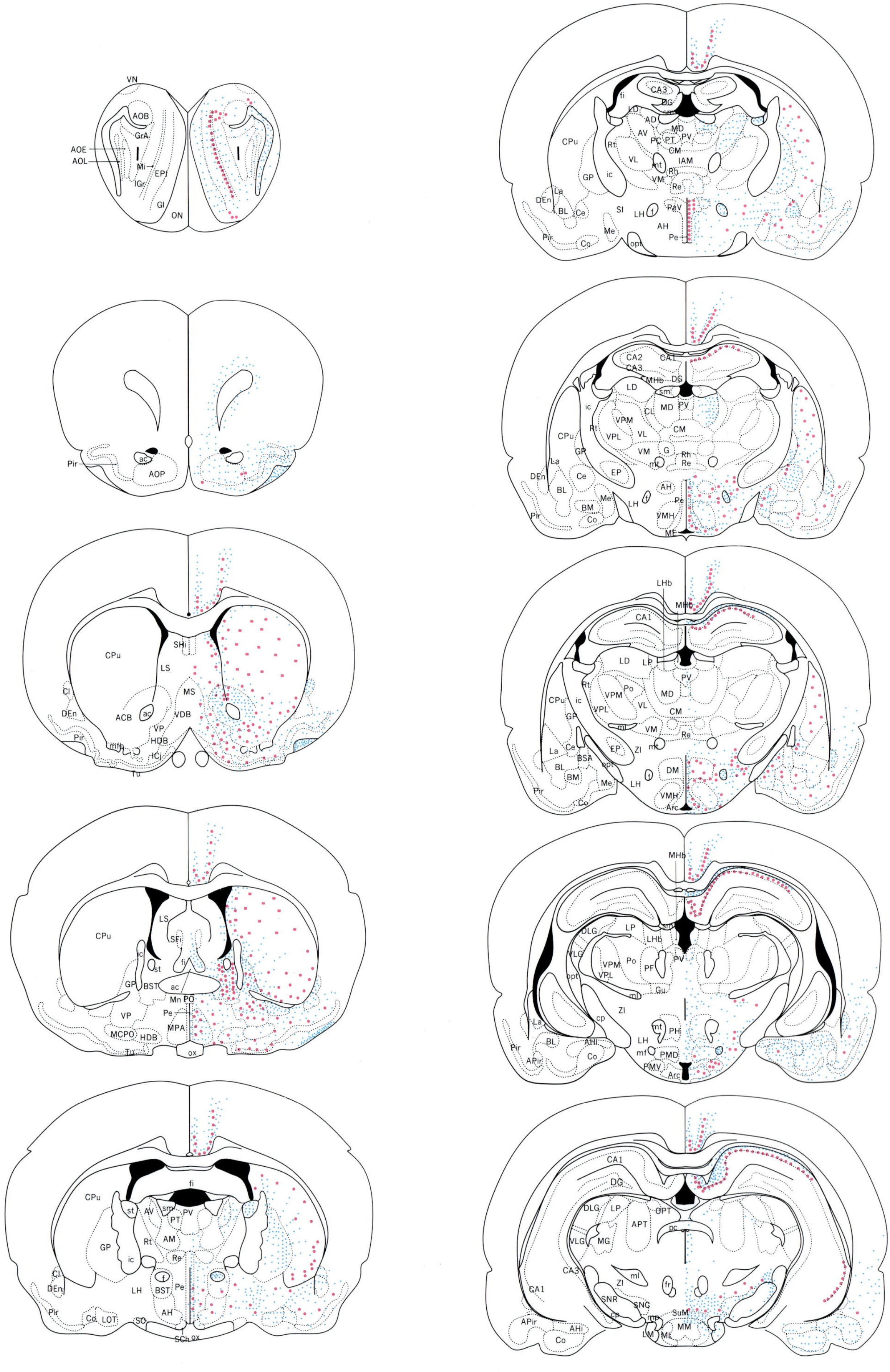

Neurotensin (immunocytochemistry)

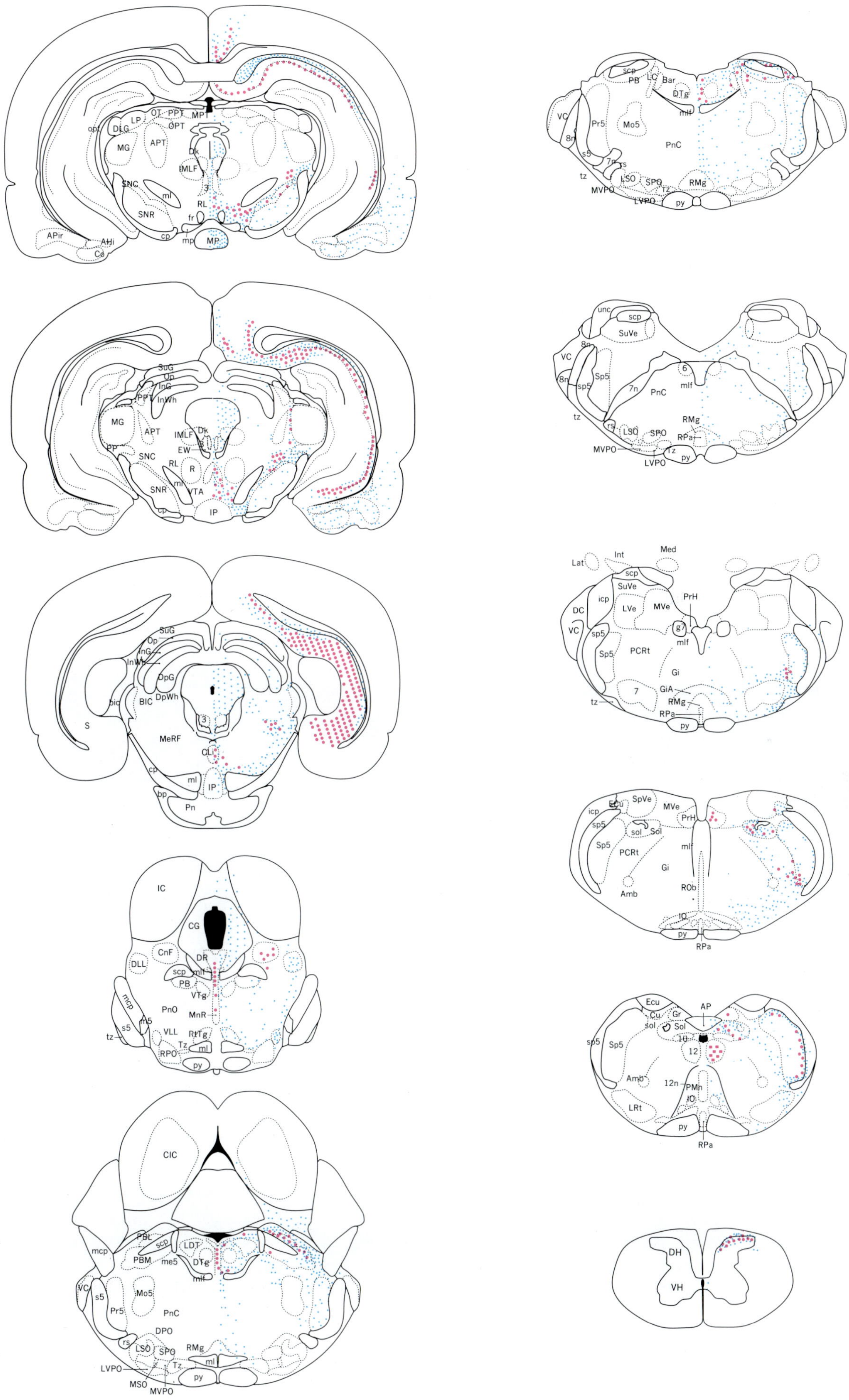

Neurotensin receptor

Structure and signalling

At present, binding assays have shown that the neurotensin receptor is known to exist in two types; high affinity ($K_d < 1\,\text{nM}$) and low affinity ($K_d > 1\,\text{nM}$)[5,6,9,10.] The low-affinity receptor can bind selectively to levocabastine ($IC_{50} = 10\,\text{nM}$) histamine blocker. When observing the brain as a whole, it is known that the low-affinity receptor increases gradually in proportion as it matures; in comparison, the high affinity receptor expresses transiently on a large scale at a juvenile phase[7] (Figs. 17.3, 4, 5). At present only the high-affinity type has been cloned. It consists of 424 amino acid residues and belongs to a G-protein coupled receptor type that has seven transmembrane domains, and is known to activate IP_3/DG and then to raise intracellular calcium concentration as well as activation of PKC(Fig. 17.5)[7]. There is also a report of the inhibition of adenylate cyclase activity.

The binding site of neurotensin to its receptor is located at the C-terminal part (8–13) of neurotensin, and in fact, the intracellular second messenger system does not change in neurotensin (1–8)[4]. The binding site of neurotensin is highly analogous to neuromedin N in its amino acid sequence. Thus neuromedin N is also considered to use the neurotensin receptor. At present, the existence of a neuromedin N specific receptor is still unclear. Neurotensin raises the intracellular Ca^{2+} concentration by intermediating IP_3, which has become visible by a recent calcium imaging technique that used a fluorochrome of Fura 2, etc. Similarly, neuromedin N also raises the intracellular calcium concentration in the same cells that respond to neurotensin (Fig. 17.6). In the neurons of the cerebral cortex or a part of neuroblast cell line (NG 108), a transient rise of calcium concentration was observed after stimulation by neurotensin or neuromedin N, and then the neurotensin receptor was strongly desensitized. This strong desensitization is one characteristic of the neurotensin receptor: the desensitization caused by neurotensin blocks the response to neuromedin N. Conversely, desensitization by neuromedin N blocks the response against subsequent neurotensin N. These ligands are therefore considered to use the same receptor, or at least the action site of desensitization is located at the same point on the information transduction path.

Distribution

Since no specific antibodies against neurotensin receptors had existed, an *in situ* hybridization method based on the basic sequence of high affinity receptors which had been cloned, was the best available method of investigating the distribution of receptor-producing cells. *In vitro* receptor binding autoradiography using an isotope-labelled ligand is a method of making the binding sites themselves visible. However, discrepancies between results derived from receptor binding and those from immunohistochemistry

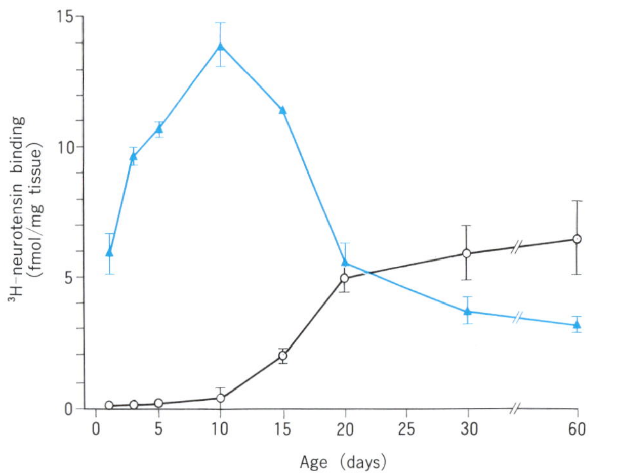

Fig. 17.3 Ontogeny of high- and low-affinity neurotensin receptors. [^{3}H]-neurotensin binding was measured with total brain homogenate. The high- and low-affinity receptors were distinguished by the presence or absense of levocavastin. The filled triangles show the high-affinity type receptor, while open circles show the low-affinity bindings (from ref. 7).

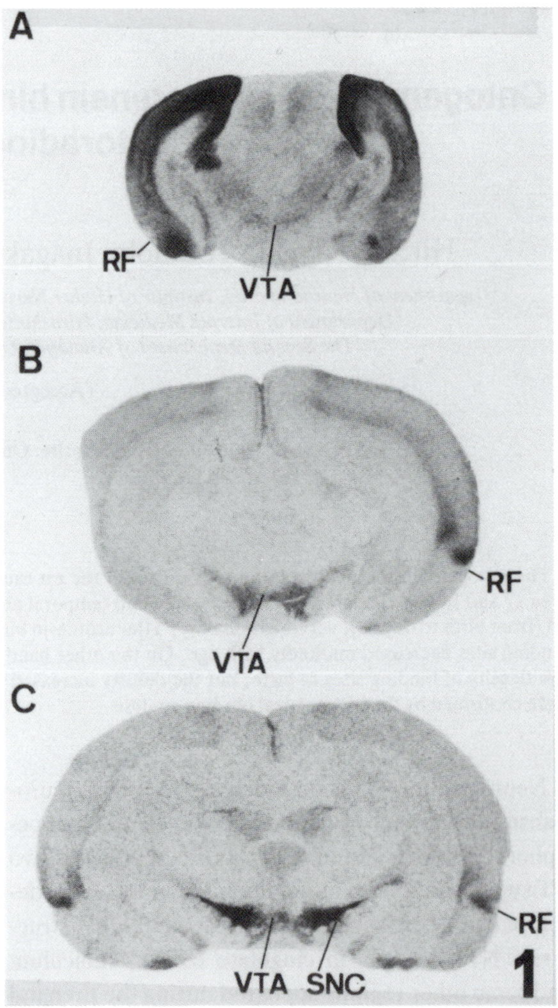

Fig. 17.4 *A*, Autoradiogram of 5-day-old rat brain showing [^{3}H]NT binding sites. Dense [^{3}H]NT binding sites were observed in the cortical area. RF, rhinal fissure. *B*, autoradiogram of 15-day-old rat brain showing [^{3}H]NT binding sites. The binding sites decreased in the cortical area, though in the area surrounding RF, VTA and SNC come to have denser binding sites. *C*, autoradiogram of 50-day-old rat brain showing [^{3}H]NT neurotensin binding sites. Very dense binding sites were observed in the VTA and SNC (from ref. 2).

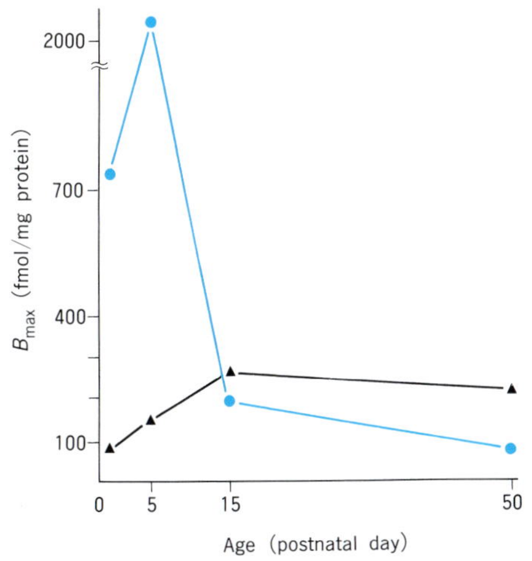

Fig. 17.5 The ontogenetic pattern of B_{max} values using slide-mounted sections; closed circles indicated cortex section and closed triangles indicated the midbrain section containing both VTA and SNC (from ref. 2).

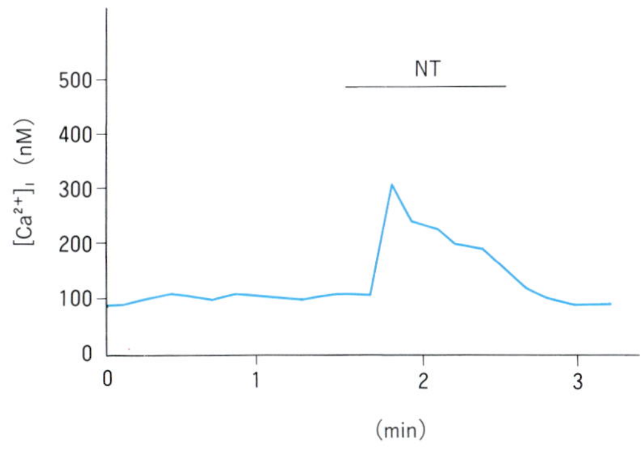

Fig. 17.6 Neurotensin (100 nM) elevates the cytosolic calcium concentration via transiently appearing neurotensin binding sites in cultured neonatal cortical cells. Intracellular calcium concentration $[Ca^{2+}]_i$ was measured by a calcium-sensitive dye, Fura-2 AM (from ref. 4).

using receptor antibody have sometimes been reported. So, in the distribution of binding sites presented here, it is not always clear whether the distribution of receptors is shown correctly or not. However, the areas that show results similar to the distribution of mRNA can be considered as reliable indicators of the distribution of receptors. The distribution of binding sites that is shown at present is the one in which ^{3}H-labelled neurotensin has been used as a ligand, and it has been considered that all show high-affinity binding sites[1–3,8].

The biggest major characteristic of the neurotensin receptor is that, in the cerebral cortex, there are extremely abundant high-affinity binding sites, and its receptor mRNA expresses transiently at a juvenile phase but is not observed in adults[5]. Considering the abundant expression of ligand simultaneously at a juvenile period and transiently[2,3], it can easily be inferred that the activities of neurotensin in the brain are associated with development. The results obtained from the binding site and *in situ* hybridization methods show that the high-affinity neurotensin receptors increase in a part of the area as the brain grows, in addition to the one that expresses abundantly but transiently at a juvenile period. The high-affinity receptors are also classified into two types, based on ontogeny. Type 1 is the one that expresses temporarily until a few weeks postnatally, as noted above for the cerebral cortex. Type 2 is the one that increases gradually after birth and then reaches its maximum concentration in adults. Broca's diagonal band nucleus, the substantia nigra, and ventral tegmental area are examples of the type 2. When observing the amount of expression of neurotensin receptors of the whole brain, the expression of type 1 is overwhelmingly higher than that of type 2 (Figs. 17.3, 17.4). To comprehend the physiological functions of neurotensin in the nervous system, we need further information on the exact distribution of receptor localization. Recently, the high affinity receptor has been localized immunohistochemically[11].

References

1. Kislauskis, E., Bullock, B., McNeil, S., and Dobner, P.R. (1988). *J. Biol. Chem.*, **263**, 4963–8.
2. Kiyama, H., Inagaki, S., Kito, S., and Tohyama, M. (1987). *Dev. Brain Res.*, **31**, 303–6.
3. Palacios, J.M., Pazos, A., Dietl, M.M., and Lichtensteiger, W. (1988). *Neuroscience*, **25**, 307–17.
4. Sato, M., Shiosaka, S., and Tohyama, M. (1991). *Dev. Brain Res.*, **58**, 97–103.
5. Sato, M., Kiyama, H., and Tohyama, M. (1992). *Neuroscience*, **48**, 137–49.
6. Schotte, A., Leysen, J.E., and Laduron, P.M. (1986). *Naunyn-schmiedebergs Arch. Pharmacol.*, **333**, 400–5.
7. Schotte, A. and Laduron, P.M. (1987). *Brain Res.*, **408**, 326–8.
8. Tanaka, K., Masu, M., and Nakanishi, S. (1990). *Neuron*, **4**, 847–54.
9. Uhl, G.R. (1991). In *Handbook of chemical neuroanatomy* (ed. A. Björklund, T. Hökfelt, and M. Kuhar). Vol. 9, pp. 443–51. Elsevier, Amsterdam.
10. Young, W.S. III and Kuhar, M.J. (1979). *Eur. J. Pharmacol.*, **59**, 161–3.
11. Boudin, H. *et al.* (1996). *J. Comp. Neurol.*, **373**, 76–89.

Neurotensin receptor (binding sites)

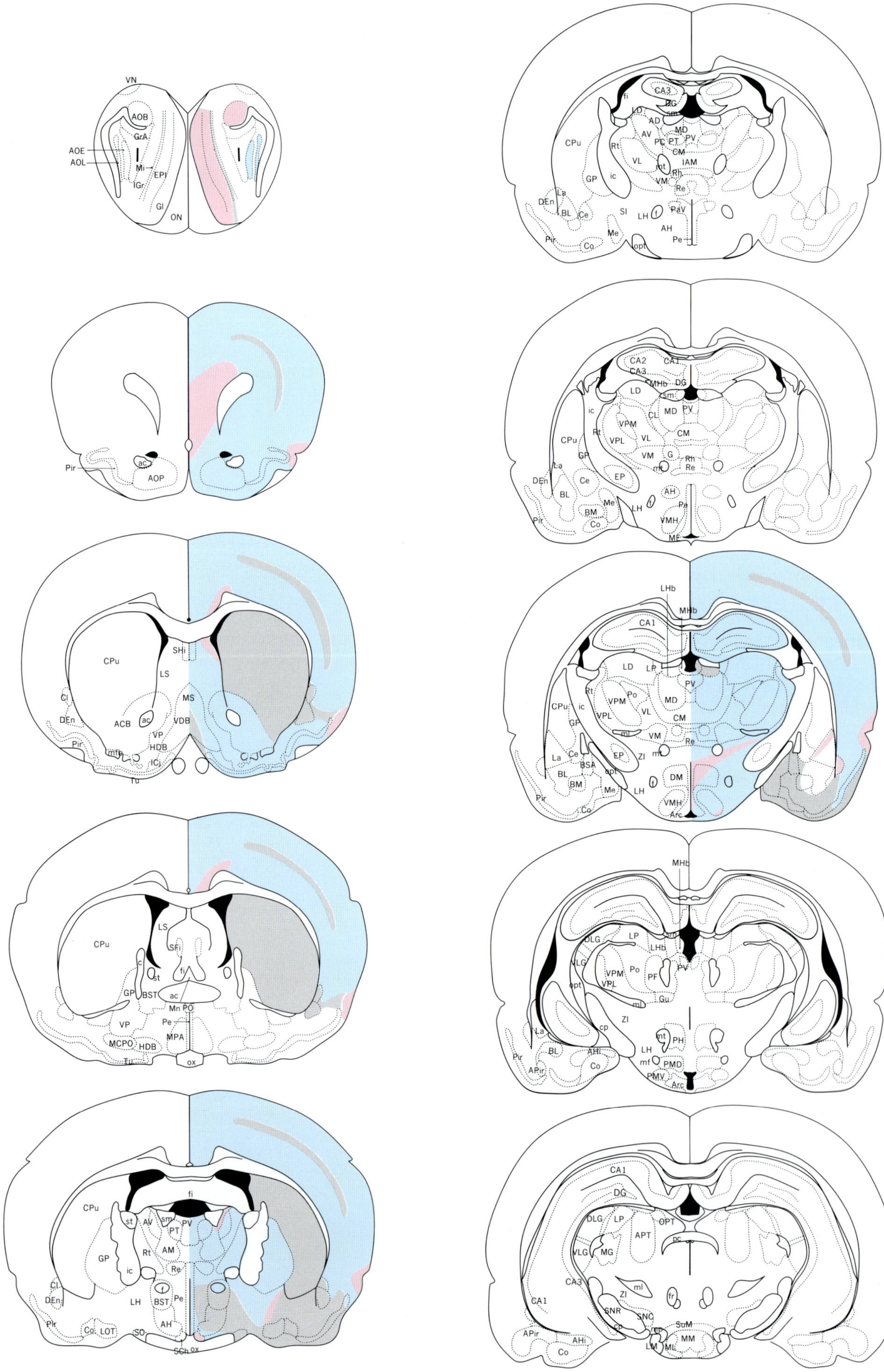

Neurotensin receptor (binding sites)

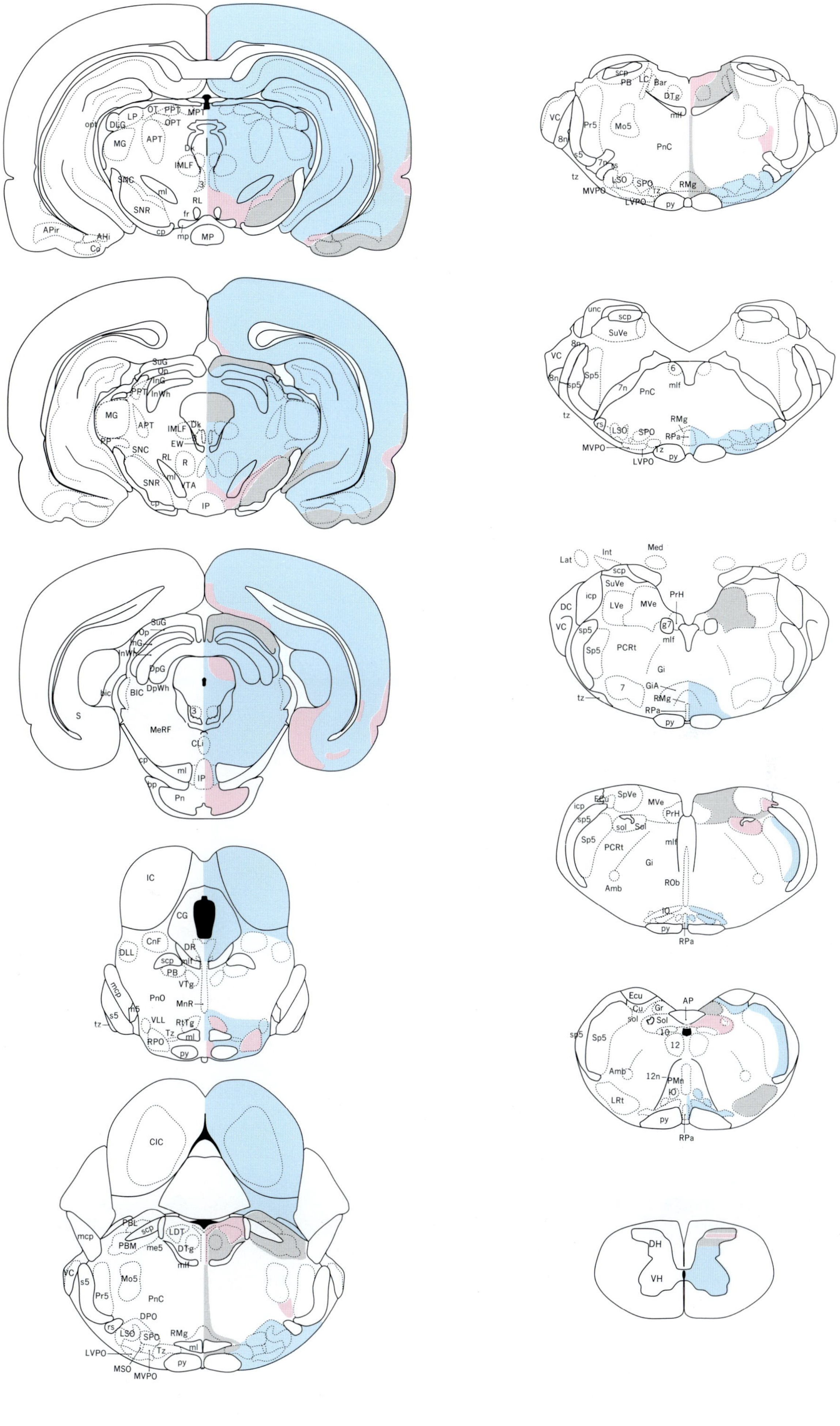

Neurotensin receptor (ISH)

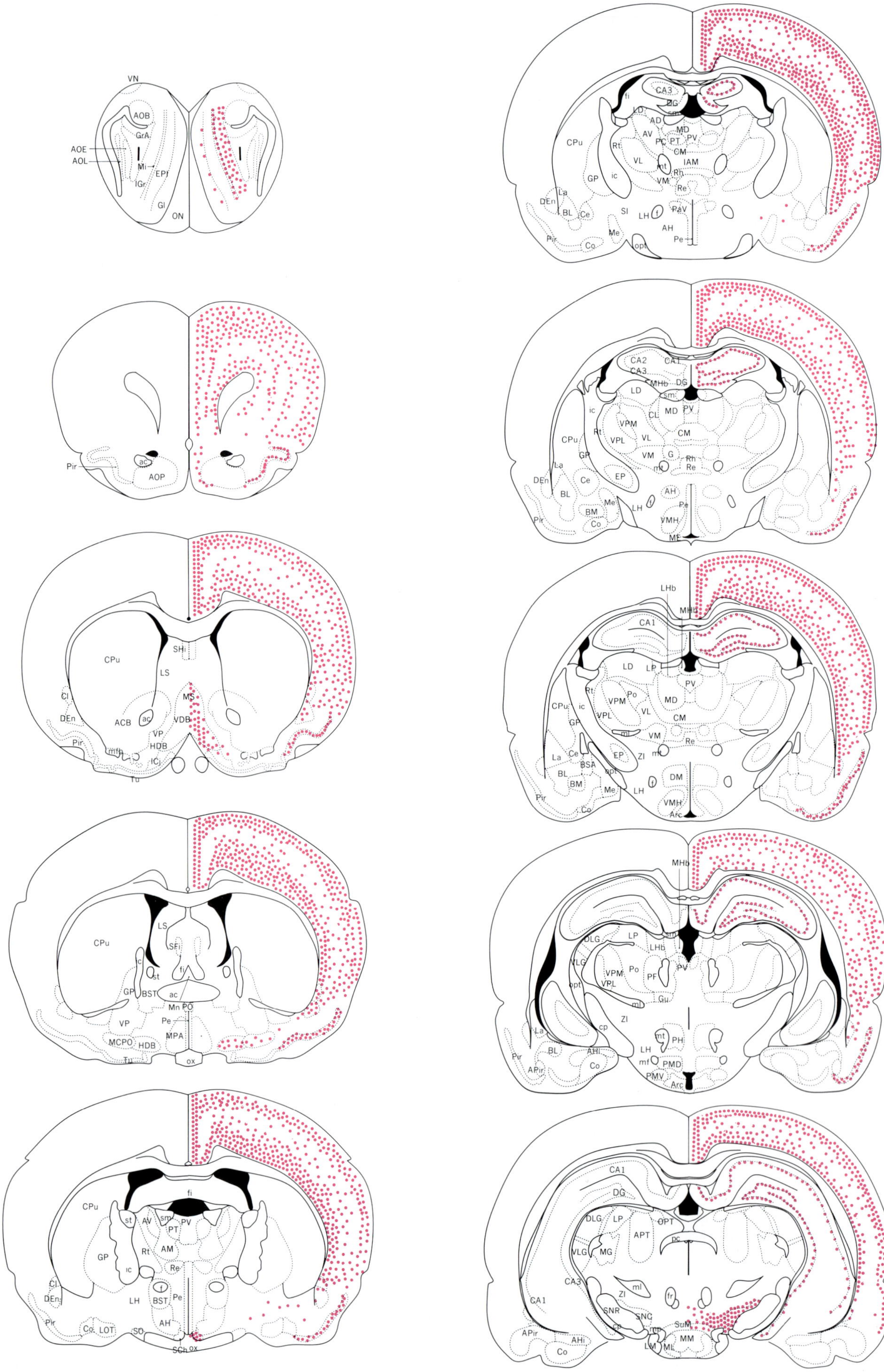

Neurotensin receptor (ISH)

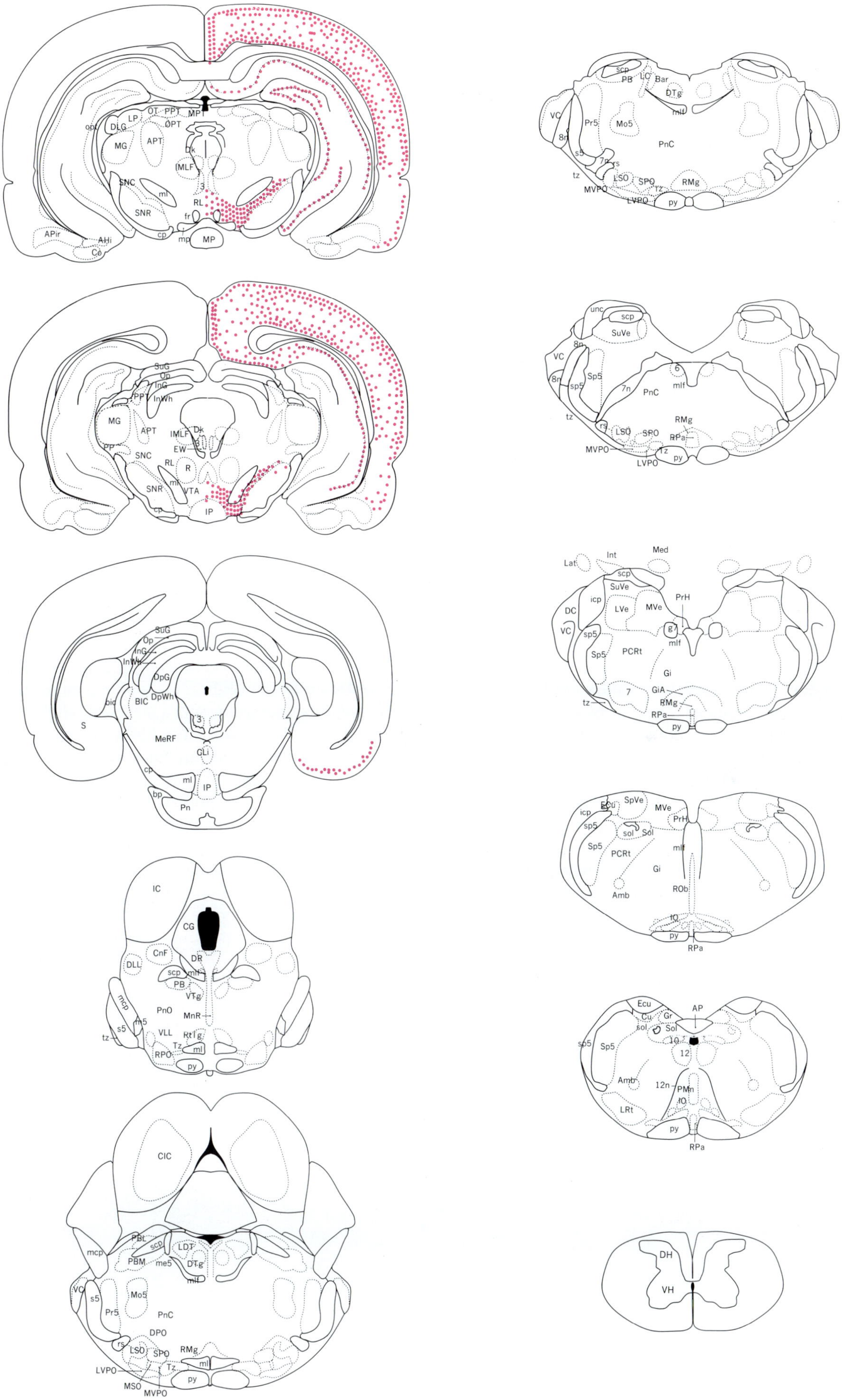

18 The angiotensin II system

Angiotensin II

Structure and physiological function

Angiotensin II (Ang II) is a peptide hormone that consists of eight amino acid residues and is the main physiologically active substance of the renin–angiotensin line (R–A line). Figure 18.1 shows a profile of the metabolic pathway of angiotensin. Ang II influences increased haemal resistance by a strong vasopressor action in peripheral tissues and stimulates the secretion of aldosterone from the adrenal cortex. In the central nervous system, Ang II is suggested to participate in the regulation and maintenance of blood pressure and water electrolyte balance as well as controlling other functions. These physiological actions of Ang II are considered to be expressed by Ang II binding to the specific receptor which exists at the surface of the cell membrane of the target cell.

The mechanism of blood pressure regulation by the intracerebral R–A line as deduced from the distribution of rat intracerebral Ang II fibres, is shown in Fig. 18.2[8]. Ang II infused into the rat cerebral ventricles acts on the medial preoptic area, the ventral anterior part of the circumventricular region including the organovasculosum lamina terminalis, and the paraventricular nucleus, and causes sympathetic nerve activation and vasopressin secretion which results in a rise in blood pressure. This vasopressor reaction is inhibited if α-receptor blocker is administered in the cerebral ventricles. Therefore, at least part of the physiological function of Ang II is mediated by noradrenaline-containing neurons. A population of Ang II neurons in the hypothalamus project into the solitary tract nucleus (NTS), and in this region Ang II is considered to inhibit the pressoreceptive reflex and then to cause vasopressor action. Practically, it has been reported that blood pressure rises when Ang II is infused in microdoses into the NTS. When sararasin, an antagonist of the Ang II receptor, is infused, vasopressor sensitivity of pressoreceptors enhances and then blood pressure decreases[2].

Meanwhile, when low doses of Ang II are infused into the ventrolateral part of the medulla oblongata (VLM), considered to be one of the vasomotor centres, it acts as a vasopressor at the rostral and a depressor at the caudal part respectively. In contrast, infusion of Ang II antagonist results in depressor action at the rostral part and vasopressor action at the caudal part[5,6,9]. These facts suggest that Ang II contributes to the regulation of the cardiovascular function in the VLM. When Ang II is infused into certain regions in the brain, amnesia is caused which is inhibited by captopril, an ACE inhibitor[7,10]. This finding also suggests that Ang II has physiological activities in the central nervous system.

Distribution and projection tract

Each constitutive element of the R–A line is also produced in the brain, and has been clarified by molecular biological, biochemical, and morphological methods[1,3,4]. In this atlas we have focused on mapping the central distribution of Ang II with potent biological activities.[3]

Ang II-containing cells are widely distributed in regions such as the brainstem, hypothalamus, limbic system, medulla oblongata, and spinal cord. High concentrations of Ang II-positive nerve fibres are observed in the median eminence, paraventricular hypothalamic nucleus, periventricular nucleus, dorsomedial hypothalamic nucleus, supraoptic hypothalamic nucleus, amygdaloid complex, bed nucleus of the stria terminalis, subfornical organ, lateral septum, parabrachial nucleus, solitary tract nucleus, nucleus of the spinal tract of the trigeminal nerve, anterolateral column of the spinal cord, and dorsal column of the spinal cord. The distribution of these Ang II-positive structures does not always correspond to the

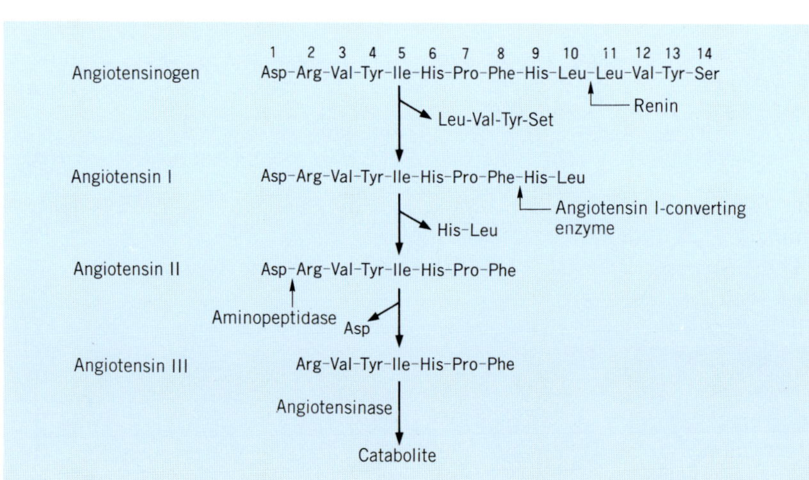

Fig. 18.1 Metabolism of angiotensinogen and angiotensins.

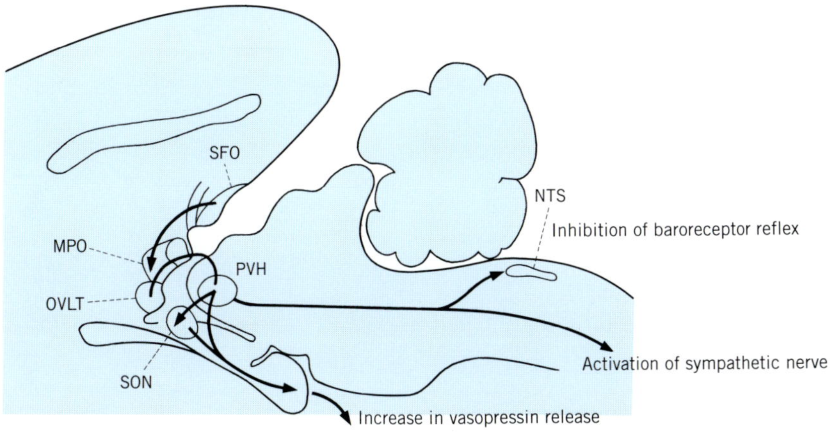

Fig. 18.2 Mechanism underlying the regulation of blood pressure by the brain renin–angiotensin system. SFO, subfornical organ; MPO, medial preoptic area; OVLT, organ vasculosum lamina terminalis; SON, supraoptic nucleus; PVH, paraventricular hypothalamic nucleus; NTS, solitary tract nucleus (from ref. 8).

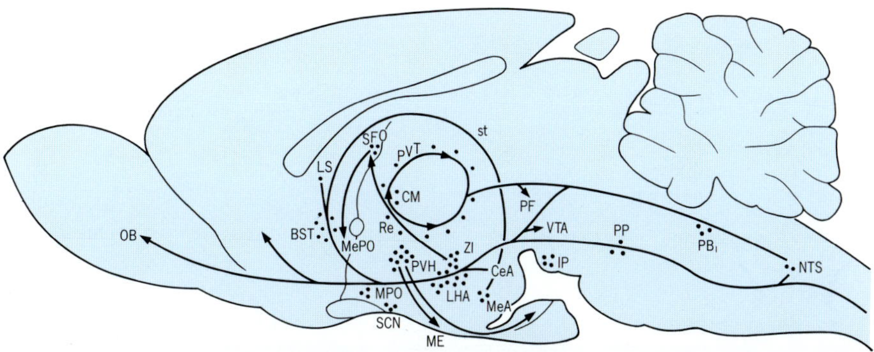

Fig. 18.3 Summary of angiotensin II-containing neuronal pathways in the rat brain. OB, olfactory bulb; BST, bed nucleus of the stria terminalis; LS, lateral septum; SFO, subfornical organ, MePO, median preoptic nucleus; MPO, medial preoptic area; SCN, suprachiasmatic nucleus; RE, reuniens thalamic nucleus; CM, centromedian thalamic nucleus; PVT, paraventricular thalamic nucleus; PVH, paraventricular hypothalamic nucleus; ME, median eminence; LHA, lateral hypothalamic area; ZI, zona incerta; MeA, medial amygdaloid nucleus (anterior division); CeA, central amygdaloid nucleus; PF, parafascicular nucleus; VTA, ventral tegmental area; IP, interpeduncular nucleus; PP, peripeduncular nucleus; PB_l, lateral parabrachial nucleus; NTS, solitary tract nucleus (from ref. 4).

locations of Ang II receptors, but in the hypothalamic area, subfornical organs, solitary tract nucleus, and area postrema, the local distributions of both are similar.

An outline of projection tracts of Ang II-containing nerve components is shown in Fig. 18.3, but one must take into account that not all of these have been certified[10]. Representative projection tracts are the ones that lead to the medial preoptic area from the subfornical organ, to the parabrachial nucleus from the solitary tract nucleus, and to the median eminence from the paraventricular hypothalamic nucleus. Thus, Ang II is produced in central neurons and transported with the axonal flow in nerve processes, and so it is likely that Ang II acts as a neurotransmitter.

Cerebral Ang II is also reported to exhibit complex interactions with other neurotransmitters, neuromodulators such as neuropeptides, monoamines, bradykinin, and prostaglandins. Details are given in Printz *et al.*[7].

References

1. Akaishi, T., Negoro, H., and Kobayashi, S. (1981). *Brain Res.*, **220**, 386–90.
2. Campagnole-Santos, M.J., Diz, D.I., and Ferrario, C.M. (1988). *Hypertension*, **11** (Suppl.), I167–I171.
3. Harding, J.W., Stone, L.P., and Wright, J.W. (1981). *Brain Res.*, **205**, 265–74.
4. Lind, R.W., Swanson, L.W., and Ganten, D. (1985). *Neuroendocrinology*, **40**, 2–24.
5. Muratani, H., Averill, D.B., and Ferrario, C.M. (1991). *Am. J. Physiol.*, **260**, R977–R984.
6. Muratani, H., Averill, D.B., and Ferrario, C.M. (1991). *Hypertension*, **17**, 422–3.
7. Printz, M.P., Ganten, D., Unger, T., and Phillips, M.I. (1982). In *Experimental brain research. The renin angiotensin system: a model for the synthesis of peptides in the brain* (ed. D. Ganten, *et al.*) Suppl. 4. Springer, Heidelberg.
8. Phillips, M.I. (1987). *Annu. Rev. Physiol.*, **49**, 413–35.
9. Sasaki, S. and Dampeny, R.A.L. (1990). *Hypertension*, **15**, 274–83.
10. Unger, T., Badoer, E., Ganten, D. *et al.* (1981). *Eur. J. Pharmacol.*, **70**, 467–8.

Angiotensin II (immunocytochemistry)

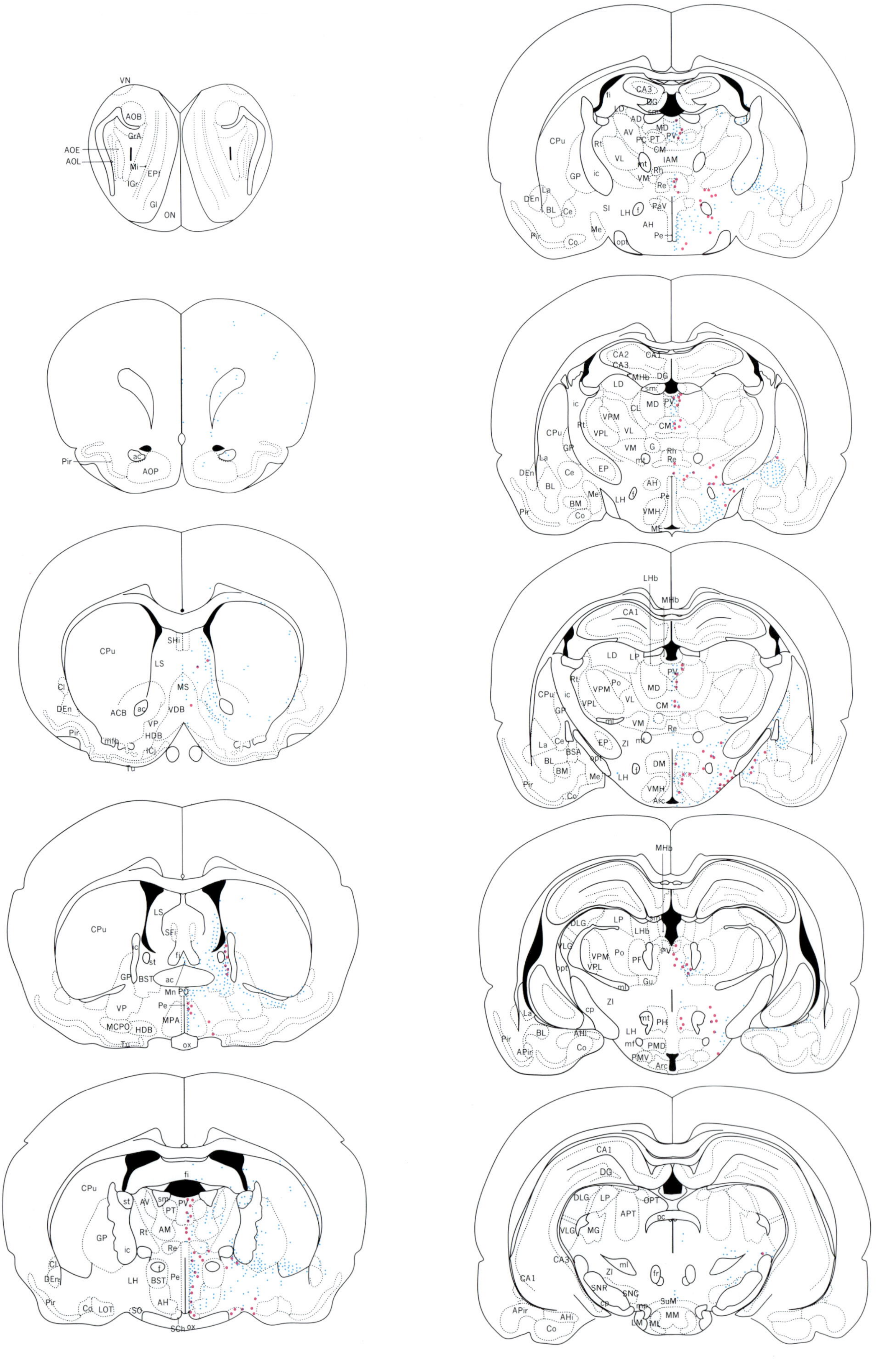

Angiotensin II (immunocytochemistry)

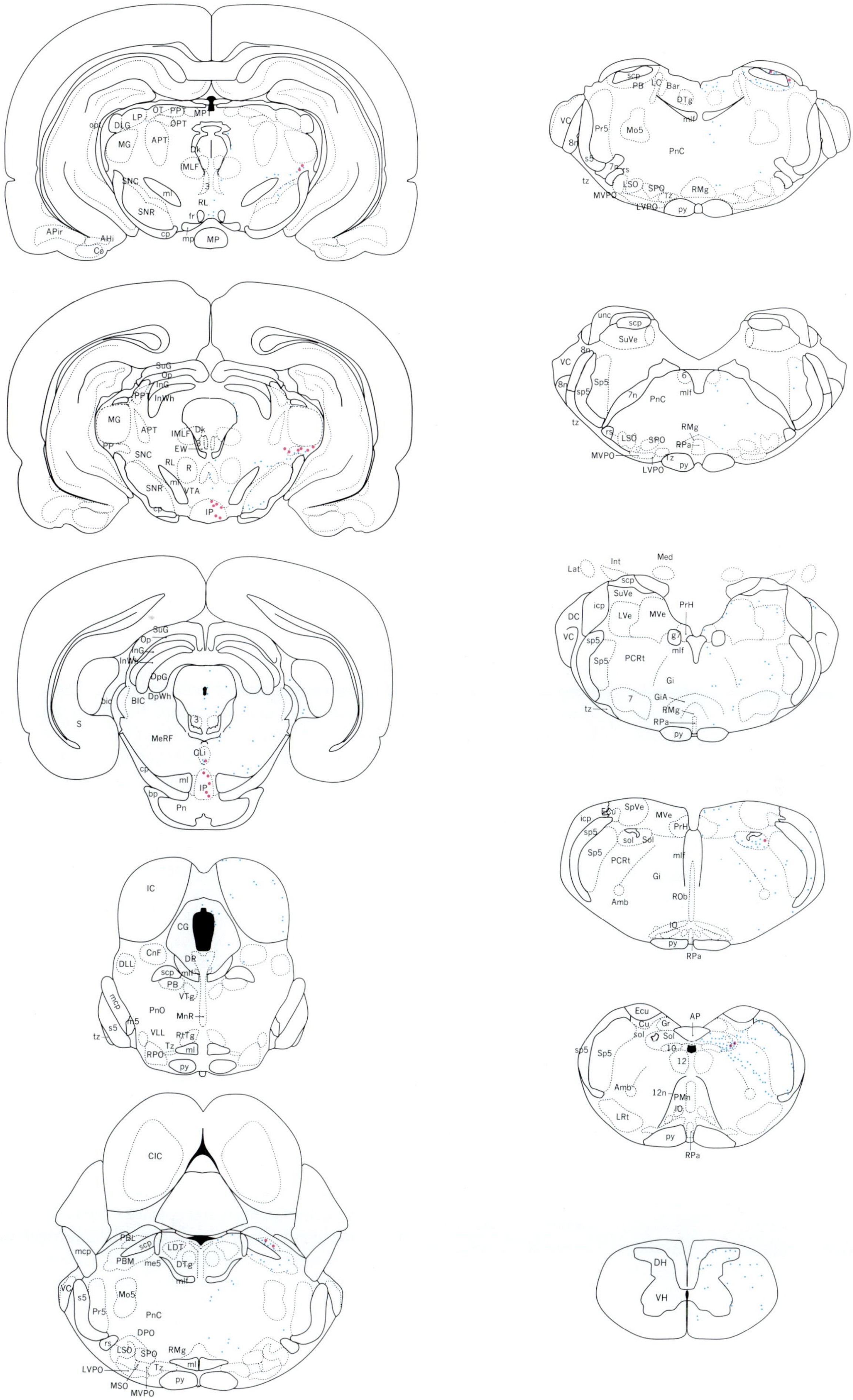

Angiotensin II receptor

Subtypes and structure

Angiotensin receptors have been classified into two subtypes (AT_1 and AT_2) based on their chemical properties (Table 18.1)[1,4,7,8]. AT_1 has affinity to Dup 753, one of the antagonists of Ang II; it is the binding site that is inhibited by sulfhydryl reducing agents such as dithiothreitol (DTT). AT_2 does not have an affinity to Dup 753, but binds with PD 123177, one of the antagonists of Ang II, and is not inhibited by DTT.

The molecular weight of AT_2 (dimer or tetramer) is believed to be around 65 kDa, although variations could occur due to differences in species, organs, ligands, and isolation methods used. With a molecualr biological technique, cDNA endocing AT_1 has been recently determined (Fig. 18.4)[3] and the primary structure of the receptor protein has been deduced (Fig. 18.5). Ang II receptor is a protein consisting of 359 amino acid residues and is considered to be a G-protein coupled type receptor with seven transmembrane sites. Details about AT_2 are not well understood and its structure and physiological significance have not been established.

1	MALNSSAEDG	IKRIQDDCPK	AGRHSYIFVM	IPTLYSIIFV	VGIFGNSLVV
51	IVIYFYMKLK	TVASVFLLNL	ALADLCFLLT	CPLWAVYTAM	EYRWPFGNHL
101	CKIASASVTF	NLYASVFLLT	CLSIDRYLAI	VHPMKSRLRR	TMLVAKVTCI
151	IIWLMAGLAS	LPAVIHRNVY	FIENTNITVC	AFHYESRNST	LPIGLGLTKN
201	ILGFLFPFLI	ILTSYTLIWK	ALKKAYEIQK	NKPRNDDIFR	IIMAIVLFFF
251	FSWVPHQIFT	FLDVLIQLGV	IHDCKISDIV	DTAMPITICI	AYFNNCLNPL
301	FYGFLGKKFK	KYFLQLLKYI	PPKAKSHSSL	STKMSTLSYR	PSDNMSSSAK
351	KPASCFEVE*				

Fig. 18.5 Deduced amino acid sequence of angiotensin II receptor. (I+II))–VII, possible transmembrane sequences; *, possible site for *N*-glycosylation; •, G-protein-binding site; ○, sequence identical with human mas protein; –, possible site of phosphorylation (from ref. 3).

Receptor and second messenger

The main intracellular signal transduction pathway mediated by AT_1 is shown in Fig. 18.6. When Ang II binds to the G-protein coupled specific receptor, phospholipase C is activated and this accelerates the decomposition of phosphatidyl inositol diphosphoric acid (PIP_2). As a result, inositol 1,4,5-triphosphoric acid (IP_3) and diacylglycerol (DG) are produced and this leads to activation of various phosphorylation enzymes[2].

1	CGGGGACTGG	ATGATGCTGT	CCCGCTGGAG	AGGATTCGTG	GCTTGAGTCC
51	TGTTCCACCC	GATCACCGGG	GTGGCCGAGG	CGGTACCAGC	GCTGCGCTCT
101	CTCAGCTCCT	GCCACATTCC	CTGAGTTAAC	ATATGAGAGA	TTGGTATAGA
151	AATGGCTGGC	ATTTTGTCTG	GATAAATCAC	ACAACCCTCC	CAGAAAGTGA
201	TCACCAGGTC	AAGTGGATTT	CGAATAGTGT	CTGAGACCAA	CTCAACCCAG
251	AAAAACAAAA	TGGCCCTTAA	CTCTTCTGCT	GAAGATGGTA	TCAAAAGAAT
301	CCAAGATGAC	TGCCCCAAGG	CTGGCAGGCA	CAGTTACATA	TTTGTCATGA
351	TCCCTACCCT primer-1	CTACAGCATC	ATCTTTGTGG	TGGGAATATT	TGGAAACAGC
401	TTGGTGGTGA	TTGTCATTTA	CTTTTACATG	AAGCTGAAGA	CTGTGGCCAG
451	CGTCTTTCTT	CTCAATCTCG	CCTTGGCTGA	CTTATGCTTT	TTGCTGACTT
501	GTCCCCTGTG	GGCAGTCTAT	ACCGCTATGG	AGTACCGCTG	GCCCTTCGGC
551	AATCACCTAT	GTAAGATCGC	TTCGGCCAGC	GTGACGTTCA	ACCTCTACGC
601	CAGTGTGTTC	CTTCTCACGT	GTCTCAGCAT	CGACCGCTAC	CTGGCCATCG
651	TCCACCCAAT	GAAGTCTCGC	CTTCGCCGCA	CGATGCTGGT	GGCCAAAGTC
701	ACCTGCATCA	TCATCTGGCT	GATGGCTGGC	TTGGCCAGTT	TGCCAGCTGT
751	CATCCACCGA	AATGTATACT	TCATCGAGAA	CACCAATATC	ACAGTGTGCG
801	CGTTTCATTA primer-2	TGAGTCTCGG	AATTCGACGC	TCCCCATAGG	GCTGGGCCTT
851	ACCAAGAATA	TTCTGGGCTT	CTTGTTCCCT	TTCCTTATCA	TTCTCACCAG
901	CTATACCCTT	ATTTGGAAAG	CTCTAAAGAA	GGCTTATGAA	ATTCAAAAGA
951	ACAAACCAAG	AAACGATGAC	ATCTTTAGGA	TAATTATGGC	GATTGTGCTT
1001	TTCTTCTTCT	TTTCCTGGGT	CCCCCACCAA	ATATTCACTT	TCCTGGATGT
1051	GCTGATTCAG	CTGGGCGTCA	TCCATGACTG	TAAAATTTCT	GACATCGTGG
1101	ACACTGCCAT	GCCCATCACC	ATCTGCATAG	CGTATTTTAA	CAACTGCCTG
1151	AACCCTCTGT	TCTACGGCTT	TCTGGGGAAG	AAATTTAAAA	AGTATTTCCT
1201	CCAGCTCCTG	AAATATATTC	CCCCAAAGGC	CAAGTCCCAC	TCAAGCCTGT
1251	CTACGAAAAT	GAGCACGCTT	TCTTACCGGC	CTTCGGATAA	CATGAGCTCA
1301	TCGGCCAAAA	AGCCTGCGTC	TTGTTTTGAG	GTGGAGTGAC	AGGTTCAAAG
1351	CACACTGGCA	ATGTAATGCC	CTGACAGAAG	CCAGGGAGCC	TCTGACTAAA
1401	TGGCTTACGA	CCAAAGGACC	ATTCACCCTG	CCTCAGGATC	TAAGCAGAAA

Fig. 18.4 DNA sequence encoding angiotensin II receptor. Initiation and stop codons are indicated by double underlines.

Distribution

In this receptor atlas, binding sites of Ang II (including AT_1 and AT_2) are presented, based on the reports of Gehlert *et al.*[1] and Tsutsumi *et al.*[8]. Furthermore, besides the areas shown in this atlas, the OVLT, infundibular stalk, and subfornical organ are also mentioned as sites of high receptor concentration.

The distributions of AT_1 and AT_2 in the brain are somewhat different. AT_1 is distributed mainly in the paraventricular hypothalamic nucleus, solitary tract nucleus, dorsal nucleus of the vagus nerve, subfornical organ, supraoptic hypothalamic nucleus, suprachiasmatic nucleus, piriform cortex, choroid plexus, etc. AT_2 is distributed in the thalamus, medial amygdaloid nucleus, inferior olive, lateral septum, superior colliculus, etc. (Table 18.2)[7].

Physiological function

Cerebral Ang II receptors are abundant in the subfornical organ, suprachiasmastic nucleus, and OVLT, which are involved in water-drinking behaviour or pressor action, and especially in dehydration an increase in the receptors of the subfornical organ has been reported[5]. High densities of Ang II receptors exist in the paraventricular hypothalmic nucleus, solitary tract nucleus, area postrema, dorsal vagus nucleus, and supraoptic hypothalamic nucleus, which regulate secretion of vasopressin, ACTH, prolactin, or autonomic functions[5]. Since the distribution of Ang II receptor is partially analogous to that of catecholamine-containing nerve components, in such sites Ang II is suggested to stimulate noradrenaline secretion presynaptically. Central action of Ang II is believed to depend on, at least, the activities of the sympathetic nervous system.

Table 18.1. Subtypes of angiotensin receptor

subtypes	AT_1	AT_2
Affinity to Ang II and Ang III	Ang II > Ang III	Ang I = Ang III
Selective antagonists	Dup 753	PD 123177 PD 123319 CGP 42112 A
Second messengers	IP_3/DG cAMP ↓	

Ang, angiotensin; IP3, inositol triphosphate; DG, diacyl glycerol

Table 18.2. Distribution of angiotensin receptor subtypes in the rat brain (from ref. 6)

$AT_1 \gg AT_2$	$AT_1 \geq AT_2$	$AT_2 > AT_1$
Anterior pituitary	Parabrachial nucleus	Lateral septum
Periventricular organs		
Subfornical organ	Central grey matter	Medial amygdaloid nucleus
Organ vasculosum lamina terminalis	Raphe nuclei	Thalamus
Area postrema		Reuniens thalmic nucleus
Median eminence		
Medial preoptic area		Medial geniculate body
Hypothalamus		
Suprachiasmatic nucleus		Lateral habenular nucleus
Paraventricular hypothalamic nucleus		Hypothalamus
		Superior colliculus
Periventricular hypothalmic nucleus		Inferior colliculus
		Locus coeruleus
Dorsomedial hypothalamic nucleus		Inferior olive
Supraoptic nucleus		
Piriform cortex		
Ventral hippocampus		
Solitary tract nucleus		
Dorsal motor nucleus of vagus		
Spinal nucleus of trigeminal nerve		

Ang II receptors are classified pharmacologically into AT_1 and AT_2 and their distributions also differ. The differences in their physiological actions have also been reported. The expression of AT_1 has been identified selectively in various brain nuclei that regulate blood pressure, water intake, salt ingestion, and vasopressin secretion[9]. The functions of intracerebral AT_2 receptors have not yet been clarified. Since adult rats possess only a small amount of AT_2 receptors and in juvenile rats AT_2 receptors show a different distribution from AT_1, it is believed that AT_2 receptors play important roles in the development of visual transmission, motion, sensation, and limbic function, or in the structural organization of the central nervous system[8].

Recently, Mukoyama *et al.*[6] have cloned a cDNA, encoding a unique 363-amino-acid protein with pharmacological specificity, tissue distribution and developmental pattern of the AT_2 receptor. It is 34 per cent identical in sequence to the AT_1 receptor, sharing a seven-transmembrane domain topology possibly without G-protein coupling.

References

1. Gehlert, D.R., Gackenheimer, S.L., and Schober, D.A. (1991). *Brain Res.*, **546**, 161–5.
2. Griedling, K.K., Berk, B.C., Socorro, L. *et al.* (1988). *Clin. Exp. Pharmacol. Physiol.*, **15**, 105–12.
3. Iwai, N., Yamano, Y., Chaki, S. *et al.* (1991). *Biochem. Biophys. Res. Commun.*, **177**, 299–304.
4. Leung, K.H., Smith, R.D., Timmermans, P.B.M.W.M., and Chiu, A.T. (1991). *Neuroscience*, **123**, 95–8.
5. Mendelsohn, F.A.O. (1985). *J. Hypertens.*, **3**, 307–16.
6. Mukoyama, M., Nakajima, M, Horiuchi, M. *et al.* (1993). *J. Biol. Chem.*, **268**, 24539–42.
7. Speth, R.C., Rowe, B.P., Grove, K.L. *et al.* (1991). *Brain Res.*, **548**, 1–8.
8. Tsutsumi, K. and Saavedra, J.M. (1991). *Am. J. Physiol.*, **261**, R209–R216.
9. Sutsumi, K. and Saavedra, J.M. (1991). *J. Neurochem.*, **56**, 348–51.

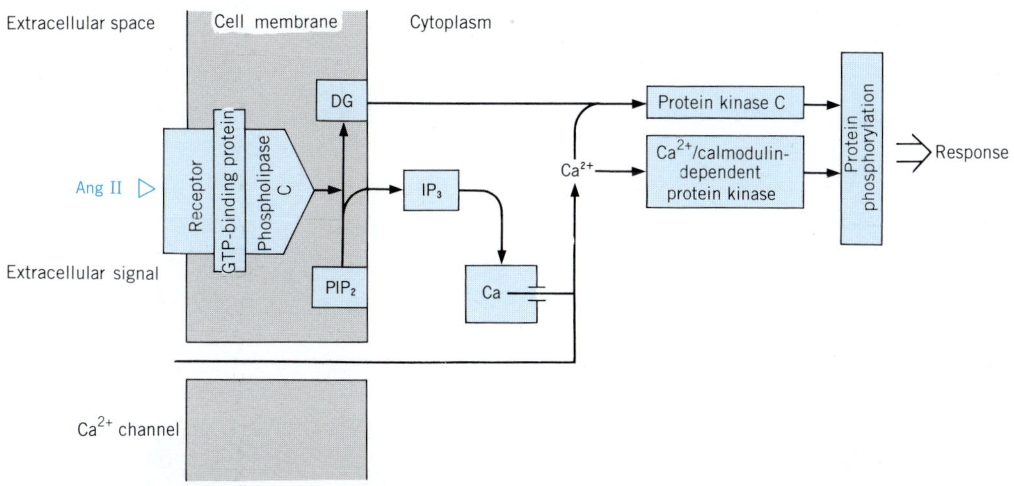

Fig. 18.6 Intracellular signal transduction mediated by angiotensin II receptor. PIP_2, phosphatidylinositol diphosphate; DG, diacyl glycerol; IP_3, inositol triphosphate; Ang II, angiotensin II.

Angiotensin II receptor (binding sites)

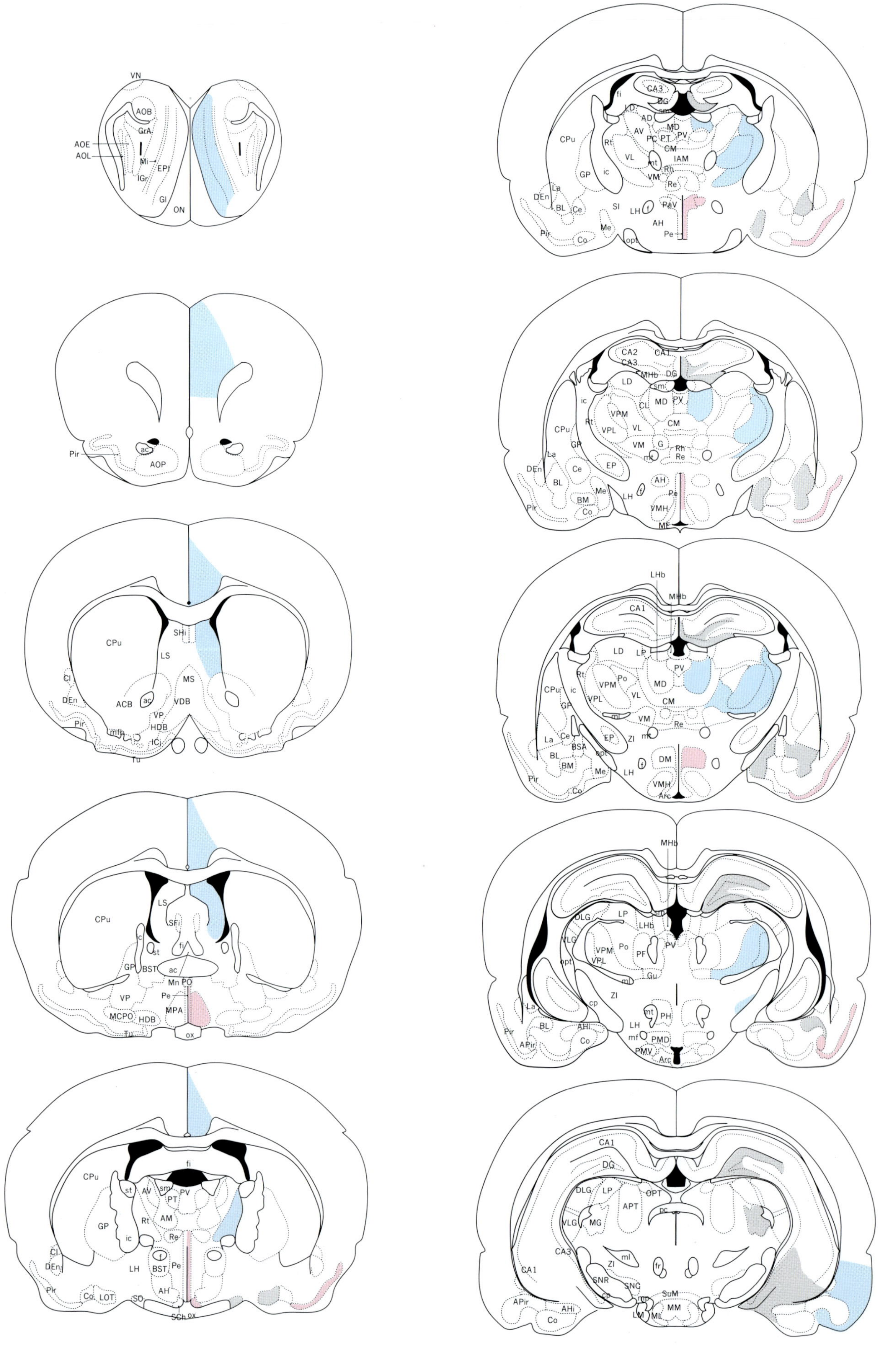

Angiotensin II receptor (binding sites)

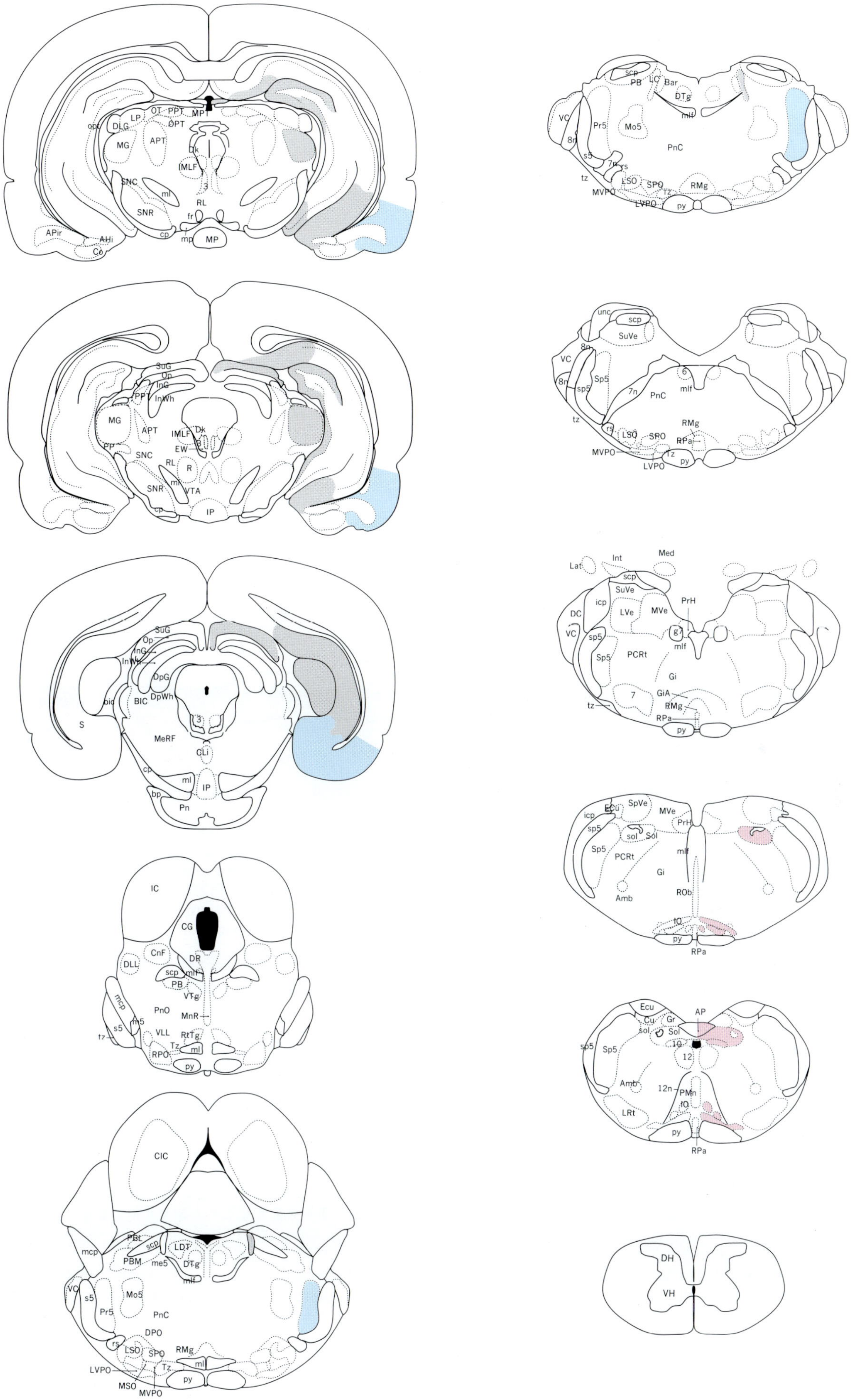

19 The corticotropin releasing factor system

Corticotropin releasing factor

Physiological actions

Corticotropin releasing factor (CRF) is a peptide that was isolated and identified from the ovine brain by Vale *et al.*[6] in 1981. CRF is secreted into hypothalamohypophyseal portal vessels and then accelerates liberation of ACTH as well as β-endorphin from the anterior lobe of the hypophysis, and thus CRF is believed to participate mainly in neuroendocrine functions *in vivo*. However, later immunohistochemical studies[2,8,11,15] clarified that CRF exists widely in the parvicellular group of the paraventricular hypothalamic nucleus as well as in extrahypothalamic regions. Thus CRF has been suggested to function as a neurotransmitter or neuromodulator in the brain. When CRF is injected into the cerebral ventricles, various autonomic reactions and/or behavioural activation are observed. These reactions are inhibited by an intraventricular administration of CRF antiserum or CRF antagonist, so CRF is considered to play an active role in the central nervous system[1,9,14].

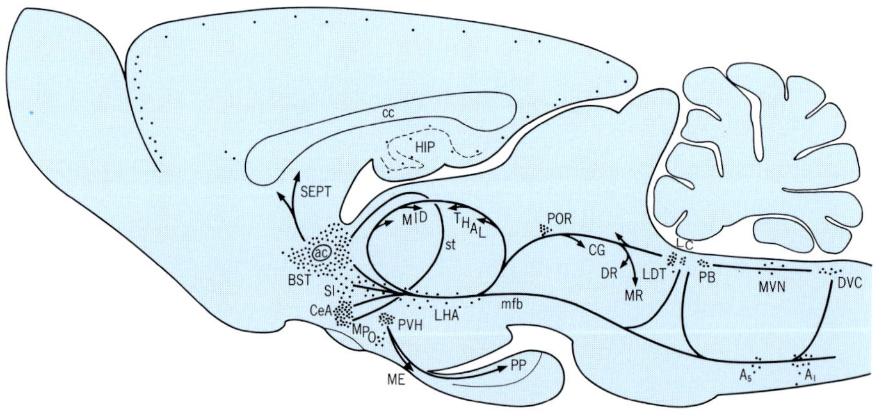

Fig. 19.1 Summary of CRF-containing neuronal pathways. A_1, A_5, noradrenergic cell groups; BST, bed nucleus of the stria terminalis; CeA, central amygdaloid nucleus; CG, central grey matter; DR, dorsal raphe nucleus; DVC, dorsal motor nucleus of vagus; HIP, hippocampus; LC, locus ceruleus; LDT, laterodorsal tegmental nucleus of pons; LHA, lateral hypothalamic area; ME, median eminence; MIDTHAL, midline thalamic nuclei; MPO, medial preoptic area; MR, midline raphe nuclei; MVN, medial vestibular nucleus; PB, parabrachial nucleus; POR, perioculomotor nucleus; PVH, paraventricular hypothalamic nucleus; SEPT, septum; SI, substantia innominata; ac, anterior commissure; cc, corpus callosum; PP, posterior pituitary; st, stria terminalis (from ref. 15).

```
                    5         10        15        20        25        30        35        40
Rat CRF     S E E P P I S L D L T F H L L R E V L E M A R A E Q L A Q Q A H S N R K L M E I I ■

Ovine CRF   S Q E P P I S L D L T F H L L R E V L E M T K A D Q L A Q Q A H S N R K L L D I A ■

Sucker U-I  N D D P P I S I D L T F H L L R N M I E M A R I E N E R E Q A G L N R K Y L D E V ■

Carp U-I    N D D P P I S I D L T F H L L R N M I E M A R N E N Q R E Q A G L N R K Y L D E V ■
                      5         10        15        20        25        30        35        40
Sauvagine     Z G P P I S I D L S L E L L R K M I E I E K Q E K E K Q Q A A N N R L L L D T I ■
```

Fig. 19.2 Amino acid sequence of CRF-related peptides (single-letter code). Amino acids common to all peptides are indicated in italics. Z, pyrrolidone carboxylic acid; ■, aminated C-terminus (from ref. 9).

Distribution

Among the pathways of CRF-containing neurons, there are descending fibres that lead to the brain stem from the central amygdaloid nucleus, ascending fibres that lead to the lateral septum from the hypothalamus, and nerve fibres that lead to the hypothalamus or the spinal cord from the brain stem[10,13]. Details can be found in the published reports. A scheme of CRF projection tracts is shown in Fig. 19.1[2]. Not all of these projection tracts have been experimentally verified. Studies on the chemical neuroanatomy of CRF projection tracts strongly suggest that CRF functions as a neurotransmitter.

Since CRF was first isolated and its structure determined from the sheep brain (referred to as ovine CRF, oCRF), peptides that show neuroendocrine activities similar to ovine CRF have been identified from various species and these have been gathered into a CRF family (see Fig. 19.2). Among these, sauvagine was isolated and identified from the cutis of frog[7] and urotensin I was characterized from the urophysis of fish[6].

The distribution of CRF-positive cells and fibres presented in this atlas has been drawn mainly from our data based on the method of cobalt-enhanced immuno-histochemistry[11,12]. *In situ* hybridization studies have not shown the overall distribution of CRF mRNA-producing neurons, but in the olfactory bulb, hippocampus, cerebral cortex, paraventricular hypothalamic nucleus, solitary tract nucleus, inferior olive and auditory system, the data obtained from *in situ* hybridization and immunohistochemistry agree except for a few regions[3,4,5].

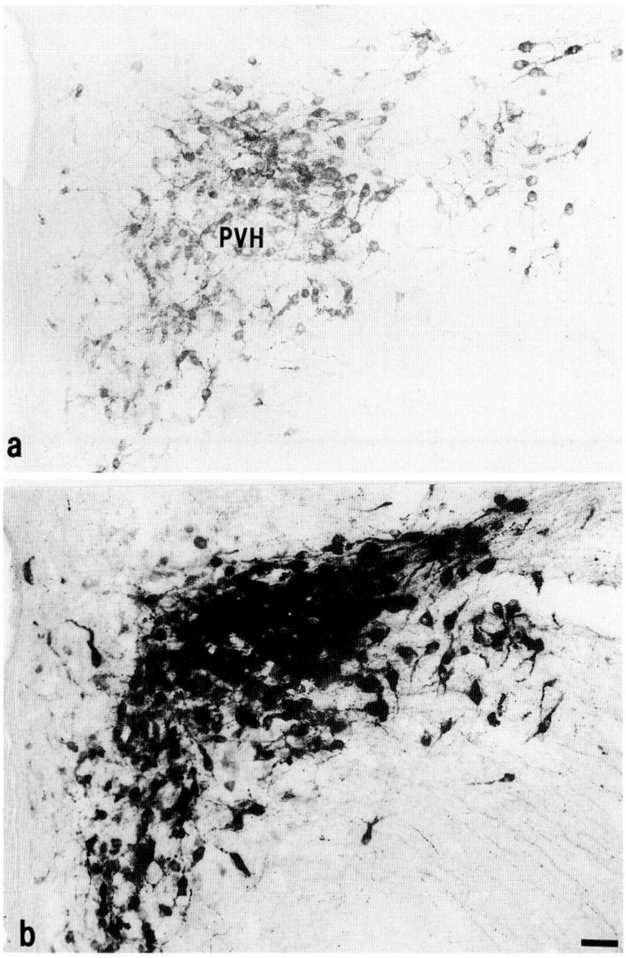

Fig. 19.3 CRF-immunoreactive cells in the paraventricular hypothalamic nucleus (PVH). (a) Stained with conventional immunohistochemistry (PAP method); (b) stained with the modified cobalt-glucose oxidase–diaminobenzidine method. Note a marked intensification of CRF immunoreaction in (b).

References

1. Britton, D.R., Koob, G.F., Rivier, J., and Vale, W. (1982). *Life Sci.*, **31**, 363–8.
2. Cummings, S., Elde, R., and Lindall, A. (1983). *J. Neurosci.*, **3**, 1355–68.
3. Cummings, S.L., Young, W.S., Bishop, G.A. *et al.* (1989). *J. Comp. Neurol.*, **280**, 501–21.
4. Imaki, T., Nahon, J.L., Sawchenko, P.E., and Vale, W. (1989). *Brain Res.*, **496**, 35–44.
5. Imaki, J., Imaki, T., Vale, W., and Sawchenko, P.E. (1991). *Brain Res.*, **547**, 28–36.
6. Lederis, K., Letter, A., McMaster, D. *et al.* (1982). *Science*, **218**, 162–5.
7. Montecucchi, P.C., Anastasi, A., De Castiglione, R., and Erspamer, V. (1980). *Int. J. Peptide Protein Res.*, **16**, 191–9.
8. Olschowka, J.A., O'Donohue, T.L. Mueller, G.P., and Jacobowitz, D.M. (1982). *Peptides*, **3**, 995–1015.
9. Rivier, C., Rivier, J., and Vale, W. (1986). *Science*, **231**, 607–9.
10. Sakanaka, M., Shibasaki, T., and Lederis, K. (1986). *Brain Res.*, **382**, 213–38.
11. Sakanaka, M., Shibasaki, T., and Lederis, K. (1987). *J. Comp. Neurol.*, **260**, 256–98.
12. Sakanaka, M., Shibasaki, T., and Lederis, K. (1987). *J. Histochem. Cytochem.*, **35**, 207–12.
13. Sakanaka, M., Magari, S., Shibasaki, T., and Lederis, K. (1988). *J. Comp. Neurol.*, **270**, 404–15.
14. Sutton, R.E., Koob, G.F., LeMoal, M. *et al.* (1982). *Nature*, **297**, 331–3.
15. Swanson, L.W., Sawchenko, P.E., Rivier, J., and Vale, W.W. (1983). *Neuroendocrinology*, **36**, 165–86.
16. Vale, W., Spiess, J., Rivier, C., and Rivier, J. (1981). *Science*, **213**, 1394–7.

Corticotropin releasing factor (immunocytochemistry)

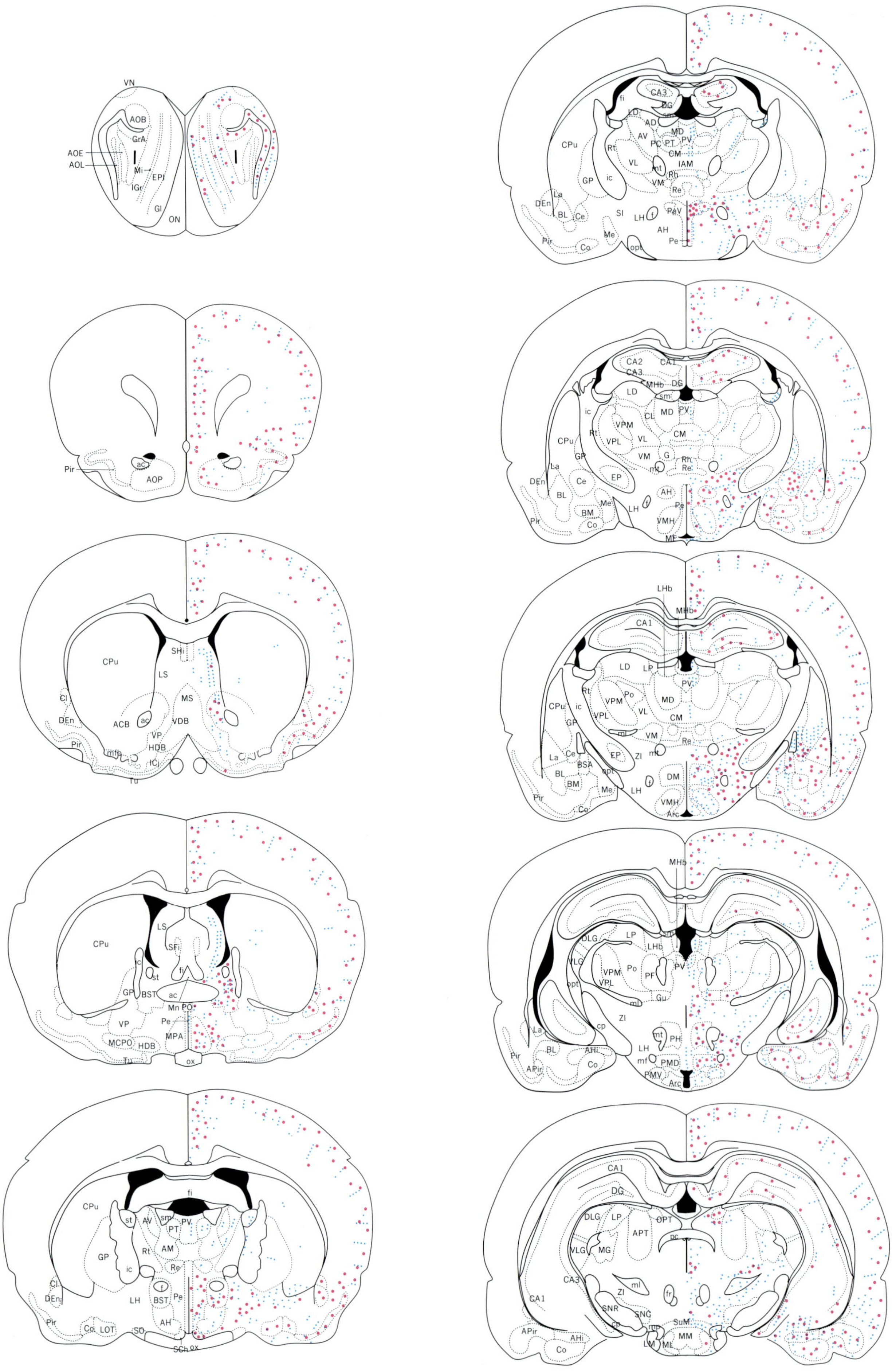

Corticotropin releasing factor (immunocytochemistry)

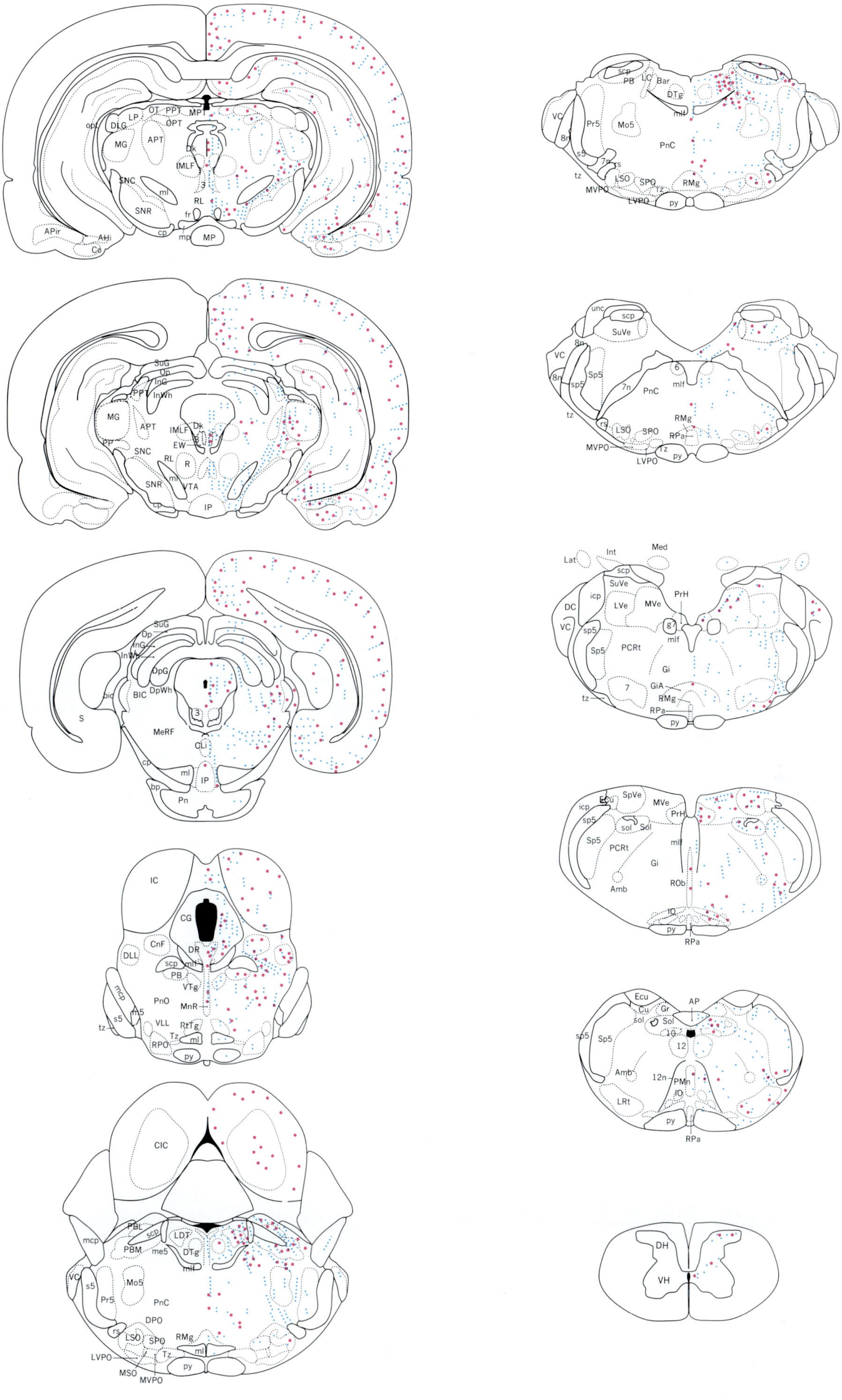

Corticotropin releasing factor receptor

Chemical properties

The nature, regulation mechanism, and distribution of CRF receptors have been studied using ^{125}I-Tyr-CRF. Scatchard plot analysis suggests that the ^{125}I-Tyr-rat CRF binding to CRF receptors depends on temperature, time, pH, ionic strength, and cation composition in buffer solution; under optimum conditions (10 mg magnesium chloride containing 50 mM Tris hydrochloric acid buffer solution, pH 7.2, 23° C, reaction time 60 min), K_d is approximately 0.2 nM[7]. B_{max} differs from brain nucleus to brain nucleus, with the ratio of binding sites at the olfactory bulb (B_{max} 243 fmol/mg protein) set at 100 (Fig. 19.4). A second group of receptors with K_d value 20 nM can also be detected[7]. CRF receptors bind with CRF as well as its analogues (α-helical oCRF, Ac oCRF etc.), but they do not bind with angiotensin II, arginine vasopressin, VIP, or GRF analogues; thus they show high hormone specificity (Table 19.1).

Receptor and second messenger

After binding to the specific receptor on the cell membrane of a target cell, CRF activates the adenylate cyclase–cyclic AMP–protein kinase line with intermediating guanine nucleotide protein (G protein)[1]. Binding with its receptor, CRF is taken into the target cell within several minutes following signal transduction (receptor mediated internalization) and is transported to lysosomes and the Golgi apparatus, and is then metabolized[3,4,5,8].

Distribution

CRF receptors exist in high concentration in the anterior and intermediate lobes of the hypophysis[6]. In the brain, they are widely distributed in the grey matter of cerebral cortex, the cerebrum, basal ganglia, paraventricular hypothalamic nucleus, external layer of median eminence, various nuclei of cranial nerves, olivary nuclei, deep cerebellar nuclei, cerebellar cortex, olfactory bulb, amygdaloid complex, claustrum, and spinal cord[4,9]. Since intracerebral distribution of these receptors corresponds approximately to that of CRF-positive structures which have been identified by immunostaining, CRF is believed to function as a neurotransmitter or neuromodulator in the brain. The CRF receptor atlas presented here is drawn from the data of binding sites of ^{125}I-Tyr^{32}-oCRF as a probe (K_d 6.2± 0.8 nM, 100 nM)[4]. Specific binding in individual brain areas is presented in Table 19.2.

In addition to the hypophysis and brain, CRF receptors exist in the adrenal medulla, sympathetic ganglia, prostate, spleen, liver, kidney, testis, erythrocyte membrane, spleen macrophages, and lymphocytes. In times of stress, CRF is believed not only to accelerate ACTH/β-endorphin secretion from the anterior pituitary but also to participate in stress-induced behavioural, autonomic, and immune responses by means of its binding to the receptors. Moreover, CRF is thought to facilitate ACTH/β-endorphin production in tissues (i.e. lymphocytes) other than the hypophysis[14].

CRF receptor has been known to vary in patients with certain neurological diseases. B_{max} rises in the cerebral cortex of patients suffering from Alzheimer's disease[5] and it decreases in the frontal lobe of suicidal patients with depression[10]. The significance of these variations remains to be determined.

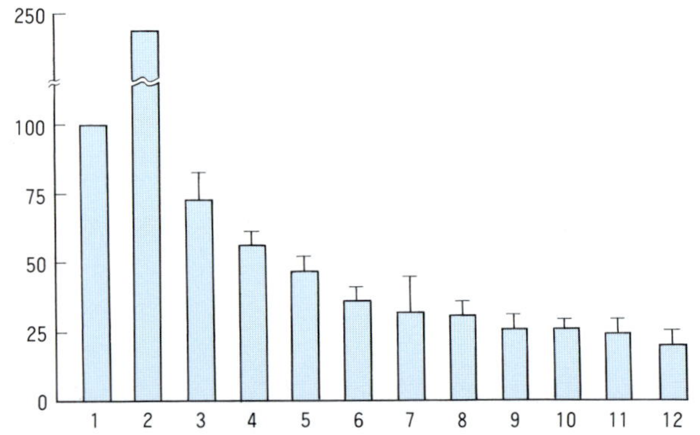

Fig. 19.4 Relative binding of ^{125}I-Tyr-rat CRF to individual brain regions; the binding of ^{125}I-Tyr-rat CRF to the olfactory bulb (1) is designated as 100. 2, anterior pituitary; 3, cerebellum; 4, cerebral cortex; 5, striatum; 6, cervical cord; 7, hypothalamus; 8, medulla oblongata; 9, midbrain; 10, thalamus; 11, pons; 12, hippocampus.

Receptor and binding protein

The cDNA of CRF binding protein has been described and its primary structure has been clarified[12]. Additional investigations are needed to find out whether this binding protein is a real receptor or a protein which simply inactivates CRF.

Recently, two types of CRF receptor with seven putative membrane-spanning domains have been cloned and they appear to increase intracellular cAMP content in response to ligand binding[2,11]. *In situ* hybridization histochemistry revealed that CRF receptor mRNA is expressed abundantly in all areas of the neo-, olfactory, hippocampal, and cerebellar cortices and in subcortical limbic structures[13].

References

1. Aguilera, G., Harwood, J.P., Wilson, J.X. *et al.* (1983). *J. Biol. Chem.*, **258**, 8039–45.
2 Chen, R., Lewis, K.A., Perrin, M.H., and Vale, W.W. (1993). *Proc. Natl. Acad. Sci.*, **90**, 8967–71.

Table 19.1. Inhibition by CRH-related peptides of ^{125}I-Tyr rat/human CRF binding

Peptides	Ki (nM)	Relative potency
rat CRF (rCRF)	1.01 ± 0.05	1.0
pCRF	2.39 ± 0.11	0.4
α-Helical oCRF (9–41)	16.73 ± 1.67	0.06
Ac oCRF (4–11)	0.47 ± 0.08	2.1
oCRF (1–39)	828.2	0.001
rCRF (1–20)	>1000	<0.001
rCRF (21–41)	>1000	<0.001
rCRF (6–33)	>1000	<0.001
Angiotensin II	>1000	<0.001
Vasopressin	>1000	<0.001
His^1, Nle^{27} GRF	>1000	<0.001

Table 19.2. CRF-binding sites in the rat brain

Brain regions	Specific binding	Brain regions	Specific binding
	fmol/mg protein ± SEM		fmol/mg protein ± SEM
Sensory cortex		Whole hippocampus	19.9 ± 4.1
Layers I, II	68.7 ± 6.2	dentate gyrus	22.4 ± 4.9
Layers III, IV	98.6 ± 6.4	Subiculum	31.3
Layers V, VI	71.2 ± 5.5		
		Midbrain	
Motor cortex		Inferior colliculus	43.1 ± 9.2
Layers I, II	48.3 ± 3.9	Dorsal naphe nucleus	8.2 ± 6.2
Layers III, IV	74.3 ± 3.9	Central grey matter	2.24 ± 1.6
Layers V, VI	58.0 ± 4.1		
		Medula oblongata, pons	
Visual cortex		Inferior olive	75.8 ± 8.4
Layers I, II	35.3 ± 7.4	Superior olive	71.3 ± 13.4
Layers III, IV	82.6 ± 6.5	Lateral reticular nucleus	61.8 ± 9.2
Layers V, VI	53.9 ± 5.5	Paragigantocellular reticular nucleus	49.7
		Spinal nucleus of the trigeminal nerve	96.1 ± 14.1
Cingulate cortex (whole layer)	43.6 ± 4.1	Cuneate nucleus	88.1 ± 6.7
Internal olfactory area (whole layer)	76.1 ± 9.3	Gracile nucleus	86.1 ± 10.1
Olfactory tubercle	65.6 ± 11.6	Solitary tract nucleus	42.4 ± 23.1
Basal ganglia		Solitary tract	58.6 ± 10.3
Nucleus caudatus putamen	47.2 ± 3.7	Lateral vestivular nucleus	29.1
Globus pallidus	16.5 ± 0.5	Medial vestibular nucleus	73.9 ± 4.7
Claustrum	78.8 ± 13.3	Facial nucleus	150.6 ± 28.4
		Cochlear nucleus	107.4 ± 24.2
Septal regions		Dorsal tegmental nucleus	83.9
Nucleus accumbens	65.2 ± 8.0	Ventral tegmental nucleus	43.5 ± 13.0
Diagonal band	34.5 ± 2.6	Parabrachial nucleus	51.1
Lateral septum	19.6 ± 2.6	Pontine nucleus	59.4
Medial septum	24.3 ± 4.2	Hypoglossal nucleus	33.2
Bed nucleus of the stria terminalis	21.5	Pyramidal tract	65.4 ± 6.9
		Locus coeruleus	9.1 ± 7.1
Amygdala			
Basolateral nucleus	50.5 ± 3.8	Whole cerebellum	98.5 ± 12.7
Central nucleus	21.6 ± 2.8	Intermediate nucleus	71.3
Medial nucleus	25.6 ± 2.7	Medial nucleus	92.2
		Paraflocculus	118.1 ± 20.0
Thalamus			
Lateral nuclei	37.5 ± 4.5	Cervical cord	
Medial nuclei	21.2 ± 3.7	Dorsal horn (layers 1–4)	23.5 ± 0.9
Anterodorsal nuclei	29.7	Dorsal horn (layers 4–6)	44.3
		Anterior horn	40.3 ± 1.4
Habenular nucleus	33.5 ± 3.3		
		White matter tracts	
Hypothalamus		Corpus callosum	7.5 ± 4.9
Lateral nucleus	24.0 ± 4.8	Fornix	0
Medial nuclei	24.5 ± 4.5	Anterior commissure	25.0 ± 2.7
Anterior nucleus	27.8 ± 1.1	Optic chiasma	0
Paraventricular nucleus	32.9 ± 1.6		
Arcuate nucleus	0		
Mammillary peduncle	54.7 ± 5.9		

Specific binding more than 80 fmol/mg protein, 40–80 fmol/mg protein and less than 40 fmol/mg protein is designated as high density, moderate density, and low density, respectively, in the atlas.

3. Childs, G.V., Morell, J.L., Niendorf, A., and Aguilera, G. (1986). *Endocrinology*, **119**, 2129–42.
4. De Souza, E.B., Insel, T.R., Perrin, M.H. *et al.* (1985). *J. Neurosci.*, **5**, 3189–203.
5. De Souza, E.B., Whitehouse, P.J., Kuhar, M.J. *et al.* (1986). *Nature*, **319**, 593–5.
6. De Souza, E.B. and Kuhar, M.J. (1986). *Methods Enzymol.*, **124**, 560–90.
7. De Souza, E.B. (1987). *J. Neurosci.*, **7**, 88–100.
8. Leroux, P., Vaudry, H., Fournier, A. *et al.* (1984). *Endocrinology*, **114**, 1506–12.
9. Millan, M.A., Jacobowitz, D.M., Hauger, R.L. *et al.* (1986). *Proc. Natl Acad. Sci.*, **83**, 1921–5.
10. Nemeroff, C.B., Owens, M.J., Bissette, G. *et al.* (1988). *Arch. Gen. Psychiat.*, **45**, 577–9.
11. Perrin, M., Donaldson, C., Chen, R. *et al.* (1995). *Proc. Natl. Acad. Sci.*, **92**, 2969–73.
12. Potter, E., Behan, D.P., Fischer, W.H. *et al.* (1991). *Nature*, **349**, 423–6.
13. Potter, E., Sutton, S., Donaldson, C. *et al.* (1994). *Proc. Natl. Acad. Sci.*, **91**, 8777–81.
14. Smith, E.M., Morrill, A.C., Meyer, W.J., and Blalock, J.E. (1986). *Nature*, **321**, 881–2.

Corticotropin releasing factor receptor (binding sites)

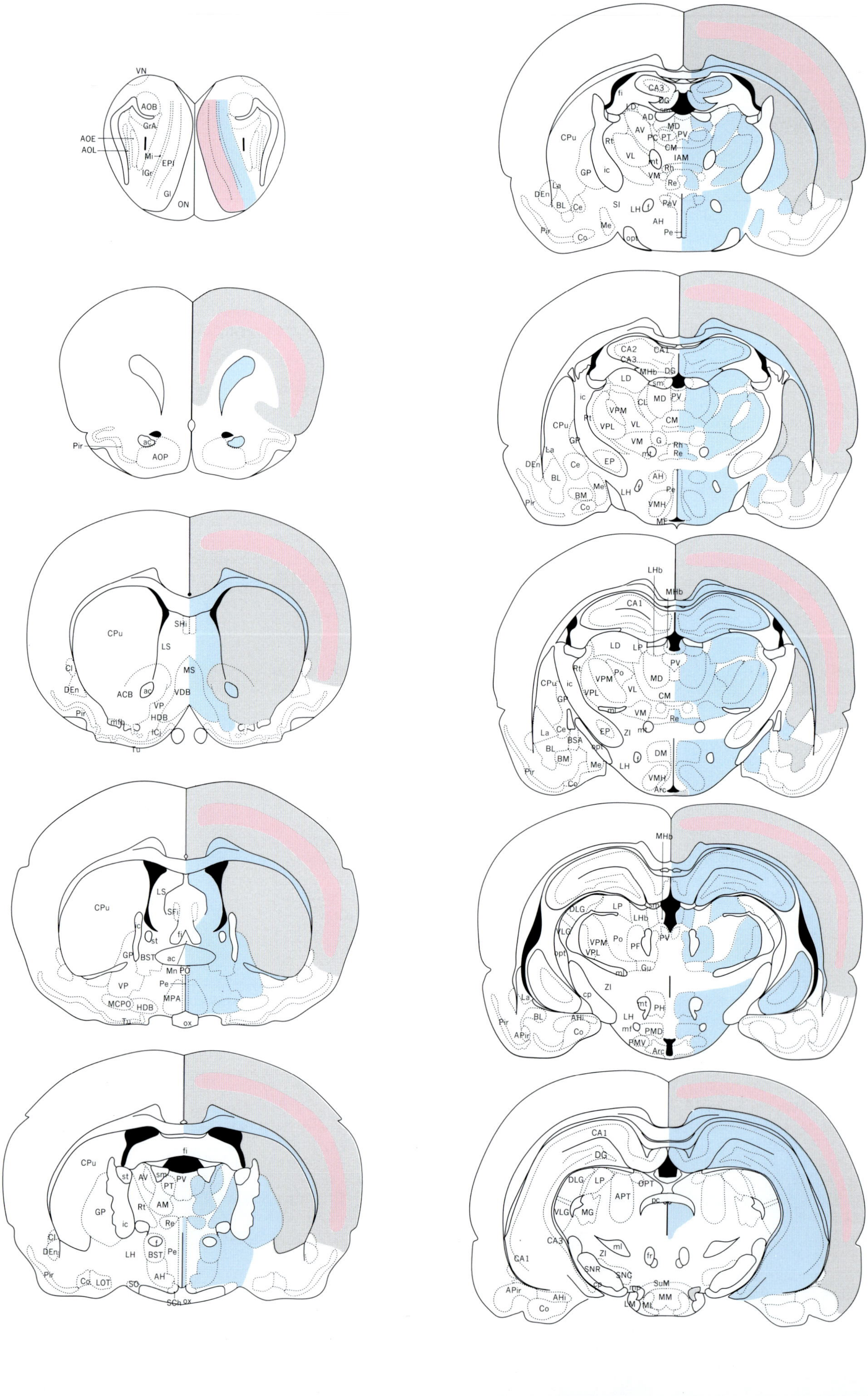

Corticotropin releasing factor receptor (binding sites)

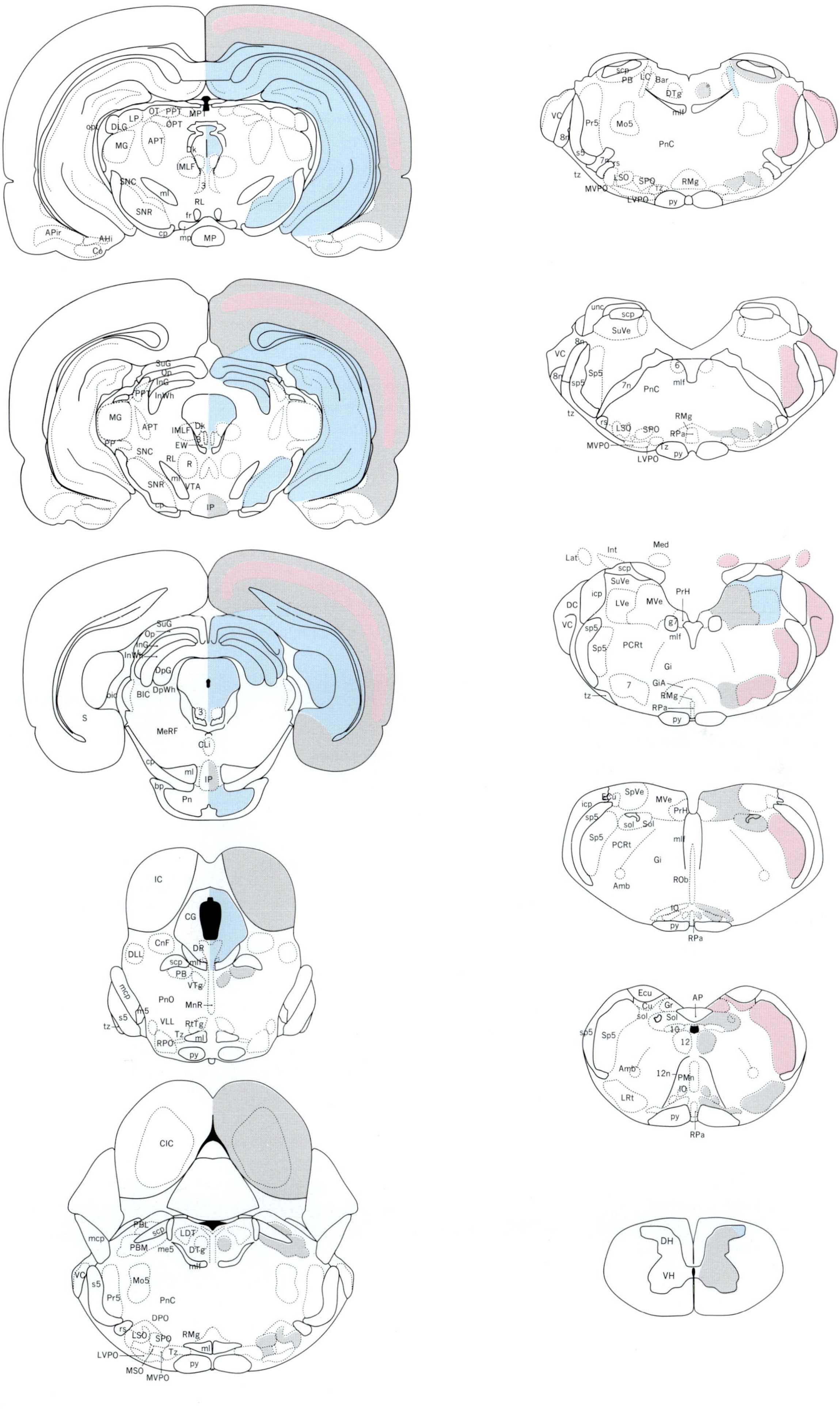

20 The bombesin-like peptide system

Bombesin-like peptides

Type and structure

The structures of principal bombesin (BN)-like peptides reported to date are shown in Fig. 20.1[8]. BN-like peptides are classified into bombesin, ranatensin, and phyllolitorin subfamilies, based on the structure of the C-terminal ends, which contribute to the main physiological activities of these peptides.

Distribution and projection tract

Each of the bombesin-like peptides has many common amino acid residues at the C-terminal end, so the antiserum to each of these peptides shows significant cross-reactivity with other peptides. Recent immunohistochemical investigations with specific antibodies to BN and a member of ranatensin subfamily show different distribution patterns of the peptides in the central nervous system of rat. For space reasons, this chapter provides information only on BN and gastrin-releasing peptide (GRP). Results on ranatensin/neuromedin (NMB) immunostaining can be found in Chronwall *et al.*[3].

BN/GRP-containing nerve cells are mainly observed in the preoptic hypothalamic area, paraventricular hypothalamic nucleus, suprachiasmatic nucleus, and solitary tract nucleus. BN/GRP-immunoreactive nerve fibres are observed in the preoptic hypothalamic area, suprachiasmatic nucleus, solitary tract nucleus, gigantocellular reticular nucleus, and dorsal column of the spinal cord (see the atlas)[2]. The distribution of these bombesin-like substances thus resembles that of distribution of intracerebral binding sites of BN.

Recent *in situ* hybridization studies have also presented GRP mRNA expression sites in the rat brain. These reveal that the distribution of BN/GRP positive cells is much wider than that obtained by immunohistochemical staining[9]. The expression sites of GRP mRNA and NMB mRNA are shown in Table 20.1.

Among the projection tracts of BN/GRP-containing nerve components, one leading to the medulla oblongata from the hypothalamic paraventricularis nucleus and one leading to the surface layer of dorsal column of spinal cord from the dorsal root ganglia have been experimentally demonstrated[7].

Physiological function

From the central distribution and the projection tracts of BN-like substances, BN is thought to participate actively in neuroendocrine function, regulation of autonomic nerve activity, and sensory conduction. Intraventricular injection of BN inhibits sleep and intake of food and water. It also influences certain behaviours such as locomotion, rearing, and grooming[5]. One report[4] suggests that a peritoneal injection of BN affects memory and learning. The results of intracerebral BN injection are commented upon further in the section on receptors.

BN-like peptides have other physiological actions as well. These include constriction of smooth muscle, stimulation of the secretion of other neuropeptides or hormones, acceleration of growth of fibroblast or lung tumour, thermoregulation, and control of sugar metabolism[1,2,6].

References

1. Alexander, R.W., Upp, J.J., Poston, G.J. *et al.* (1988). *Cancer Res.*, **48**, 1439–41.
2. Battey, J. and Wada, E. (1991). *Trends Neurosci.*, **14**, 524–8.
3. Chronwall, B., Pisano, J., Bishop, J.F. *et al.* (1985). *Brain Res.*, **338**, 97–113.
4. Flood, J.F. and Morley, J.E. (1988). *Brain Res.*, **460**, 314–22.
5. Kulkosky, P.J. (1988). *Ann. New York Acad. Sci.*, **525**, 201–18.

Bombesin family–		
Gastrin-releasing peptide (GRP)	Gly·Asn·His·Trp·Ala·Val·Gly·	His·Leu·Met.NH2
Bombesin	pGlu·Gln·Arg·Leu·Gly·Asn·Gln·Trp·Ala·Val·Gly·	His·Leu·Met.NH2
Alytesin	pGlu·Gly·Arg·Leu·Gly·Thr·Gln·Trp·Ala·Val·Gly·	His·Leu·Met.NH2
Ranatensin family –		
Porcine neuromedin B (PMB)	Gly·Asn·Leu·Trp·Ala·Thr·Gly·	His·Phe·Met.NH2
Ranatensin C	pGlu·Thr·Pro·Gln·Trp·Ala·Thr·Gly·	His·Phe·Met.NH2
Ranatensin	pGlu·Val·Pro·Gln·Trp·Ala·Val·Gly·	His·Phe·Met.NH2
Liktorin	pGlu·Gln·Trp·Ala·Val·Gly·	His·Phe·Met.NH2
Phyllolitorin family–		
Leu^8-phyllolitorin	pGlu·Leu·Trp·Ala·Val·Gly·	Ser·Leu·Met.NH2
Phe^8-phyllolitorin	pGlu·Leu·Trp·Ala·Val·Gly·	Ser·Phe·Met.NH2

Fig. 20.1 Amino acid sequence comparison of bombesin-like peptides. Other than neuromedin B and human GRP, all peptides shown are amphibian (from ref. 8).

Table 20.1. Distributions of NMB and GRP mRNAs in rat nervous system (from ref. 9).

Tissue	NMB	GRP
I. Forebrain		
A. Isocortex		
II, III	–	++
V, VI	–	+
B. Olfactory regions		
1. Main olfactory bulb		
Periglomerular layer	+	–
External plexiform layer	+++	–
Mitral cell layer	+++	–
2. Accessory olfactory bulb		
Glomerular layer	+	–
Mitral cell layer	++	–
3. Anterior olfactory nucleus	–	++
4. Olfactory tubercle	(+)	–
5. Tenia tecta	–	+
C. Hippocampal formation		
1. Entorhinal area (medial)		
II	–	++
IV–VI	–	+
2. Entorhinal area (lateral)		
II	–	+
IV–VI	–	(+)
3. Presubiculum		
II	–	++
IV	–	+++
V, VI	–	+
4. Parasubiculum		
II	–	+++
IV–VI	–	+
5. Subiculum, pyramidal layer	–	+++
6. CA1, pyramidal layer	–	+
7. CA2, pyramidal layer	–	++
8. CA3, pyramidal layer	++	++
9. Dentate gyrus		
granular layer	–	+++(v)
polymorph layer	+++	–
D. Amygdala		
1. Medial n.	–	++
2. Amygdalohippocampal area	–	++
3. Cortical n. posterior part	–	+++
4. Anteiror amygdaloid area	–	+
5. Central n.	++	–
6. Lateral n.	–	++
7. Basolateral n.	–	+
8. Basomedial n.	–	++
E. Septum		
1. Medial n.	–	+
2. N. diagonal band	–	+
3. Bed n. stria terminalis		
dorsal lateral subnucleus	+	–
F. Basal ganglia and related regions		
1. N. accumbens, shell	+	–
2. Magnocellular preoptic n.	–	+
3. Substantia innominata	–	(+)
4. Substantia nigra[a]		
compact part	++	–
5. Ventral tegmental area[a]	++	–
G. Thalamus		
1. Lateral habenula	+	–
2. Anteroventral n.		
dorsomedial ventrolateral	–	+
3. Medial geniculate n.	–	++
4. Posterior intralaminar n.	–	+
5. Reticular n.	(+)	–
6. Zona incerta	–	(+)
H. Hypothalamus		
1. Median preoptic n.	–	+
2. Anteroventral		
periventricular n.	–	+
3. Suprachiasmatic n.	–	++
4. Supraoptic n.	(+)	–
5. Paraventricular n.		
parvicellular part	–	+
magnocellular part	–	(+)
6. Arcuate n.	+	–
7. Medial preoptic n.		
lateral, medial part	+	++
central aprt	(+)	+
8. Dorsomedial nucleus	–	(+)
9. Supramammillary n.	+	(+)
10. Posterior hypothalamic area	–	+
II. Brain stem		
A. Sensory		
1. Somatosensory		
Principal sensory n.	++	–
Dorsal column n.	–	+
Spinal trigeminal n.	(+)	+
2. Auditory		
Cochlear n.	(+)	–
N. trapezoid body	+	–
N. lateral lemniscus, ventral	+	–
Inferior colliculus		
external, dorsal part	–	++
3. Visceral		
N. solitary tract		
medial part	–	++
lateral pat	–	(+)
Area postrema	+	–
Parabrachial n.		
central lateral	+	++
dorsal lateral	+	++
external lateral	–	+
internal lateral	(+)	++
superior lateral	+	+
ventral lateral	–	+
Kölliker-Fuse n.	+	(+)
B. motor		
1. Facial n. (VII)	+	–
2. Dorsal motor n. X	++	–
C. Reticular core (including central grey matter and raphe)		
1. Central grey	–	+
2. Peridorsal tegmental n.	++(p)	–(p)
3. N. incertus	++	–
4. Locus ceruleus	(+)	+
5. Interfascicular n.	–	+
6. Paranigral n.	–	+
7. Median raphe	+	–
8. Pontine reticular n.	(+)	–
9. Hindbrain raphe	++	(+)
10. Gigantocellular reticular field	+	–
11. Rostroventrolateral reticular n.	–	+
12. Lateral reticular n.	+	–
III. Spinal cord		
1. Substantia gelatinosa	(+)	++
IV. Peripheral nervous system		
1. Trigeminal ganglion	+++	–
2. Dorsal root ganglion	+++	–

[a] The substantis nigra and ventral tegmental area are brainstem structures.

6. Moody, T.W., Getz, R., O'Donohue, T.L., and Rosenstein, J.M. (1988). *Ann. New York Acad. Sci.*, **547**, 114–30.

7. Panula, P., Nieminen, O., Falkenberg, M., and Auvinen, S. (1988). *Ann. New York Acad. Sci.*, **547**, 54–69.

8. Spindel, F.R. and Krane, I.M. (1988). *Ann. New York Acad. Sci.*, **547**, 10–20.

9. Wada, E., Way, J., Lebacq-Verheyden, A.M., and Battey, J.F. (1990). *J. Neurosci.*, **10**, 2917–30.

Bombesin (immunocytochemistry)

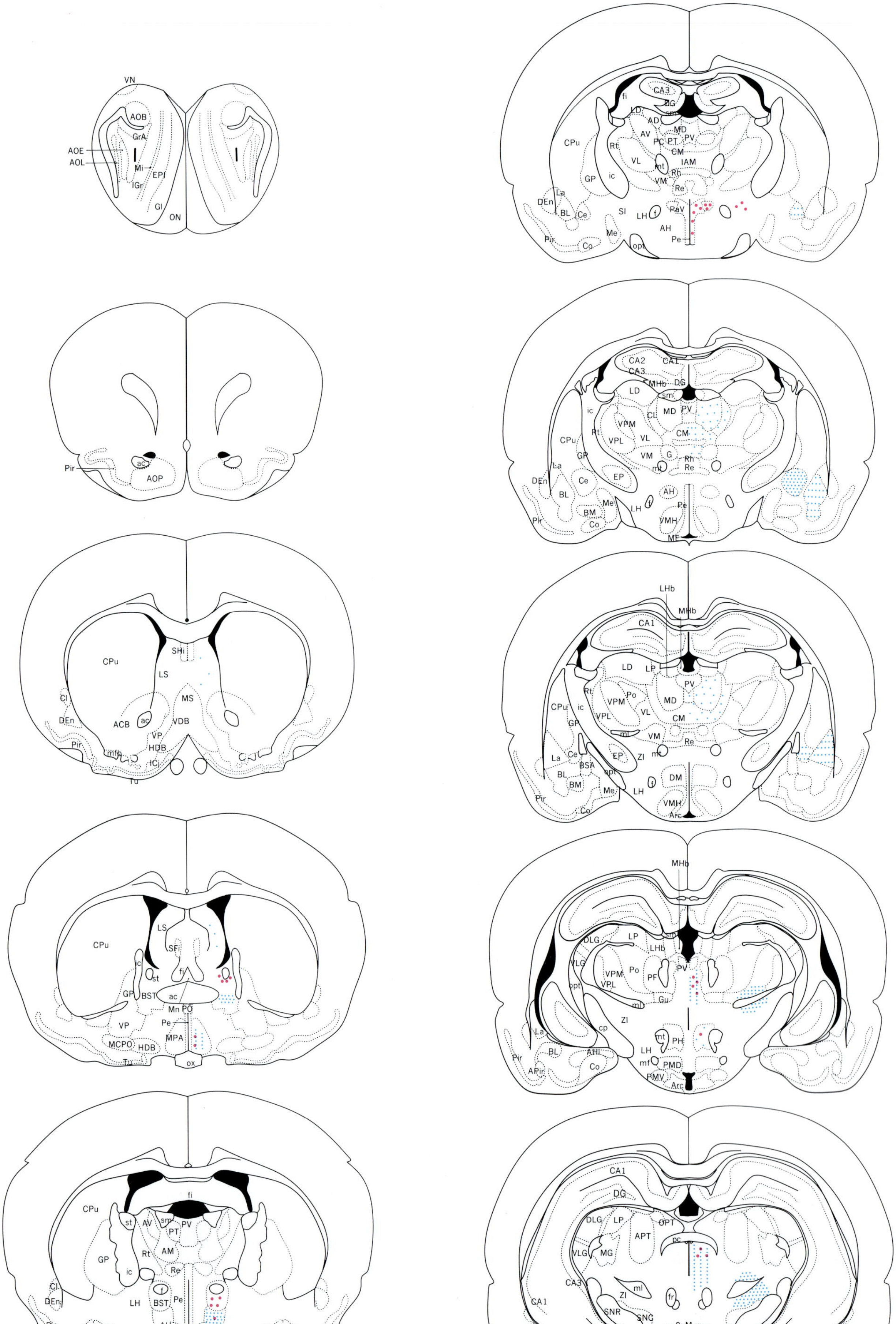

Bombesin (immunocytochemistry)

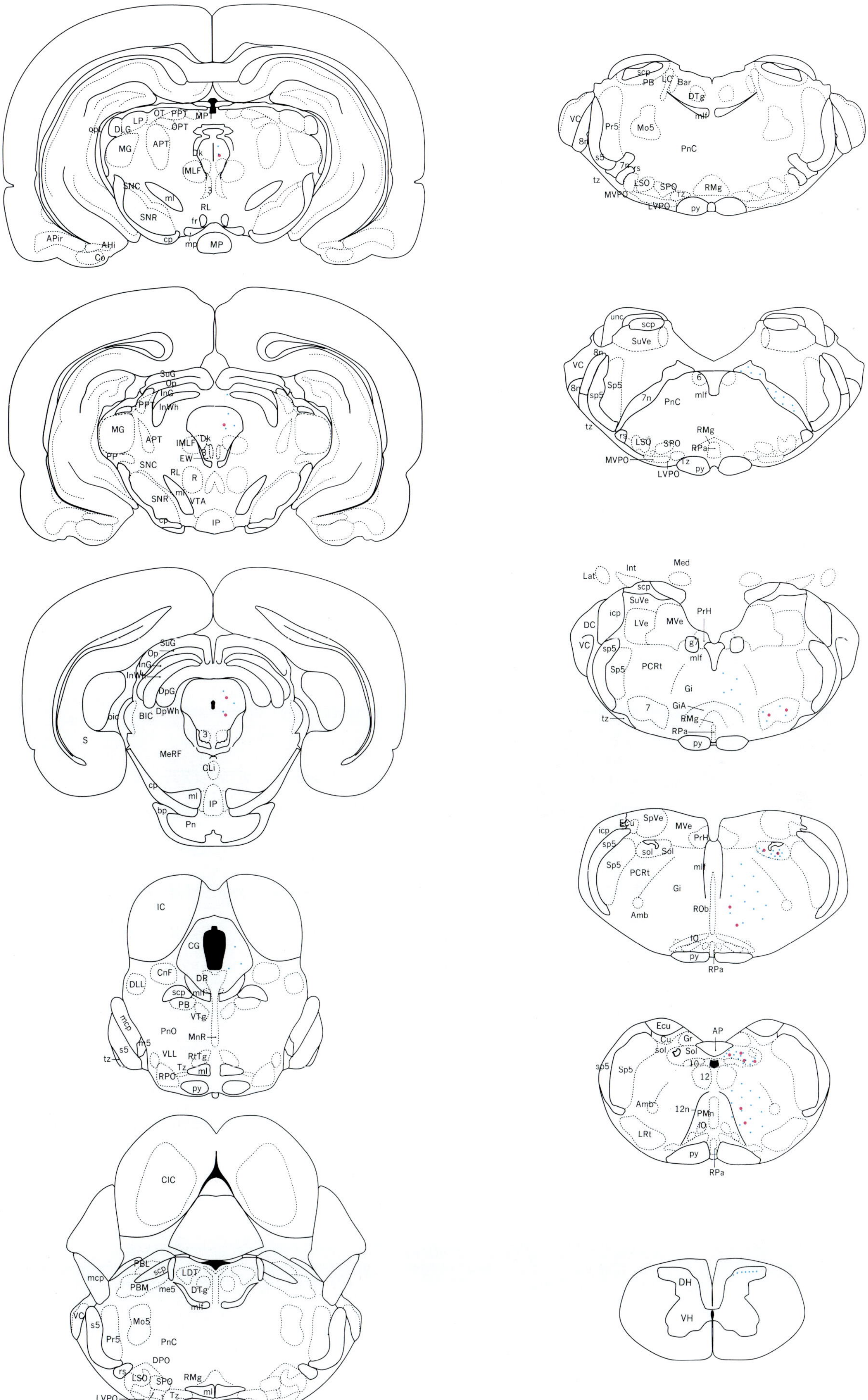

Bombesin-like peptide receptors

Subtypes

As shown in Fig. 20.1, two kinds of GRP NMB have been identified as mammalian BN-like peptides. These have specific binding sites, and exhibit different distribution patterns in the central nervous system[5]. Therefore, from receptor binding experiments and autoradiography data, BN-like peptide receptors are believed to be present in at least two subtypes[5,6]. The primary structure of BN receptor protein (GRP receptor, NMB receptor) that has high affinity to GRP and NMB respectively, has recently been clarified[2] (Fig. 20.2). This has brought a new focus to research on BN receptors.

Structure

Both the GRP receptor (GRP-R) and the NMB receptor (NMB-R) have a molecular weight of 43 kDa, and they share 56 per cent homology in the amino acid residues[6,9,10]. As a result of the analysis on the primary structure, seven transmembrane structures are known to be present in G-protein-coupled receptors, have been discovered to date[2]. Binding studies with receptor protein expressed on *Xenopus* oocyte have demonstrated that GRP selectively inhibits the binding between (^{125}I)Tyr^{4}BN and GRP-R, and that NMB selectively inhibits the binding between (^{125}I)Tyr^{4}BN and NMB-R. With the same experimental procedures, a BN receptor antagonist 2-(D-Phe^6)BN$_{6\text{–}14}$ ethyl ester has been shown to exhibit high affinity with GRP-R as demonstrated previously by *in vitro* binding experiments (Table 20–2).

Distribution

According to previous *in situ* hybridization studies, GRP-R mRNA is expressed abundantly in the hypothalamus, and specially in the suprachiasmatic nucleus and supraoptic nucleus. The expression of NMBR-mRNA is intense in the anterior olfactory nucleus and central thalamic nucleus (Table 20.3)[2]. In some locations, including the solitary tract nucleus, both types of receptors are present, but it is unclear whether GRP-R mRNA and NMB-R mRNA are expressed in the same neurons or not. Data on the BN receptor mRNA expression sites which were obtained from *in situ* hybridization are similar to those obtained previously from receptor-binding autoradiographic experiments. Thus, receptor binding sites based on the studies of Moody *et al.*[7] are shown in this atlas.

The distribution of intracerebral BN/GRP and GRP-R is presented in Table 20.3. Special attention may be focused on the suprachiasmatic nucleus where both

Table 20.2. Displacement of [^{125}I]Try^4 bombesin binding by GRP, NMB and the antagonist [D-Phe^6] bombesin$_{6\text{–}13}$ ethyl ester (from ref. 2).

Subtype	GRP	K_1 (nM) NMB	Antagonist
GRP receptor	2	62	1.6
NMB receptor	43	2	>1000

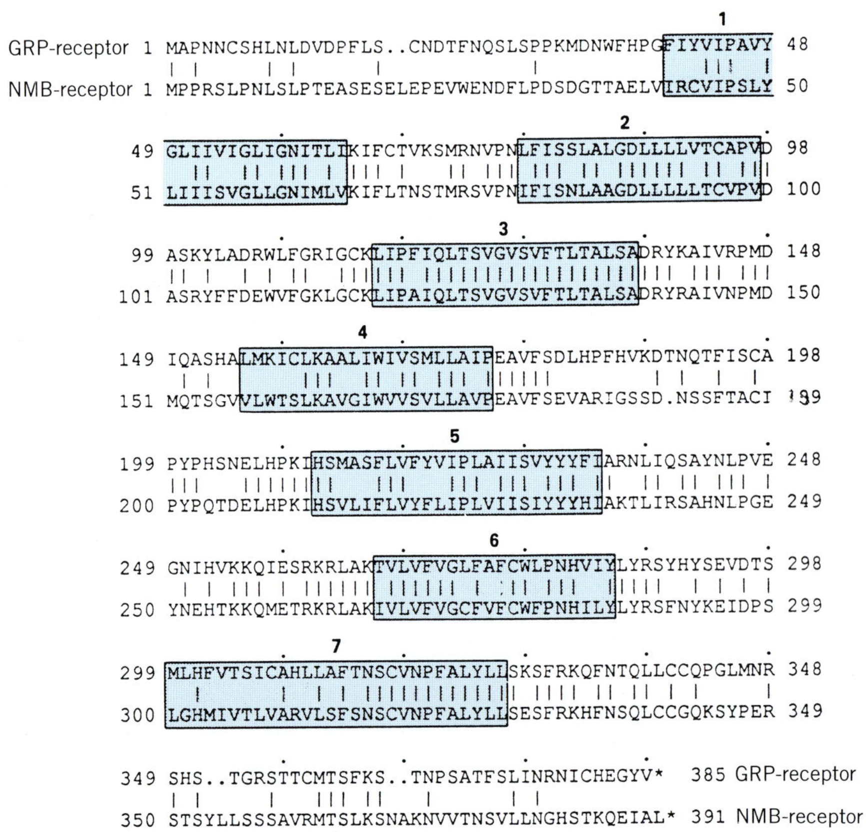

Fig. 20.2 Comparison of amino acid sequences for the murine GRP-R and the rat NMB-R subtypes (from ref. 2).

Table 20.3. Distribution of GRP, NMB, GRP-R, and NMB-R mRNAs in the CNS (from ref. 2)

Tissue	GRP	NMB	GRP-R	NMB-R
I Forebrain				
A Isocortex				
II	++	–	++	–
III	++	–	+	–
V	+	–	–	++
VI	+	–	–	+
B Olfactory regions				
Main bulb	–	+++	–	–
Anteiror olfactory n.	++	–	(+)	+++
C Hippocampal formation				
Dentate gyrus	++	++	++	++
D Amygdala				
Amygdalohippocampal area	++	–	(+)	+++
N. lateral olfactory tract	–	–	+++	(+)
E Basal ganglia				
N. accumbens	–	+	+	–
Magnocellular preoptic n.	+	–	+++	+
F Thalamus				
Paraventricular n.	–	–	–	+++
Central medial n. and central lateral n.	–	–	–	+++
G Hypothalamus				
Suprachiasmatic n.	++	–	+++	–
Supraoptic n.	–	(+)	++	–
Paraventricular n.	+	–	++	(+)
Medial preoptic n.	++	–	++	++
Lateral mammillary n.	–	–	+++	–
II Brainstem				
N. Solitary tract	++	–	+	+
N. ambiguus	–	–	++	++
Central grey matter	+	–	+	+

The strength of hybridization signal is graded based on the grains per positive cell: +++, strong signal; ++, moderate signal; +, weak signal; (+), very weak signal; –, no signal over the background. (Data from refs 7, 8.) Abbreviation: n., nucleus.

BN/GRP and GRP-R are produced simultaneously. BN/GRP secreted from the short axons of BN/GRP-containing neurons binds to local GRP-R and thereby probably influences the circadian rhythm. Indeed when GRP, vasoactive intestinal peptide, and peptide histidine-isoleucine were simultaneously injected into the suprachiasmatic nucleus, the phase of the circadian rhythm was altered[1]. In addition, BN-R-mediated inhibition of cold-sensitive neurons in the preoptic area of hypothalamus has been reported in the preoptic area of hypothalamus. The effects of BN injection in different brain locations are presented in Table 20.4. Most of these locations contain BN-like peptide receptor (Table 20.3).

Table 20.4. Effect of direct BN injection into the central nervous system (from ref. 7)

Injection sites	Response after injection
Olfactory tubercle	Dopamine metabolism activate
Caudate putamen	Dopamine metabolism activate
Nucleus accumbens	Grooming behaviour
	Locomotor, floor, and rearing activity induced
Hypothalamus	
anterior	Hypothermia
preoptic area	Hypothermia
dorsal	Plasma adrenaline level and blood sugar level elevated, satiety
paraventricular	Gastric acid secretion dieceased, satiety
lateral	satiety
Hippocampus	Locomotor, floor, and rearing activity induced, satiety
Solitary tract nucleus	Facial scratching, facial and body licking

Recently, another two types of bombesin receptor (BRS-3[3,4] and BB4[8]) have been cloned[8]. BRS-3 is expressed in uterus[4], testis, and lung carcinoma cells[3]; BB4 is expressed only in the brain[8].

References

1. Albers, H.E., Liou, S.Y., Stopa, E.G., and Zoeller, T. (1991). *J. Neurosci.*, 846–51.
2. Battey, J. and Wada, E. (1991). *Trends Neurosci.*, **14**, 524–8.
3. Fathi, Z., Corjay, M.H., Shapira, H. *et al* (1993). *J. Biol. Chem.*, **268**, 5979–84.
4. Gorbulev, V., Akhundova, A., Büchner, H., and Fahrenholz, F. (1992). *Eur. J. Biochem.*, **208**, 405–10.
5. Ladenheim, E.E., Jensen, R.T., Mantey, S.A., McHugh, P.R., and Moran, T.H. (1990). *Brain Res.*, **537**, 233–40.
6. Lee, M., Jensen, R.T., Coy, D.H., and Moody, T.W. (1990). *Mol. Cell Neurosci.*, **1**, 161–7.
7. Moody, T.W., Getz, R., O'Donohue, T.L., and Rosenstein, J.M. (1988). *Ann. New York Acad. Sci.*, **547**, 114–30.
8. Nagalla, S.R., Barry, B.J., Creswick, K.C. *et al* (1995). *Proc. Natl. Acad. Sci. USA*, **92**, 6205–9.
9. Spindel, E. (1986). *Trends Neurosci.*, **9**, 130–3.
10. Wada, E., Way, J., Lebacq-Verheyden, A.M., and Battey, J. (1990). *J. Neurosci.*, **10**, 2917–30.
11. Wada, E., Way, J., Shapira, H. *et al.* (1991). *Neuron*, **6** 421–30.

Bombesin receptor (binding sites)

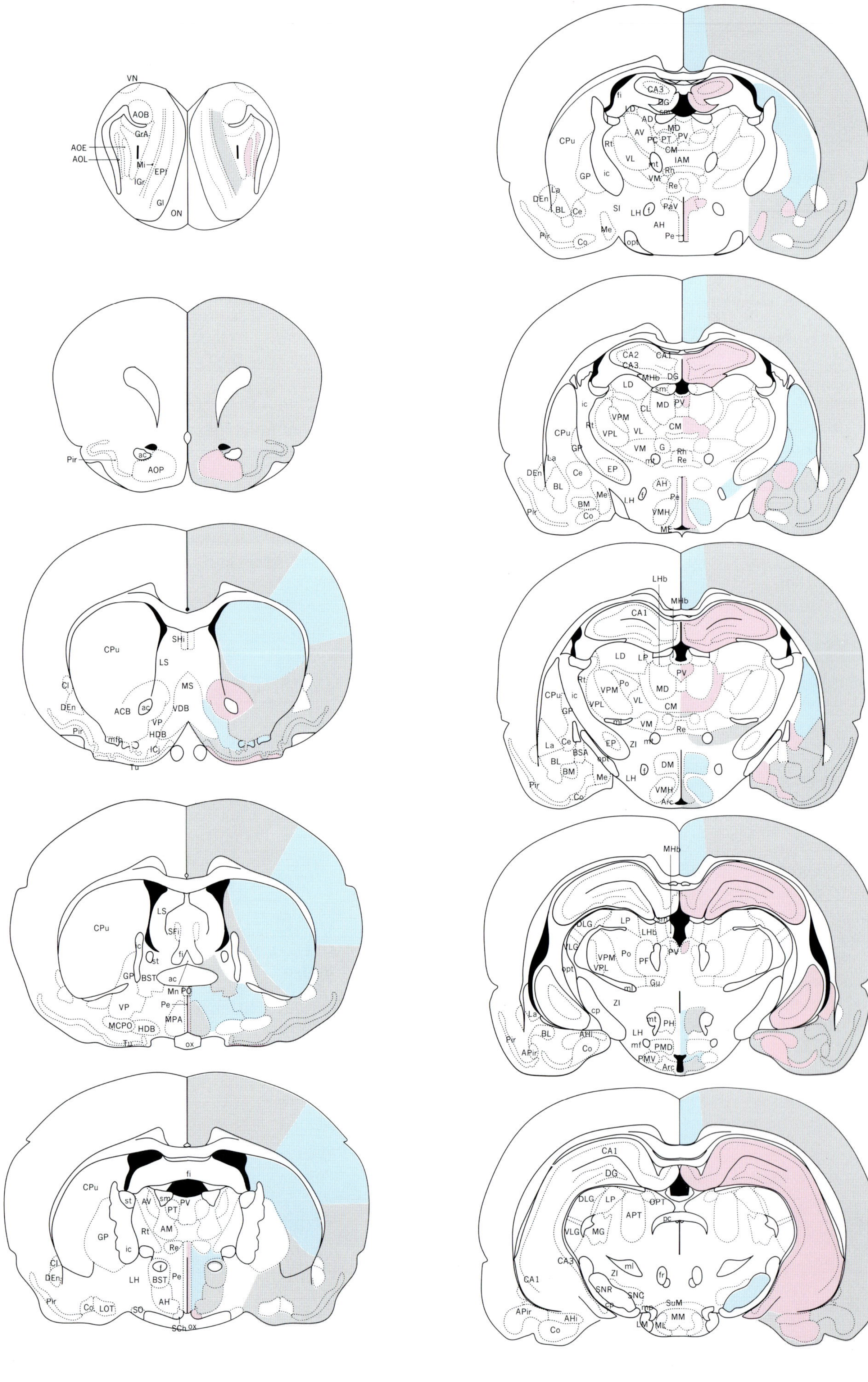

Bombesin receptor (binding sites)

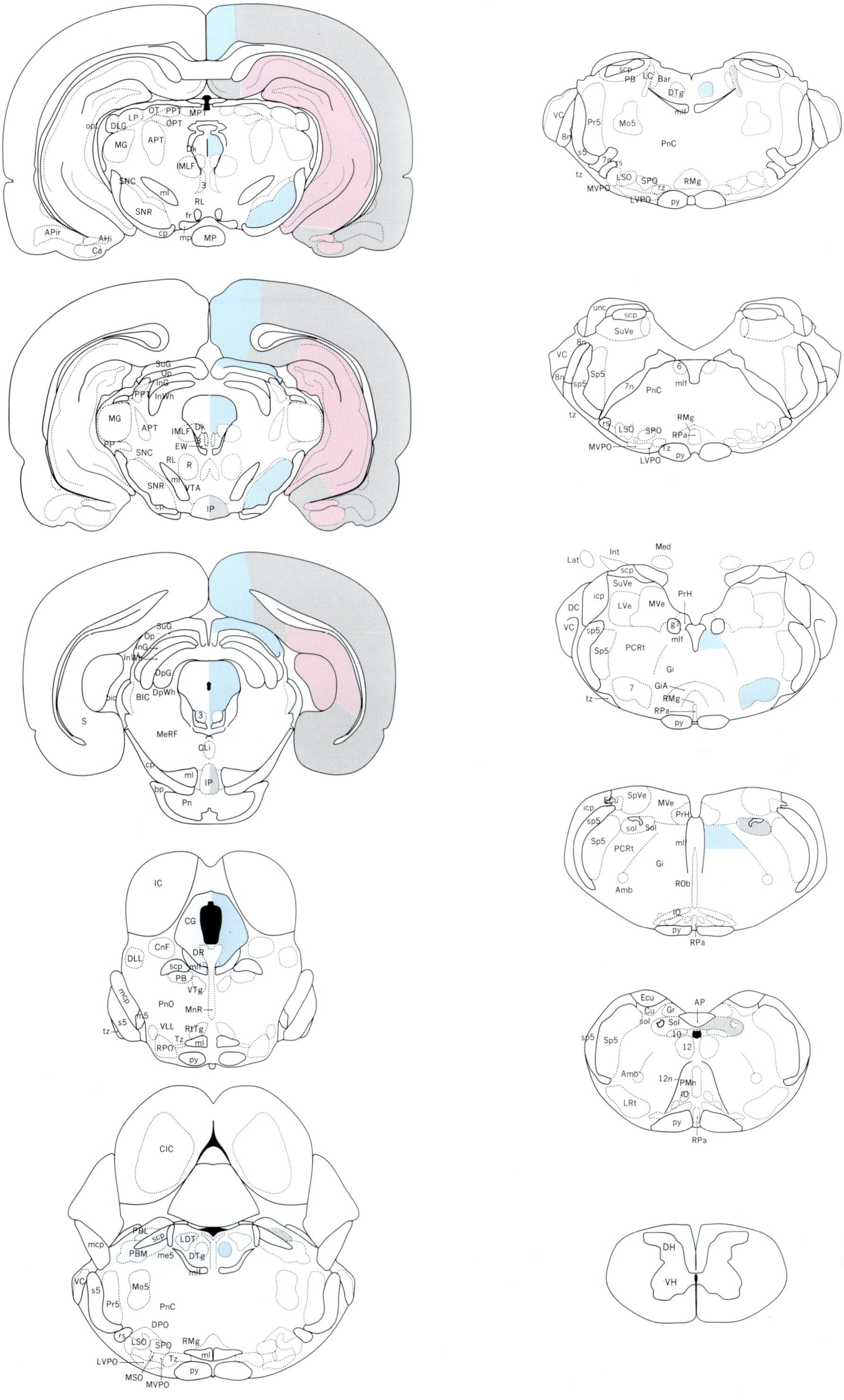

21 The opioid peptide system

Opioid peptides

Classification

For the expression of analgesic activity and other various physiological activities of opioid, the existence of endogenous ligands of the opioid receptor had been pharmacologically predicted from early days. Its intrinsic ligands, two kinds of pentapeptide as met-enkephalin and leu-enkephalin, were discovered by Hughes *et al.* 1975. Following that, dynorphin (DYN) and β-endorphin were discovered. Since then, various opioid peptides have been shown to originate from three kinds of large molecule precursors (proenkephalin, prodynorphin, and proopiomelanocortin) (Fig. 21.1).

Proopiomelanocortin (POMC) series peptides include β-endorphin as an opiod peptide and α-MSH, γ-MSH, ACTH as nonopioid peptides. Neurons which produce the POMC peptides are contained in the arcuate nuclei of the hypothelamus and the commissural nucleus of the medula oblongata in the brain[1,4,7] so that the projection system is comparatively simple. But the α-MSH synthesizing cells are distributed in various brain regions including arcuate nuclei and the commissural nucleus. Thus the synthetic pathway of α-MSH may differ from that of other POMC series polypeptides. In the acruate nucleus, neurons that contain β-endorphin, γ-MSH, and ACTH are distributed and their coexistence is also observed, so their projection areas in the brain are similar. Hence, in this section the collective pattern of the POMC series peptides is presented.

Distribution

Immunoreactive fibres and terminals for POMC series peptides are widely distributed in the brain except in the corpus striatum, hippocampus, and cerebral cortex. These fibres and terminals are distributed in hypothalamic nuclei such as the medial preoptic area, paraventricularis hypothalamic nucleus, nucleus periventricularis, dorsomedial hypothalamic nucleus, and arcuate nuclei. Besides, additional structures with abundant fibres include the bed nucleus of stria terminalis, medial nucleus of amygdaloid complex, central gray, nucleus tractus solitarii, and commissural nucleus[1,4]. In the cases of the preproenkephalin-A (PPEA) series and preprodynorphin (=preproenkephalin-B) (PPD) series, the distribution area of producing neurons is very wide, and their projection patterns are also somewhat complex. The distribution of PPEA series peptides is based on reports of Fallon *et al.*[2], Harlan *et al.*[3], Khachaturian *et al.*[6], and Murakami *et al.*[9]. The distribution of PPD series is adopted from the information provided in the reports of Fallon *et al.*[2], Khachaturian *et al.*[5], and Vincent *et al.*[10].

Cerebral cortex

Enkephalin (Enk) applied electrophoretically into the cerebral cortex shows remarkable inhibitory effects on cortical neurons. Opioid receptors are distributed in various cortical regions, with abundance in the frontal region or cingulate gyrus. Opioid peptide containing neurons are also distributed in the cerebral cortex. DYN positive neurons are observed in both the neocortex and paleocortex, and enriched in layers II, III and IV[2,10]. These are non-pyramidal cells.Enk positive neurons are less numerous in the neocortex, but present in piriform lobe or lateral entorhinal cortex[2,6,8].

Hippocampus

The opioid peptide participates in two important fibre systems of the hippocampus. One is mossy fibres which originate from granules cells, pass through dentate gyrus obex and terminate as pyramidal cells of CA3, CA4 areas. The other system is one whch originates from the lateral entorhinal area pass through perforans path and lead to hippocampus. The mossy fibre system contains more DYN fibres than Enk, while the lateral entorhinal cortex and perforans path contain more Enk fibres[2].

Corpus striatum

Enk-positive neurons in the corpus striatum are medium spiny cells and are concentrated in the 'patch' which receives projections from the thalamus, but there are few in matrix. Opioid receptors are also concentration in patch. DYN and Enk positive neurons show similar distributions.

Corpus striatum–substantia nigra tract. Both Enk- and DYN-positive neurons of the corpus striatum project into globus pallidus (GP) but only DYN-positive neurons project to the substantia nigra. Leu-Enk in the corpus striatum–substantia nigra tract, is derived from PPD and not from PPEA. Enk-positive fibres are observed regionally in the substantia nigra pars compacta and in the ventral tegmental area which is located dorsally to the substantia nigra. But the DYN positive fibres are distributed densely in substantia nigra pars reticulata[2].

Corpus striatum–globus pallidus tract. The distribution of DYN and Enk differs in GP also. DYN positive fibres are seen in entopeduncular nucleus and ventral pallidum (VP), while Enk fibres are confined to GP[2]. While Enk fibres form the 'dorsal' corpus striatum–globus pallidus tract, DYN fibres form the 'ventral' corpus striatum–globus pallidus tract, and the above mentioned corpus striatum–substantia nigra tract. Furthermore, DYN fibres of VP are also supplied from interneurons[2].

Paraventricularis hypothalamic nucleus; nucleus supraopticus hypothalami

Numerous DYN positive neurons are observed in the anterior, posterior, and medial regions of the paraventricularis hypothalamic nucleus as well as the nucleus supraopticus hypothalami, which are called magnocellular neurons. The hypothalamo-pituitary tract as well as the posterior lobe of the pituitary also carry DYN-positive fibres. Meanwhile Enk positive cells are observed mostly in parvocellular neurons.

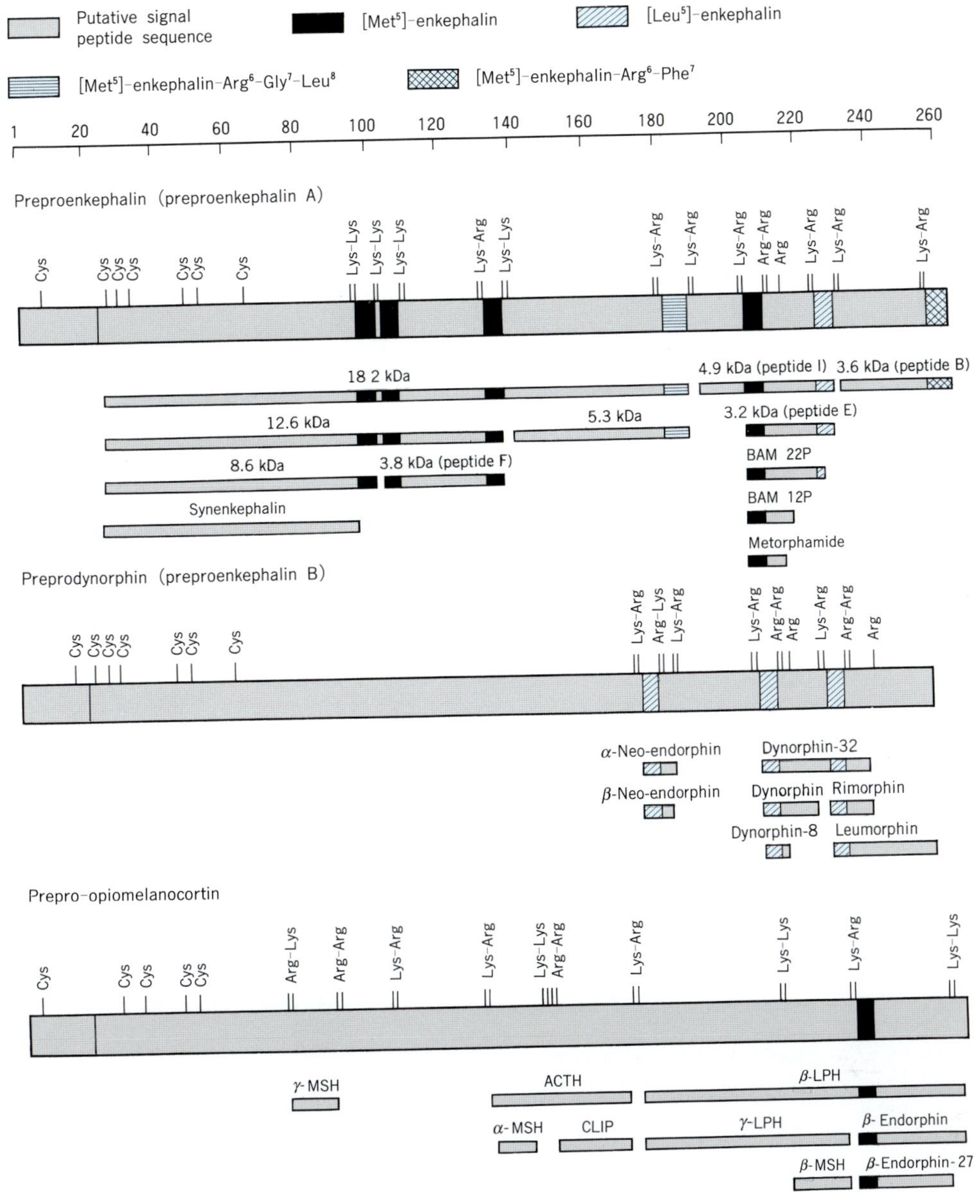

Fig. 21.1 Schematic representation of the proteolytic processing of the three opioid peptide-containing prohormones (modified from ref. 8). ACTH, adrenocorcotropin; CLIP, corticotropin-like intermediate-lobe peptide; LPH, lipotropin; MSH, melanotropin.

Spinal cord

In the spinal cord, opioid peptide is distributed densely in the dorsal horn intermedio-lateral nucleus and area surrounding the central canal, and it is supposed to play an important role in pain transmission and the autonomic nervous systems. However, its distribution or projection differs between Enk, DYN, and POMC series. For example, in the dorsal horn, Enk-positive neurons are densely observed in the layer II but DYN-positive neurons are predominant in layer I. In the interrmedio-lateral nucleus, Enk-positive neurons are observed, but DYN-positive ones are not. Conversely, DYN-positive neurons are present in the lateral spinal nucleus, but Enk neurons are not. Positive neurons for POMC series are not observed in the spinal cord of the adult rat, but descending fibres immunoreactive for β-endorphin originated from upper brain regions, terminate in the area surrounding the central canal.

References

1. Bloom, F., Batenberg, E., Rossier, J.K. *et al.* (1978). *Proc. Natl Acad. Sci. USA*, **75**, 1591–5.
2. Fallon, J.H. and Leslie, F.M. (1986). *J. Comp. Neurol.*, **249**, 293–336.
3. Harlan, R.E., Shivers, B.D., Romano, G.J. *et al.* (1987). *J. Comp. Neurol.*, **258**, 159–84.
4. Kawai, Y., Inagaki, S., Shiosaka, S. *et al.* (1984). *Brain Res.*, **297**, 21–32.
5. Khachaturian, H., Watson, S.J., Lewis, M.E. *et al.* (1982). *Peptides*, **3**, 941–54.
6. Khachaturian, H., Lewis, M.E., Hollt, V., and Watson, S.J. (1983). *J. Neurosci.*, **3**, 844–55.
7. Khachaturian, H., Lewis, M.E., Schaffer, M.K.H., and Watson, S.J. (1985). *Trends Neurosci.*, **8**, 111–19.
8. Merchenthaler, I., Maderdrut, J.L., Altschuler, A., and Petrusz, P. (1986). *Neuroscience*, **17**, 325–48.
9. Murakami, S., Okamura, H., Yanaihara, C. *et al.* (1987). *J. Comp. Neurol.*, **261**, 193–208.
10. Vincent, S.R., Hokfelt, T., Christernsson, I., and Terenius, L. (1982). *Neurosci. Lett.*, **33**, 185–90.

Opioid peptides (immunocytochemistry)

Pro-opiomelanocortin

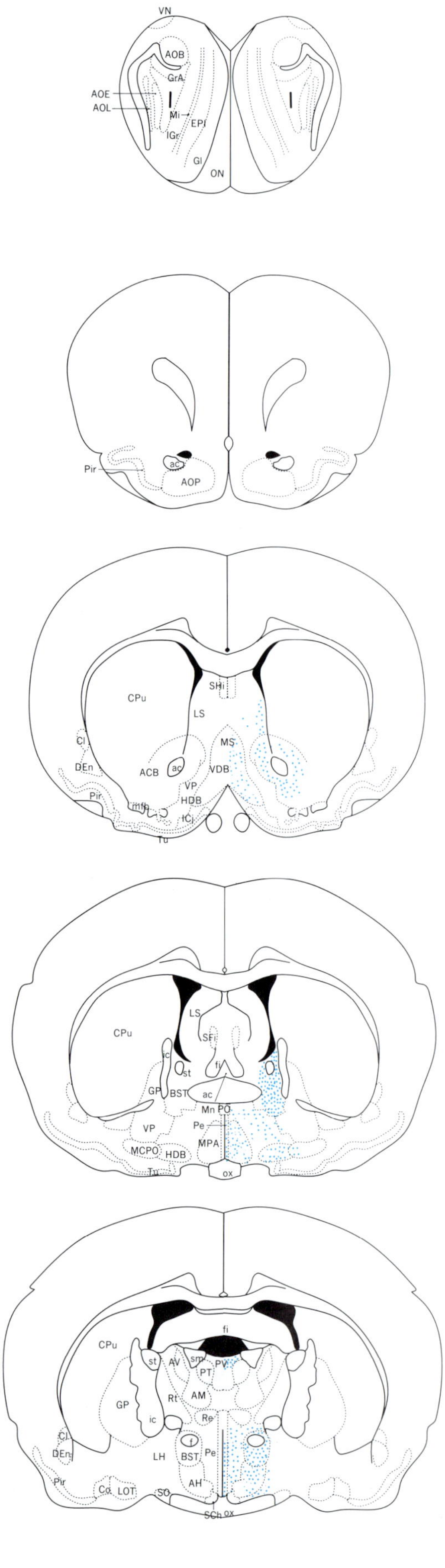

Pro-opiomelanocortin

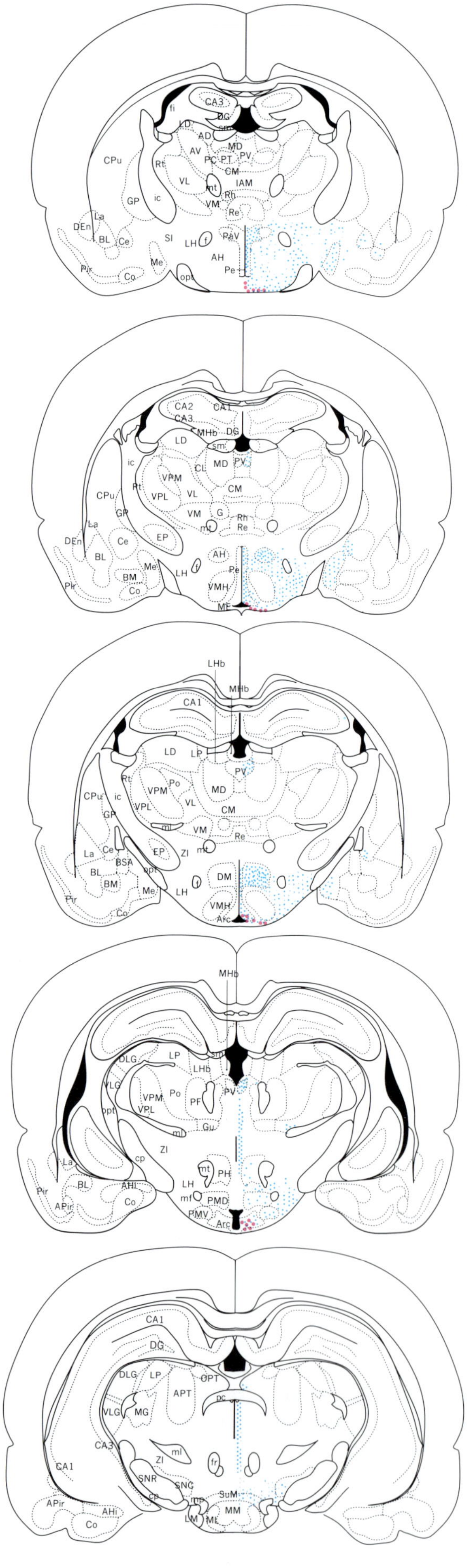

Opioid peptides (immunocytochemistry)

Pro-opiomelanocortin

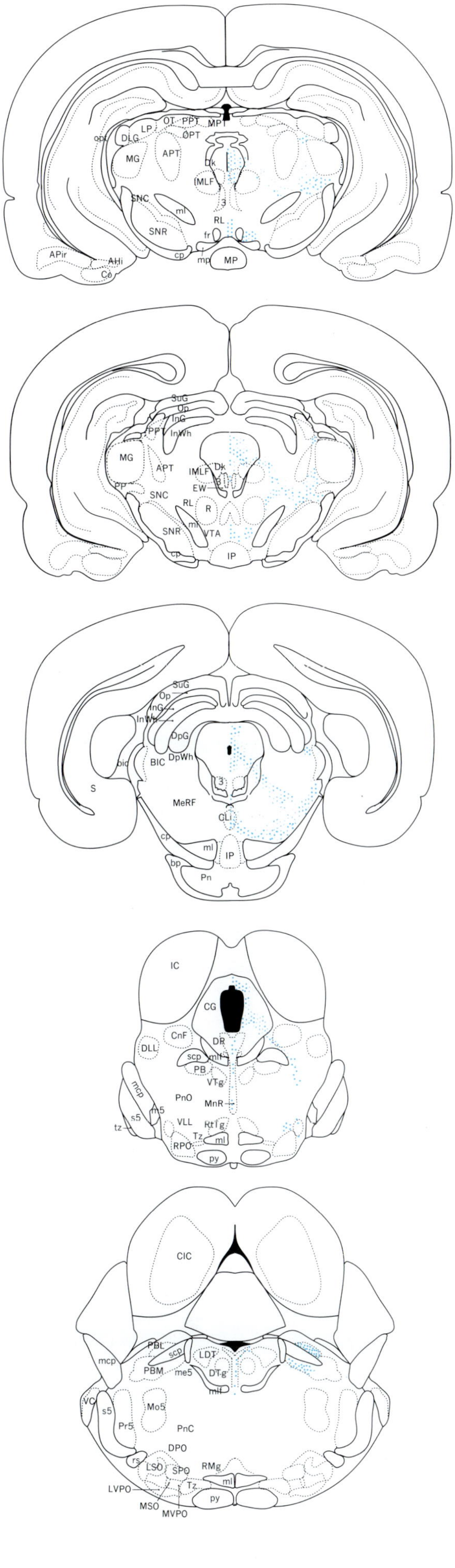

Pro-opiomelanocortin

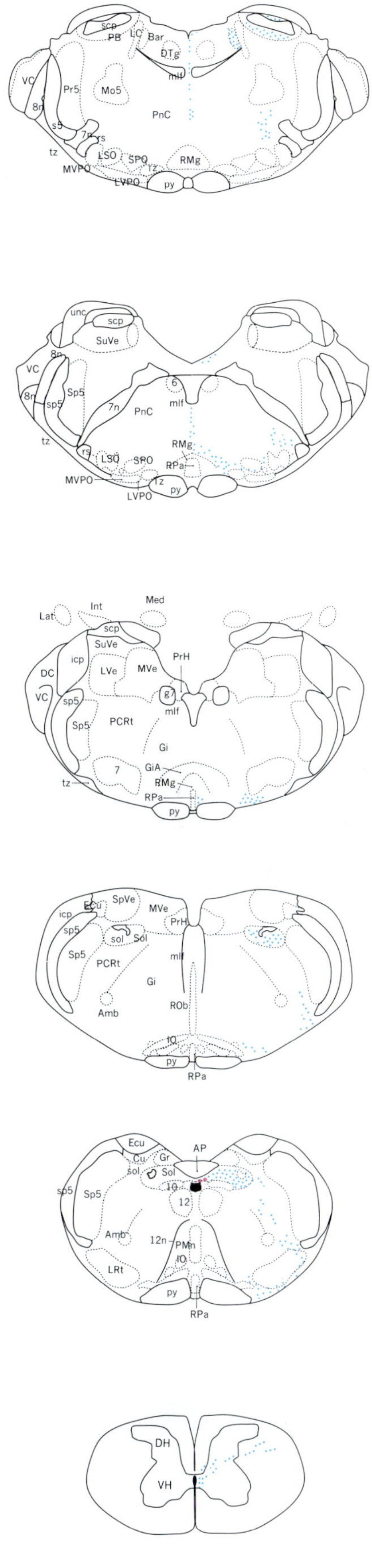

Opioid peptides (immunocytochemistry)

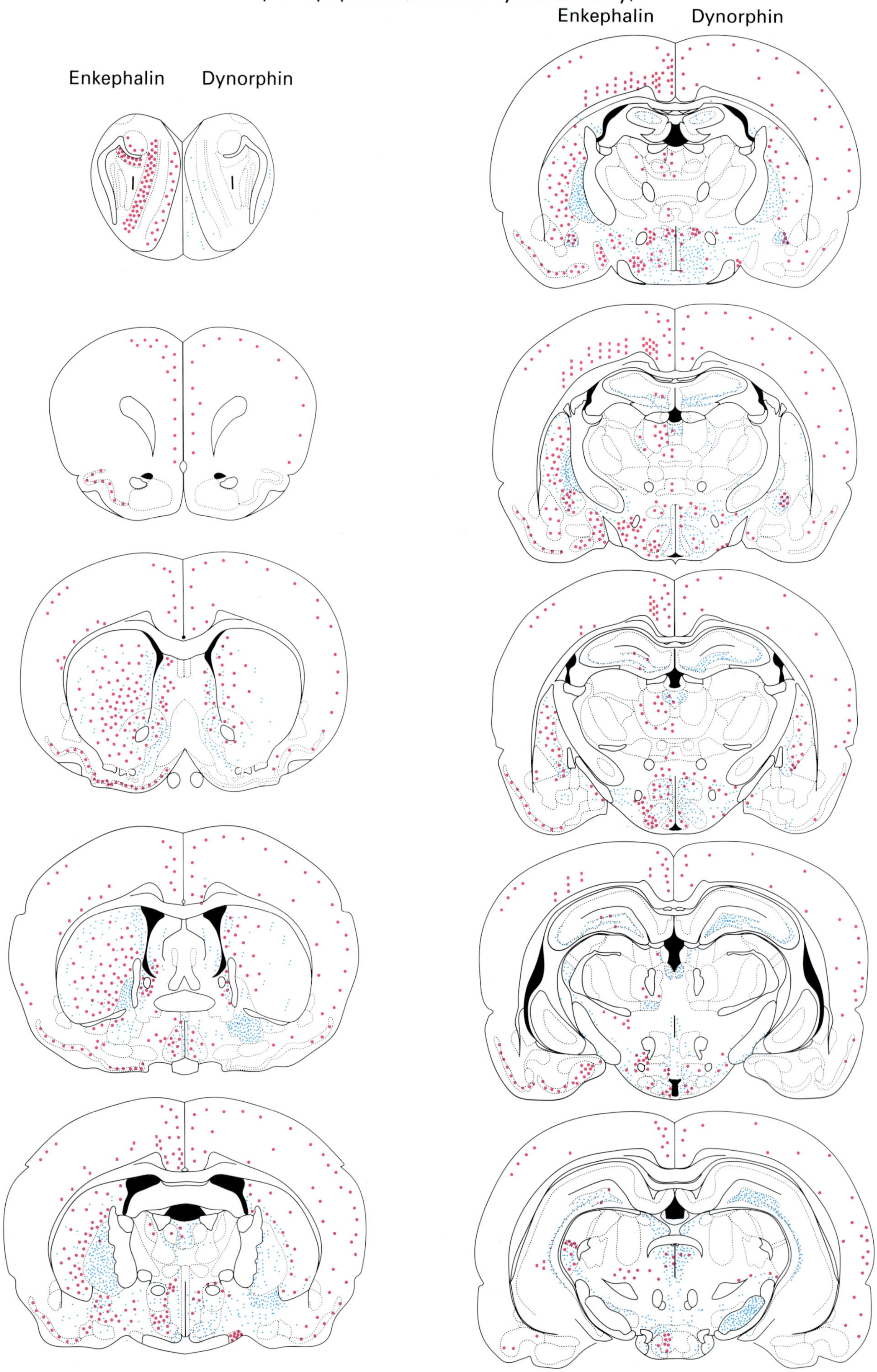

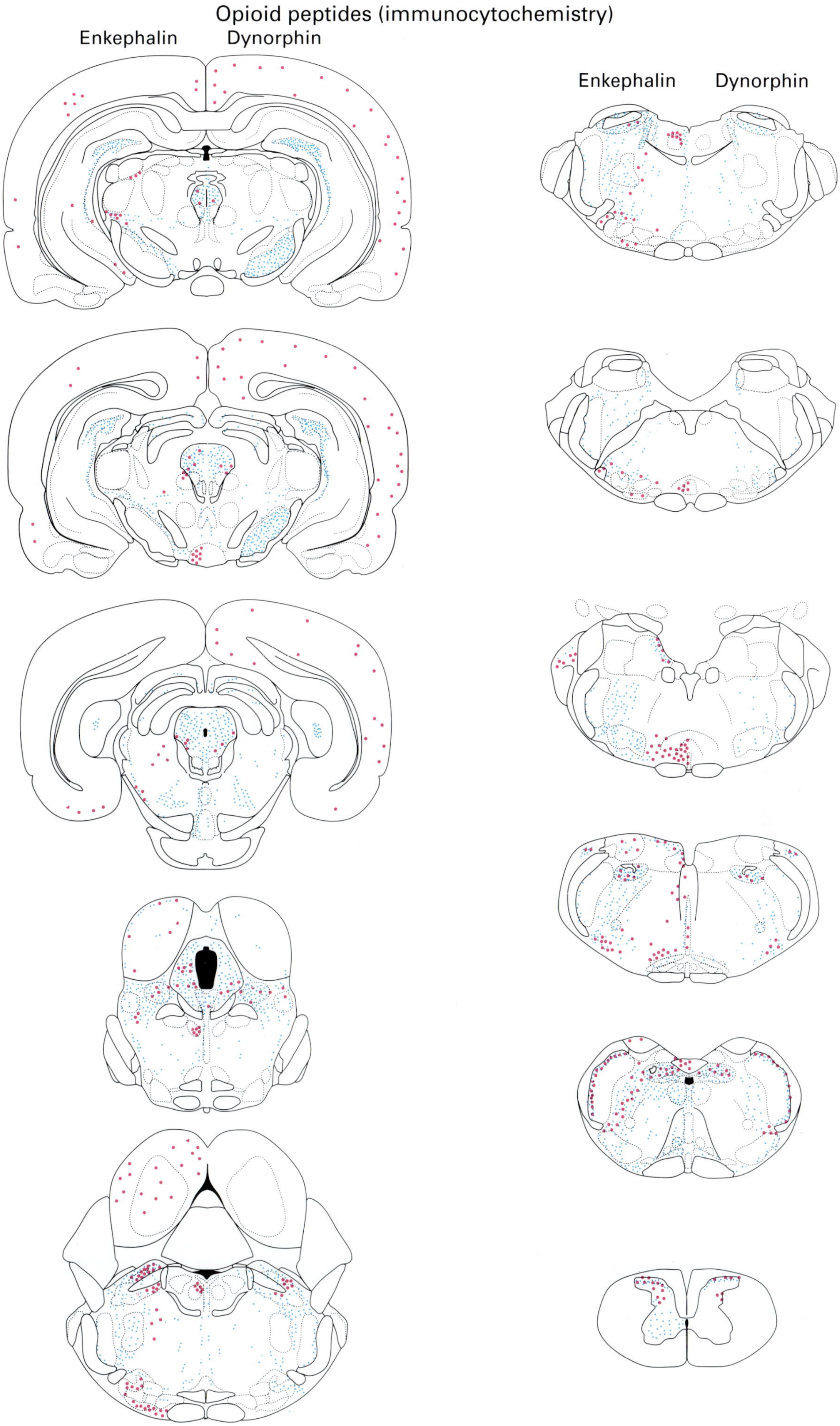
Opioid peptides (immunocytochemistry)
Enkephalin
Dynorphin
Enkephalin
Dynorphin

Opioid receptors

Subtypes

Opioid peptide initiates its pharmacological activities at specific recognition sites, i.e., by binding with receptors. Even in 1950s, a model of a specific receptor which could recognize the chemical structure of morphine was postulated[5]. Since then, the development of the receptor-assay method using radioactive ligand[7] has shown the heterogenous distribution of opioid receptors in the brain. Following the identification of two types of enkephalin as endogenous ligand for the opioid receptors in 1975, a large number of opioid peptides were discovered.

Several lines of research, including the work by Martin *et al.*[14], have demonstrated that multiple opioid receptors exist in the brain which can be classified into three main subtypes: mu (μ), delta (δ), and kappa (κ). These early studies have been extended by many anatomical, pharmacological, biochemical, and behavioural studies demonstrating the existence of distinct opioid receptor subtypes in the brain and periphery. The μ subtype acts with high selectivity on morphine-like compounds. Similarly, δ subtype shows an affinity to enkephalin-like peptides and κ subtype is selective for dynorphin-like peptides. Based on pharmacological and receptor binding studies, some researchers have provided further detailed classification, as μ_1 and μ_2[15], or κ_1 and κ_2[4]. in addition, ϵ subtypes, which has a high selectivity to β-endorphin, has been postulated[16].

Recently, the cDNA cloning of the mouse δ-opioid receptor has been reported using an expression cloning strategy[6,9]. This success facilitated the cloning of other members of the opioid receptor family. The three cloned opioid receptors are members of the seven transmembrane superfamily of G-protein coupled receptors and are homologous to one another at both the amino acid and the nucleic acid levels.

Distribution

One of the lines of evidence in support of the multiplicity of opioid receptors has been their differential localization in the nervous system. Numerous inconsistencies of differential distribution of μ and δ receptors have appeared in the literature. For example, some researchers have suggested that δ receptors are diffusely localized in rat cortex, while others reported that their density varies with cortical lamina. Similarly, some studies have suggested that the hypothalamus contains both μ and δ sites, while others reported either a predominance of μ sites or sparse, if any, opioid binding sites. In the case of κ receptors, there has been even less agreement concerning their distribution.

One of the major sources of confusion concerning the distribution of the opioid receptor subtypes may have been the lack of selective opioid ligands. The recent advent of selective ligands, such as the μ agonist ^{3}H-DAGO and the δ agonist ^{3}H-DPDPE has greatly reduced this problem (Table 21.1). There continues to be a problem, however, in selectively labelling κ receptors. Another possible source of problem may have been the varying concentrations of radioligand used to label the receptor sites. Because of the varying affinities of the ligands for each of the opioid receptor subtypes, the labelling concentration is of primary importance when comparing these subtypes.

A recent study[11] has examined the distribution of the μ, δ, and κ opioid receptor subtypes in the rat forebrain and midbrain, using DAMGO as a ligand for μ receptor, DPDPE for δ receptor, and bremazocine for κ receptor in the presence of unlabelled DAGO and DPDPE. In this study, μ receptor is distributed most widely, with high concentrations in the anterior cingulate cortex, neocortex, amygdala, hippocampus, ventral dentate gyrus, presubiculum, nucleus accumbens, caudate putamen, thalamus, habenula, interpeduncular nucleus, pars compacta of the substantia nigra, superior and inferior colliculi, and raphe nuclei. In contrast, δ binding was restricted to only a few brain areas, including the anterior cingulate cortex, neocortex, amygdala, olfactor tubercle, nucleus accumbens, and caudate putamen. κ receptor was densely distributed in the amydala, olfactory tubercle, nucleus accumbens, caudata putamen, medial preoptic area, hypothalamus, median eminence, periventricular thalamus, and interpeduncular nucleus. This atlas is based on the data of Adler *et al.*[1] and Mansour *et al.*[11].

It is now possible to determine more precisely the anatomical localization of opioid receptors using *in situ* hybridization since the opioid receptor family has recently been cloned. These studies[12,13] examining the distribution of μ, δ, κ receptor mRNAs in the CNS and comparing opioid receptor mRNA and receptor binding, complemented previous studies demonstrating a differential

Table 21.1. Receptor classification of opiates and opioid peptides (modified from ref. 1).

Receptor subtype	Agonist	Antagonist
μ	Morphine dihydromorphine Oxymorphone DAMGO Morphiceptin β-casomorphin PL-017	Naloxone Naltrexone
δ	[Met] Enkephalin [Leu] Enkephalin DADLE DSTLE DAMEA DPDPE	Naloxone ICI 154, 129 ICI 171, 864
κ	Dynorphin Ketocyclazocine Tifuadome Bremazocine U 50488 H Mr 2034	Naloxone Mr 2266 WIN 44441–3
ε	β-Emdorphin	Naloxone
σ	*N*-Allylnormetazocine (SKF 10, 047)	

DAMGO [D-ala2-MePhe4, Gly(ol)5]-Enkephalin H-Tyr-D-Ala-Gly-MePhe-Gly(ol)
DADLE [D-ala2, D-leu5]-Enkephalin H-Tyr-D-Ala-Gly-Phe-D-Leu-OH
DSTLE [D-Ser2, Leu5, Thr6]-Enkephalin H-hyr-D-Ser-Gly-Phe-Key-Thr
DAMEA [D-ala2, met5]-Enkephalinamide H-tyr-D-Ala-Gly-Phe-Met-CONH2
DPDPE [D-Pen2, D-Pen5]-Enkephalin H-tyr-D-Pen-Gly-Phe-D-Pen

Table 21.2. The effect of U50,488H and dynorphin A on the rat dorsal horn cells (from ref. 10). The numbers refer to the numbers of cells influenced in a particular direction by the κ-agonists. Three cells were excited by U50,488H at low doses and then inhibited by higher doses.

Effect on C-fibre evoked activity	U 50488 H	Dynorphin A_{1-13}
Excitation	4	12
Inhibition	6	8
Excitation then inhibition	3	6
No effect	0	6
Total number of cells	13	26

distribution of the μ, δ, κ binding sites in the CNS. They also provided new insights into the cellular localization of the opioid receptor mRNA encoding these receptors. Other studies[2,3] using specific antisera which recognize epitopes of the cloned δ and μ-opioid receptor reported a precise immunohistochemical localization of the δ-opioid receptor in brain and spinal cord.

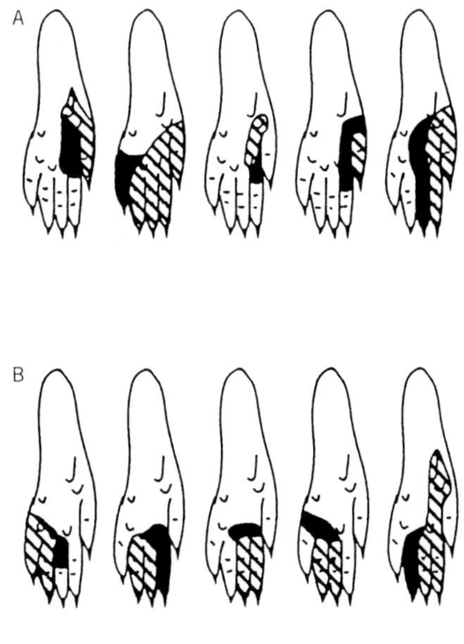

Fig. 21.3 Receptive fields of superficial dorsal horn cells that exhibited expansion after spinal dynorphin (*A*) or spinal U-50,488H (*B*). Shaded areas represent receptive fields prior to manipulation. Filled areas indicate the maximum extent of the expanded portion of the receptive field (modified from ref. 8).

Physiological functions

A variety of controversial proposals concerning the function of each receptor subtype have been made, concerning several areas in the CNS. For example, it has been reported that in the dorsal horn of the spinal cord, dynorphin like substances act mainly through κ receptor, as compared to enkephalin-like substances which act mainly via the δ receptor. In animal models of peripheral inflammation and hyperalgesia, μ and δ agonists injected into the intrathecal or peripheral tissues bring about an analgesis effect by inhibiting the hyperexcitability of dorsal horn neurons. However, in the case of κ agonist, a dual effect on dorsal horn neurons has been reported[10]. Dynorphin of U-50488 of κ agonist, when intrathecally administered into the spinal cord, produced facilitation, or inhibition, or a combination of both effects simultaneously of the nociceptive dorsal horn neurons (Table 21.2). Another study[8] reported that one-third of the dorsal horn neurons, when κ receptor agonist was administrated spinally, exhibited the expansion of their receptive field (Fig. 21.3). These two types of responses may represent activation of functionally different subpopulations of dorsal horn neurons, or, alternatively, the two responses may reflect interaction of the agonists at the multiple receptor types mentioned above. Studies to clarify these controversial points will be helped by the information on the molecular structure of several opioid receptor mRNAs that is now available.

References

1. Adler, B., Goodman, R.R., and Pasternak, G.W. (1990). In *Handbook of chemical neuroanatomy* (ed. A. Björklund, T. Hökfelt, and M.J. Kuhar) Vol. 9, pp. 359–93. Elsevier, Amsterdam.
2. Arvidsson, U., Dado, R.J., Riedl, M., Lee, J.-H. *et al.* (1995) *J. Neurosci.*, **15**, 1215–35.
3. Arvidsson, U., Riedl, M., Chakrabarti, S. *et al.* (1995). *J. Neurosci.*, **15**, 3328–41.
4. Attali, B., Gouarderes, C., Mazarguil, H. *et al.* (1982). *Neuropeptides*, **3**, 53–64.
5. Beckett, A.H. and Casy, A.F. (1954). *J. Pharm. Pharmacol.*, **6**, 986–1001.
6. Evans, C.J., Keith, D.E. Jr, Morrison, H. *et al.* (1992). *Science*, **258**, 1952–5.
7. Goldstein, A., Lowney, L.I., and Pal, B.K. (1971). *Proc. Natl Acad. Sci. USA*, **68**, 1742–7.
8. Hylden, J.L.K., Nahin, R.L., Traub, R.J., and Dubner, R. (1991). *Pain*, **44**, 187–93.
9. Kieffer, B.L., Befort, K., Gaveriaux-Ruff, C., and Hirth, C.G. (1982). *Proc. Natl Acad. Sci. USA*, **89**, 12048–52.
10. Knox, R.J. and Dickenson, A.H. (1987). *Brain Res.*, **415**, 21–9.
11. Mansour, A., Khachaturian, H., Lewis, M.E. *et al.* (1987). *J. Neurosci*, **7**, 2445–64.
12. Mansour, A., Fox, C.A., Thompson, H., and Watson, S.J. (1994). *Brain Res.*, **643**, 245–65.
13. Mansour, A., Fox, C.A., Burke, S. *et al.* (1994). *J. Comp. Neurol.*, **350**, 412–38.
14. Martin, W.R., Eades, C.G., Thompson, J.A. *et al.* (1976) *J. Pharmacol Exp. Ther,.* **197**, 517–32.
15. Pasternak, G.W., Gintzler, A.R., Houghten, R.A. *et al.* (1983). *Life Sci.*, **33**, Suppl. 1, 167–73.
16. Schulz, R., Wuster, M., and Herz, A. (1981). *J. Pharmacol. Exp. Ther.*, **216**, 604–6.

Opioid peptides (binding sites)

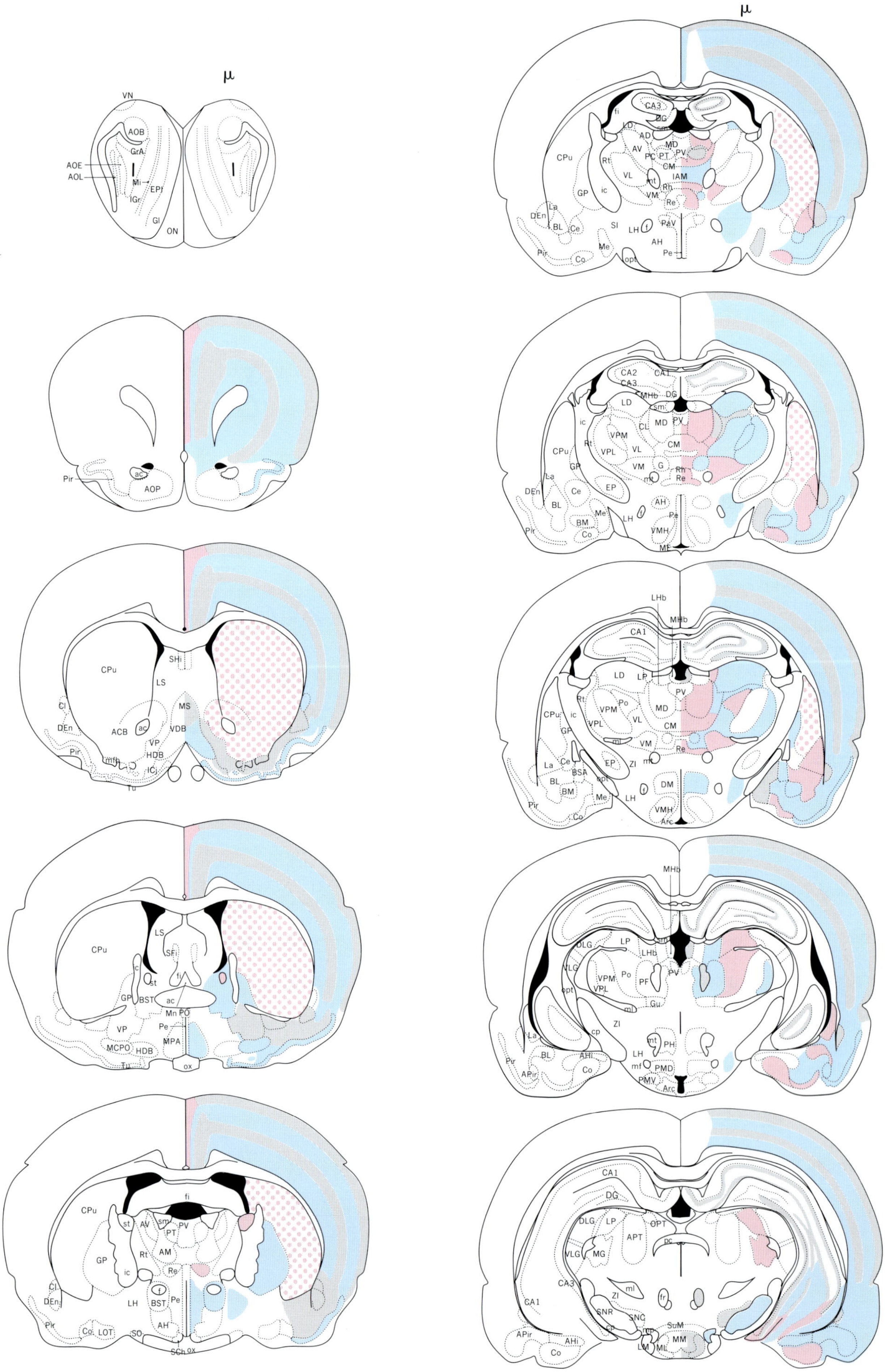

Opioid peptides (binding sites)

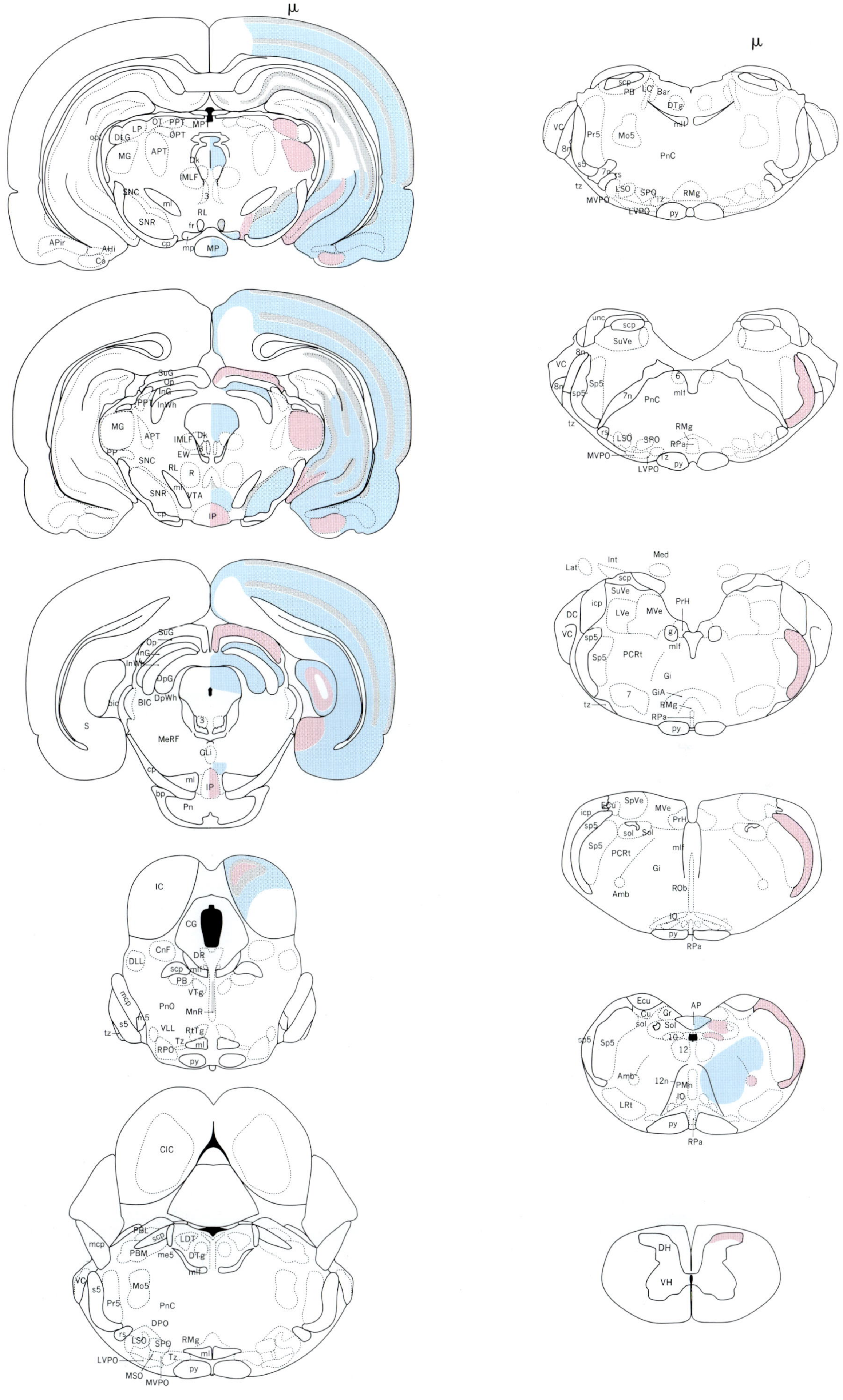

Opioid peptides (binding sites)

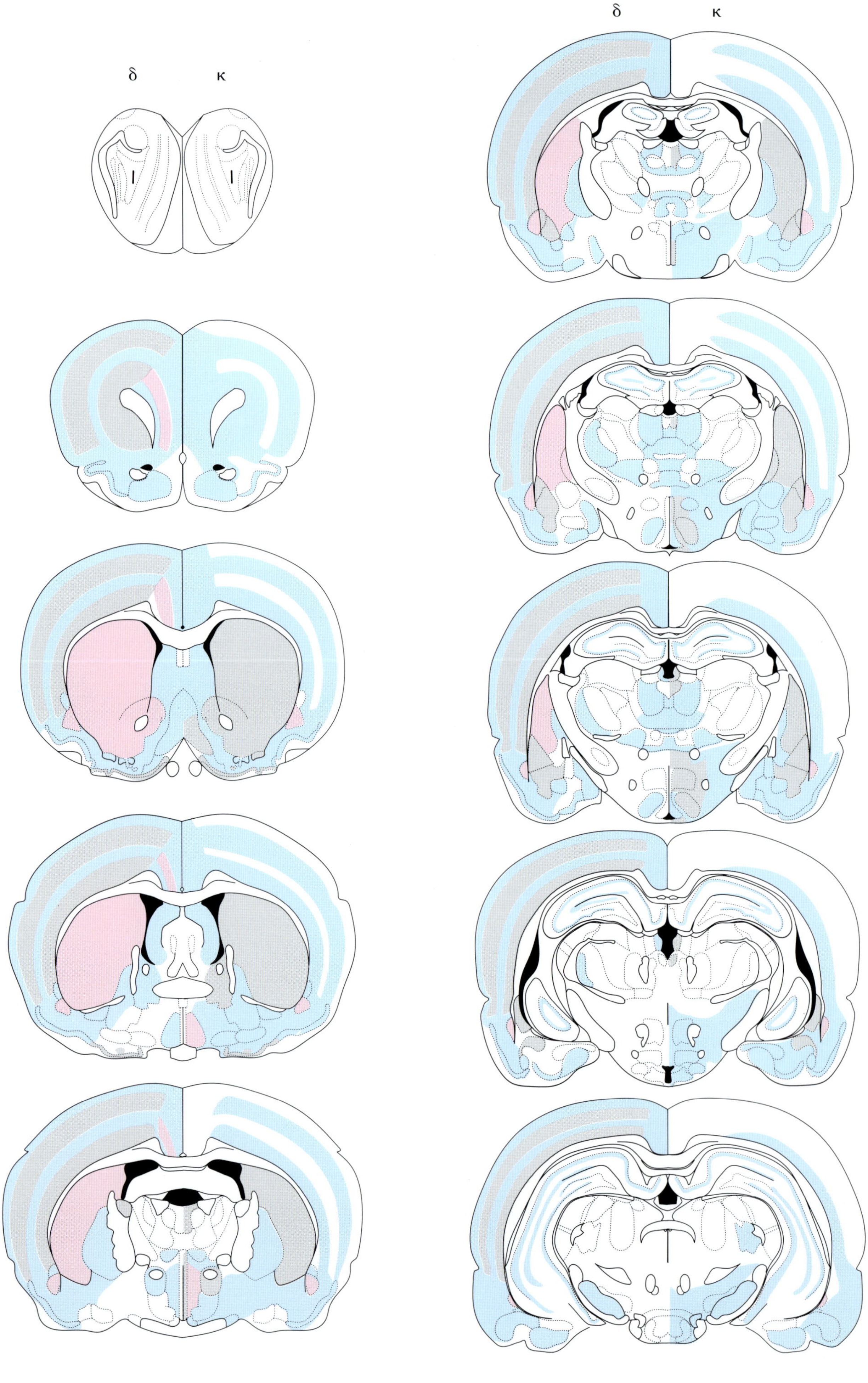

Opioid peptides (binding sites)

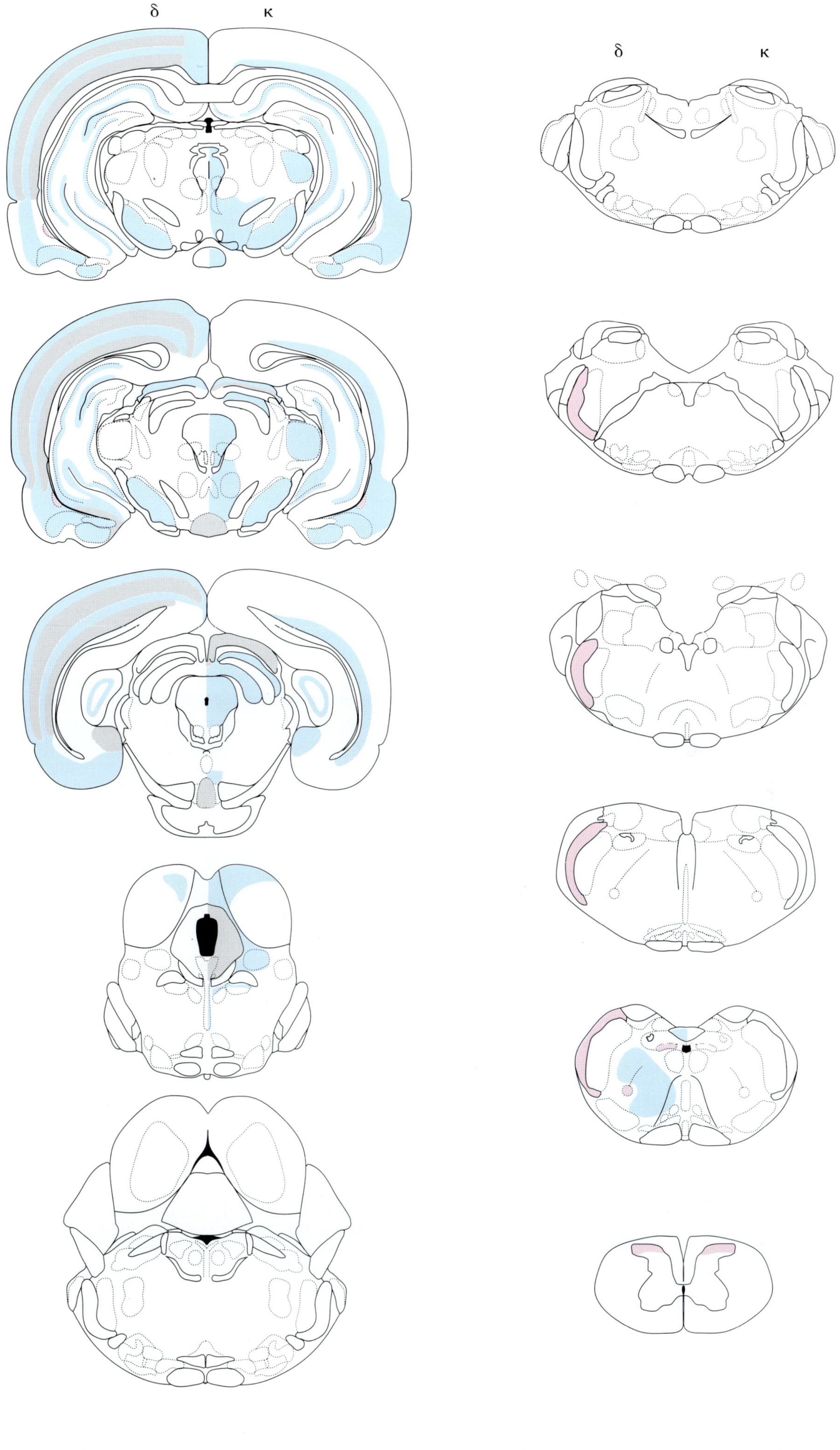

22 The galanin system

Galanin

Structure

Galanin is a peptide originally purified from swine intestine and consisting of 29 amino acid residues[7] (Fig. 22.1) with a C-terminal alanine amide. The five amino acid sequence from the C-terminal end of galanine (-Tyr-Gly-Leu-Ala-NH_2), shows structural analogy to physalaemin (-Tyr-Gly-Leu-Met-NH_2), substance P (-Phe-Gly-Leu-Met-NH_2), and gonadoliberin (Gly-His-Trp-Ser-Tyr-Gly-Leu-Arg-Pro-Gly-NH_2).

The N-terminal sequence of galanin (Gly-Trp-Thr-Leu-Asn-Ser) resembles that of propiomelanocortin (Gly-Trp-Cys-Leu-Glu-Ser), and the (-Gly-Tyr-Leu-Leu-Gly-) sequence of the intermediate region of galanin resembles the (-Gly-Tyr- Ser-Leu-Gly-) sequence of the phosphate acceptor peptide. Though galanin shows partial homology to the amino acid sequence of other peptides, it does not appear to belong to any of the peptide families described to date.

Physiological function

The reported physiological actions of galanin include inhibition of insulin secretion[3], induction of hyperglycemia[7], and, in the nervous system, inhibition of dopamine release from the median eminence[5] and inhibition of acetylcholine release in the ventral hippocampus[2].

From experiments using the smooth muscle of small intestine, the physiological activities of galanin have been attributed to its N-terminal end and not to the C-terminal end. In the nervous system also the N-terminal fragment (1–16) of galanin is an agonist of high affinity ($K_d \approx 3$ nM) of galanin receptor in rat hippocampus and has been found to inhibit acetylcholine release from the hippocampus nerve cells and break down inositol phosphate stimulated by muscarinic agonists. Thus, for reference of distribution of galanin receptors, the N-terminal peptide fragment is employed (see also the next section and Tables 22.2 and 22.3). Galanin accelerates transcription remarkably during nerve regeneration of peripheral sensory and motor nerves. A rise in galanin expression in the locus ceruleus, caused by depletion of catecholamine, as well as regulation in the expression of galanin in hypophysis due to steroid stimulus, are also known.

1	2	3	4	5	6	7	8	9	10
Gly-	Trp-	Thr-	Leu-	Asn-	Ser-	Ala-	Gly-	Tyr-	Leu-
11	**12**	**13**	**14**	**15**	**16**	**17**	**18**	**19**	**20**
Leu-	Gly-	Pro-	His-	Ala-	Ile-	Asp-	Asn-	His-	Arg-
21	**22**	**23**	**24**	**25**	**26**	**27**	**28**	**29**	
Ser-	Phe-	His-	Asp-	Lys-	Tyr-	Gly-	Leu-	Ala-NH_2	

Fig. 22.1 Amino acid sequence of galanin.

Distribution

Immunohistochemical investigations have shown that galanin is distributed widely in the central and peripheral nervous systems[1,4,6].

The highest concentrations of galanin-containing nerve cells are seen in the hypothalamus, medulla oblongata, and spinal cord. However, a large number of positive signals are detected widely in the telencephalon, thalamus, mesencephalon, and pons as well. Galanin seems to be absent in the olfactory bulb and cerebellum.

Similar to the distribution pattern of other neuropeptides, the distribution of galanin often deviates from the ranges of nerve nuclei.

Diffused galanin-positive cells have been observed in alignment with the following nerve nuclei: diagonal band nucleus, medial septal nucleus, bed nucleus of stria terminalis, central nucleus of amygdaloid complex, supraoptic nucleus, paraventricular hypothalamic nucleus, medial preoptic area, dorsomedial hypothalamic nucleus, arcuate nucleus, magnocellular and tuberal hypothalamic nuclei, dorsal raphe nucleus and raphe nuclei of medulla oblongata, locus ceruleus, solitary tract nucleus, and ventral reticular nucleus of medulla oblongata. Diffused fibre distribution of high–medium concentration is observed in ventral and caudal regions of the bed nucleus of the stria terminalis, medial preoptic area, paraventricular hypothalamic nucleus, median eminence (very high concentration), dorsomedial nucleus of hypothalamus, central grey matter of pons, parabrachial nucleus, solitary tract nucleus, spinal tract nucleus of the trigeminal nerve, and the dorsal column of spinal cord.

References

1. Ch'ng, J.L.C., Christofides, N.D., Anand, P. *et al.* (1985). *Neuroscience*, **16**, 343–54.
2. Fisone, G., Wu, C.F., Consolo, F. *et al.* (1987). *Proc Natl Acad. Sci. USA*, **88**, 7339–43.
3. McDonald, T.J., Dupre, J., Tatemoto, K. *et al.* (1985). *Diabetes*, **34**, 192–6.
4. Melander, T., Hökfelt, T., and Rökaeus, Å. (1986). *J. Comp. Neurol.*, **248**, 475–517.
5. Nordström, Ö., Melander, T., Hökfelt, T. *et al.* (1987). *Neurosci. Lett.*, **73**, 21–6.
6. Skofitsch, G. and Jacobowitz, D.M. (1985). *Peptides*, **6**, 509–46.
7. Tatemoto, K., Rökaeus, Å., Jörnvall, H. *et al.* (1983). *FEBS Lett.*, **164**, 124–8.

Galanin receptor

The structure of galanin receptor has not yet been clarified. Whether a specific agonist of the galanin receptor exists or whether there are receptor subtypes are also not known. Reports on the distribution of galanin receptor are therefore based on the whole or a fragment of ^{125}I-labelled galanin. The quantitative distribution of the galanin receptor in the nervous system is shown in Table 22.1.

The N-terminal (1–15) fragment of galanin is conserved in the rat, porcine, and bovine amino acid sequences[1]. The literature on the distribution of the galanin receptor is limited to studies on the tissue binding distribution by autoradiography of radioactively labelled galanin or its partial peptide[1,2,3].

High concentrations of binding sites are observed in the cerebral cortex, hippocampus, septal area, amygdaloid complex, thalamus, hypothalamus, sensory regions of hindbrain, and spinal cord. The results obtained from the immunohistochemical studies of galanin correspond to high concentration of binding sites in the anterior frontal cortex, septal area, ventral pallidum, bed nucleus of stria terminalis, stria terminalis, median forebrain bundle, amygdaloid complex, periamygdaloid cortex, hippocampus, dorsal mammillary nucleus of the hypothalamus, lateral and periventricular thalamic nuclei, internal medullary lamina of the thalamus, interpeduncular nucleus, substantia nigra zona compacta, parabrachial nucleus, sensory nucleus of trigeminus, and dorsal column of spinal cord. Contrastingly, the olfactory tubercle, piriform cortex, and superior colliculus show a low concentration of galanin binding sites.

From the distribution of galanin and its binding sites, one can infer that galanin participates in the somatic endocrine functions in the amygdaloid complex and hypothalamus region. Since galanin binding sites exist abundantly in sensory regions of hindbrain and spinal cord where primary afferent sensory fibres terminate, galanin is suggested to play a role in the formation or regulation of nociceptive information.

References

1. Land, T., Langel, Ü., Fisone, G., Bedecs, K., and Bartfai, T. (1991). In *Methods in neurosciences*, Vol. 5 (ed. P.M. Conn) pp. 225–34. Academic Press, San Diego.
2. Melander, T., Köhler, C., Nilsson, S. *et al.* (1988). *J. Chem. Neuroanat.*, **1**, 213–33.
3. Skofitsch, G., Sills, M.A., and Jacobowitz, D.M. (1986). *Peptides*, **7**, 1029–42.

Table 22.1. Distribution of galanin receptors or ^{125}I-labelled galanin-binding sites in rat tissues (from ref. 1)

Region	B_{max} (fmol/mg protein)
Brain	
Cerebral cortex	0.96 ± 0.28
Hypothalamus	7.46 ± 0.30
Hippocampus dorsalis	1.7 ± 0.30
Hippocampus ventralis	3.6 ± 0.4
Smooth muscle	
Gastric	58.2 ± 11.0
Jejunal	32.4 ± 5.7
Spinal cord (dorsal lumbar)	5.9 ± 1.5

Table 22.2. Galanin and galanin receptor ligand effects on acetylcholine release and displacement of ^{125}I-labelled galanin (1–29) in rat ventral hippocampus (from ref. 1)

Galanin receptor ligand	Inhibition (%) of acetylcholine release (10 μg/μl, intracerebroventricularly)	Displacement of ^{125}I-labelled galanin, IC_{50} (nM)
Galanin (1–29)	100	1
Galanin (1–16)	50	3
[D-Trp^2]-Galanin (1–16)	35	500

Table 22.3. Galanin and galanin receptor ligand effects on feeding behaviour and displacement of ^{125}I-labelled galanin (1–29) in rat hypothalamus (from ref. 1)

Galanin receptor ligand	Effect on food consumption	Displacement of ^{125}I-labelled galanin, IC_{50} (nM)
Galanin (1–29)	Increase	1
Galanin (1–16)	Increase	6.3
[D-Trp^2]-Galanin (1–16)	No effect	500
Galanin (17–29)	No effect	>10 000
Galanin (1–9)	No effect	>1000
Galanin (10–20)	No effect	Not determined
Galanin (12–29)	No effect	Not determined
Galanin (21–29)	No effect	Not determined

Galanin (immunocytochemistry)

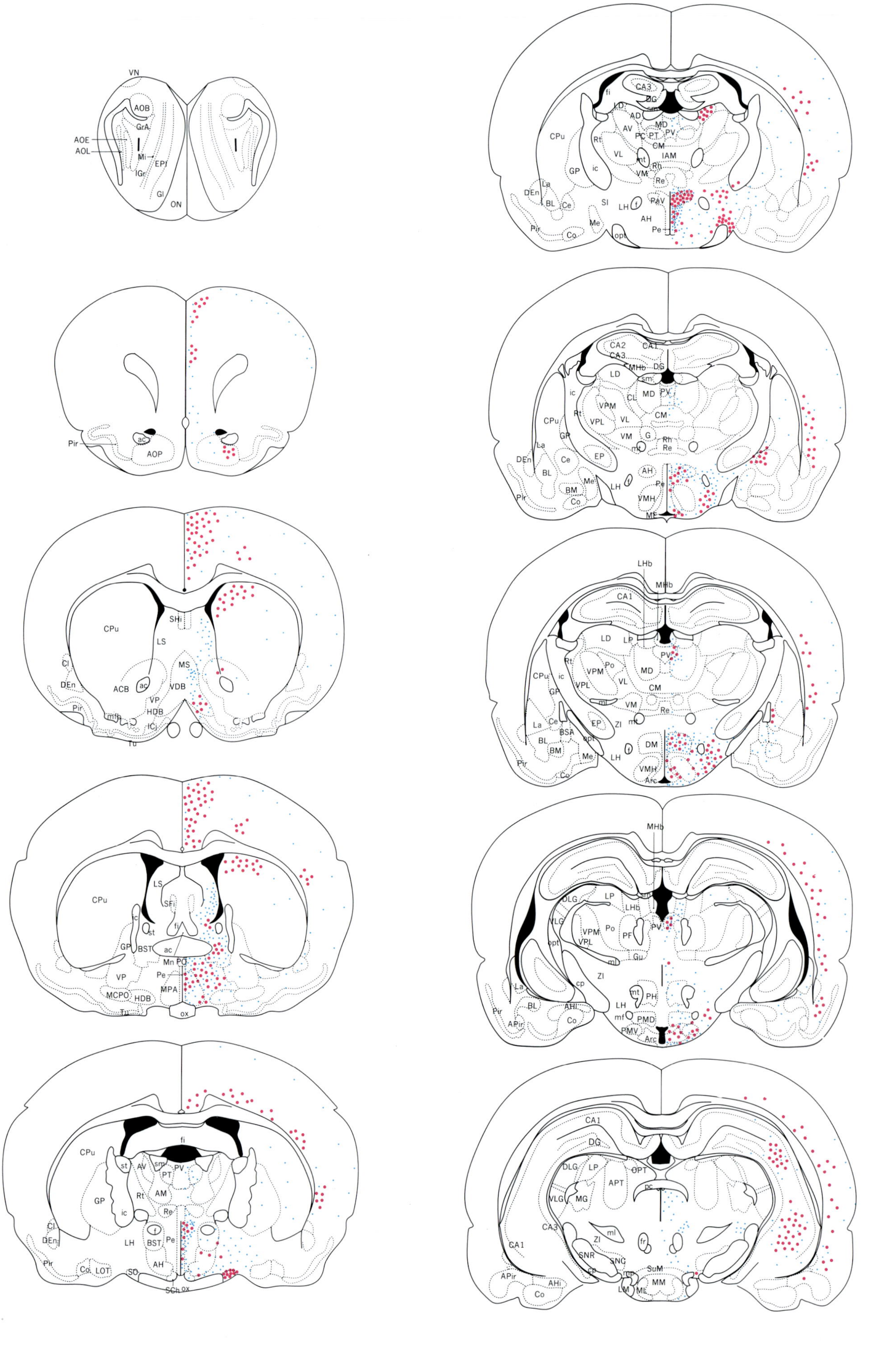

Galanin (immunocytochemistry)

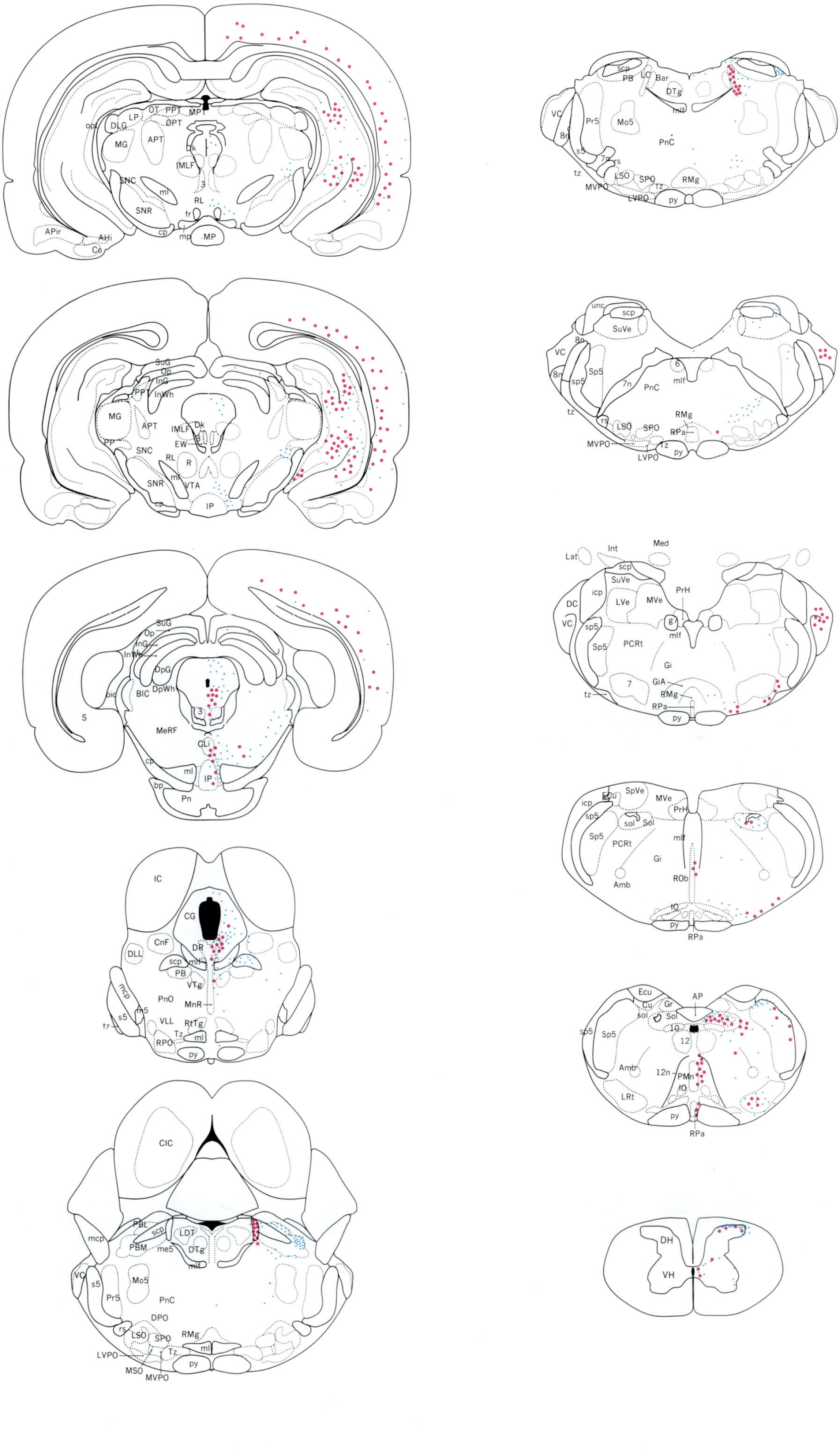

Galanin receptor (binding sites)

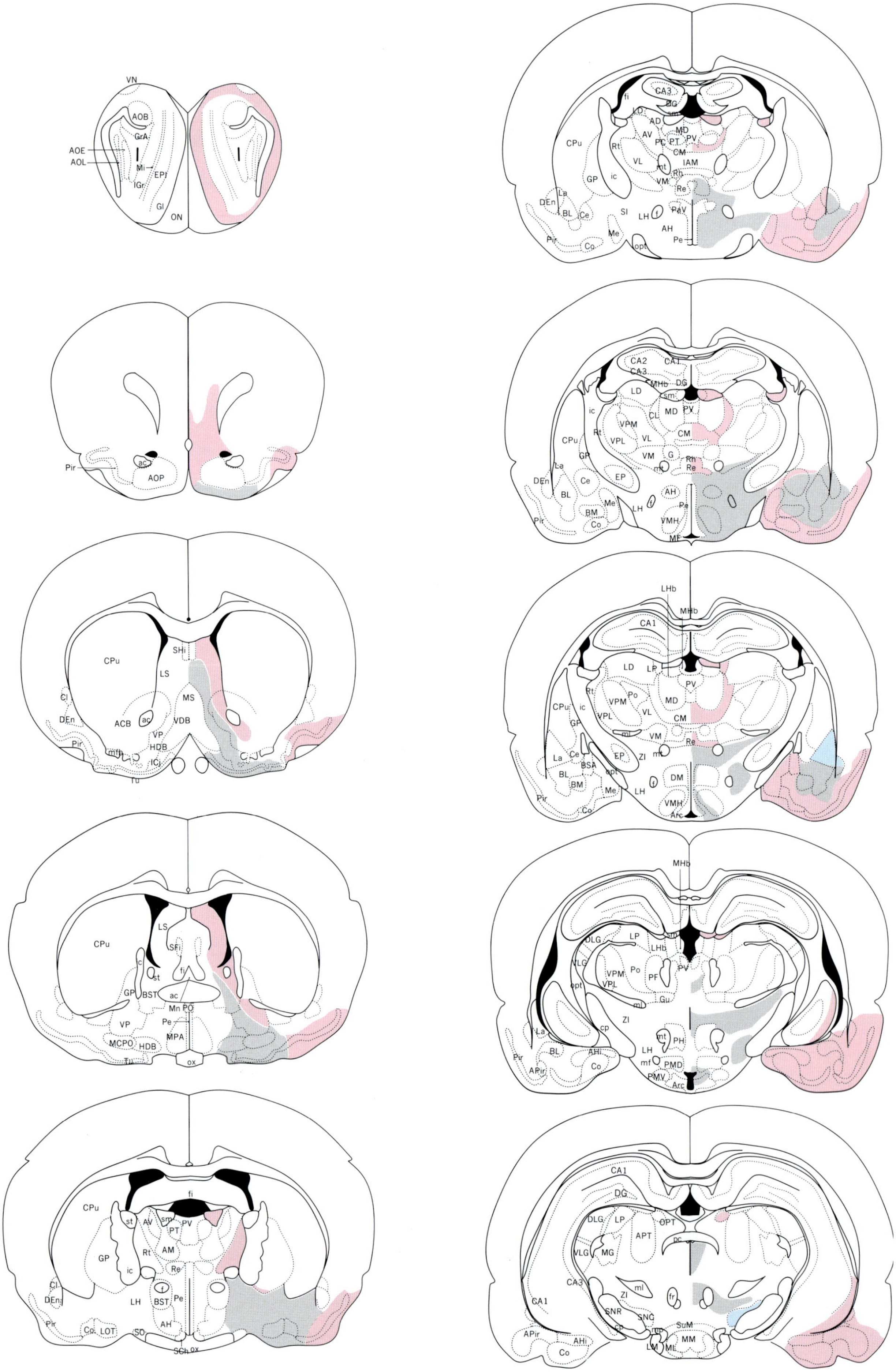

Galanin receptor (binding sites)

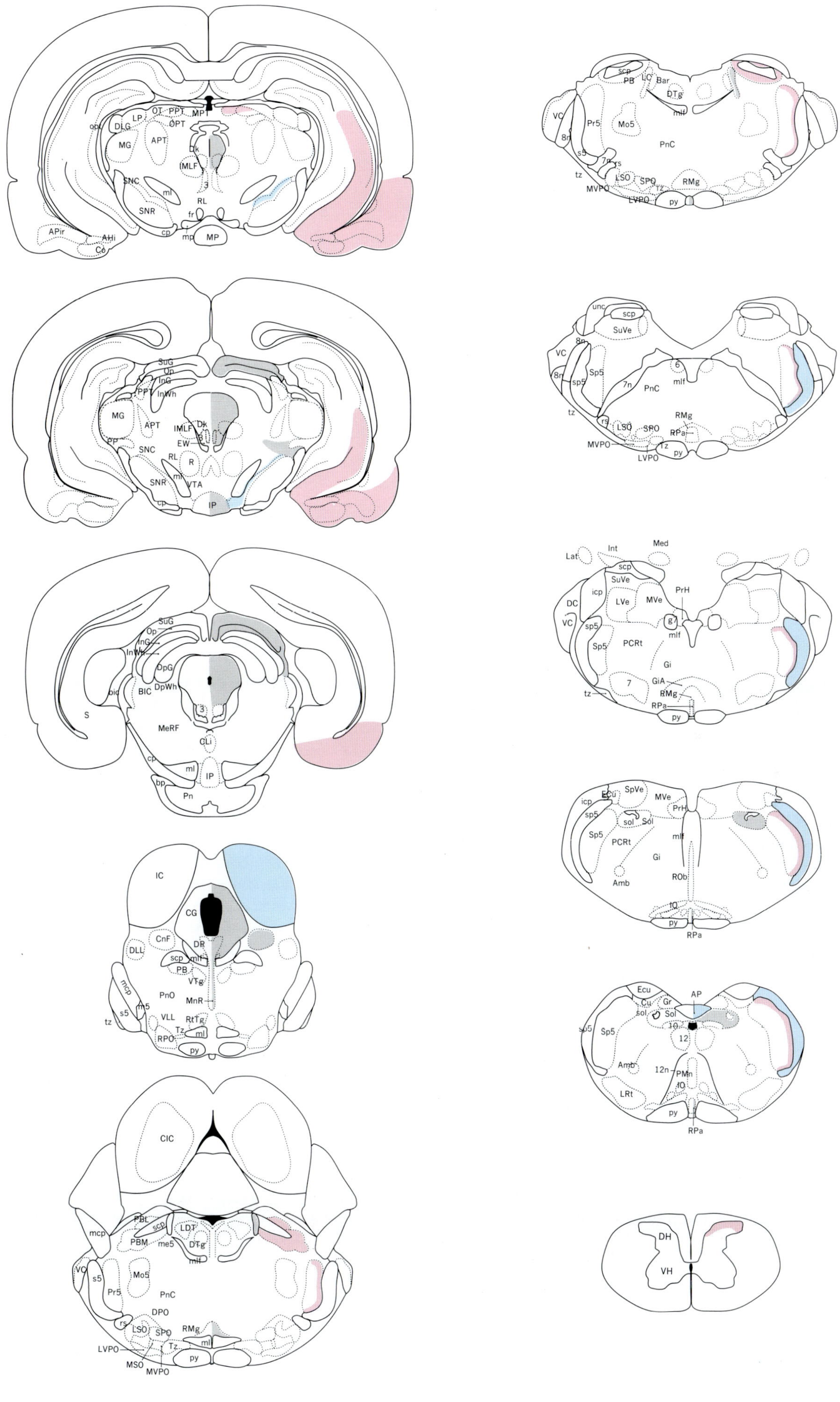

23 The somatostatin system

Somatostatin

Somatostatin was first isolated as a growth hormone release inhibiting factor (somatotropin release inhibiting factor) by Brazeau *et al.*[1] in 1973. It was also known to inhibit the secretion of thyroid stimulating hormone (TSH). Apart from these actions, somatostatin is also considered to function as a neurotransmitter or a neuromodulator. Its presence is identified not only in the hypothalamus but also in the olfactory bulb, cerebral cortex, corpus striatum, claustrum, hippocampus, bed nucleus of stria terminalis, amygdaloid complex, inferior colliculus, central grey matter of mesencephalon, dorsal and lateral nuclei tegmenti, cochlear nuclei, and various nuclei in the reticular formation. Studies on the ontogeny of this peptide suggest that somatostatin has neurotrophic activity[2,4].

Biosynthesis and structure

Somatostatin has 14 amino acid residues, with an intramolecular disulfide bond between the two cysteine residues giving it a circular form. Somatostatin-28 is a form in which the N-terminal ends are extended. Somatostatin-14 is synthesized as follows: preprosomatostatin is processed to somatostatin-28, which is further processed to somatostatin-14 by somatostatin-converting enzyme. The 116 amino acid precursor protein, called preprosomatostatin, of human, monkey, rat, and mouse has been isolated (Fig. 23.1).

Distribution[4,5]

In the hypothalamus, somatostatinergic neurons are distributed most abundantly in the periventricular nucleus. Some other somatostatin-containing cells in the hypothalamus exist in the zona incerta, dorsomedial hypothalamic nucleus, and arcuate nucleus. Somatostatin-containing nerve fibres that originate in these cells terminate in the outer lamina of the median eminence and the somatostatin is released into the portal blood of the pituitary gland. From here, the released somatostatin reaches the anterior lobe and regulates the secretin of growth hormone and TSH from the anterior lobe cells.

Apart from the hypothalamus, somatostatin-containing cells are present in abundance in the cerebral cortex and hippocampus. In the brains of patients with Alzheimer's disease, the amount of somatostatin in the cerebral cortex is decreased and somatostatin-immunopositive deposition is observed in senile plaque as well. Thus, a role for this polypeptide in Alzheimer's disease has also been suggested. Furthermore, in the forebrain abundant somatostatin-positive cells in the limbic system (bed nucleus of stria terminalis, amygdaloid complex, etc.) have been observed.

In the brain stem region, somatostatin-positive cells are concentrated in the components of the auditory system, such as the dorsal and ventral cochlear nuclei, lateral lemniscus, inferior colliculus, etc. Furthermore, somatostatin cell bodies exist in large numbers in structures such as the dorsolateral tegmental nucleus, reticular formation at the facial nucleus, and solitary tract nucleus. In the spinal trigeminal nucleus and dorsal horn of spinal cord, positive interneurons have been observed. However, these cell bodies are covered by exceedingly dense somatostatin terminals. Injection of colchicine into the fourth ventricle or central canal can render the cell bodies easily immunostainable. Somatostatin-positive neurons are present in the trigeminal ganglion and dorsal root ganglion, which project a large number of primary sensory fibres to the nuclei mentioned above.

The main somatostatin projection tracts are shown in Fig. 23.2. Somatostatin cells that exist in the amygdaloid complex project their axons through stria terminalis, to the bed nucleus of stria terminalis, and to the auxiliary brain stem regions. The auditory system, such as pathways from the cochlear nuclei to the lateral lemniscus nucleus and from the lateral lemniscal nucleus to the inferior colliculus, contain somatostatin. The somatostatinergic primary sensory fibres or hypothalamus–pituitary gland system are a comparatively long projection tract. However, the somatostatin cells in the cerebral cortex, hippocampus, and corpus striatum are considered to be inter-neurons.

Physiological function

The functions of somatostatin, except for its role in the hypothalamus–pituitary hormonal axis, are somewhat unclear. Many studies based on animal models and clinical data suggest that the somatostatinergic system may be involved in cognitive functions. This involvement may be supported by the finding that somatostatin content is decreased in the brains of Alzheimer's disease patients. In the hippocampus, it is known that somatostatin modifies cholinergic transmission, and it has therefore been hypothesized that somatostatin may be involved in cognitive functions via acetylcholine[2].

The action of somatostatin as a neurotransmitter is probably carried out through a specific receptor (described on p. 264) and is believed to be due to the inhibition of cAMP production. Accumulated data suggest that somatostatin is an inhibitory neurotransmitter (or neuromodulator), but some studies provide evidence that it can act as an excitatory neurotransmitter in certain brain locations[3].

References

1. Brazeau, P., Vale, W., Burgus, R. *et al.* (1973). *Science*, **179**, 77–9.
2. Deft, I.R. and Dicher, M.A. (1985). In *Neurotransmitter actions in the vertebrate nervous system* (ed. M.A. Rogawski, and I.L. Barker) pp. 411–38. Plenum Press, New York.
3. Dodd, J. and Kelly, I.S. (1978). *Nature*, **273**, 674–5.
4. Shiosaka, S. (1992). In *Handbook of chemical neuroanatomy*. Vol. 10: *Ontogeny of transmitter and peptides in the CNS* (ed. A. Björklund, T. Hökfelt, and M. Tohyama) pp. 369–97. Elsevier, New York.
5. Vincent, S.R., Mcintosh, C.H.S., Buchan, A.M.J., and Brown, J.C. (1985). *J. Comp. Neurol.*, **238**, 169–86.

Sequence	Form	
Ala-Gly-Cys-Lys-Asn-Phe-Phe-Trp-Lys-Thr-Phe-Ser-Ser-Cys	SS-14(Sea lamprey)	fish
Ala-Leu-Arg-Ala-Ala-Ala-Val-Ala-Gly-Ser-Pro-Gln-Gln-Leu-Leu-Pro-Leu-Gly-Gln-Arg-Glu-Arg-Lys- -Ala-Gly-Cys-Lys-Asn-Phe-Phe-Trp-Lys-Thr-Phe-Ser-Ser-Cys	SS-34(Sea lamprey)	fish
Ala-Ala-Ala-Val-Ala-Gly-Ser-Pro-Gln-Gln-Leu-Leu-Pro-Leu-Gly-Gln-Arg-Glu-Arg-Lys- -Ala-Gly-Cys-Lys-Asn-Phe-Phe-Trp-Lys-Thr-Phe-Ser-Ser-Cys	SS-37(Sea lamprey)	fish
Ala-Gly-Cys-Lys-Asn-Phe-Phe-Trp-Lys-Thr-Phe-Thr-Ser-Cys	SS-I(American goosefish)	fish
Ala-Gly-Cys-Lys-Asn-Phe-Phe-Trp-Lys-Thr-Phe-Thr-Ser-Cys	SS-14(Channel catfish)	fish
Ala-Gly-Cys-Lys-Asn-Pro-Phe-Tyr-Lys-Thr-Phe-Thr-Ser-Cys	SS-14(Shorthorn sculpin)	fish
Ala-Gly-Cys-Lys-Asn-Phe-Tyr-Trp-Lys-Gly-Phe-Thr-Ser-Cys	SS-14(European flounder)	fish
Ser-Ile-Glu-Pro-Pro-Asn-Asn-Leu-Pro-Pro-Arg-Glu-Arg-Lys-Ala-Gly-Cys-Lys-Asn-Phe-Tyr-Trp-Lys-Gly-Phe-Thr-Ser-Cys	SS-28(European flounder)	fish
Ala-Gly-Cys-Lys-Asn-Phe-Phe-Trp-Lys-Thr-Phe-Thr-Ser-Cys	SS-14(Human,Macaque,Pig,Rat,Mouse)	
Ser-Ala-Asn-Ser-Asn-Pro-Ala-Met-Ala-Pro-Arg-Glu-Arg-Lys-Ala-Gly-Cys-Lys-Asn-Phe-Phe-Trp-Lys-Thr-Phe-Thr-Ser-Cys	SS-28(Human,Macaque,Pig,Rat,Mouse)	

Fig. 23.1 Amino acid sequences of somatostatins. Somatostatin contains an intramolecular disulfide bond linking two cystein residues. S-28, S-34, and S-37 are larger forms of S-14 with an extended N-terminal.

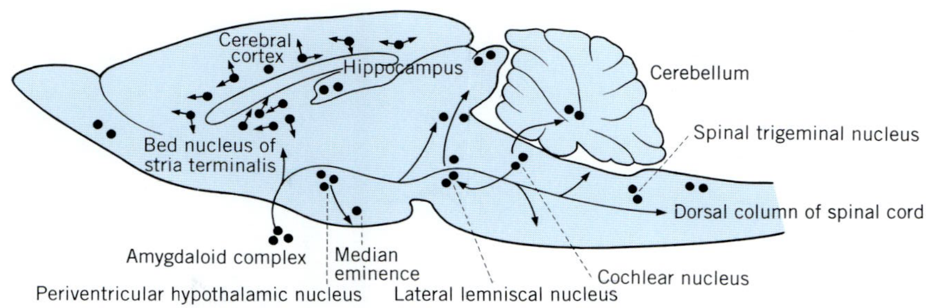

Fig 23.2 Somatostatinergic pathways.

Somatostatin (immunocytochemistry)

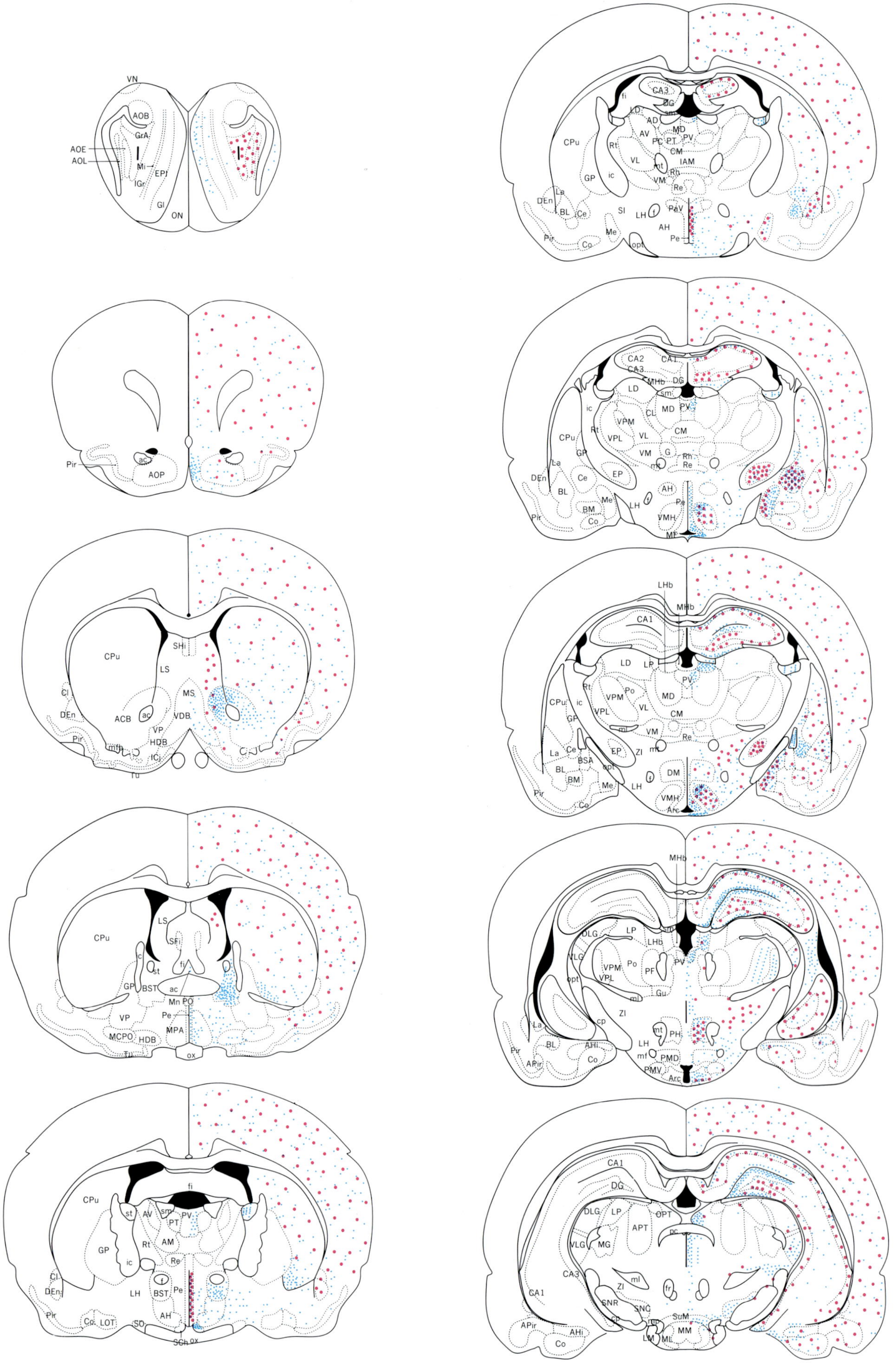

Somatostatin (immunocytochemistry)

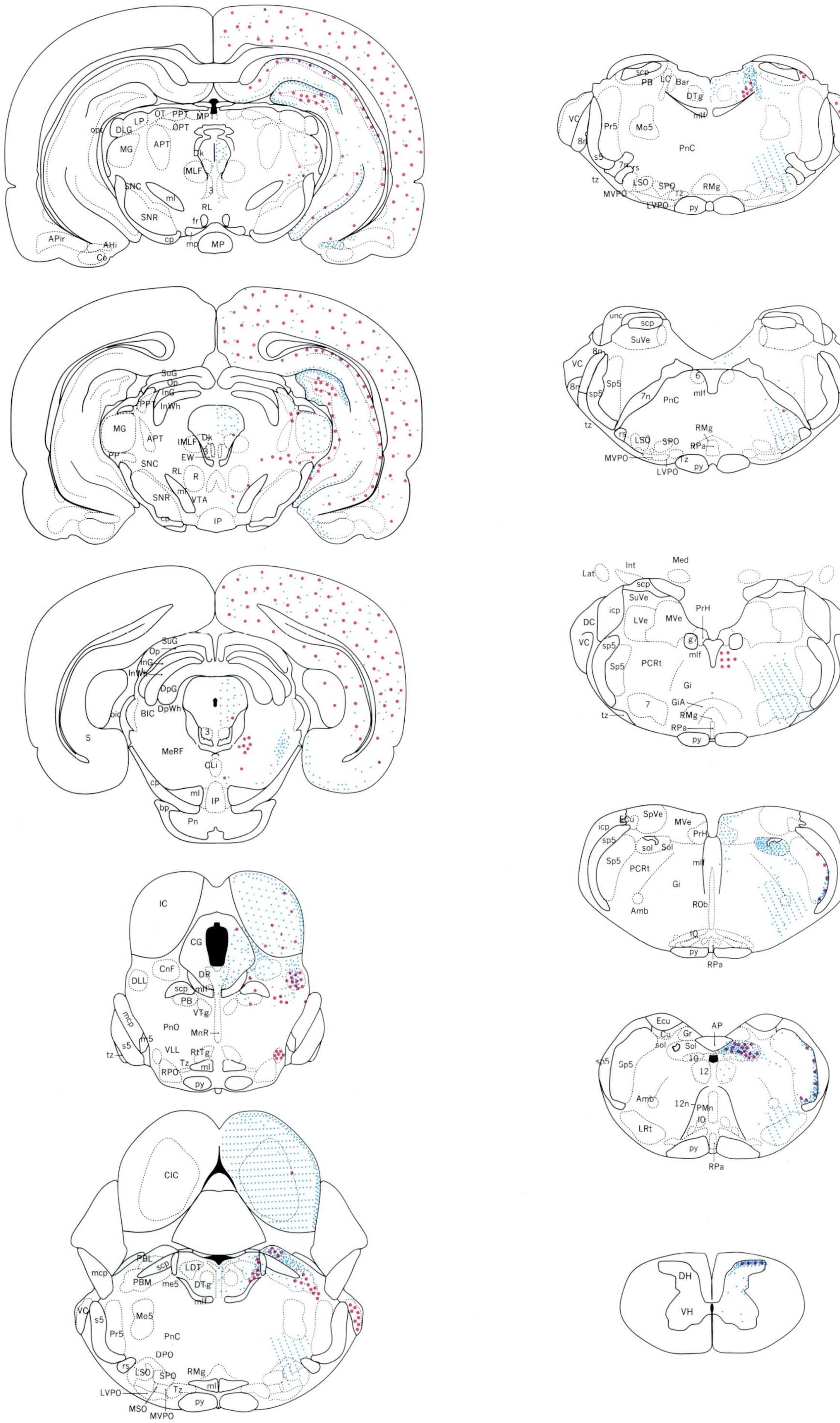

Somatostatin receptor

Subtypes

Among mammals, two types of somatostatin are present: somatostatin-14, with 14 amino acid residues, and somatostatin-28, with 28 amino acid residues. Both types show physiological activity, such as inhibiting the secretion of growth hormone. At least two types of receptor for somatostatin also exist. One type, existing widely in the brain, binds to both somatostatin-14 and somatostatin-28. The second type of receptor binds only to somatostatin-28. Both receptor types are present in the cerebral cortex, hippocampus, mesencephalon, and corpus striatum[2,4,5].

Compared to somatostatin-14, somatostatin-28 has stronger inhibition potency for secretion of growth hormone, insulin, and glucagon. The binding potency of somatostatin-28 is 3.2 fold higher than somatostatin-14 in pituitary gland. On the other hand, in the central nervous system, the affinity of somatostatin-28 to the receptor is weaker in comparison to somatostatin-14. Therefore, somatostatin-14 is considered to exhibit stronger physiological actions in the regulation of brain functions.

Somatostatin receptors have been shown to act by coupling in an inhibitory way with adenylate cyclase. Changes in Ca^{2+} concentration are also implicated, in an independent manner[1,3] (Fig. 23.3).

Cloning

Recently, two different somatostatin receptors have been cloned in human, rat, and mouse[5]. These contain 391 and 369 amino acids and possess higher affinity to somatostatin-14 than to somatostatin-28. The amino acid sequences of both receptors show high similarity, and belong to the receptor superfamily possessing seven transmembrane domains (Fig. 23.4).

Table 23.1. Localization of somatostatin receptors in the rat brain (from ref. 2)

Structure	S-14	S-28
Olfactory bulb	++	+++
Nucleus accumbens	–	–
Cerebral cortex	+++	+++
Cingulate cortex	+++	+++
Claustrum	+++	+++
Central amygdaloid nucleus	+	+
Lateral amygdaloid nucleus	+++	+++
Caudate-putamen	++	++
Septum	+++	+++
Lateral olfactory tract nucleus	+++	+++
Medial habenula	+	++
Periventricular nucleus of thalamus	++	++
Subfornical organ	++	++
Corticotectal tract	–	+
Hippocampus	+++	+++
Dentate gyrus	++	++
Hypothalamus	–	+C
Zona incerta	–	+C
Lateral geniculate body	–	++
Substantia nigra	–	++
Inferior colliculus	+++	+++
Superior colliculus	+	+++
Interpeduncular nucleus	+	+++
Periventricular grey matter	+	++
Locus ceruleus	+++	+++
Interposate nucleus	+++	+++
Dentate nucleus	–	+++
Fastigial nucleus	–	+++
Ventral cochlear nucleus	–	++
Dorsal cochlear nucleus	–	++
Medial vestibular nucleus	–	+
Nucleus of the solitary tract	+	+++
Solitary tract	–	+
Nucleus of the nerve vagus	–	+++
Nucleus of the hypoglossal nerve	–	++
Substantia gelatinosa of medulla oblongata and spinal cord	++	++++

C, cysteamine-treated rats.

Distribution

Distribution[6] of somatostatin-14 and somatostatin-28 receptors in the brain is shown in Table 23.1. In adults, regions such as the amygdaloid complex and lateral olfactory tract nucleus show dense binding sites to both receptors. Both these receptors are densely present in various nuclei of the limbic system such as the subiculum, claustrum, hippocampus, and layers III–V of the cerebral cortex. In the lower brain stem, both somatostatin-14 and somatostatin-28 receptors are present in the inferior colliculus and locus ceruleus. On the other hand, in regions such as the deep cerebellar nuclei and cochlear nuclei, only the somatostatin-28 receptor exists. The cerebellar cortex of an adult animal is not rich in somatostatin receptors. However, in newborn animals, somatostatin-14 and somatostatin-28 receptors are expressed in high concentrations, and disappear in parallel with the loss of the external germinal layer during maturation.

References

1. Patel, Y.C., Murthy, K.K., Escher, E.E. *et al* (1990). *Metabolism*, **39**, 63–9.
2. Pelletier, G., Leuroux, P. and Morel, G. (1986). In *Methods in enzymology*, Vol. 124, pp. 607–17. Academic Press, New York.
3. Schettini, G., Florio,T., Meucci, O. *et al.* (1989). *Brain Res.*, **492**, 65–71.
4. Srikant, G.B. and Patel, Y.C. (1981). *Nature*, **294**, 259–60.
5. Yamada, Y., Post, S.R., Wang, K. *et al.* (1992). *Proc. Natl Acad. Sci. USA*, **89**, 251–5.
6. Whitford, Y.A., Candy, J.M., Snell, C.R. *et al.* (1987). *Eur. J. Pharmacol.*, **138**, 327–33.
7. Patel, Y.C., Greenwood, M.T., Panetta, R *et al.* (1995). *Life Sci.*, **57**, 1249–65.

Appendix

Five somatostatin receptors (SSTR1–5) which are all of the G-protein coupled type have been identified. mRNA for SSTR1–5 is widely distributed in the brain and peripheral organs[7]. The highest expression of SSTR mRNAs is found in the cerebral cortex and amygdala (SSTR1), in the cerebral cortex (SSTR2), in the cerebellum (SSTR3) and in the hypothalamus (SSTR5). SSTR4 mRNA is relatively poorly expressed in the brain. In the pituitary, the most prominent somatostatin receptor is SSTR5; SSTR5 mRNA and SSTR1–3 are moderately expressed.

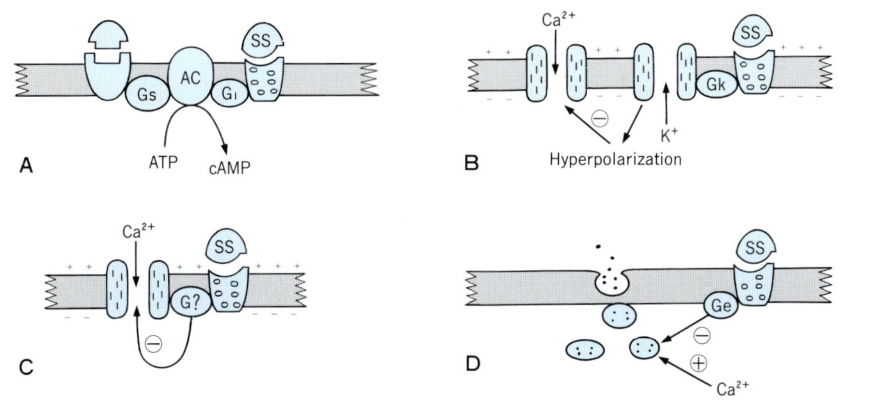

Fig. 23.3 Transmembrane signalling pathways linked to the somatostatin receptor: (*A*), receptor coupling to adenyl cyclase via Gi; (*B*), receptor coupling to K^+ channel via Gk; (*C*), receptor coupling to Ca^{2+} channels via putative G protein (G?); (*D*), receptor coupling to exocytosis via putative Ge (from ref. 1).

A

```
             MELTSEOFNGSQVWIPSPFDLNGSLGPSNGSNQTEPYYDMTSNA
VLTFIYFVVCVVGLCGNTLVIYVILRYAKMKTITNIYILNLAIADELFMLGLPFLAMQ
VALVHWPFGKAICRVVMTVDGINQFTSIFCLTVMSIDRYLAVVHPIKSAKWRRPRTAK
MINVAVWGVSLLVILPIMIYAGLRSNQWGRSSCTINWPGESGAWYTGFIIYAFILGFL
VPLTIICLCYLFIIIKVKSSGIRVGSSKRKKSEKKVTRMVSIVVAVFIFCWLPFYIFN
VSSVSVAISPTPALKGMFDFVVILTYANSCANPILYAFLSDNFKKSFQNVLCLVKVSG
AEDGERSDSKQDKSRLNETTETQRTLLNGDLQTSI
```

B

```
   1 gccaccggca cgctggcgag gccaccggcc ctggagcacc agtccgccgc tgggcgtcga
  61 tgatctacag gccagggtag ctctactggg gcccaggcaa gctctctcag acgccaggag
 121 ggccagcacg agccagactg ggaagctgcg agcccgagag ctactgcgga gcgccaaaca
 181 ccccctaaac ctgctgcgct cccgggcgcc cggctgggta aaggacagct tctgggagct
 241 agagaacaca gagaagcgag tgctcgtgga aaagcaagat gtcacgatag acccttggcc
 301 ccagagtcca ctgaggtgag aggaagatct ctgggctgct tggttctagg cggactgaag
 361 agcagccatg gagttgacct ctgagcagtt caatgggagc caagtgtgga taccttctcc
 421 ctttgacctc aacggctcac tggggccaag caatggctcc aaccagacag agccatacta
 481 cgacatgaca agcaacgcgg tcctcacgtt catctacttc gtggtgtgcg tggtggggct
 541 gtgcggcaac acgctcgtca tctacgtcat cctccgctac gccaagatga aaaccatcac
 601 caacatttac atcctcaacc tggccatcgc agatgaactc ttcatgctgg ggctgccctt
 661 cttggccatg caggtggcgc tggtccactg gccttttggc aaggccatct gccgggtggt
 721 catgactgtg gacggtatca accagttcac cagtatcttc tgcttgacgg tcatgagcat
 781 cgaccgttac ctggccgtgg tccaccccat taagtcagcc aaatggaggc gaccccggac
 841 agccaagatg atcaacgtgg ctgtgtgggg tgtgtccctg cttgtcattt tgcccatcat
 901 gatatacgct ggcctccgga gcaaccagtg ggtaggagc agctgcacca tcaactggcc
 961 gggcgaatcc ggggcatggt acacgggttt cattatctat gccttcatcc tggggttcct
1021 ggtaccccta accatcatct gtctctgcta cctgttcatc atcatcaagg tgaagtcctc
1081 tgggatccga gtggggtcgt ccaagaggaa aaagtcagag aaaaaggtga cccgaatggt
1141 atccatcgtg gtggctgtct tcatcttctg ctggctcccc ttctatatct tcaatgtctc
1201 gtccgtgtct gtggccatca gccccacccc tgccctgaaa ggcatgtttg actttgtggt
1261 tatcctcacc tacgccaaca gctgcgccaa ccccatcctg tacgccttct tgtccgacaa
1321 cttcaagaag agcttccaga atgttctttg cttggtcaag gtgagtggtg cggaggatgg
1381 ggagcggagc gacagtaagc aggacaaatc ccggctgaat gagaccacgg agacccagag
1441 gaccctcctc aatggagacc tccaaaccag tatctgaaac aacccgggaa cgcaacgtgc
1501 acacgcacta gccaagcccc gcctcctggc agtgcgagcc ccattcaccc gcttcctgcc
1561 tcccctaccc atcacacccg gcttttctag agcagagcgg atttgagtct ggcttgtccg
1621 aaagtatacc cctctggtca catctacccc taaagtgaac gttttcgtgc aggcagacaa
1681 ttcaaagact ggagaagagg acacgatggc ctgggtgtga cccggtggaa agcagctacc
1741 cggcagaaac cggaaaaacc aaaactaaaa tcaaagttcc gcgcgtgtac gtgtgcttgc
1801 ccgctatgta atctcgtgat ctgatatttc cgtttgtaca tcacctcccc acccccaccc
1861 cggtctctgc ggagccagta tacacgtgtc ctgtgtttgt aaacccaagt agctagttca
1921 tgtgcgtcta gtataggtgg acatttacca cagcgctgaa cctgacgaca aggactcacc
1981 atgtcagagt caatctaatc taagcttcca gcatccctct tgcatgggcc tttcccagac
2041 ccaggaggag catgagcagt atgttcatat aataatacat ttttgtaaaa agaaaaaaaa
2101 aaaaaaaaaa aaaaaa
```

Fig. 23.4 Amino acid (A, single-letter code) and nucleic acid (B) sequences of somatostatin receptor from rat brain (DDBJ Accession No. M96817).

Somatostatin receptor (binding sites)

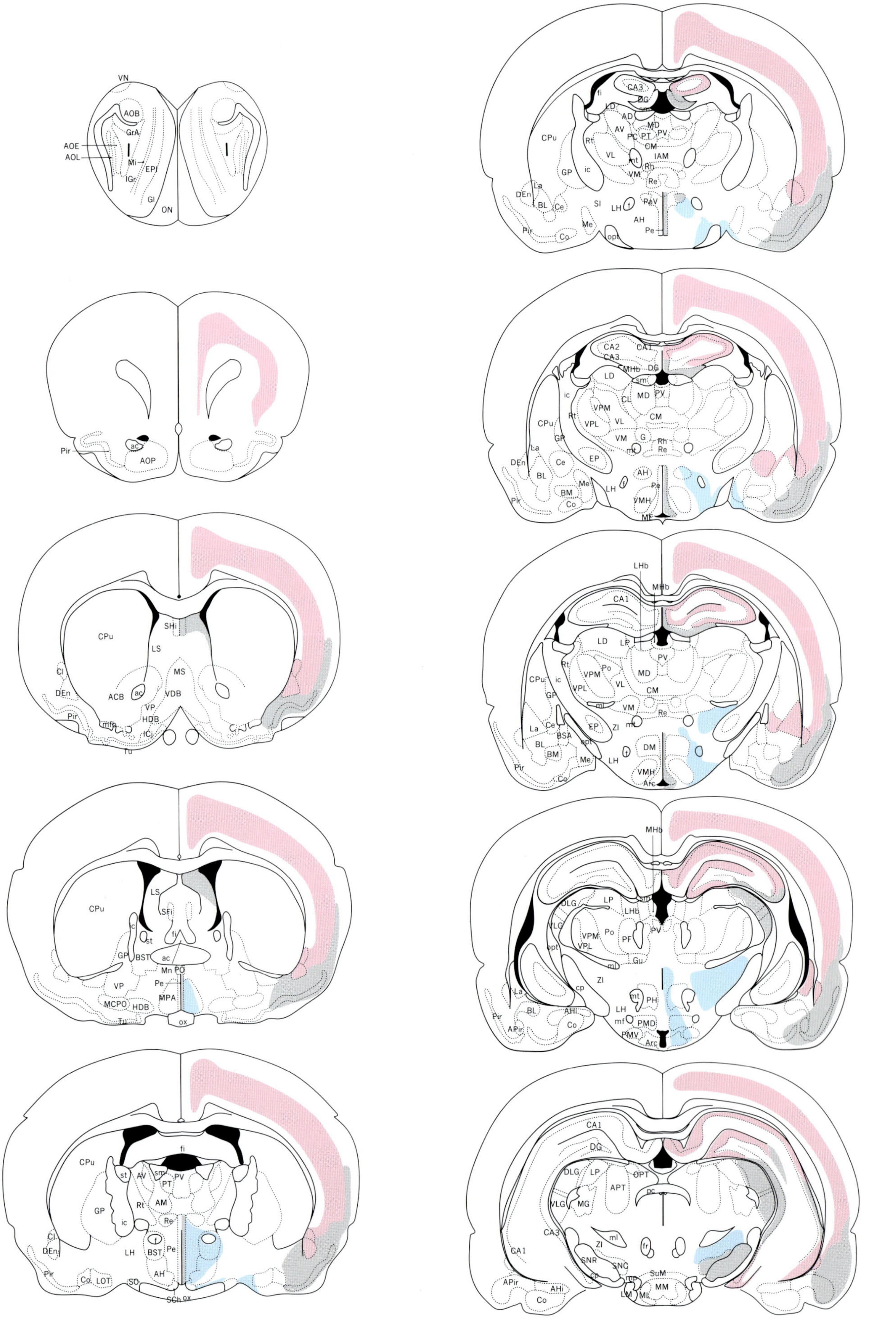

Somatostatin receptor (binding sites)

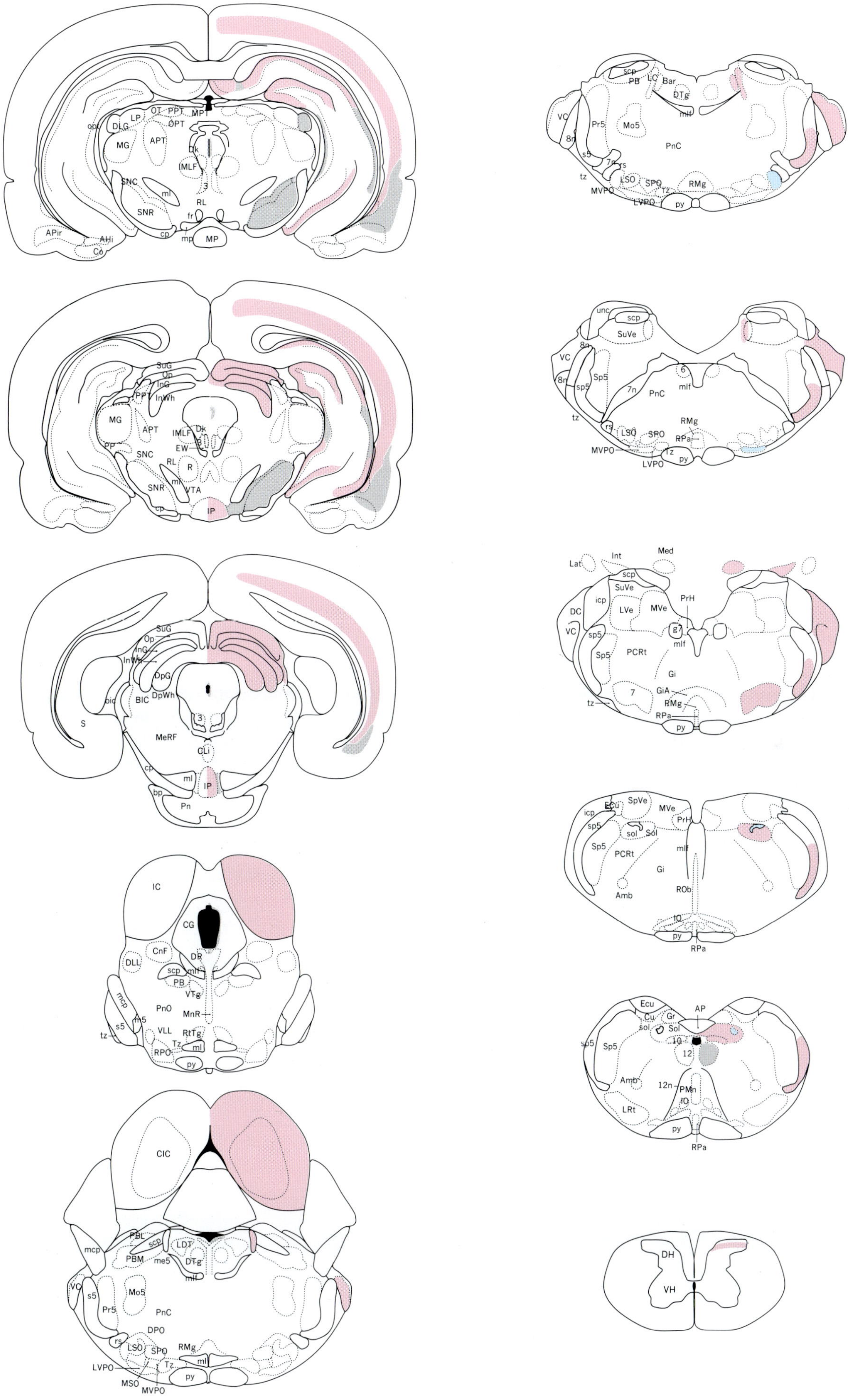

24 The vasopressin–oxytocin system

Vasopressin and oxytocin

In 1895, Oliver and Schafer discovered that an extract from the posterior lobe of hypophysis has a strong vasopressor action. Sixty years later, du Vigneaud *et al.* isolated this substance and it was subsequently determined to be made up of vasopressin and oxytocin each consisting of nine amino acid residues. Vasopressin exhibited vasopressor action and antidiuretic action; oxytocin showed uterine contraction action. These were the first peptides to be isolated from the nervous tissue and identified. As demonstrated by Scharrers and Bargmann, vasopressin and oxytocin are synthesized by the magnocellular neurons of the supraoptic and paraventricular hypothalamic nuclei. These neurons then project fibres into the posterior lobe of hypophysis and secrete these peptides into the blood as hormones. The concept that vasopressin and oxytocin function as hormones had been supported for a long time. However, the fact that these peptides are distributed widely in the central nervous system of mammals, like other peptides, has been revealed only within the last decade or so[1,2,3,6,7].

Structure

Vasopressin and oxytocin are peptides of nine amino acid residues. The structure of oxytocin remains the same among all the mammals examined; species difference is absent for this peptide. However, depending on the species, the eighth amino acid from the N-terminus of vasopressin is either arginine (Arg-vasopressin) or lysine (Lys-vasopressin). Birds, reptiles, amphibians, and fishes have a highly analogous peptide, termed vasotocin. As shown in Fig. 24.1, all of them have a circular structure consisting of six amino acid residues. Vasotocin is believed to be phylogenetically the oldest peptide, and the other hormones may be considered as derivatives of it.

Biosynthesis

The biosynthetic pathways for vasopressin and oxytocin have prominent characteristics. In the case of vasopressin synthesis, vasopressin mRNA is first transcribed from the vasopressin gene, and from this mRNA the precursor of vasopressin (166 amino acid residues) is translated. This precursor is made up of vasopressin, neurophysin I, and a glycopeptide of 39 amino acid residues. Neurophysin II forms secretory granules as a carrier of vasopressin. In the case of oxytocin biosynthesis, after the mRNA for oxytocin is transcribed the precursor of oxytocin is translated. This precursor consists of oxytocin, neurophysin I, and signal peptide. As in the case of vasopressin, neurophysin I becomes a carrier and forms a secretory granule with oxytocin[6].

Physiological function

The best-known function of vasopressin is its antidiuretic hormonal action of raising the water permeability of the nephritic collecting tubule. This is the only known function of vasopressin at physiological concentrations in blood. Above physiological concentration, vasopressin has vasoconstriction activity, stimulates adrenocorticotropic hormone secretion, and oxytocin-like action. In the central nervous system, vasopressin is also linked to thermoregulation and cardiovascular regulation. Several studies have been done on the action of vasopressin in relation to memory. As shown in Fig. 24.2, rats injected with vasopressin significantly improved their memory retention in comparison with the control group. Brattleboro rats which lack the ability to synthesize vasopressin show remarkable memory disability[4].

The best-known actions of oxytocin are lactation and uterine contraction. Some vasopressin-like actions also exist. In the central nervous system, oxytocin is involved in the cardiovascular regulation as well. Interestingly, oxytocin is also reported to enhance amnesia, an opposite action to that of vasopressin (Fig. 24.2).

Distribution and projection tract

Large amounts of vasopressin and oxytocin exist in magnocellular neurons in the hypothalamic supraoptic nucleus and paraventricular nuclei. These neurons project fibres into the posterior lobe of hypophysis. The secretory magnocellular neurons form a large cluster at the level of

A Cys-Try-Phe-Gln-Asn-Cys-Pro-Arg-Gly (bovine, guinea pig, human, rat, ovine)

B Cys-Try-Phe-Gln-Asn-Cys-Pro-Lys-Gly (porcine)

C Cys-Try-Ile -Gln-Asn-Cys-Pro-Arg-Gly

D Cys-Try-Ile -Gln-Asn-Cys-Pro-Leu-Gly

Fig. 24.1 Amino acid sequences of arginine-vasopressin (A), lysine-vasopressin (B), vasotocin (C), and oxytocin (D).

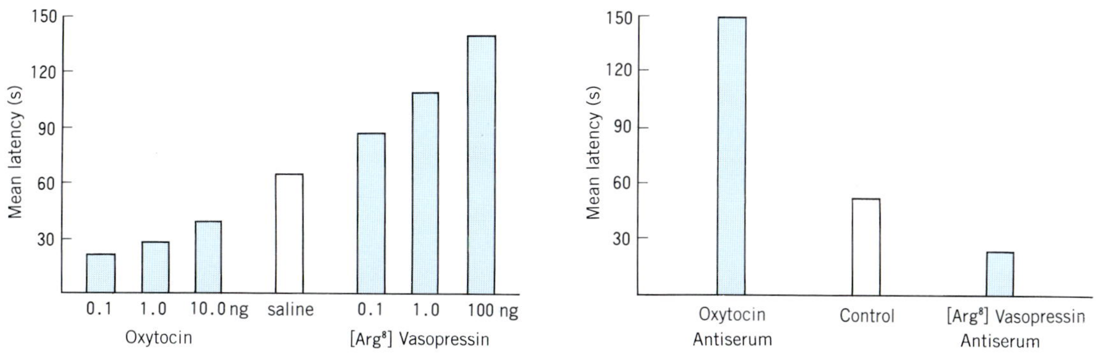

Fig. 24.2 Effect of [Arg^8] vasopressin, oxytocin, and antisera to these peptides on the retention of passive avoidance behaviour. Mean latency pariods before entrance into the dark chamber during retention testing of these materials-treated rats (from ref. 8).

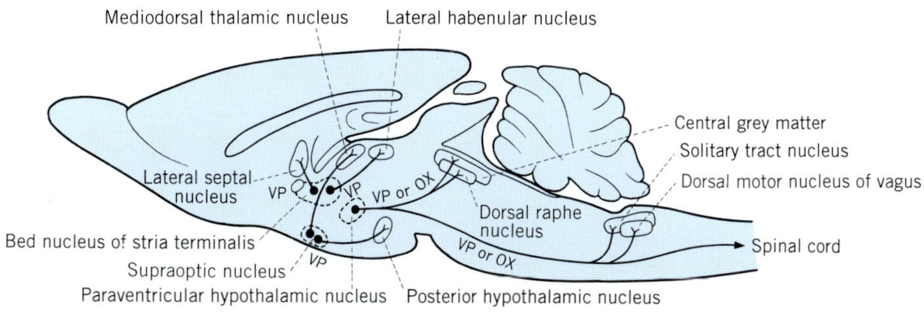

Fig. 24.3 Extrahypothalamo-hypophyseal pathways of vasopressin (VP) and oxytocin (OX) in brain.

the anterior hypothalamic nucleus. In the more posterior part, at the level where the optic tract separates in to right and left fibre bundles, these neurons form small clusters, and these clusters are observed up to the level of the median eminence. It has been clarified that vasopressin neurons and oxytocin neurons are widely distributed in areas other than the hypothalamic magnocellular regions mentioned above. Both these types of neuron are mostly observed in the suprachiasmatic nucleus, paraventricular parvocellular nucleus, lateral nucleus of hypothalamus, posterior hypothalamic nucleus, septal area, stria terminalis, amygdaloid complex, and locus ceruleus. The oxytocin neurons are distributed in supraoptic nucleus, hypothalamic parvocellular nucleus, and lateral nucleus of hypothalamus.

Apart from neuroendocrine axi, vasopressin and oxytocin-producing neurons project their fibres to broad areas between the preoptic areas and the mesencephalic central grey matter such as the periventricular nucleus and the median part of the superior nucleus of the mammillary body. Major projections of magnocellular neurons producing oxytocin and vasopressin are to the posterior lobe of pituitary through the inner lamina of the median eminence. In the paraventricular and subfornical organs, abundant vasopressin and oxytocin fibres are observed, and these seem to pass through the organum vasculosum of the lamina terminalis. Although the origin of these fibres is unclear, they may originate from parvocellular neurons in the paraventricular nucleus. Vasopressin neurons in the supraoptic nucleus are considered to project to the thalamic dorsomedial nucleus and hypothalamic posterior nucleus, and the vasopressin neurons in the bed nucleus of stria terminalis are considered to project to the lateral nucleus of the septulum and the lateral habenular nucleus. Furthermore, vasopressin and oxytocin fibres in lower brain stem or spinal cord have been thought to project from parvocellular neurons in the paraventricular nucleus. These are outlined in Fig. 24.3[5].

References

1. Boer, G.J., Buijs, R.M., Swaab, D.F., and De Vries, G.J. (1980). *Peptides*, **1**, Suppl. 1, 203–9.
2. Buijs, R.M., Velis, D.N., and Swaab, D.F. (1980). *Peptides*, **1**, 315–24.
3. Burbach, J.P.H., Voohuis, T.A.M., van Tol, H.H.M., and Ivell, R. (1987). *Biochem. Biophys. Res. Commun.*, **145**, 10–4.
4. de Wied, D. (1983). In *Progress in brain research*, Vol. 60 (ed. B.A. Cross and G. Leng), pp. 155–67. Elsevier, Amsterdam.
5. Gainer, H. (1983). In *Progress in brain research*, Vol. 60 (ed. B.A. Cross and G. Leng), pp. 205–15. Elsevier, Amsterdam.
6. Laurent, F.M., Hindelang, C., Klein, M.J. *et al.* (1989). *Dev. Brain Res.*, **46**, 145–54.
7. Sofroniew, M.V. (1985). In *Handbook of chemical neuroanatomy* (ed. A. Björklund and T. Hökfelt), Vol. 4, pp. 93–146. Elsevier, Amsterdam.
8. de Wied, D. and Bohus, B. (1979). In *Brain mechanisms in memory and learning from a single neuron to man.* IBRO Monogr. (ed. M.A.B. Brazier) Vol. 4, pp. 139–49. Raven Press, New York.

Vasopressin (immunocytochemistry)

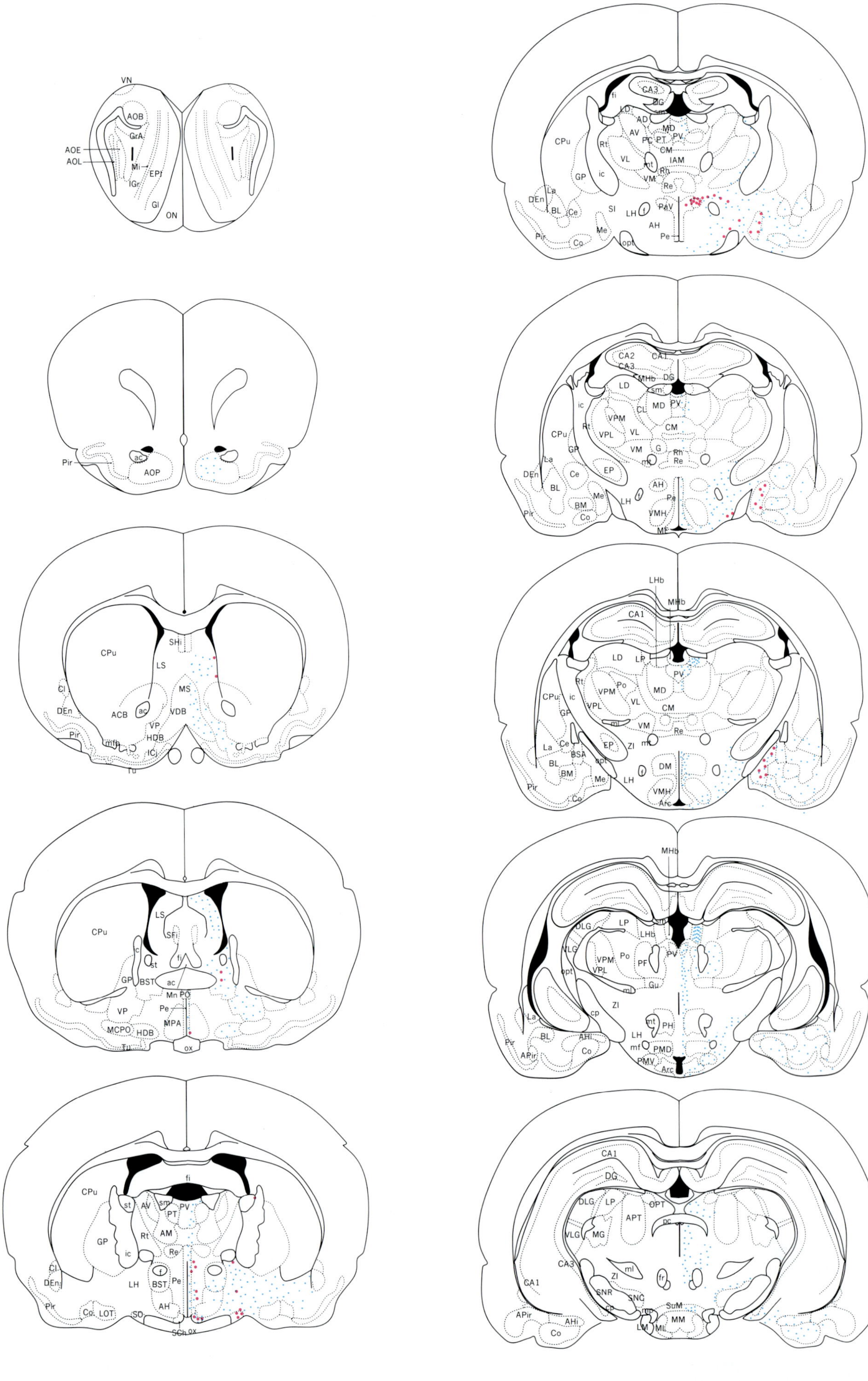

Vasopressin (immunocytochemistry)

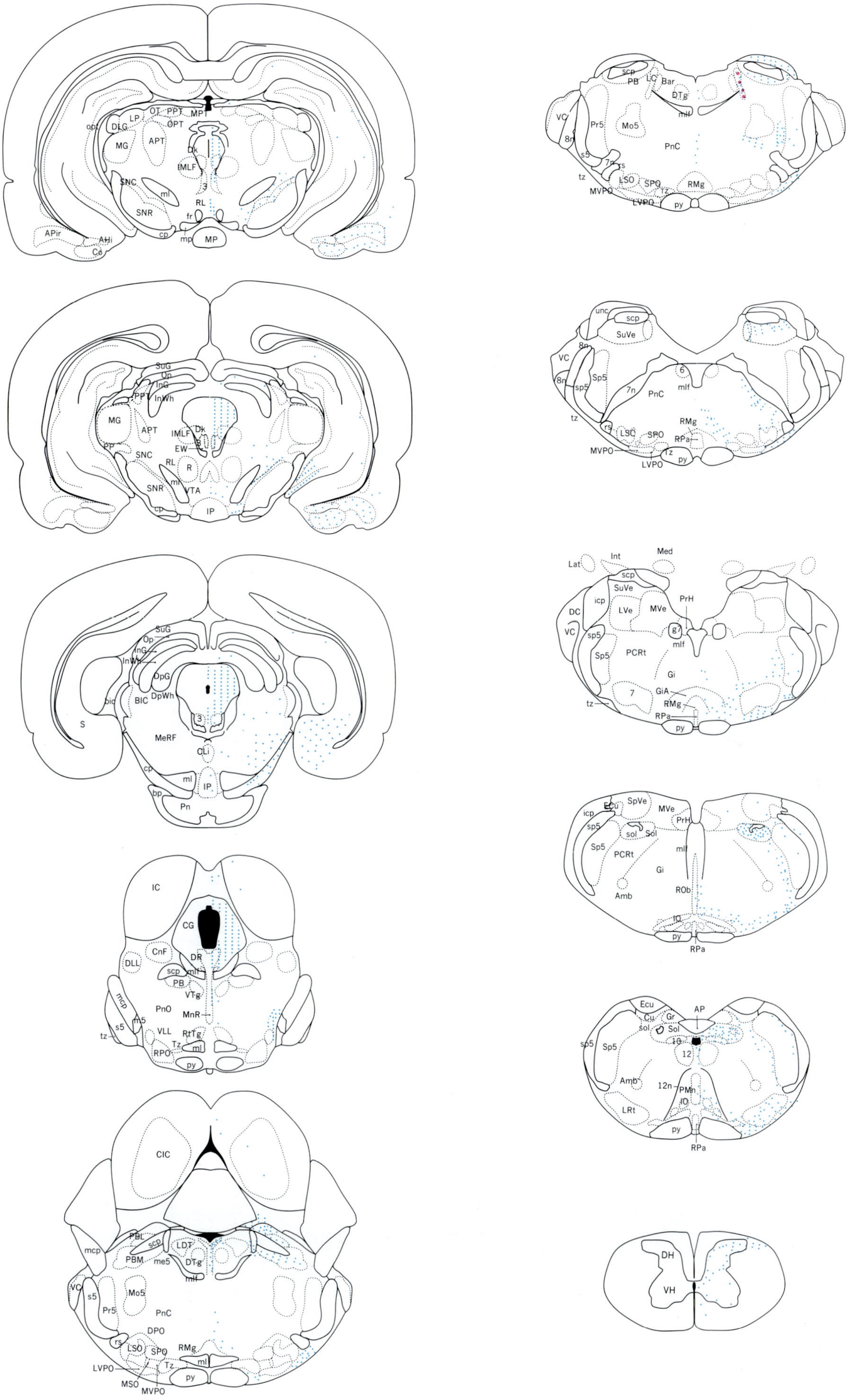

Vasopressin receptor (binding sites)

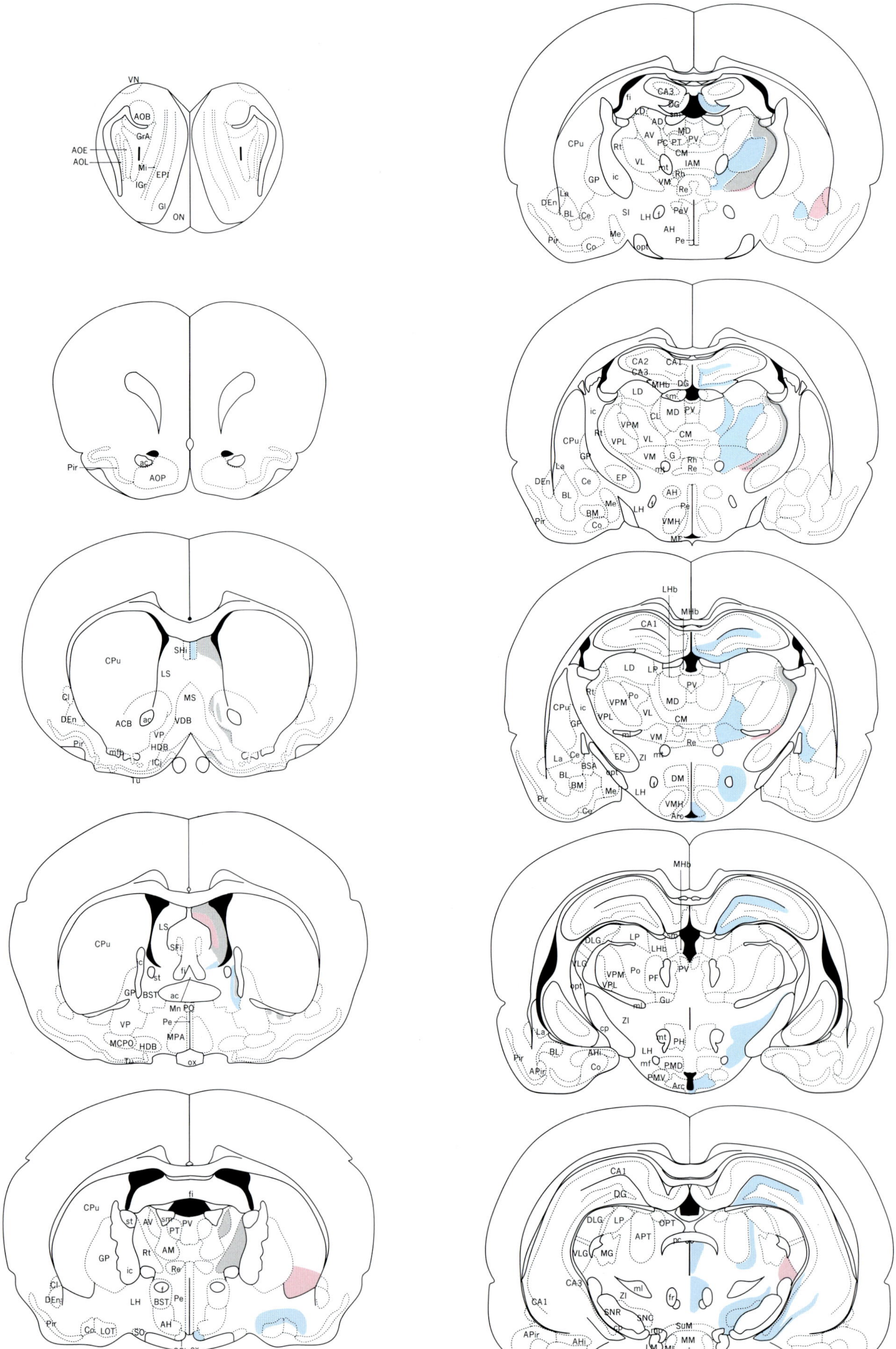

Vasopressin receptor (binding sites)

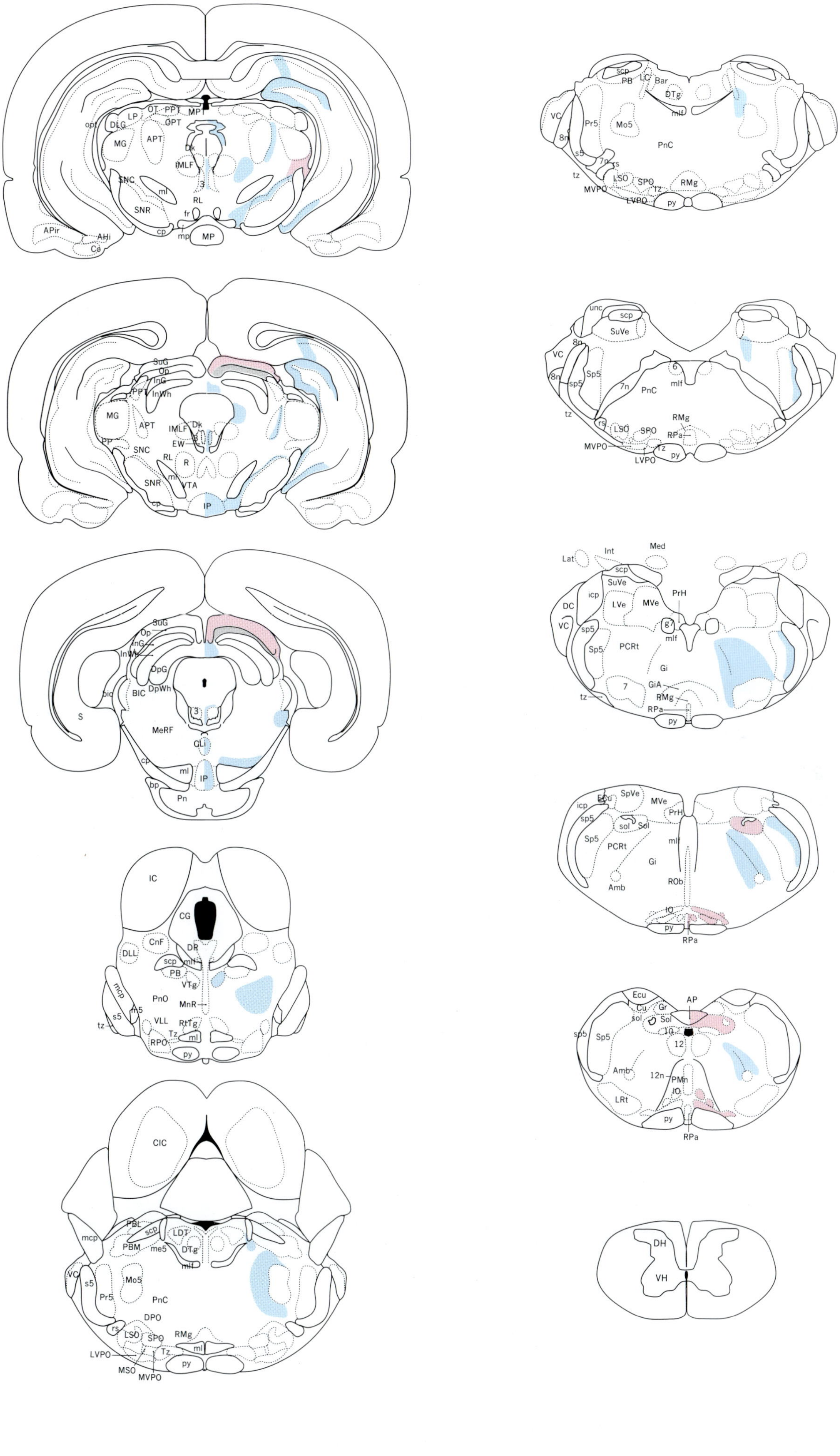

Vasopressin and oxytocin receptors

Subtypes

Based on the studies using the specific agonists and antagonists to vasopressin receptors, it is believed that two types of receptors exist in peripheral tissues; vascular-type receptor (V_1) and renal-type receptor (V_2). Vasopressin receptors of V_1 type are further distinguished into V_{1a} (blood vessel and liver) and V_{1b} (anterior pituitary). V_2 receptors exist in the medulla of the kidney and cause oliguria in response to vasopressin (Table 24.1, Fig. 24.4)[3]. The V_1 type is believed to exist in the brain[2,5]. Whereas V_1 associates phospholipase C, V_2 associates adenylate cyclase. Electrophysiological evidence also show that at least some V_1 receptor exists in neurons. This is because, when vasopressin is administered into neurons containing vasopressin binding sites, firing of neurons has been observed (Fig. 24.5)[7].

So far no report has suggested that more than one type of oxytocin receptor exists. Intracerebral oxytocin binding sites have been considered to be the same type of receptors as those of uterus and mammary gland. Intracerebral oxytocin receptors are regulated by the ovarian hormone. Rats injected with 17β-oestradiol (E2) showed increased oxytocin binding in ventramedial nucleus of hypothalamus[1].

There is also a suggestion that vasopressin and oxytocin bind at the same sites. Audigier and Barberis have shown two types of binding sites. Vasopressin binding sites showed a high affinity (K_d = 1.7 nM) to vasopressin and a low affinity (K_d = 56 nM) to oxytocin. Both oxytocin and vasopressin showed a high affinity to oxytocin binding sites (K_d = 2.0 nM, 3.6 nM) (Table 24.1).

Distribution

In embryonic rats, from day 16, vasopressin binding sites are expressed in the ventral part of the reticular formation of the pons. From day 17 to postnatal day 5, vasopressin binding sites are expressed in the facial nucleus, suprachiasmatic nucleus, septal area, solitary tract nucleus, trigeminal spinal tract nucleus, grey matter of spinal cord, anterior olfactory nucleus, nucleus accumbens, amygdaloid complex, thalamus, hippocampus, dorsal raphe nucleus, area postrema, hypoglossal nucleus, inferior olive nucleus, superior colliculus, inferior colliculus, and zona incerta. After the adult phase is reached, a decrease in binding sites is observed in hypoglossal nucleus, zona incerta, dorsal nuclei raphe, locus ceruleus, trigeminal spinal tract nucleus, grey matter of spinal cord and intermediate parts of facial nucleus. In the Brattle-

Table 24.1. K_d (nM) of vasopressin, oxytocin and structural analogues for vasopressin binding sites[6]

Peptide	Liver	Brain	Pituitary
OH[Thr^4, Gly^7]OT	n.d.	13,500	n.d.
dDAVP	100	250	444
d[Tyr(me)2]VDAVP	128	135	n.d.
OT	78	56	364
[Phe^2, Orn8]VT	n.d.	3.5.	5.7
AVP	3	1.7	4
desGly^{9}d(CH_2)$_5$AVP	0.23	n.d.	1905

n.d., not determined.

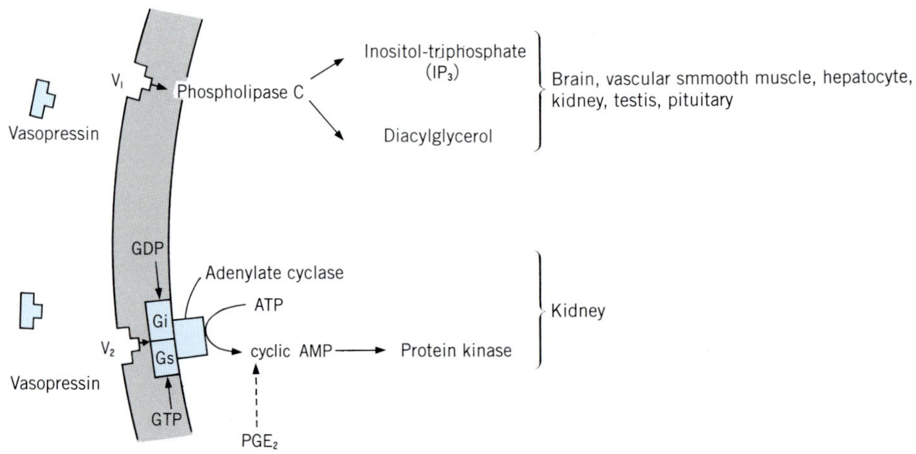

Fig. 24.4 V_1 and V_2 receptors and their localizations.

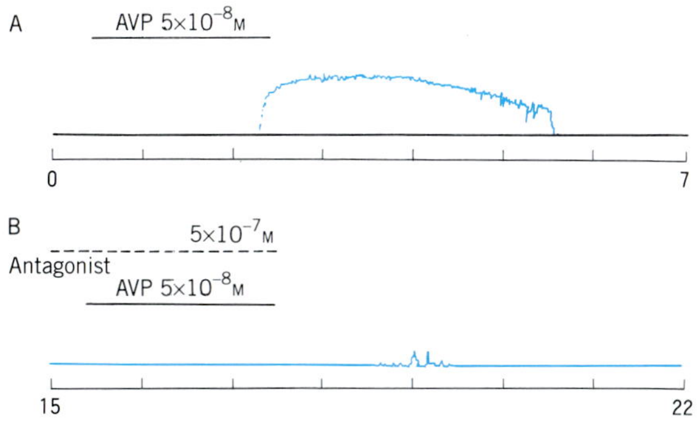

Fig. 24.5 Effects of vasopressin (AVP) and of a synthetic structural analogue on a facial neuron from a 5-day-old rat. The neuron, which had no spontaneous activity, was exited by AVP (A). The antagonist, dTyr(Me) VDAVP, caused a reversible blockade of the action of AVP (B) (from ref. 6).

boro rat, which does not show expression of vasopressin, binding sits of vasopressin are stronger than in normal rats in the lateral nucleus of reticularis formation, anterior commissure, and suprachiasmatic nucleus, and weaker in amygdalod complex, hippocampus, and cingulate gyrus. In other regions there is little difference from normal rats. It is therefore considered that most vasopressin receptors are not regulated by vasopressin[4].

In rats, the oxytocin binding sites show a different distribution between the perinatal period and adult phase. From perinatal day 14, expression of oxytocin binding is seen in the dorsal vagal nucleus; from embryonic day 20 to postnatal day 5, expression is observed in the anterior olfactory nucleus, striatum, accumbens nucleus, mammillary body, subiculum, cingulate gyrus, reticular formation, trigeminal spinal tract nucleus, substantia gelatinosa of spinal cord, stria terminalis, septum, thalamus, amygdaloid complex, and hypothalamus. However, from postnatal day 17, oxytocin binding sites disappear from the cingulate gyrus, spinal tract of trigeminal nerve, substantia gelatinosa of spinal cord, accumbens nucleus, striatum, hypothalamus, mammillary body, reticular formation, and amygdaloid complex. In the hippocampus, as the receptors in the dorsal subiculum region disappear, the receptors appear in the ventral subiculum. This reversal is completed between postnatal days 19 and 22[2].

cDNA cloning

Three vasopressin receptors ($V1_a$, $V1_b$, and V_2) and one oxytocin receptor have been successfully cloned (ref. 8 and references therin). All of these four receptors belong to G-protein coupled receptors which have seven transmembrane domains. Rat V_{1a} receptor consists of 394 amino acids and the first extracellular loop might interact its ligand. By *in situ* hybridization and immunohistochemical studies, V_{1b} receptor has been found to be widely and abundantly expressed in the brain. V_{1a} and oxytocin receptors are also expressed in many brain areas, but they are less abundant than V_{1a} receptor Some discrepancies between the binding results and the immunohistochemical or *in situ* hybridization results were reported[8,9], suggesting that there is another unknown type of oxytocin receptor in the brain.

References

1. Johnson, A.E., Coirini, H., Ball, G.E. and McEwen, B.S. (1989). *Endocrinology*, **124**, 207–11.
2. Moratalla, R., Vallejo, M., and Lightman, S.L. (1988). *Brain Res.*, **450**, 398–402.
3. Phillips, P.A., Abrahams, J.M., Kelly, J.M. *et al.* (1990). *Endocrinology*, **126**, 1478–84.
4. Snijdewint, F.G.M., Van Leeuwen, L.W., and Boor, G.J. (1989). *J. Chem. Neuroanat.*, **2**, 3–17.
5. Stephens, L.R. and Logan, S.D. (1986). *J. Neurochem.*, **46**, 649–51.
6. Tribollet, E., Barberis, C., Jard, S. *et al.* (1988). *Brain Res.*, 442, 105–18.
7. Tribollet, E., Goumaz, E., Raggenbass, M. *et al.* (1991). *Dev. Brain Res.*, **58**, 13–24.
8. Burbach, J.P.H., Adan, R.A.H., Lolait, S.J. *et al.* (1995). *Cell. Mol. Neurobiol.*, **15**, 573–95.
9. Yoshimura, R., Kiyama, H., Kimura, T. *et al.* (1993). *Endocrinology*, **133**, 1239–46.

Oxcytocin (immunocytochemistry)

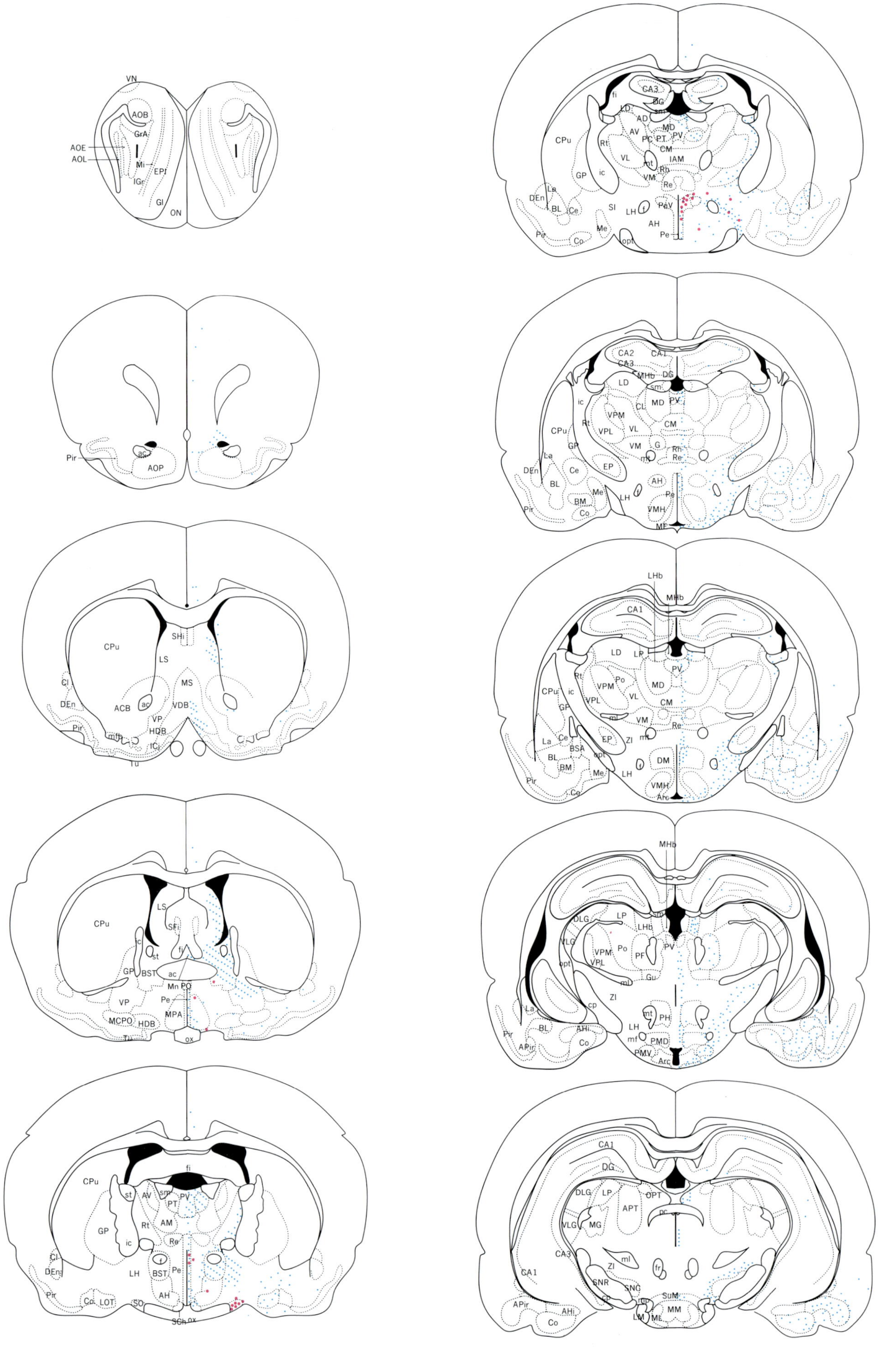

Oxcytocin (immunocytochemistry)

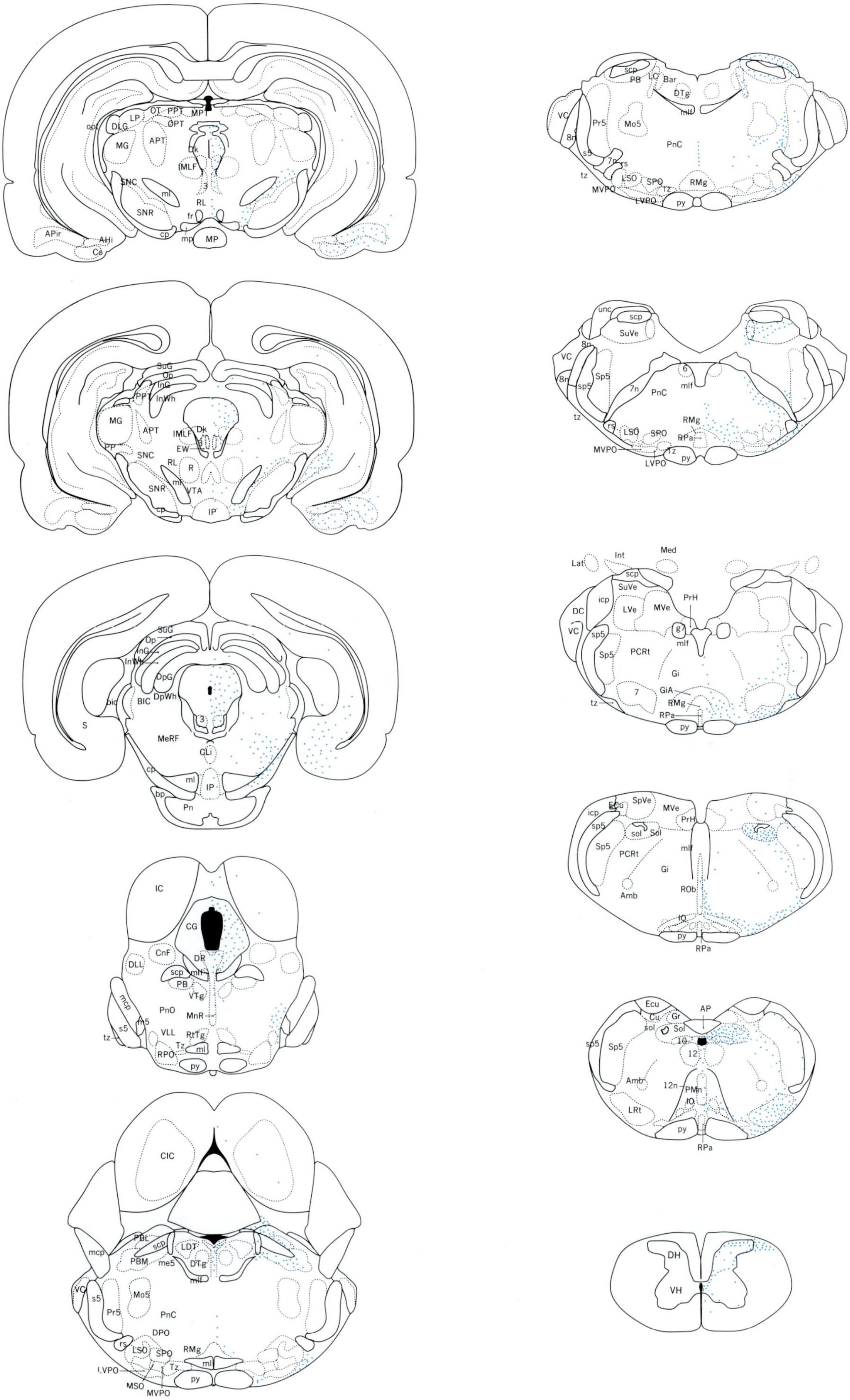

Oxcytocin receptor (binding sites)

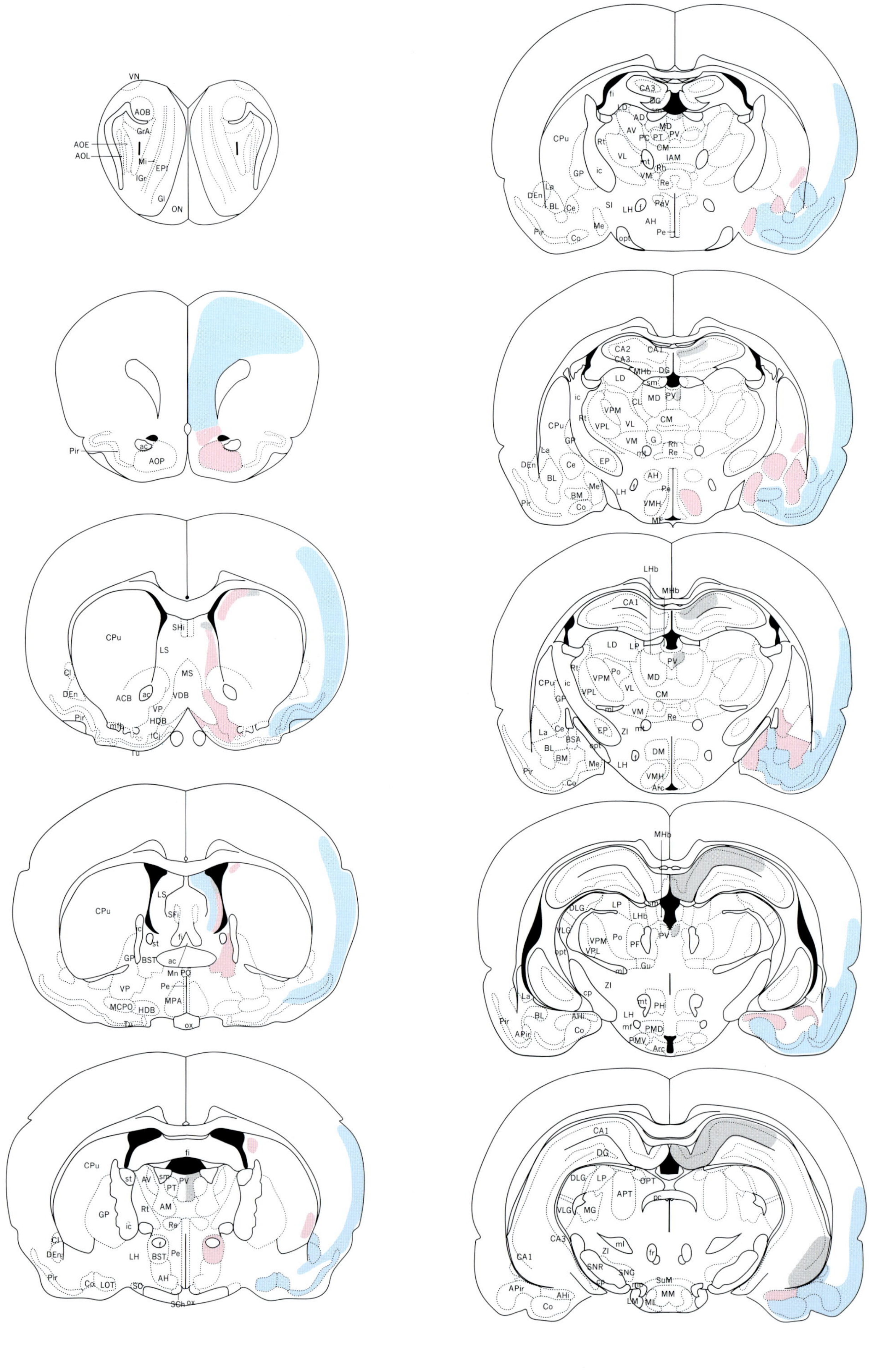

Oxcytocin receptor (binding sites)

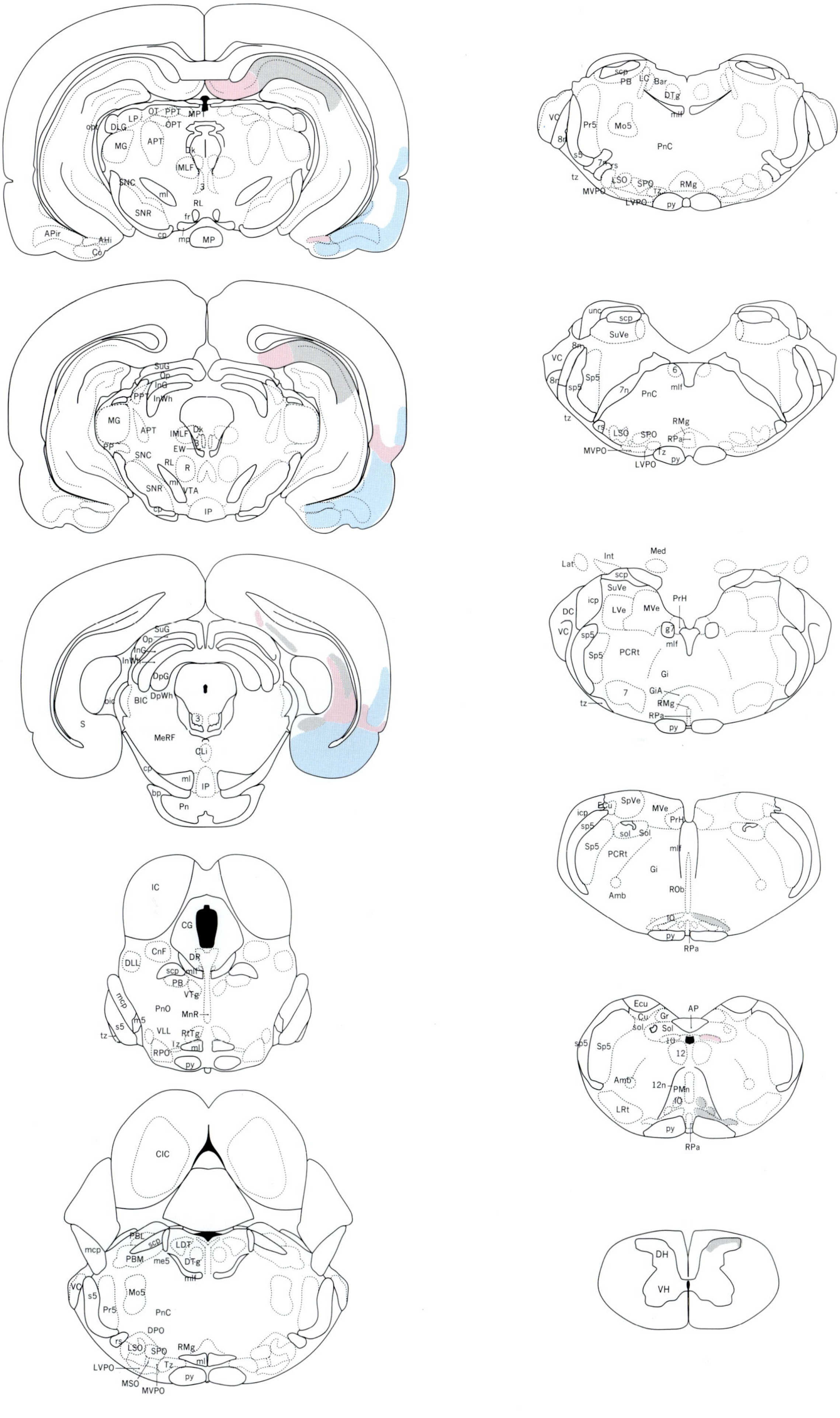

25 The luteinizing hormone releasing hormone system

Luteinizing hormone releasing hormone (gonadotropin releasing hormone)

Luteinizing hormone releasing hormone (LHRH) is an important hormone or releasing factor which constitutes the diencephalohypophyseal system. The diencephalohypophyseal system is composed of two endocrine systems, namely hypothalamohypophysis (posterior lobe) and hypothalamoinfundibulum. Vasopressin and oxytocin are the regulatory factors of the former system. The hypothalamoinfundibulum system regulates by LHRH, growth hormone release inhibiting factor (GRIF) or somatostatin, growth hormone releasing hormone (GRH), thyroid stimulating hormone (TSH), and thyrotropin releasing hormone (TRH).

Structure

Between 1971 and 1973, LHRH was isolated and purified from the hypothalami of sheep and swine by Guillemin[5] and Schally[6] respectively. It is a decapeptide and the presence of pyroglutamic acid (glutamic acid with an annular structure) at the N-terminal end and amidized glycine at the C-terminal has been clarified. The amino acid sequence of LHRH is shown in Fig. 25.1. This sequence is the same in mammals (human, sheep swine, rat, and mouse), but one amino acid varies in the chicken[1,2,4,5,6].

Physiological function

The LHRH is transported through the axon processes from hypothalamus, to the outer lamina of the median eminence, and is released into the blood capillaries which branch away from the superior hypophyseal artery. These blood capillaries are connected to the pituitary gland portal vein, which leads into the anterior lobe of hypophysis. Thus, the LHRH which is produced in the the hypothalamus can reach the endocrine cells which are located in the anterior lobe (Fig. 25.2). At this location, LHRH accelerates the secretion of luteinizing hormone (LH) and follicle stimulating hormone (FSH)[2,6].

LHRH influences many important sexual and reproduction-related behaviours in mammals. Excision of the ovary or pituitary gland of female rats makes them sexually inactive toward males. Oestrogen injection into the ovarectomized female rats, followed by a subcutaneous injection of LHRH, leads to hyperactive sexual stimulation lasting for several hours. Although progesterone also has a similar action, the potency of LHRH is 1000-fold higher. Since this action is observed in animals in which pituitary gland has been excised, the effect is not mediated by the pituitary hormones. The influence of LHRH on the sexual behaviour of male rats has been also observed. The sites where LHRH elicits its function on this behaviour are still being sought. It can be stated that the sites are somewhat circumscribed. When the injection site of LHRH is the medial preoptic area and arcuate nuclei of hypothalamus or the central grey matter of the mesencephalon, the sexual behaviours increases remarkably. However, when an antibody to LHRH is injected into the third ventricle, the sexual behaviour is inhibited. It has also been shown that iontophoretic injection of LHRH into the medial preoptic area or to the neurons of arcuate nuclei resulted in excitation. Thus, the actions of LHRH on sexual behaviour can be considered as central actions which affect the neurons directly[7].

Distribution

Immunohistochemical studies have revealed that the neurons that contain LHRH exist mostly in the hypothalamus, septal area, and diagonal band. The LHRH nerve fibres project mostly into the median eminence. However, a few of the LHRH nerve fibres project externally from hypothalamus into the medial nucleus of amygdaloid complex, hippocampus, medial habenular nucleus, and central grey matter of the mesencephalon.

References

1. Adelman, J.P., Mason, A.J., Haytlick, J.S., and Seeburg, P.H. (1986). *Proc. Natl. Acad. Sci. USA*, **83**, 179–83.
2. Amoss, M., Bugus, R., Blackwell, R. *et al.* (1971). *Biochem. Biophys. Res. Comm.*, **44**, 205–10.
3. Barry, J. (1985). In *Handbook of chemical neuroanatomy* (ed. A. Björklund and T. Hökfelt), Vol. 4, pp. 166–215. Elsevier, New York.
4. King, J.A. and Miller, R.P. (1982). *J. Biol. Chem.*, **257**, 10722–8.
5. Ling, N., Rivier, J., Burgus, R., and Guillemin, R. (1973). *Biochemistry*, **26**, 5305–9.
6. Schally, A.V., Arimura, A., Baba, Y. *et al.* (1971). *Biochem. Biophys. Res. Comm.*, **43**, 393–9.
7. Kandel, E.R., Schwartz, J.H., and Jessell, T.M. (1991). *Principle of neural sciences*. Elsevier, New York.

pGlu-His-Trp-Ser-Tyr-Gly-Leu-Arg-Pro-Gly-NH_2 (Human, sheep, pig, rat, mouse)

pGlu-His-Trp-Ser-Tyr-Gly-Leu-Gln-Pro-Gly-NH_2 (Chick)

Fig. 25.1 Amino acid sequence of LHRH.

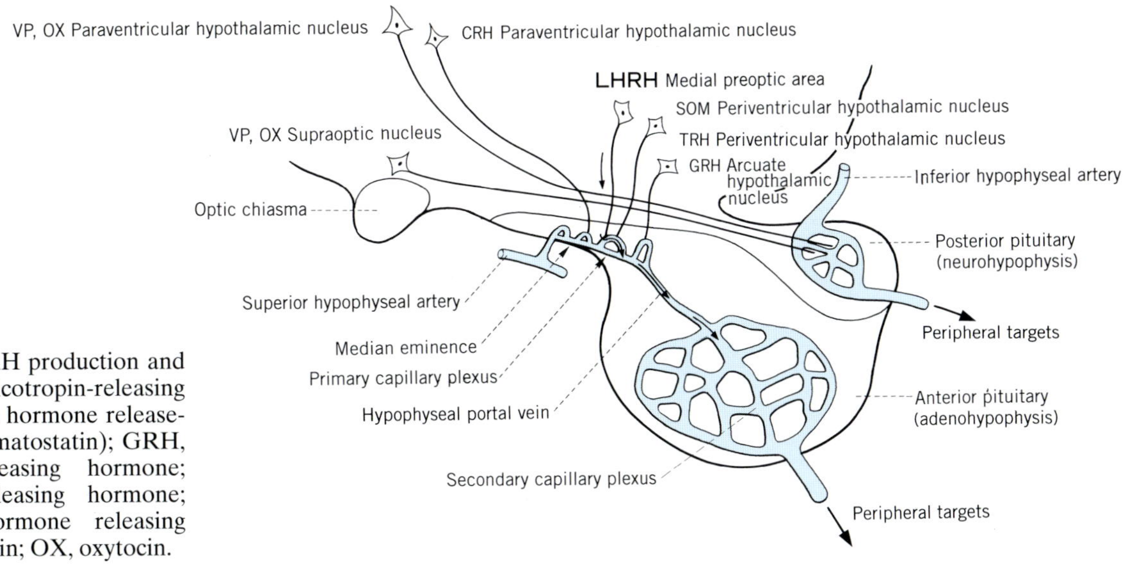

Fig. 25.2 Sites of LHRH production and release. CRH, corticotropin-releasing hormone; SOM, growth hormone release-inhibiting hormone (somatostatin); GRH, growth hormone releasing hormone; TRH, thyrotropin releasing hormone; LHRH luteinizing hormone releasing hormone; VP, vasopressin; OX, oxytocin.

Luteinizing hormone releasing hormone receptor

Distribution

Research on the distribution of LHRH receptor has been carried out by identifying the location of binding sites. High-affinity binding sites for LHRH have been found in homogenized anterior lobe of pituitary. The LHRH agonist ([DSer(TBU)^6, des-gly-NH$_2$^{10}]LHRH ethylamide) stabilizes LHRH, resulting in a conformation that facilitates stronger binding. Autoradiographic studies on sections using this agonist have shown that LHRH binding sites are present in very high density in the anterior lobe of the hypophysis. Scatchard plot analysis has clarified that LHRH binding sites are of high-affinity monomeric types. Other than hypophysis, binding sites for LHRH were also observed in the ovary and limbic system of the brain. In the brain, high density of LHRH binding sites was observed in Ammon's horn, lateral septal nucleus, and amygdaloid complex. Moderate densities of LHRH binding sites are present in the hypothalamus, entorhinal cortex, and hippocampus[1,2,3].

Physiological function

The presence of LHRH binding sites in the limbic system suggests the possibility that this peptide may function as a neurotransmitter or a neuromodulator. In the septal area and amygdaloid complex, there is a possibility of LHRH participating in the regulation of gonadotropin secretion and reproductive behaviour[2].

It has also been speculated that the binding sites in the hippocampus might be concerned with cognitive behaviour, but this has not been proved experimentally[3].

Receptor and second messenger

As a second messenger for the receptor of LHRH, cAMP or cGMP is somewhat negative. The possibility of Ca^{2+} could be considered. When LHRH binds to its receptor, the G protein that links to the receptor activates phospholipase C and leads to activation of PKC. The LHRH is secreted as a result of this cascade (Fig. 25.3).

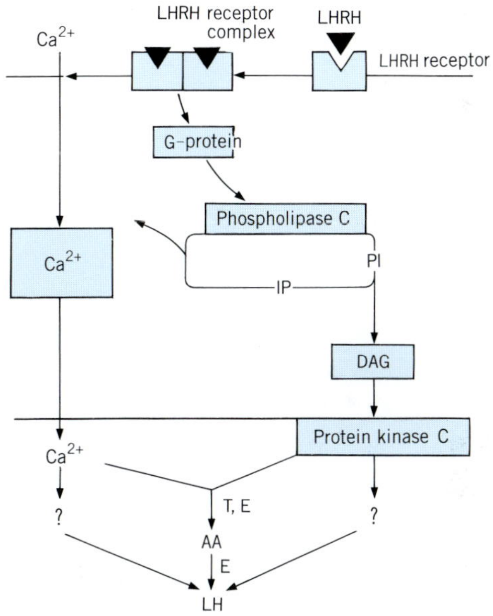

Fig. 25.3 Putative sites of steroid action for LHRH function. PI, phosphatidylinositol; T, testosterone; IP, inositolphosphate; E, oestrogen; DAG, diacylglycerol; AA, arachidonic acid; LH, luteinizing hormone (from ref. 4, modified by Y. Nakai).

References

1. Badr, M. and Pelletier, G. (1987). *Synapse*, **1**, 567–71.
2. Badr, M. and Pelletier, G. (1988). *Neuropeptides*, **11**, 7–11.
3. Badr, M., Marchetti, B., and Pelletier, G. (1989). *Devel. Brain Res.* **45**, 179–84.
4. Kamel, F. and Kubajak, C.L. (1988). *Am. J. Physiol.* **255**, E314–21.

Luteinizing hormone releasing hormone (immunocytochemistry)

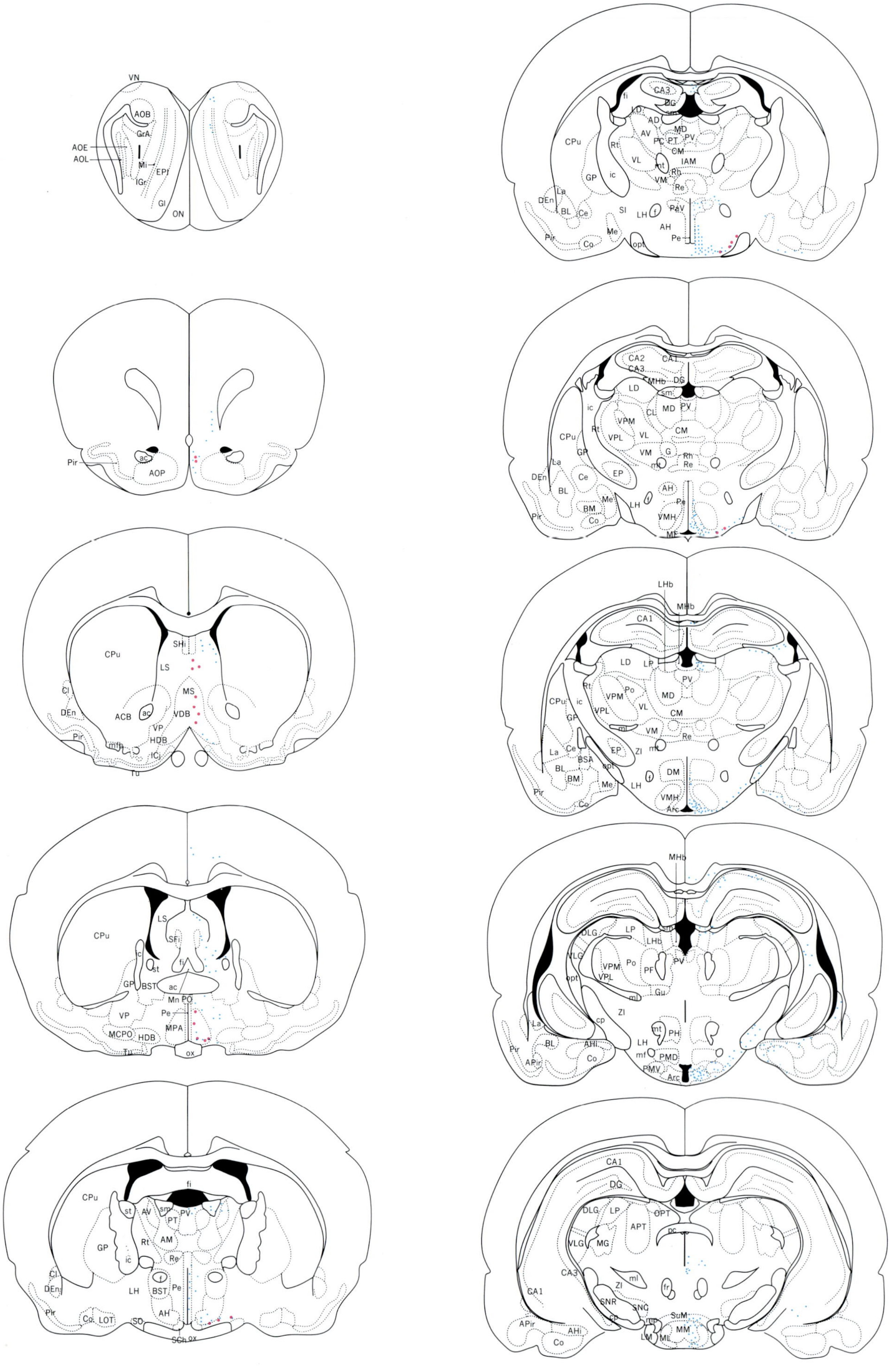

Luteinizing hormone releasing hormone (immunocytochemistry)

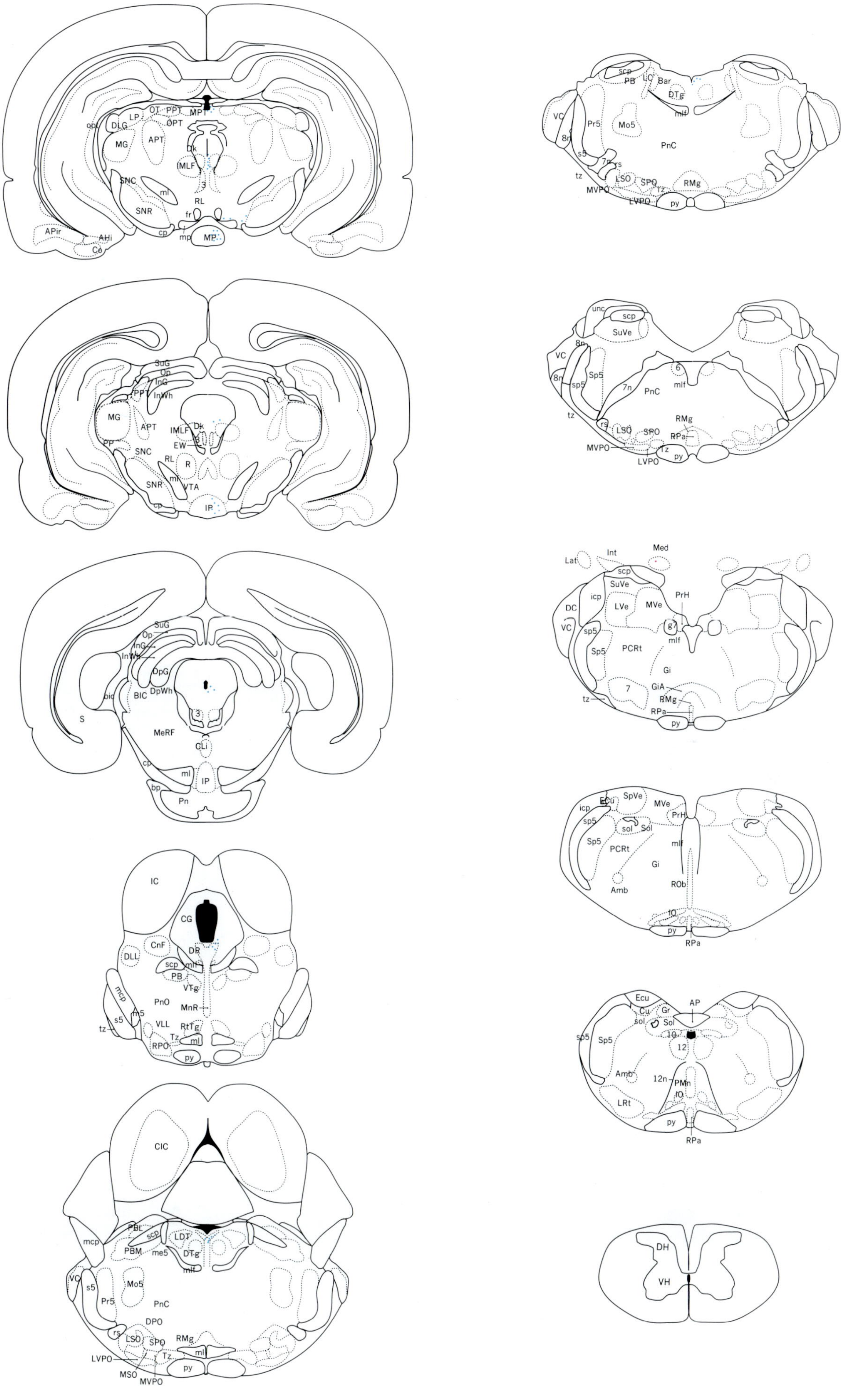

Luteinizing hormone releasing hormone receptor (binding sites)

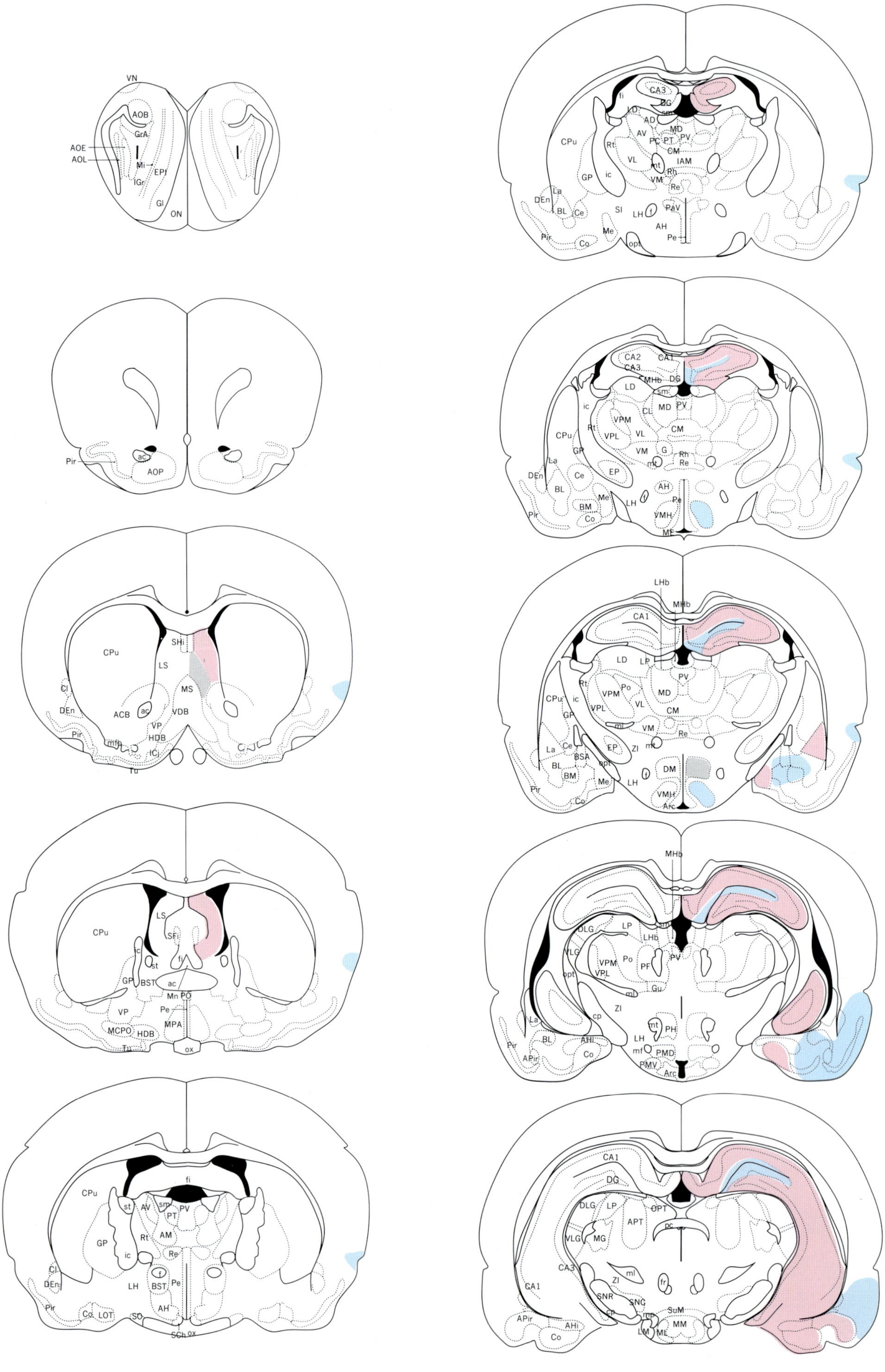

Luteinizing hormone releasing hormone receptor (binding sites)

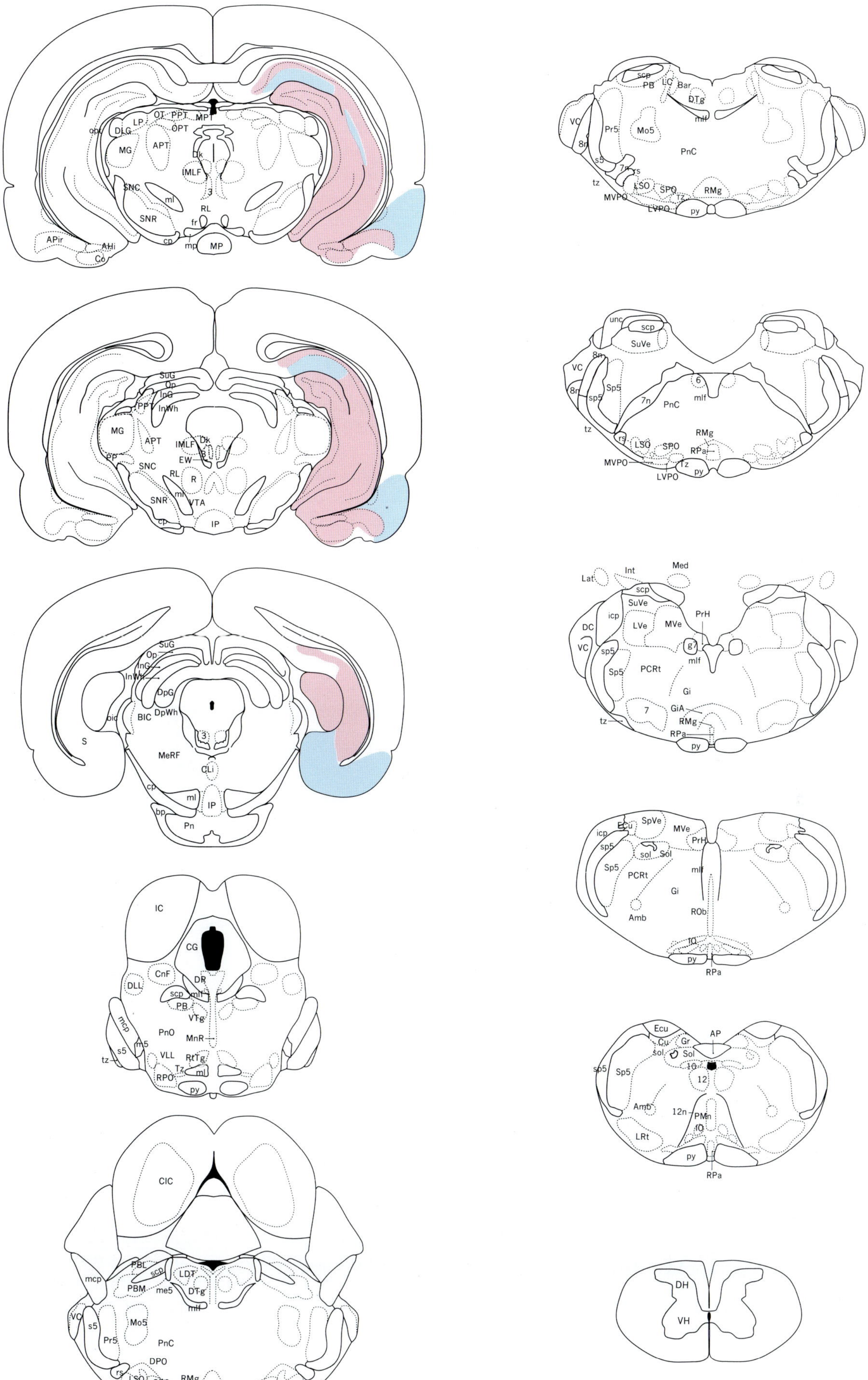

26 The thyrotropin releasing hormone system

Thyrotropin releasing hormone

Structure

Thyrotropin releasing hormone (TRH) was first isolated from the hypothalamus as a hormone accelerating thyroid stimulating hormone (TSH) secretion from the hypophysis, and its structure was determined by Schally and Guillemin in 1970[2,13] (Fig. 26.1). It has a molecular weight of 362.4 and is made up of (pyro) Glu-His-Pro-NH_2 with three amino acid residues. There are no species difference in structure among mammals. In Fig. 26.1, the principal TRH analogues and their structures are shown. [3-Met-His^2] TRH exhibit 8–10-fold higher potency in hormone action than TRH. Recently, hypophysiotropic and central neurotropic actions of TRH have been recognized, and compounds such as MK-771 and DN-1417 are much used in pharmacological studies.

The hypophysiotropic action of MK-771 is 40 times stronger than that of TRH. The central neurotropic action of DN-1417 is only 1/30–1/40 of that of TRH. The TRH precursor protein (pre-pro TRH) has a molecular weight of 29 247 and information for the generation of five molecules of TRH are coded for on DNA. Post-translation of precursor protein therefore leads to production of multiple copies of TRH simultaneously[10].

Distribution and projection tract

To identify the distribution of TRH, the localization of mRNA of the TRH precursor protein has been carried out by immunohistochemistry[7,17] and by *in situ* hybridization using the antiserum of TRH or TRH precursor protein[3,11]. Some differences exist in the distribution of TRH-containing cells and TRH precursor protein-containing cells. For example, TRH precursor protein, but not TRH, can be detected in the lateral nucleus reticularis thalami and ventral cochlear nuclei[5]. Conversely, TRH-containing cells, but not TRH precursor protein-containing cells, can be observed in the dorsal column of spinal cord and subfornical organs[10]. In the former situation, the production of TRH from TRH precursor protein seems to be inhibited by some mechanism. In the latter situation, the THR detected might have been brought from extracellular locations.

Some TRH projection tracts have been clarified. Lesion experiments have shown that TRH cells in the hypothalamus terminate in the interior of hypothalamus and immunocytochemistry combined with retrograde labelling study shows that the paraventricular nucleus TRH cells project to the median eminence and the organum vasculosum of the lamina terminalis (OVLT)[19]. The TRH cells of caudal nuclei raphe also contain substance P and serotonin, and project into the spinal cord[8]. Fig. 26.2 shows the release mechanism of these substances.

Physiological function

The well-documented physiological action of TRH is its action on the cells of anterior lobe of hypophysis, which lead to an acceleration of prolactin secretion[6]. TRH does not activate other cells of the anterior lobe of the hypophysis normally, but in disease conditions such as acromegaly, anorexia nervosa, and depression, it influences other cells of the anterior lobe and causes secretion of ACTH, LH, FSH, and GH from these cells[15].

Apart from its endocrine action, TRH also produces other actions. One of these is behavioural excitation. Intraventricular infusion of TRH in cats led to behavioural arousal. In humans also, intravenous infusion of

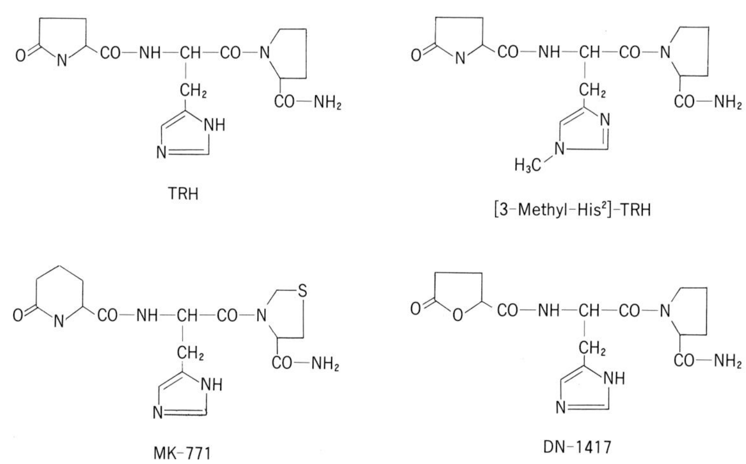

Fig. 26.1 Chemical structure of TRH and its main analogues [3-methyl-His 2]-TRH, MK-771 (pyro-2-aminoadipyl-His-thiazolidine-4-carboxyamide), DN-1417 (γ-butyrolactone-γ-carbonyl-His-Pro-NH_2).

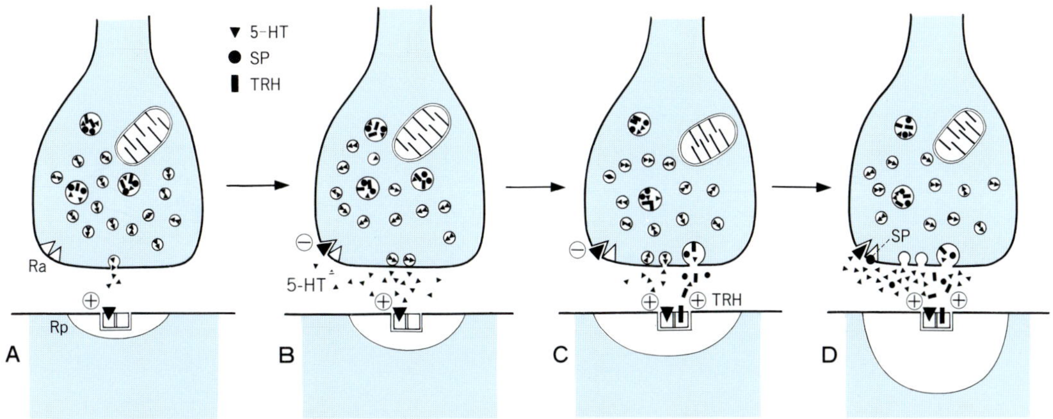

Fig. 26.2 Schematic illustration of a nerve ending in the ventral horn of spinal cord containing 5-HT, substance P (SP)-LI, and TRH-LI. (A) The two peptides are stored exclusively in the large vesicles. (B) At low-firing frequencies, only 5-HT is released, stimulating postsynaptic receptors (Rp). (C, D) With increasing activity, 5-HT may active the presynaptic autoreceptor (Ra), inhibiting further release of 5HT and limiting the postsynaptic response, At very high frequencies, the peptides are released from the large vesicles, whereby TRH synergistically activates a postsynaptic receptor (C), and substance P blocks the inhibitory 5-HT auto-receptor (D), giving rise to a maximal postsynaptic response (from ref. 5).

TRH during sleep or anesthetic conditions leads to arousal[16]. How TRH contributes to the arousal mechanism is still unclear. It can be interpreted in terms of incremental nerve excitation caused by the acceleration in the release of acetylcholine in cerebral cortex or hippocampus. Alternatively, relaxation in the inhibiting action of the GABA activation system could also be influenced by TRH. When TRH is injected into humans, it also produces euphoric stimulation, and its involvement in psychogenic activities has been documented[15]. Reports on the injection of TRH analogues into experimentally depressed rats[14] reveal the antidepressant actions of TRH as well.

Furthermore, TRH participates in motor functions such as accelerating the activities of spinal cord motoneurons[18]. The TRH of spinal cord is supplied from caudal nuclei raphe where serotonin and substance P are also colocalized[7]. The action mechanism postulated for the interactions of TRH, serotonin, and substance P in spinal cord is shown in Fig. 26.2. The participation of TRH in motor function is also supported by the clinical observation that it is used therapeutically for spinal cord trauma[4] or amyotrophic sclerosis[2,3]. In patients with Huntington's disease, a marked increase of TRH in amygdaloid complex has been reported. TRH is also implicated in the emotional imbalance seen in degenerative disease of the corpus striatum–substantia nigra tract.

Functional correlation between TRH and acetylcholine has already been mentioned. In addition, one study[9] has reported that TRH acts in the accumbent nucleus of the septum to enhance the metabolism of dopamine and increase the release of dopamine. TRH also possibly enhances the release of noradrenaline in the hypothalamus[1].

References

1. Bennett, G.W. (1983). In *Thyrotropin-releasing hormone* (ed. E.C. Griffiths and G.W. Bernett). pp. 253–69. Raven Press, New York.
2. Burgus, R., Dunn, T.F., Desiderio, D. *et al.* (1970). *Nature (Lond)*, **226**, 321–5.
3. Engel, W.K., Siddeque, T., and Nicoloff, J.T. (1983). *Lancet*, **1**, 73–5.
4. Fanden, A., Jacobs, T.P., and Holaday, J.W. (1981). *New Engl. J. Med.*, **305**, 1063–7.
5. Hökfelt, T., Tsuruo, Y., Ulfhake, B. *et al.* (1989). *Ann. New York Acad. Sci.*, **553**, 74–95.
6. Jackson, I. and Reichlin, S. (1977). *Nature*, **267**, 853–4.
7. Jackson, I. (1980). *Neurobiology of cerebrospinal fluid* **1**, 625. Plenum Press, New York.
8. Johannson, O.T., Hökfelt, B., Pernow, S.L. *et al.* (1981). *Neuroscience*, **6**, 1857–81.
9. Kalivas, P.W., Stanley, D., and Prange, J. (1987). *Neuropharmacology*, **26**, 33–8.
10. Lechan, R.M., Wu, P., Jackson, I., Wolf, H. *et al.* (1985). *Science*, **231**, 159–61.
11. Kreinder, M.S., Engber, T.M., Nilaver, G. *et al.* (1985). *Peptides*, **6**, 997–1000.
12. Mitchell, R. and Fleetwood-Walker (1981). *J. Pharmacol.*, **76**, 119–20.
13. Nair, R.M.G., Barrett, J.F., Bowers, C.Y., and Schally, A.V. (1970). *Biochemistry*, **9**, 1103.
14. Ogawa, N. (1986). *Neuroendocrinology*, 2. *Bioactive substance*, pp. 1–16. Chugai Igaku, Tokyo.
15. Prange, A. Jr, Wilson, I.C., Lala, P.P. *et al.* (1972). *Lancet*, **2**, 999–1002.
16. Prange, A.J., Breese, G.R., Cott, J.H. *et al.* (1974). *Life Sci.*, **14**, 447–55.
17. Tsuruo, Y., Hökfelt, T., and Visser, T. (1987). *Exp. Brain Res.*, **213**, 213–17.
18. White, S.R. (1985). *Brain Res.*, **335**, 63–70.
19. Merchenthaler, I., and Liposits, Z. (1994). *Acta. Biol. Hung*, **45**, 361–74.

Tyrotropin releasing hormone (immunocytochemistry)

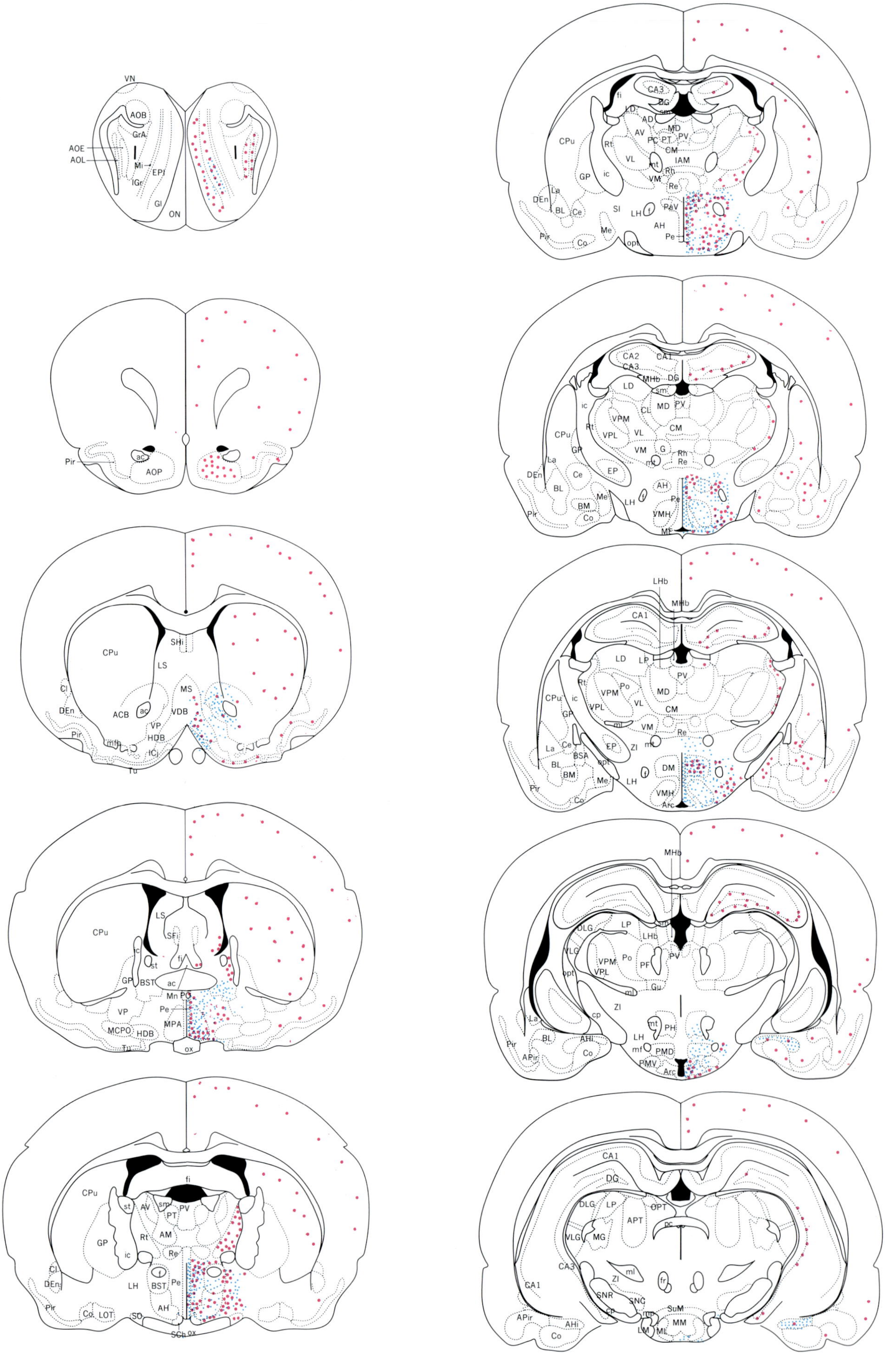

Tyrotropin releasing hormone (immunocytochemistry)

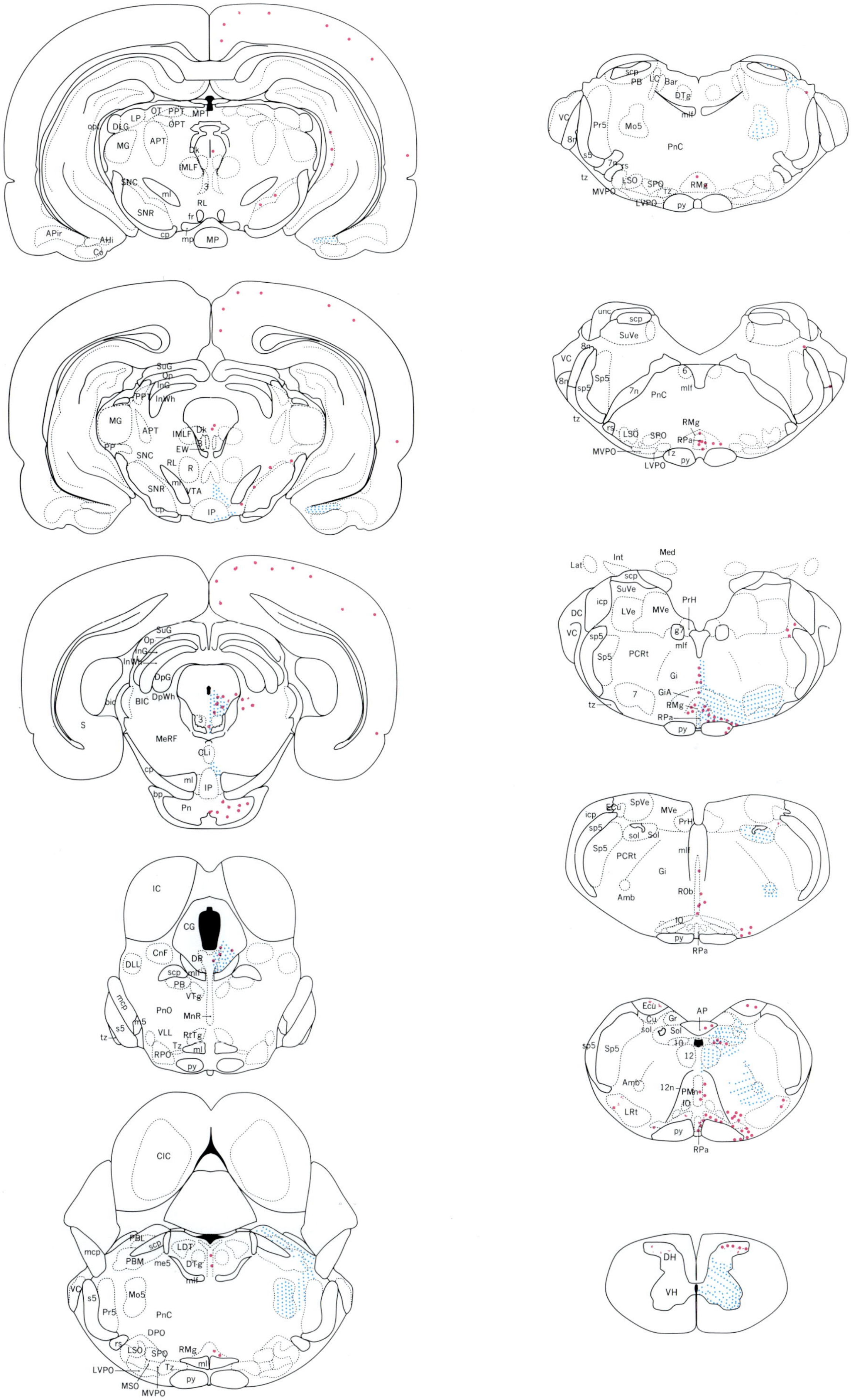

Thyrotropin releasing hormone receptor

Structure and second messenger

Recently, the cDNA sequence of the thyrotropic TRH receptor of hypophysis from mouse and rat has been clarified[10,12] (Fig. 26.3). Both are of G-protein type with seven transmembrane domains. The TRH receptor of rat has 412 amino acid residues, with 96 per cent similarity to mouse TRH receptor. The difference lies in the additional 16 amino acid residues in the C-terminus of rat TRH receptor. When transmembrane domains are compared, the TRH receptor shows the following percentage identity: 21.5 for substance P, 29.4 for rat serotonin type II receptor, 32.5 for rat dopamine D_2 receptor, 28.8 for rat muscarine M_5 receptor[11]. The second messenger system of the TRH receptor has been considered to be an intracellular Ca^{2+} mobilization system using inositol phospholipid metabolism[2,7].

Distribution

The distribution of TRH receptors has been mapped with reference to the binding sites by autoradiography using [^{3}H] (3-Me-His^2)TRH [^{3}H]MeTRH) or [^{3}H]TRH as a ligand[1,2,7]. Whereas [3H]TRH recognizes low-affinity sites, [^{3}H]MeTRH is a high-affinity analogue and is proposed to exhibit a strong specific binding[4]. The high-affinity receptor is thought to take a leading role in regulating many of the physiological actions of TRH[11]. In the brain of monkey or rat, the receptors are abundant in the limbic system, and well distributed in the cerebral cortex (frontal lobe > temporal lobe > occipital lobe), central grey matter of mesencephalon, and hypothalamus[8,9]. The binding sites were found to be weak in the anterior lobe of the hypophysis, the purported site of action of TRH. In the cerebral cortex, the binding sites are abundant in comparison to the amount of TRH, but in the hypophysis the situation is reversed. In the sheep brain, many binding sites were observed in the accumbent nucleus of the septum, caudate nucleus, and septum pellucidum. Other regions showed only few binding sites. Also, the TRH binding sites are mostly seen in the dorsal region of the spinal cord[6].

Physiological function

It is well known that TRH possesses actions related to behavioural arousal or consolidation of consciousness. In ageing rats, the amount of TRH receptors decreases by half. Despite this reduction, the amount of TRH in the cerebral cortex does not show any decrease. When dihydroergotoxine was injected into an aged rat, the amount of TRH receptor increased and the depressed discriminatory learning ability was restored to normal[5]. These actions are believed to be mediated by the intermediate TRH receptors in the cerebral cortex. TRH also has other actions, such as relieving depression and stimulating euphoria[5]. These are thought to be mediated by the intermediate TRH receptors in the limbic system. The euphoria-producing action of TRH is thought to be mediated by the TRH receptor in hypothalamus. Motor enhancement behaviour is probably mediated by the TRH receptors that exist in the substantia nigra–corpus striatum dopamine axis or mesencephalon peripheralis system or by the TRH receptors present in the accumbent nucleus

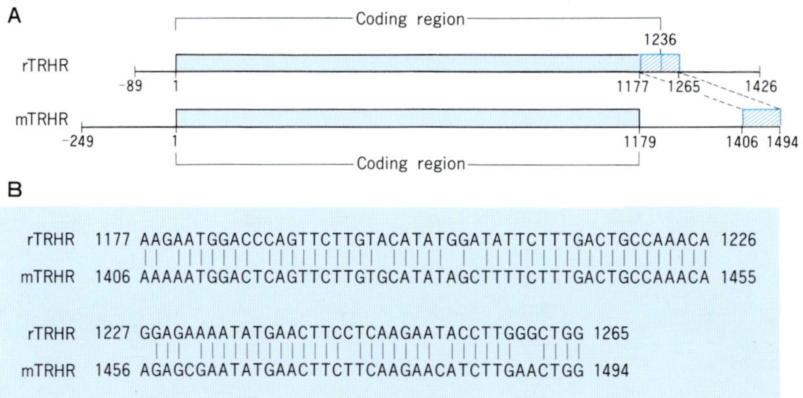

Fig. 26.3 (A) Diagrammatic comparison of the nucleated sequences of the rat (r) and mouse (m) TRHR. The coding region of the rTRHR contains 1236 bp, and that of the mTRHR contains 11 769 bp. The 3′-end of the coding sequence in teh rTRHR has 57 bp (19 amino acid residues) that are not present in the mTRHR. (B) However, begining 224 bp 3′ to the stop codon in the mTRHR is a stretch of at least 89 bases in the coding sequence, inclding the stop codon, and the first 26 bases in the 3′-untranslated region of the rTRHR cDNA. There is identity of the rTRHR and mTRHR in the 5′-untranslated region between positions −1 and −28 and some homology further upstream from the initiation codon. For clarity in this comparison, the numbering used here is different from that of the mTRHR sequence published by Straub *et al.*[10] (from ref. 12).

of the septum. It is known that irritation of the amygdaloid complex leads to spasms, during which a decrease in TRH receptors in the affected nucleus has been observed[3]. TRH therefore, has a therapeutic value for some types of epileptic patients. The prevailing knowledge of spasm production links TRH prominently with the amygdaloid complex.

Note added in proof

1. In the mouse pituitary TRHR, the amino acid sequence Cys (positions 335 or 337) in the carboxyl terminal tail are important in ligand-induced receptor internalization as determined by deletion mutagenesis. They are located near the seventh transmembrane domain[14].

2. In the human brain, TRH receptors are also concentrated in the hippocampus and amygdala. Lexow *et al.* reported that TRH receptor binding changes with several diseases; they used quantitive autoradiography in the postmortem human brain. Schizophrenic patients demonstrated an increase in the concentration of TRH receptors in the molecular layer of the dentate gyrus, while there were marked decreases in the amygdala compared to non-neurological control. In Huntington's disease, slight alterations were noted in the central and medial basal amygdala and in the cortical amygdala in Alzheimer's disease but there was no significant difference within the hippocampus in either disease[13].

References

1. Burt, D.R. and Taylor, R.L. (1980). *Endocrinology*, **106**, 1416–23.
2. Burt, D.R. and Sharif, N.A. (1990). In *Handbook of chemical neuroanatomy*, **9**, 477–500.
3. Kajita, S., Ogawa, N., and Sato, M. (1987). *Epilepsia*, **28**, 228–33.
4. Ogawa, N., Yamawaki, Y., Kuroda, H. *et al.* (1982). *Peptides*, **3**, 669–77.
5. Ogawa, N., Mizuno, S., Mori, A., and Kuroda, H. (1984). *Peptides*, **5**, 53–6.
6. Ogawa, N. (1986). In *Neuroendocrinology*, 2. *Bioactive substance* (ed. N. Yanaihara) pp. 10–16. Chugai Igaku, Tokyo.
7. Ogawa, N. (1986). *New brain receptor*. World Health Report Co., Tokyo, pp. 367.
8. Sharif, N.A. and Burt, D.R. (1983). *Regul. Pept.*, **7**, 399–411.
9. Sharif, N.A. (1989). *Ann. N.Y. Acad. Sci.*, **553**, 147–75.
10. Straub, R.E. *et al.* (1990). *Proc. Natl. Acad. Sci. USA*, **87**, 9514–8.
11. Taylor, R.L. and Burt, D.R. (1982). *J. Neurochem.*, **38**, 1649–56.
12. Zhao, D. *et al.* (1992). *Endocrinology*, **130**, 3529–36.
13. Lexow, N., Joyce, J.N., Kim, S.J. *et al.* (1994). *Synapse*, **18**, 315-27.
14. Nussenzveig, D.R., Heinflink, M., and Gershengorn, M.C. (1993). *J. Biol. Chem.*, **268**, 2389–92.

Tyrotropin releasing hormone receptor (binding sites)

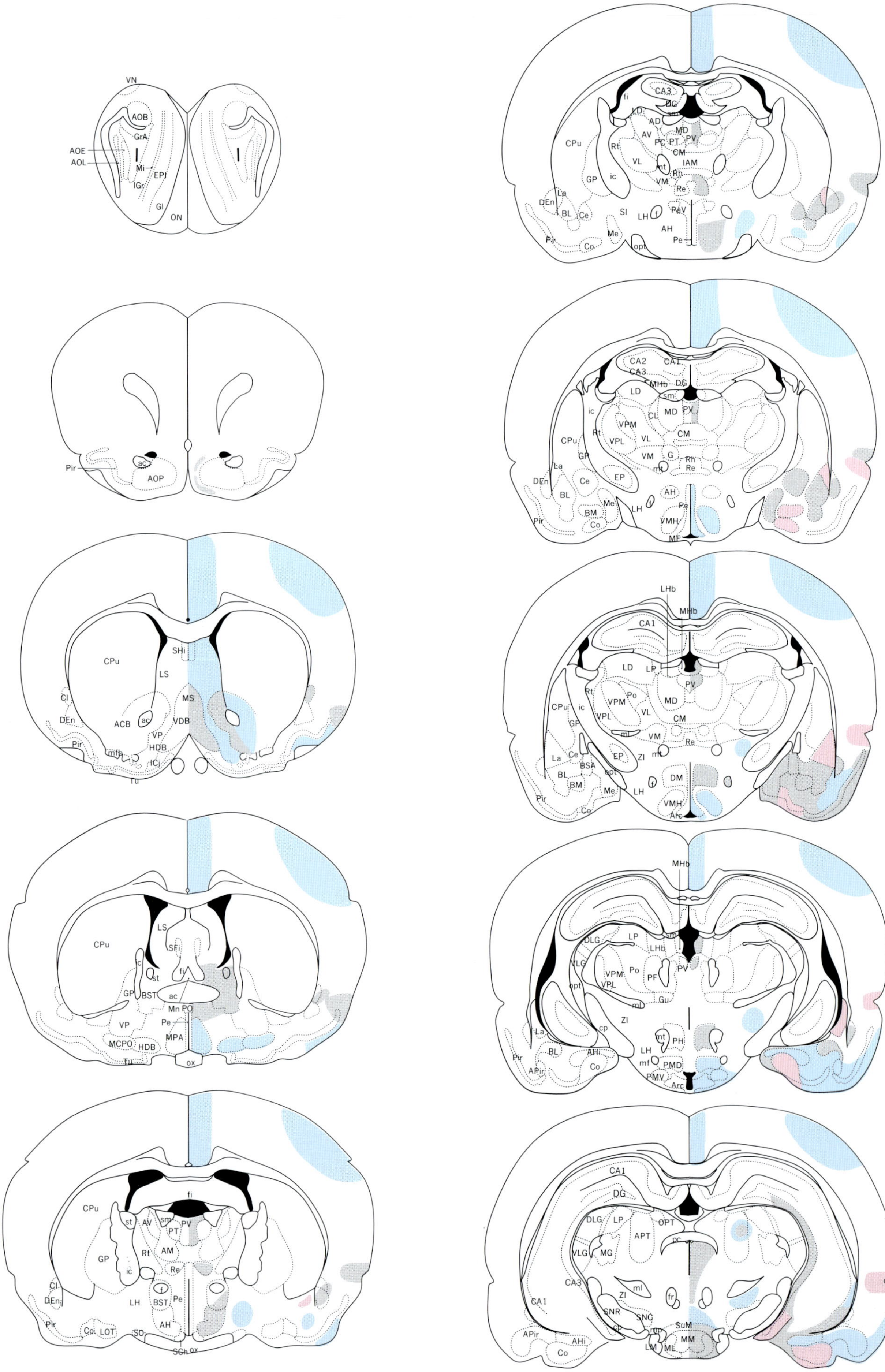

Tyrotropin releasing hormone receptor (binding sites)

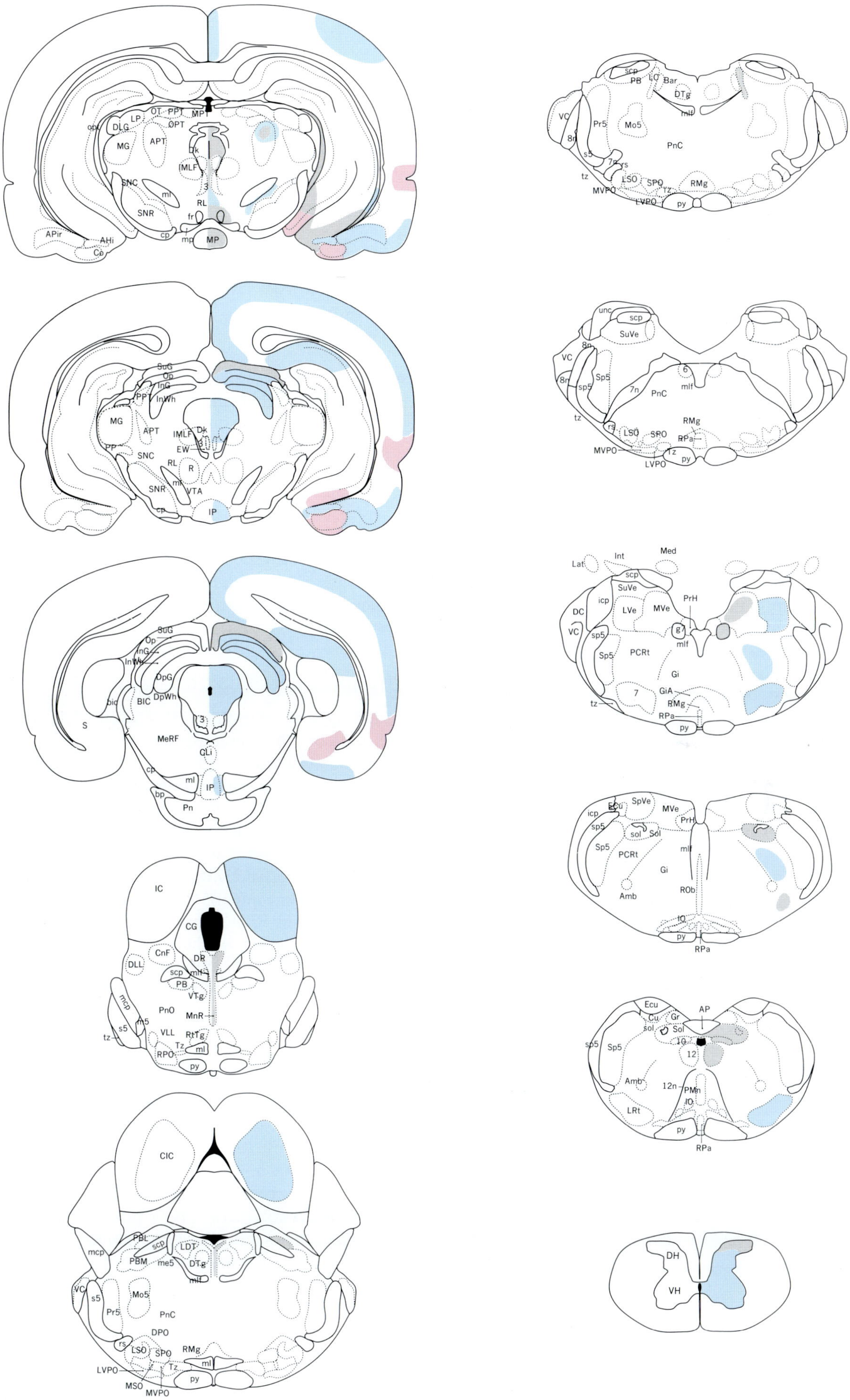

27 The neuropeptide Y system

Neuropeptide Y

Subtypes and structures of analogous peptides

Neuropeptide Y (NPY), together with pancreatic peptide (PP), and peptide YY (PYY), belongs to the pancreatic polypeptide family. Each peptide consists of 36 amino acid residues (Fig. 27.1)[8,15].

PP has been isolated from chick, cattle, sheep, dog, swine, rat, and humans; they differ in their tertiary structures and amino acid sequences. NPY has been isolated from porcine brain. PYY was first purified from the small intestine of swine, and has been purified from most mammalian species mentioned above. While NPY and PYY show approximately 70 per cent homology in their amino acid sequences, a 45–70 per cent homology is seen between NPY and PP, and between PYY and PP.

Distribution of the pancreatic peptide

More than 90 per cent of PP exists in pancreas, produced in D or F cells of the Langerhans islets[1]. NPY is abundantly present in brain and the peripheral nervous system also. Immunohistochemistry and *in situ* hybridization methods[4] revealed that they are widely distributed in the brain, with high concentrations in hypothalamus and cerebral cortex. Noradrenaline (NA) or adrenaline is co-localized in a section of NPY-containing cells of the auxiliary brain stem[10]. In the peripheral nerves, coexistence of NPY and NA has been demonstrated in the posterior neuron of sympathetic ganglia[9]. NPY also exists in chromaffin cells of adrenal gland and melanocytoma of humans. PYY is localized in the endocrine cells of the small intestine and digestive tract. Cells of a very small region of the brain (solitary tract nucleus tractus and ventral part of medulla oblongata) also contain PYY.

		PP
Avian	PP	GPSQPTTPGDDAPVEDLIRPDASLQQYLNVVTRHRY
Rat	PP	APLEPMYPGDYATHEQRAGYETQLRRYINTLTRPRY
Human	PP	APLEPVYPGDNATPEQMAGYAADLRRYINMLTRPRY
Porcine	PP	APLEPVYPGDNATPEQMAQYAAELRRYINMLTRPRY
		NPY
Rat	NPY	YPSKPDNPGEDAPAEDMARYYSALRHYINLITRQRY
Human	NPY	YPSKPDNPGEDAPAEDMARYYSALRHYINLITRQRY
Porcine	NPY	YPSKPDNPGEDAPAEDLARYYSALRHYINLITRQRY
		PYY
Rat	PYY	YPAKPEAPGEDASPEELSRTTASLRHYLNLVTRQRY
Porcine	PYY	YPAKPEAPGEDASPEELSRTTASLQQYLNLVTRQRY

Fig. 27.1 Amino acid sequence comparison of NPY family. Identical residues are shown aligned, in blue.

Neuropeptide Y projection tract

The following projection tracts have been demonstrated:

- projection from lateral geniculate body to suprachiasmatic nucleus
- coexisting A1, A3, and A6 of auxiliary brain stem NA cell group projects to forebrain
- coexisting C2 adrenaline cell group projects to paraventricularis hypothalamic nucleus[10]
- some NPY cells of the solitary tract nucleus project to the parabrachial nucleus of the pons
- a projection from arcuate nuclei to paraventricularis hypothalamic nucleus[3]
- a projection from arcuate nuclei to median eminence
- NPY cells in cerebral cortex, hippocampus, and corpus striatum, which coexist with somatostatin or GABA[4]
- NPY coexisting with NA in posterior neuron of sympathetic ganglia and projecting widely into the periphery[9]
- endogastric NPY circuit.

Physiological function

Following food ingestion, PP is released into the blood and it inhibits the secretin-stimulated pancreatic secretion in anesthetized cats[15]. PP also inhibits pancreatic juice secretion by CCK and secretin, PYY, somatostatin, and NPY possess similar actions, but the potency of action is, in decreasing order, PYY > PP, somatostatin > NPY. PP also accelerates acid and bile secretion and stimulates pepsin secretion, and thus affects stomach, intestine, and gall bladder[6]. In comparison to gastrin, the acceleration of acid secretion by PP is only 1/3 as potent and this action is inhibited by excision of the vagus nerve. This suggests a role for the parasympathetic nervous system as well. The secretion of PP is inhibited by somatostatin. PP concentration in the blood also increases with ageing. PP also shows a vasopressor action, which is somewhat weaker than that shown by PYY or NPY (PYY > NPY > PP). A higher concentration of PP is seen in chronic renal insufficiency of diabetes. NPY has an inhibitory action on vasoconstriction and on the smooth muscle motion of the digestive tract[6,10]. In the peripheral nervous system, it coexists with NA in the posterior sympathetic ganglia and exhibits a vasopressor action. While NA causes a short-lasting quicker contraction, NPY produces a tonic, slightly weak vasoconstriction. Furthermore, simultaneous injection of NA and NPY shows a synergistic effect (Fig. 27.2).

Various central effects of NPY have also been reported. For example, infusion of NPY into hypothalamus leads to an increase in food and water intake[13]. Particularly when NPY was infused into the paraventricular hypothalamic nucleus, food intake increased remarkably. The NPY of the paraventricular hypothalamic nucleus is

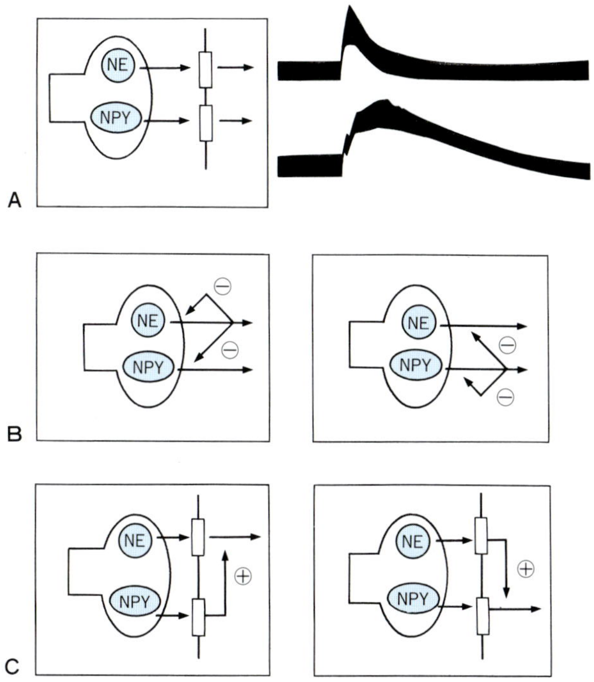

Fig. 27.2 The three principally different types of effect of NPY (or PYY) at sympathetic neuroeffector junctions. A: NE and NPY act independently, but (right panel) characteristics of response may differ (exemplified by vasopressor effects); B: Reciprocal regulation of release; C: Reciprocal enhancement at the level of vascular smooth muscle (from ref. 18).

supplied from the NA neuron of the arcuate nuclei of the hypothalamus and a part of the medulla oblongata. It is therefore highly probable that this projection tract plays a role in water intake behaviour. Infusion of NPY into the suprachiasmatic nucleus lead to phase advances in the light–dark cycle entrainment of circadian rhythm[2]. The suprachiasmatic nucleus is the centre for regulating the circadian rhythm; NPY is concentrated in the ventral part of this nucleus, and these fibres are supplied from the lateral geniculate body. It is therefore, believed that photic information from the retina acts on NPY neurons in the lateral geniculate body. The endocrine function of NPY is also of some interest. When NPY is infused into the supraoptic nucleus of the hypothalamus, it stimulates the secretion of vasopressin. A similar action is also shown by NA. NPY in the supraoptic nucleus of the hypothalamus projects from the cells in medulla oblongata A1 in which NA is co-localized. A1 is also related to the baroreceptors; it is an interesting site. NPY also accelerates the secretion of LH in hypophysis and the secretion of FSH and GH[11]. On the mechanism of how NPY accelerates LH secretion, two possibilities have been suggested. One is that NPY stimulates LHRH cells. The other is that it acts directly on hypophysis LH cells, with intermediating arcuate nuclei– median eminence tract NPY. NPY's action in accelerating the secretion of ACTH has also been reported[5]. Furthermore, NPY has been implicated in memory enhancement, probably by the actions of NPY neurons present in the hippocampus and cerebral cortex. GABA is co-localized in the NPY neurons of these areas. In contrast to NPY, GABA is believed to influence amnesia. It is therefore, suggested that NPY may inhibit the release of GABA from the same neuron (Fig. 27.3; see the next section). In addition, NPY has been demonstrated to increase in the innominate substance of Alzheimer's disease patients with amnesia[16].

References

1. Adrian, T.E., Bloom, S.R., Bryant, M.G. *et al.* (1976). *Gut*, **17**, 940–4.
2. Albers, H.E. and Ferris, C.F. (1984). *Neurosci. Lett.*, **75**, 163–8.
3. Bai, F.L., Yamano, M., Shiotani, Y. *et al.* (1985). *Brain Res.*, **331**, 172–5.
4. De Quidt, M.F. and Emson, P.C. (1990). In *Handbook of chemical neuroanatomy*, **9**, 287.
5. Harfdtrand, A.K., Uxe, K., Agnati, L.F. *et al.* (1986). *Neurochem. Int.*, **8**, 355–76.
6. Hellstrom, P.M. (1984). In *Gastrointestinal motility* (ed. C. Roman) p. 433. MTTP Press, Lancaster.
7. Kawata, M., Nako, K., Morii, N., Kiso, Y. *et al.* (1985). *Neuroscience*, **16**, 521–46.
8. Kimmel, J.R., Hayden, J., and Pollock, G. (1975). *J. Biol. Chem.*, **250**, 9369–76.
9. Lee, Y., Shiosaka, S., Emson, P.C., Powell, J.F. *et al.* (1985). *Gastoenterology*, **89**, 118–26.
10. Lundberg, J.M., Terenius, L.,Hökfelt, T. *et al.* (1982). *Acta Physiol. Scand.*, **116**, 477–80.
11. McDonald, J.K., Lumpkin, M.D., Samson, W.K., and McCann, S.M. (1985). *Proc. Natl Acad. Sci. USA*, **82**, 561–4.
12. Morley, J.E. and Food, J.F. (1990). *Ann. N.Y. Acad. Sci.*, **611**, 226–31.
13. Stanley, B.G. and Leibowitz, S.F. (1984). *Life Sci.*, **35**, 2635–42.
14. Sudou, T., Kangawa, K., Minamino, N., and Matsuo, H. (1988). *Nature*, **332**, 78–81.
15. Tatemoto, K., Carlquist, M., and Mutt, V. (1982). *Nature (Lond.)*, **296**, 659–60.
16. Takeda, Y. and Yanaihara, N. (1986). *Neuroendocrinology, 2. Bioactive substance.* pp. 182–95. Chugai Igaku, Tokyo.
17. Weiner, F.D., Makkat, A.M., Papolos, D.F., and Lachman, H.M. (1992). *Mol. Brain Res.*, **12**, 209–14.
18. Wahlestedt, C., Grandemar, L., Håkanson, R. *et al.* (1990). *Ann. N.Y. Acad. Sci.*, **611**, 7–26.

Neuropeptide Y (immunocytochemistry)

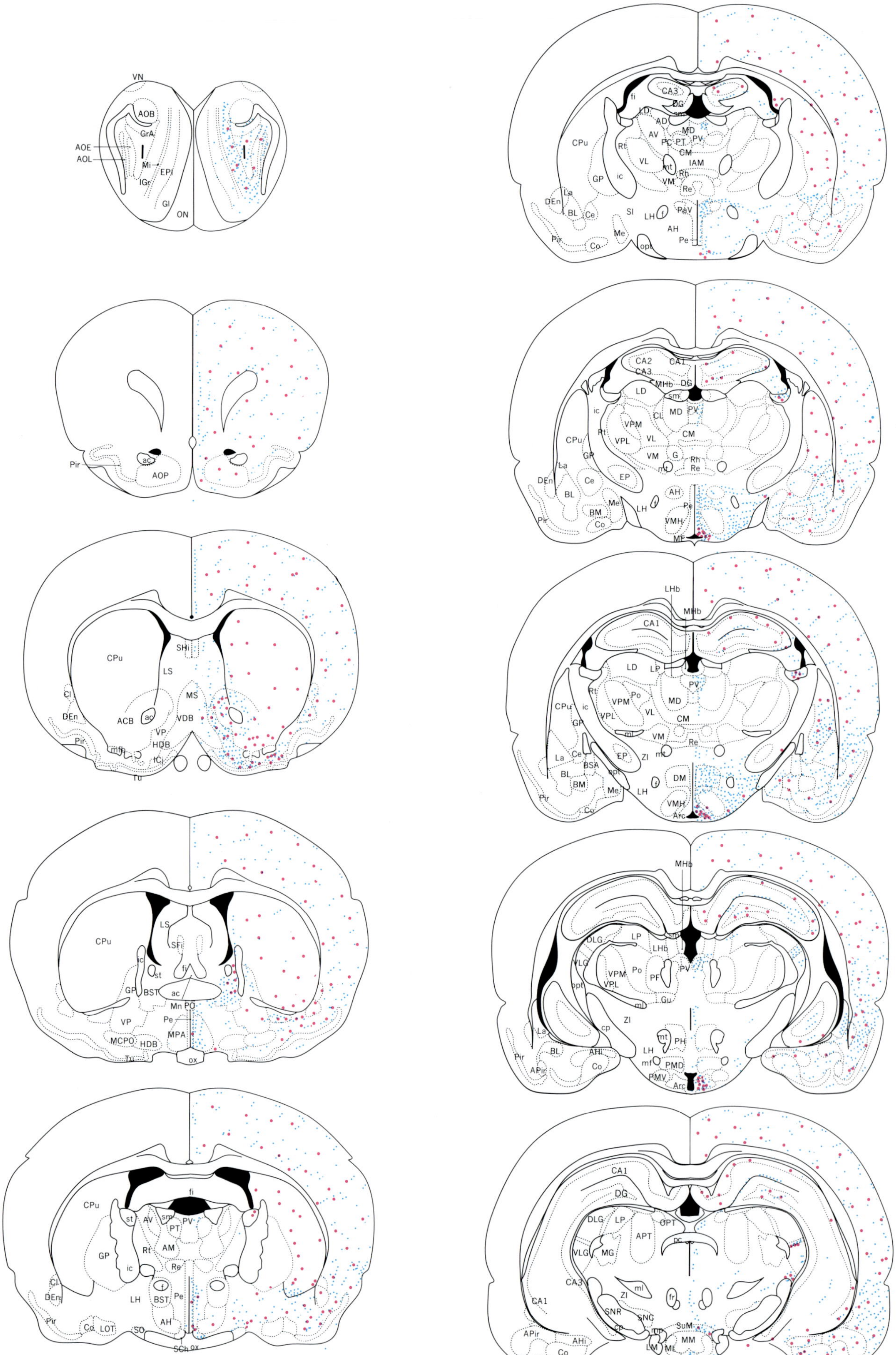

Neuropeptide Y (immunocytochemistry)

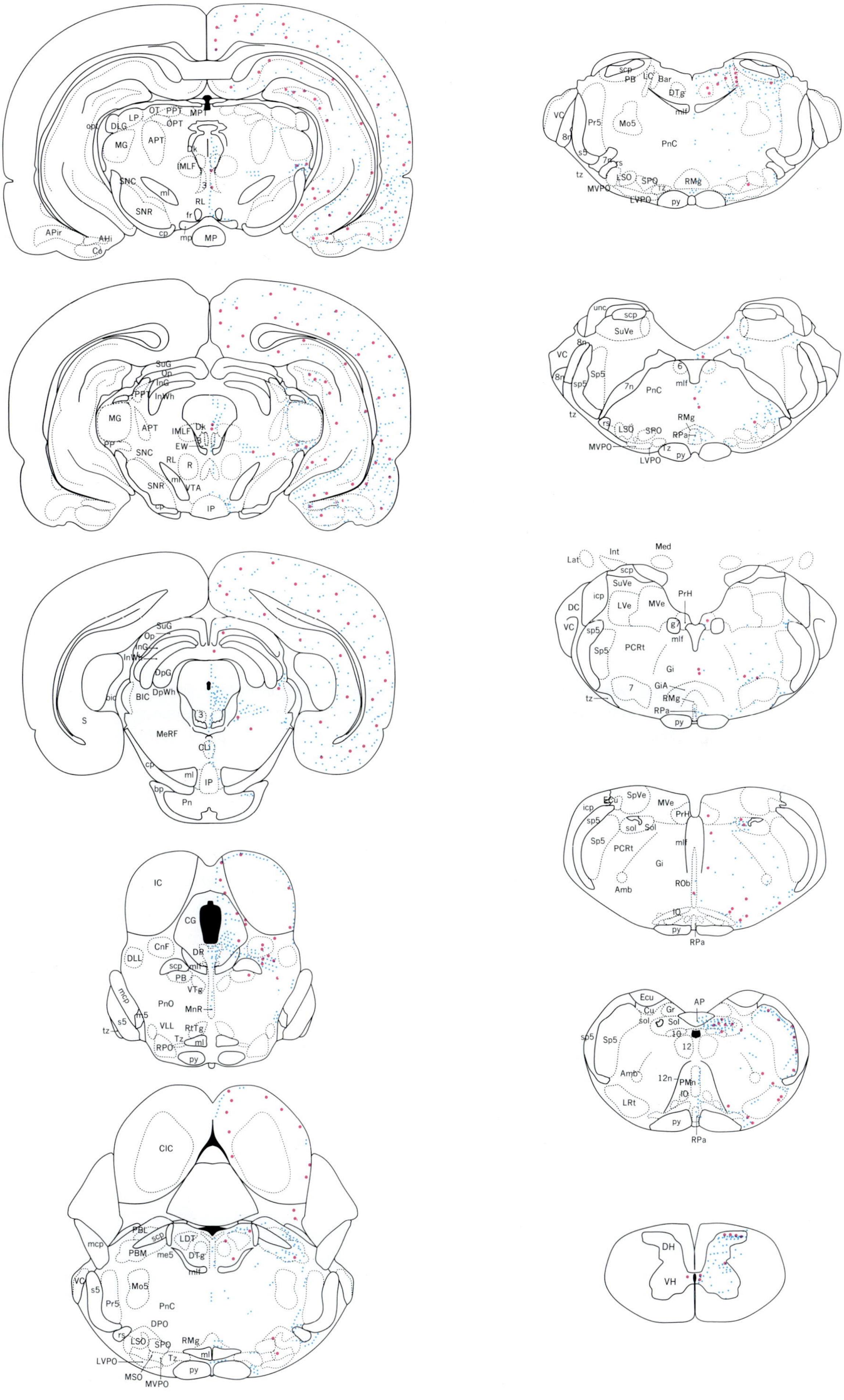

Neuropeptide Y receptor

Subtypes

On pharmacological grounds, the existence of two subtypes of NPY receptors has been suggested. These have been named Y_1 and Y_2[2,7,10,11]. Y_1 is believed to be a postsynaptic receptor[2] and Y_{2a}, a presynaptic receptor[10,11]. While Y_1 is activated in the presence of the total amino acid structure (holopeptide) of NPY or PYY, Y_2 has high affinity for the C-terminal fragment of NPY or PYY[10,11]. Since a short fragment of NPY is involved in causing memory enhancement, it is plausible that this NPY's action of memory enhancement is mediated by the Y_2 receptor[1]. For acceleration of food intake, NPY needs its complete amino acid sequence. Thus, this action is believed to be mediated by the Y_1 receptor[5]. NPY coexists with noradrenaline at many binding sites in the auxiliary brain stem and projects into the forebrain. High concentration of Y_2 receptor is seen in these terminal sites of NA and dopamine. Meanwhile, though both Y_1 and Y_2 receptors exist in the vascular smooth muscle, the main action is performed by Y_1, probably as an analogue of α-adrenoreceptor. Recently, it has also been reported that NPY and PYY have affinities for binding with sigma and phenchclidine (PCP) and that new physiological functions of NPY and PYY may be identified[6].

The main antagonists of NPY receptors are PYX-1 and PYX-2 (Fig. 27.4)[10,11]. While PYX-1 is a peptide which modifies the amino acids at the N-terminal, PYX-2 (a stronger antagonist) modifies both the N- and C-terminal ends[8]. These are selectively inhibited by ^{3}H-NPY. These antagonists are effective to the physiological actions of NPY such as altering the food intake, blood pressure, hormone secretion, pancreatic secretion, intrinsic movement of digestive tract, and circadian rhythm[8].

Distribution

The distribution of NPY binding sites has been studied by autoradiography, using [^{125}I]NPY and [^{3}H]NPY as ligands[4,6]. The distribution density of NPY fibres and the distribution of NPY binding sites show high correlation in cerebral cortex and hippocampus. NPY has low binding sites in the corpus striatum. The distribution of PYY binding sites has also been mapped by autoradiography, using [^{125}I]PYY as a ligand[3,4]. Most regions of the brain, with a few exceptions, show similar distribution of NPY binding sites[6]. Only low binding sites were observed in hypothalamus, brainstem, and cerebellum, where PYY binding sites are present in abundance[6]. However, these regions are rich in NPY fibres. It has therefore been inferred that PYY binding sites in these regions may have NPY receptors, with structure different from that of [^{125}I]NPY binding sites[9]. An autoradiographic technique which employs every part of the amino acid fragment of NPY to identify the distribution of Y_1 and Y_2 receptors exists, but the detection sensitivity limit is rather low[6].

Cloning

Recently, the genetic sequence of Y_1 receptor for NPY has been reported[12–14]. The human Y_1 receptor, based on the gene library of the human fetus, consists of 384 amino acid residues and has seven transmembrane domains similar to other G-protein conjugate type receptors. The transmembrane domains of Y_1 receptor differ only in 29 per cent of amino acids from those of human tachykinin receptor. The cDNA (LCR1) of NPY receptor based on bovine locus ceruleus nucleus has also been sequenced, but the sequence homology between the NPY receptor gene and Y_1 receptor is only 2.1 per cent.

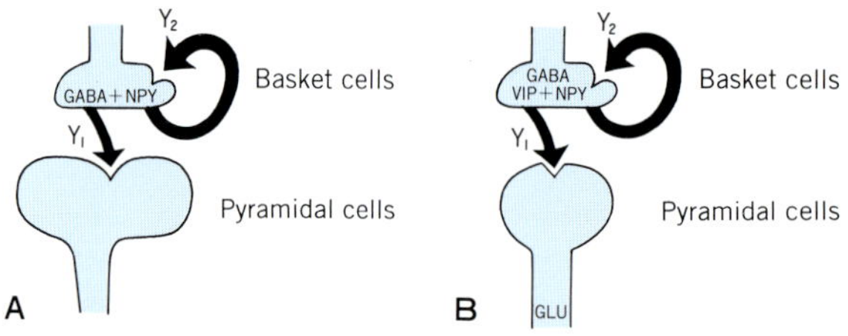

Fig. 27.3 (A) Neuropeptide Y (NPY) could be co-released with γ-aminobutyric acid (GABA) and NPY could feedback presynaptically on the basket cell to inhibit further release of GABA. Thus GABA would no longer inhibit the firing of the pyramidal cell containing glutamate. Glutamate enhance memory retention. (B) A similar scenario is postulated for NPY inhibiting the release of vasoactive intestinal peptide (VIP). Alternatively, VIP could act on a separate set of pyramidal cells (from ref. 5).

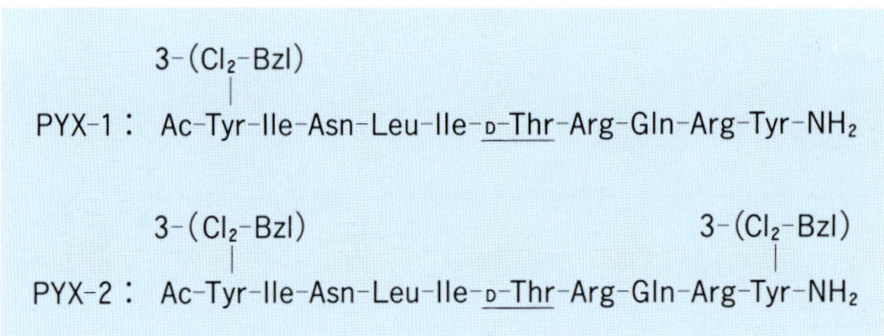

Fig. 27.4 The structures of NPY receptor antagonists PYX-1 and PYX-2. 3-(Cl_2-Bzl), 3-(2-6-dichorobenzyl); $-NH_2$, amide; Ac-, acetyl- (from ref. 8).

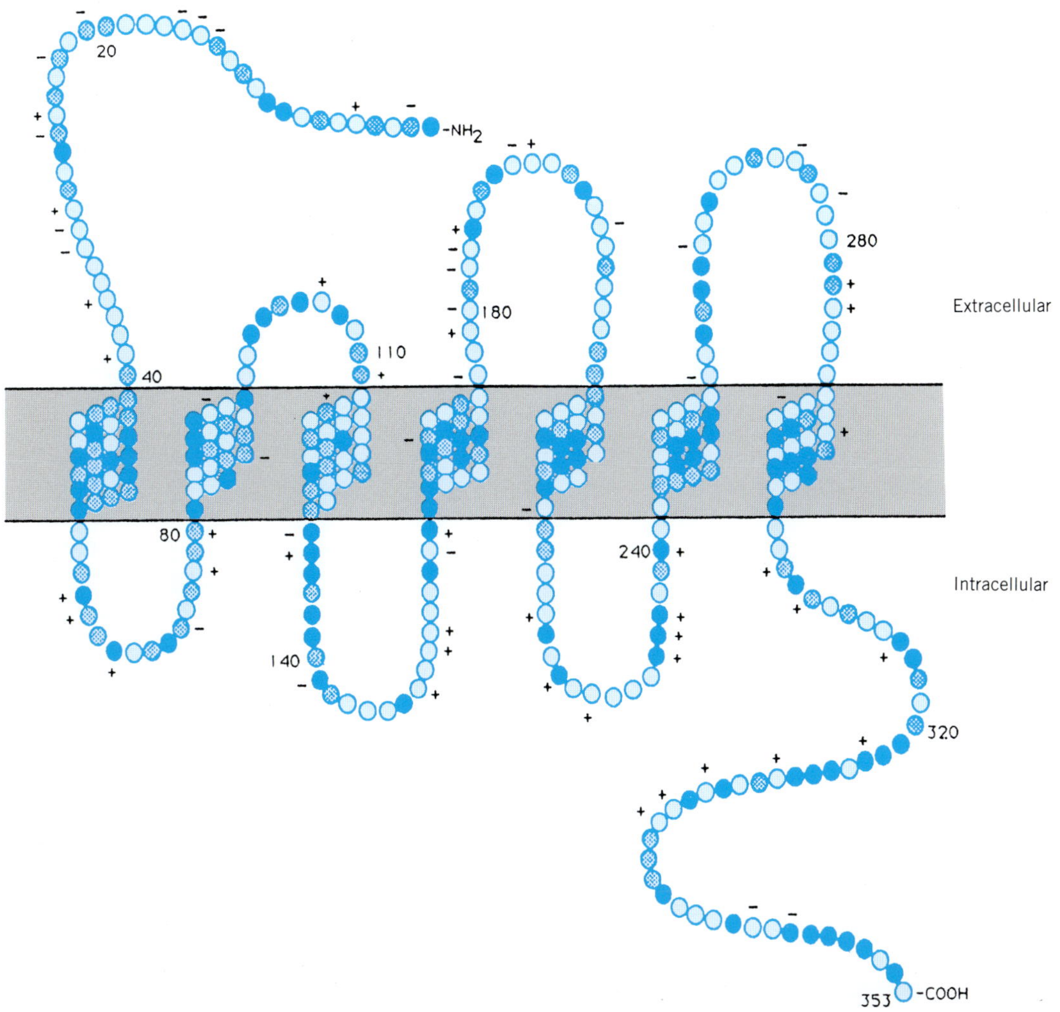

Fig. 27.5 Transmembrane model of the LCR NPY receptor. The solid or shaded residues represent amino acids that are identical to or conserved substitution of, respectively, the rat substance P receptor. The transmembrane-spanning domains were identified by hydropathy analysis and homology to other G-protein coupled receptor (from ref. 14).

Update

Now five members of NPY receptor subtypes are defined, and Y_1, Y_2, Y_4, and Y_5 receptors have been characterized. The Y_1 receptor requires the intact 36 amino acid peptides for binding; on the other hand NPY3-36, NPY13-36, and NPY18-36 seem to be Y_2 selective ligands. The Y_4 receptor exhibits the highest affinity for PP; mouse Y_5 receptor exhibits Y_1 receptor pharmacology. Y_5 receptors are expressed within discrete regions of the hypothalamus including the suprachiasmatic nucleus, anterior hypothalamus, bed nucleus stria terminalis, and the ventromedial nucleus. Y_1 receptor activation produces marked vasodepressor actions, while Y_2 receptor activation causes a vasopressor response[15].

References

1. Flood, J.F. and Morley, J.E. (1989). *Peptides*, **10**, 963–66.
2. Fuhlendorf, J.U., Gelten, U., Aakerlund, L. *et al.* (1990). *Proc. Natl Acad. Sci.*, **87**, 182–6.
3. Lynch, D.R., Walker, M.W., Miller, R.J., and Snyder, S.H. (1989). *J. Neurosci.*, **9**, 2607–19.
4. Martel, J.-C., Fournier, A., St-Pierre, S., and Quiron, R. (1990). *Neuroscience*, **36**, 255–83.
5. Morley, J.E. and Food, J.F. (1990). *Ann. N.Y. Acad. Sci.*, **611**, 226–31.
6. Quirion, R., Martel, R.Q., Dumont, Y. *et al.* (1990). *Ann. N.Y. Acad. Sci.*, **611**, 58–71.
7. Schwartz, T.W. (1989). In *Nobel Symposium 65: Neuropeptide.* (ed. Y.V. Mutt). Raven Press, New York.
8. Tatemoto, K. (1990). *Ann. N.Y. Acad. Sci.*, **611**, 1–6.
9. Taylor, I.L., Mannon, P.J., Heintz, G.G. *et al.* (1990). *Ann. N.Y. Acad. Sci.*, **611**, 48–57.
10. Wahlested, C., Grundemar, L., Hakanson, R. *et al.* (1990). *Ann. N.Y. Acad. Sci.*, **611**, 7–15.
11. Wahlested, C., Yanaihara, N., and Hakanson, R. (1985). *Regul. Pept.*, **13**, 317–28.
12. Eva, C., Keinanen, K., Monyer, H. *et al.* (1990). *FEBS Lett.*, **271**, 81–4.
13. Larhammar, D., Blomqvist, A.G., Yee, F. *et al.* (1992). *J. Biol. Chem.*, **267**, 10935–8.
14. Rimland, J., Xin, W., Sweetnam, P., Saijoh, K. *et al.* (1991). *Mol. Pharmacol.*, **40**, 869–75.
15. Weinberg, D.H., Sirinathsinghji, D.J., Tan, C.P. *et al.* (1996). *J. Biol. Chem.*, **271**, 16435–8.

Neuropeptide Y (binding sites)

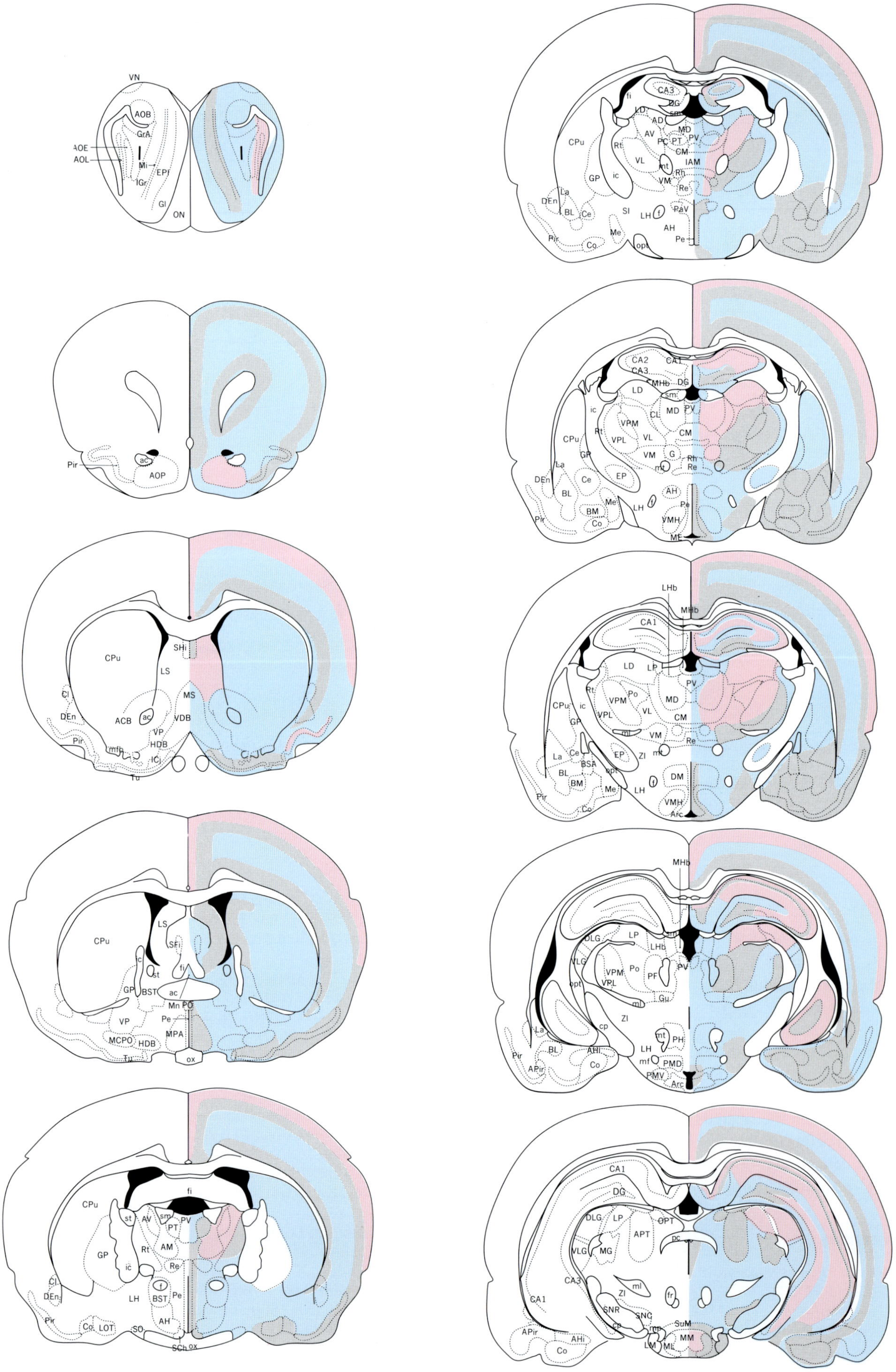

Neuropeptide Y (binding sites)

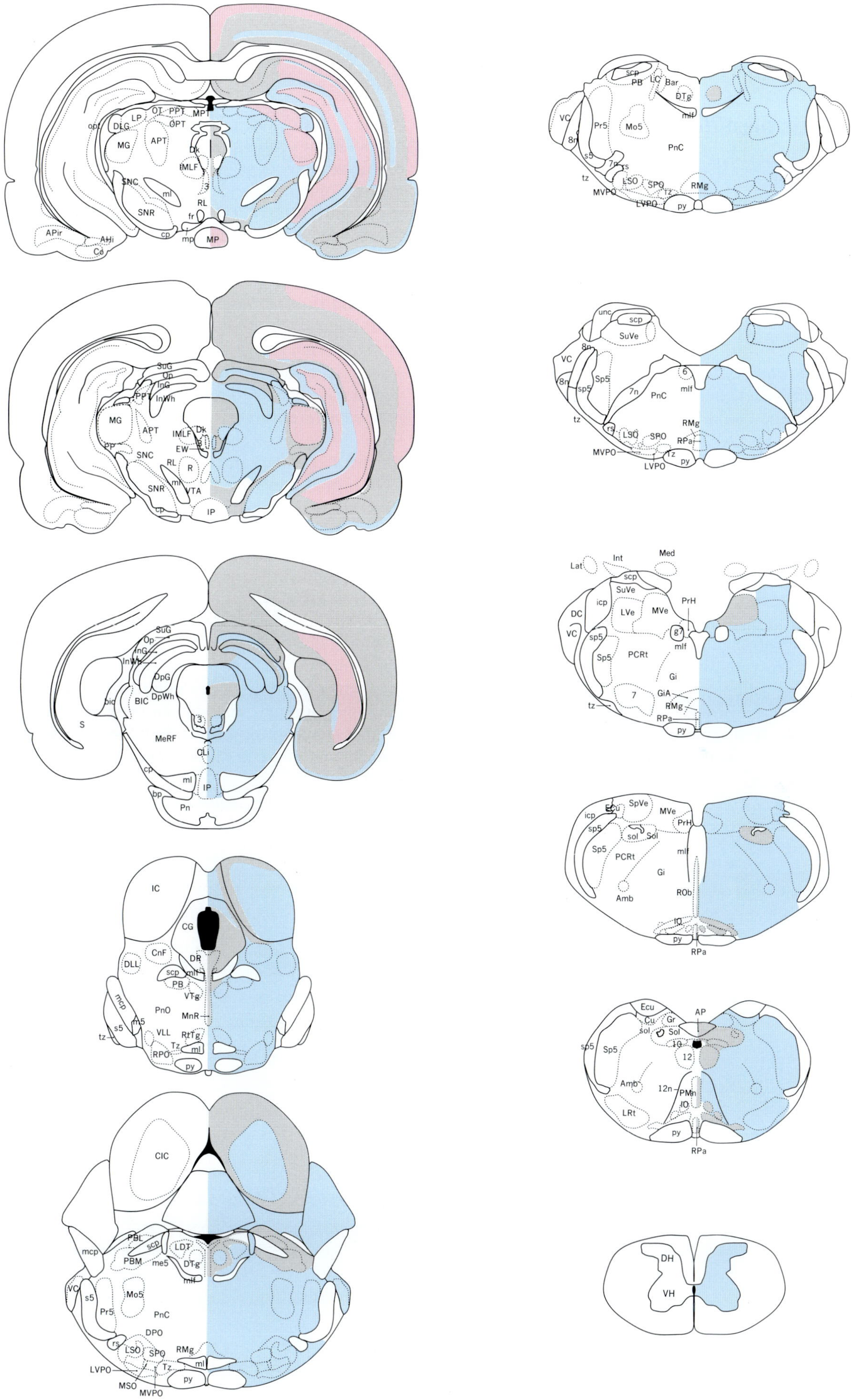

28 The natriuretic peptide system

Natriuretic peptide

Classification

The natriuretic peptide (NP) family consists of atrial NP (ANP)[6], brain NP (BNP)[14], and C-type NP (CNP)[15]. Their structures are shown in Fig. 28.1. All possess a ring structure formed by the Cys–Cys bond (Fig. 28.2)[6,14,15].

Distribution and physiological function

Atrial natriuretic peptide
ANP is synthesized and stored in cardiac atria and is released into the blood in response to a stimulus. There are three kinds of ANP, based on molecular weight variation; α-, β-, and γ-ANP. γ-ANP is considered as the precursor of α-ANP. It exists in atrium, and is transformed into α-ANP and secreted into the blood, to act on kidney, lung, and adrenal gland, influencing the regulation of fluid and electrolyte homeostasis and blood pressure[3,4]. α-ANP allows the renal blood volume and glomerular filtration rate to increase in kidney, but inhibits water and electrolyte transport in the uriniferous tubule[11]. These actions of ANP suggest a strong diuretic action. In adrenal gland, it inhibits the biosynthesis of aldosterone[3]. It also has a strong relaxing action on vascular smooth muscle[2]. These actions are directed toward reduction of blood pressure. High levels of ANP among hypertensives[3] suggest that this is a secondary defensive phenomenon, but it could possibly be related to the direct increase of systolic pressure by ANP.

Major components present in the central nervous system of rat are α-ANP [4–28] and α-ANP [5–28][13,16]. The α-ANP exists in high concentration in mesencephalon of swine, human, and monkey. Medium concentration is seen in the olfactory bulb and hypothalamus[4]. However, in the brain of rat, high concentration is present in the hypothalamus and septal nuclei, but only a little in mesencephalon[5,7]. In dogs, high concentration is seen in the medulla oblongata and olfactory bulb. Infusion of α-ANP [5–28] into the cerebral ventricle results in blood pressure depression, inhibition in the secretion of vasopressin, and uptake inhibition of salt and water[4].

Brain natriuretic peptide
BNP is biosynthesized and stored in the intra-atrial and intraventricular regions and is released into the blood, where it elicits sodium diuresis and blood pressure depression, similar to that of ANP[9,14]. There also exists a precursor form, γ-BNP, in atrium, which is released in a form of small molecular BNP, where the C-terminus was fragmented at the same sites[3]. BNP differs from ANP and CNP in that there are remarkable differences of primary structure among species. Swine and bovine BNP show a high homology of amino acids.

In the brains of swine and dog, BNP is detected in higher concentration than ANP[3,4]. In swine, BNP is observed especially in high concentration in the olfactory bulb, cerebral cortex, lenticular nucleus, caudate nucleus, and medulla oblongata, and is believed to exist in the hypothalamus as well[3,10]. In dog, BNP is observed in medium concentration in similar sites, but is not detected in corpus striatum[3]. In swine and dog, BNP is also present in high concentration in medulla oblongata. BNP is also seen in abundance in the rat brain and spinal cord[3,4,10,12].

C-type natriuretic peptide
CNP was first identified in swine brain, and shows similar physiological actions to ANP; however, its diuretic and hypotensic actions are somewhat lower in intensity than those of ANP and BNP[15]. In humans and rats, CNP-22 with 22 amino acid residues is the main component and

```
ANP
α-Rat ANP                 SLRRSSCFGGRIDRIGAQSGLGCNSFRY
α-Human ANP               SLRRSSCFGGRMDRIGAQSGLGCNSFRY

BNP
Rat     BNP   SQDSAFRIQERLRNSKMAHSSSCFGQKIDRIGAVSRLGCDGLRLF
Human   BNP                 SPKMVQGSGCFGRKMDRISSSSGLGCKVLRRH
Porcine BNP                        DSGCFGRRLDRIGSLSGLGCNVLRRY
Bovine  BNP                ALRGPKMMRDSGCFGRRLDRIGSLSGLGCNVLRRY

CNP                        GLSKGCFGLKLDRIGSMSGLGC
```

Fig. 28.1 Amino acid sequence comparison of ANP family. Identical residues are shown aligned on a blue background.

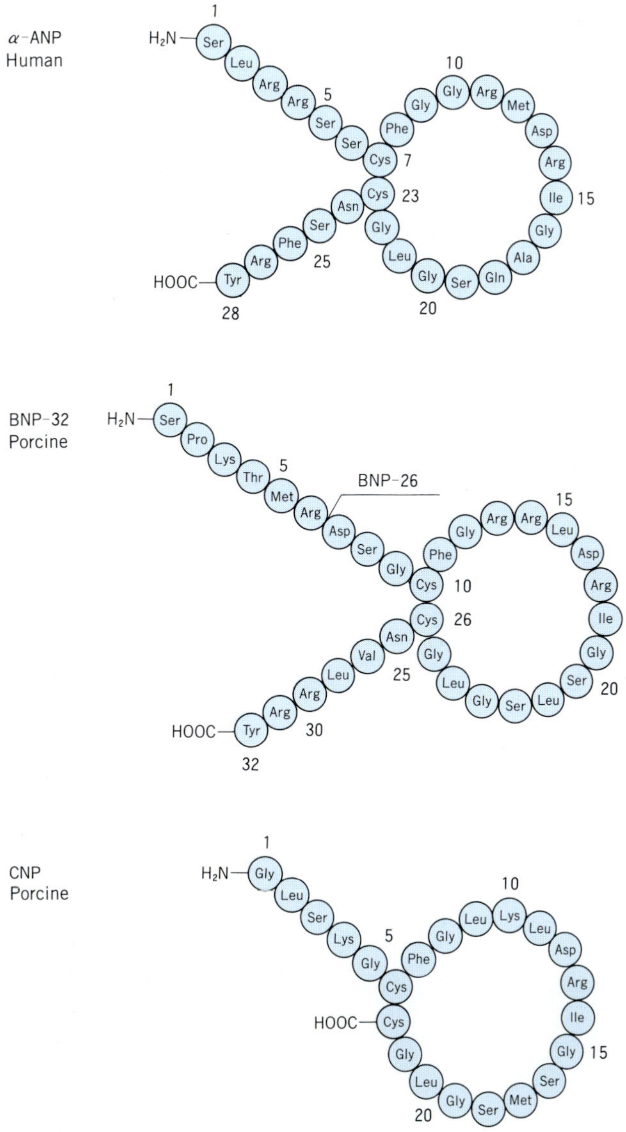

Fig. 28.2 Comparison of structure of human α ANP, porcine BNP, and porcine CNP.

the mammalian CNP is believed to be the most conserved NP in the family[8]. Porcine CNP consists of 22 amino acid residues (CNP-22) and its N-terminally elongated form of 53 amino acid residues (CNP-53)[8]. Northern blot analysis of CNP mRNA revealed that CNP is observed only in the brain, and especially in the hypothalamus, mesencephalon, and medulla oblongata, where it is detected in higher concentration than ANP and BNP[1,2,8]. It has been inferred that CNP binds with ANP receptor B type specifically and it is believed that CNP also participates in the functions of natriuretic peptide by activating this receptor.

References

1. Furuya, M. (1990). *Biochem. Biophys. Res. Commun.*, **170**, 201.
2. Inagami, T. (1989). *J. Biol. Chem.*, **264**, 3043–6.
3. Itoh, H. (1991). *Biomedica*, **6**, 1262–70.
4. Itoh, H. (1990). *Endocrinology*, **127**, 1292–300.
5. Jacobowitz, D.M., Skofitsch, G., Keiser, H.R. *et al.* (1985). *Neuroendocnology*, **40**, 92–4.
6. Kangawa, K., and Matsuo, H. (1984). *Biochem. Biophys. Res. Commun.*, **118**, 131–9.
7. Kawata, M., Nakao, K., Morii, N. *et al.* (1985). *Neuroscience*, **16**, 521–46.
8. Kojima, M., Minamino, N., Kangawa, K., and Matsuo, H. (1990). *FEBS Lett.*, **276**, 209.
9. Gunning, M., Ballermenn, B.J., Silva, P. *et al.* (1990). *Am. J. Physiol.*, **258**, F467–F472.
10. Ogawa, Y. (1990). *Endocrinology*, **126**, 2225–7.
11. Samson, W.K. (1985). *Neuroendocrinology*, **40**, 227.
12. Saper, C.B., Hurley, K.M., Moga, M.M. *et al.* (1989). *Neuroscience*, **96**, 29–34.
13. Shiono, S. (1986). *Biochem. Biophys. Res. Commun.*, **135**, 728–34.
14. Sudoh, T., Kangawa, K., Minamino, N., and Matsuo, H. (1988). *Nature*, **332**, 78–80.
15. Sudoh, T., Minamino, N., Kangawa, K., and Matsuo, H. (1990). *Biochem. Biophys. Res. Commun.*, **168**, 863–70.
16. Ueda, S., Sudoh, T., Fukuda, K. *et al.* (1987). *Biochem. Biophys. Res. Commun.*, **149**, 1055–62.

Atrial natriuretic peptide (immunocytochemistry)

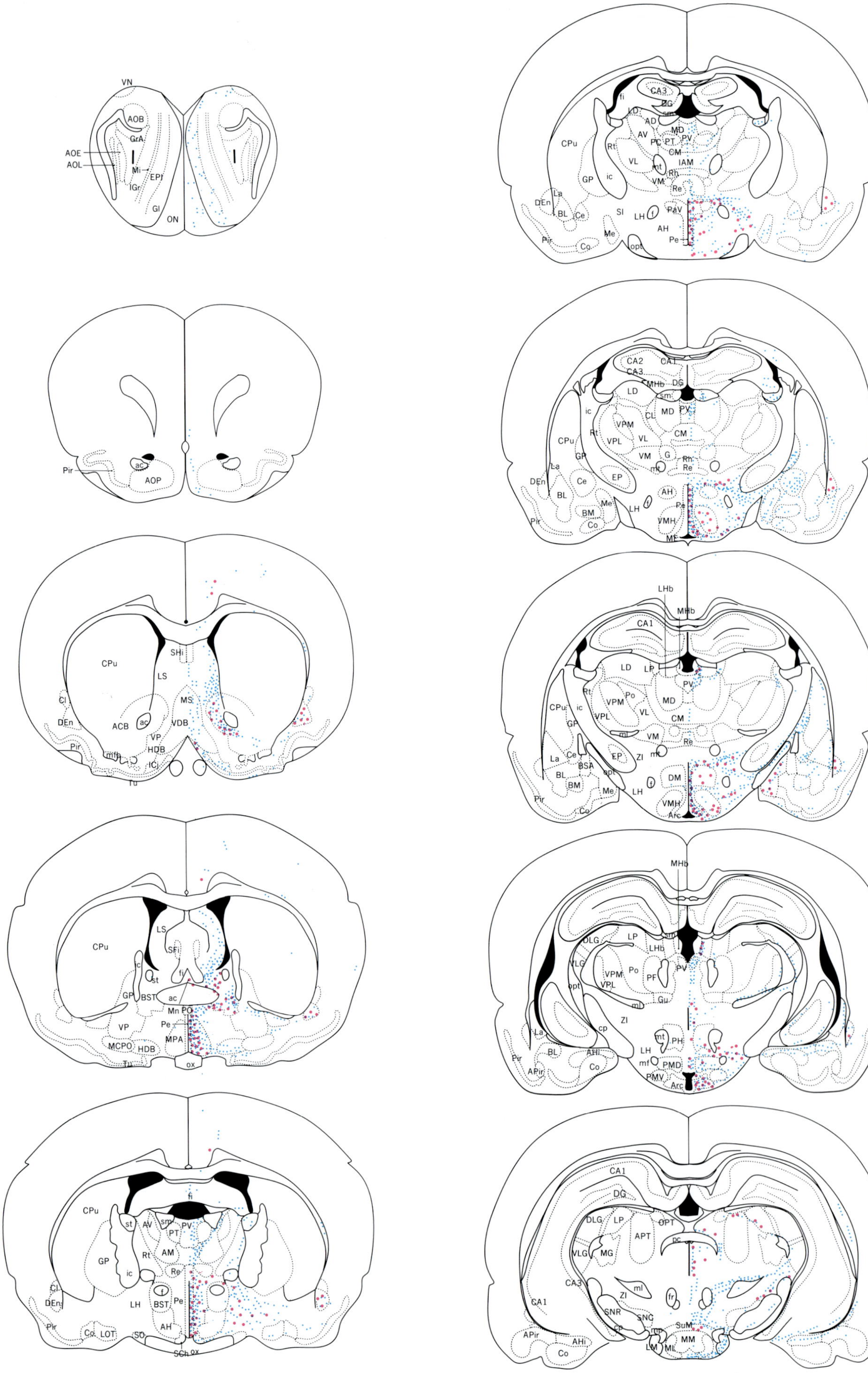

Atrial natriuretic peptide (immunocytochemistry)

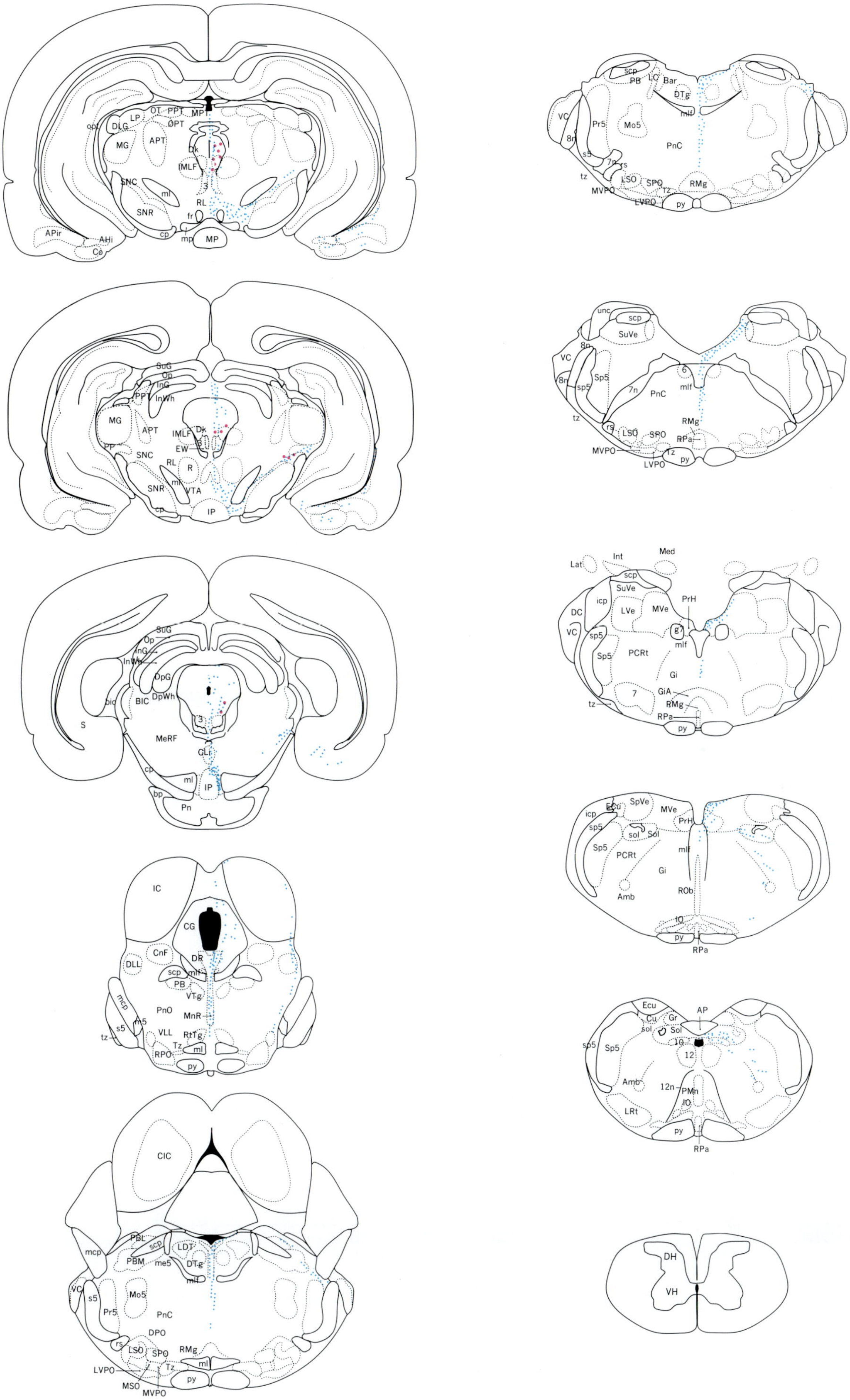

Natriuretic peptide receptor

Subtypes and structure

Atrial natriuretic peptide

For the ANP receptor, there exist at least 3–4 types of monomer of 120–135 kDa (A, B type) and a dimer of 135–140 kDa (C type)[5,6]. The monomer-type 120–135 kDa receptors (A, B type) possess guanylate cyclase activity[4,13,14], but this activity is absent in the dimer-type receptor. Since the A type that possesses guanylate cyclase activity is believed to bind ANP and thereby causes alteration in intracellular metabolism, the monomeric receptors of 120–135 kDa (A, B type) are believed to be important for the expression of physiological activities of ANP[2,5,10–11,13]. Also of interest is the fact that the ANP receptor which utilizes guanylate cyclase as an effector is a rare example, since most of the peptide receptors that are known to date use adenylic acid cyclase or phospholipase as an effector.

The cloning of the dimeric type (C type of 135–140 kDa) receptor has also been reported; this type of ANP receptor consists of 496 amino acid residues (molecular weight 55 701), and is formed from a precursor with 537 amino acid residues (molecular weight 55 701), and is formed from a precursor with 537 amino acid residues (molecular weight 59 774)[6,7] (Fig. 28.3). Although the extracellular domain of this type of receptor has a high homology with the ANP A-type and B-type receptors, the intracellular domain differs considerably and consists of only 37 amino acid residues[4]. Its intracellular signal transfer mechanism is not clear. The ANP C-type receptor binds with all ligands of ANP, BNP, and CNP[1] (Fig. 28.4). It has therefore been suggested that the ANP C-type receptor may take part in the clearance of natriuretic peptide from its binding ligands[6,8].

Brain natriuretic peptide receptor

Not much is known about the subtypes and molecular weight of the BNP receptor. Since most of the sites binding [^{125}I]-ANP are inhibited by BNP, the ANP receptor is believed to bind with BNP[5]. Furthermore, BNP is known to increase the intracellular cGMP by an involvement with ANP A-type receptor. However, in some regions of tissues such as the choroid plexus, the inhibition effect by BNP is weak, so there is a possibility that the structure of the ANP receptor is different in these regions[5].

C-type natriuretic peptide receptor

Not much is known about the CNP receptor either. CNP does not increase the cGMP of cells via the involvement of ANP A-type receptor. However, when mediated by the ANP-B receptor, CNP increases the intracellular cGMP. CMP shows 50–500 times higher affinity than ANP or BNP in this binding[9]. These facts suggest that CNP has the ability to influence physiological actions via the mediation of a receptor similar to ANP-B, or to ANP-B type receptor than ANP-A type receptor.

Distribution

ANP receptor is distributed in the brain, lung, kidney, heart, and vasculature[2,3]. The distribution of ANP receptor in brain has been traced from the binding sites. Although the distribution of binding sites between ANP and BNP is analogous, they differ widely with the distribution of ANP itself, of being a ligand[2,3].

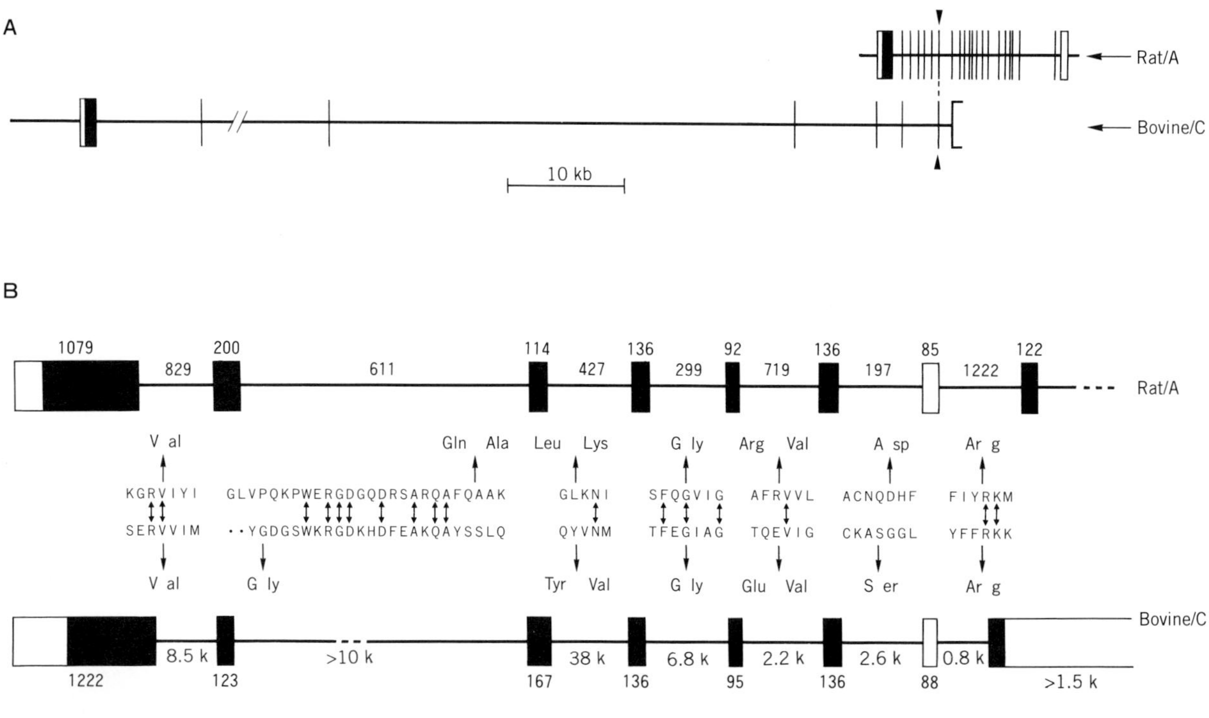

Fig. 28.3 Comparison of the gene structures of rat type A (Rat/A) and bovine type C (Bovine/C) ANP receptors. (A) Comparison of length of the type A and C receptor genes. The exons corresponding to the membrane-spanning domains of the receptor are aligned and indicated by arrow-heads. (B) Comparison of relative exon–intron organizations of types A and C receptor genes with special emphasis on intron positioning in the ligand-binding coding regions. Numbers indicate the sizes of exons and introns in bp. Introns are indicated by horizontal lines and exons by boxes. The filled and open boxes show the coding and non-coding regions, respectively; stippled boxes represent the membrane spanning domains. Between the maps is shown sequence homology around the intron positions. Intron positions, indicated by vertical arrows, are similar between the two subtypes except the second intron (from ref. 12).

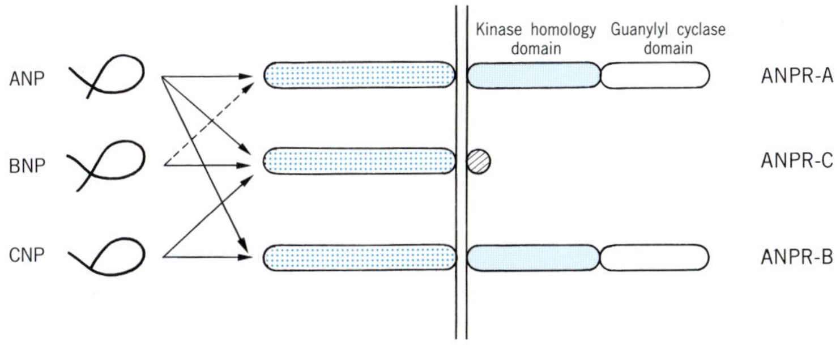

Fig. 28.4 Human specificity for the human natriuretic peptide receptors. A schematic representation of the ability of the human natriuretic peptides to specifically activate the guanylyl cyclase of human ANPR-A or ANPR-B or bind to human ANPR-C. The solid lines connect the receptors with their depreferred ligand. The ambiguity of BNP's role in activation of ANPR-A is represented by a dashed line, as described in the text (from ref. 9).

Update

ANP-C type receptors coupled to adenylyl cyclase/cAMP signal transduction system through Gi-regulatory protein differ from -A or -B type receptors. Their inhibition was altered in hypertension in the heart, aorta, brain, and adrenal glands of rats. CNP and BNP also interact with the ANP-C type receptor and are coupled to adenylyl cyclase inhibition[15].

References

1. Bovy, P.R. (1990). *Med. Res. Rev.*, **10**, 115–42.
2. Brown, J. and Czarnckl, A. (1990). *Am. J. Physiol.*, **258**, R57–R63.
3. Brown, J. and Czarneckl, A. (1990). *Am. J. Physiol.*, **258**, R1078–R1083.
4. Chang, M., Lowe, D.G., Lewis, M. *et al.* (1989). *Nature (Lond.)*, **341**, 68–72.
5. Chinkers, M., Garrbers, D.L., Chang, M.S. *et al.* (1989). *Nature (Lond.)*, **338**, 78–83.
6. Lowe, O.G., Chang, M.-S., Hellmiss, R. *et al.* (1989). *EMBO J.*, **8**, 1377–84.
7. Fuller, F.J.G., Porter, J.G., Arfsten, A.E. *et al.* (1988). *J. Biol. Chem.*, **263**, 9395–401.
8. Itoh, H. (1991). *Biomedica*, **66**, 1262–70.
9. Koller, K.J. (1991). *Science*, **252**, 120–3.
10. Kuno, T. (1986). *J. Biol. Chem.*, **261**, 5817–23.
11. Paul, A.K., Marala, AR.B., Jaiswal, R.K., and Sharma, R.K. (1987). *Science*, **235**, 1224–6.
12. Saheki, T., Mizuno, T., Iwata, T. *et al.* (1991). *J. Biol. Chem.*, **266**, 11122–5.
13. Takayanagi, R., Inagami, T., Snajdar, R.M. *et al.* (1986). *J. Biol. Chem.*, **262**, 12104–13.
14. Wildey, G.M. and Glembotski, C.C. (1986). *J. Neurosci.*, **6**, 3767–76.
15. Marcil, J. and Anand-Srivastava, M.B. (1995). *Mol. Cell Biochem*, **149**, 223–31.

Atrial natriuretic peptide receptor (binding sites)

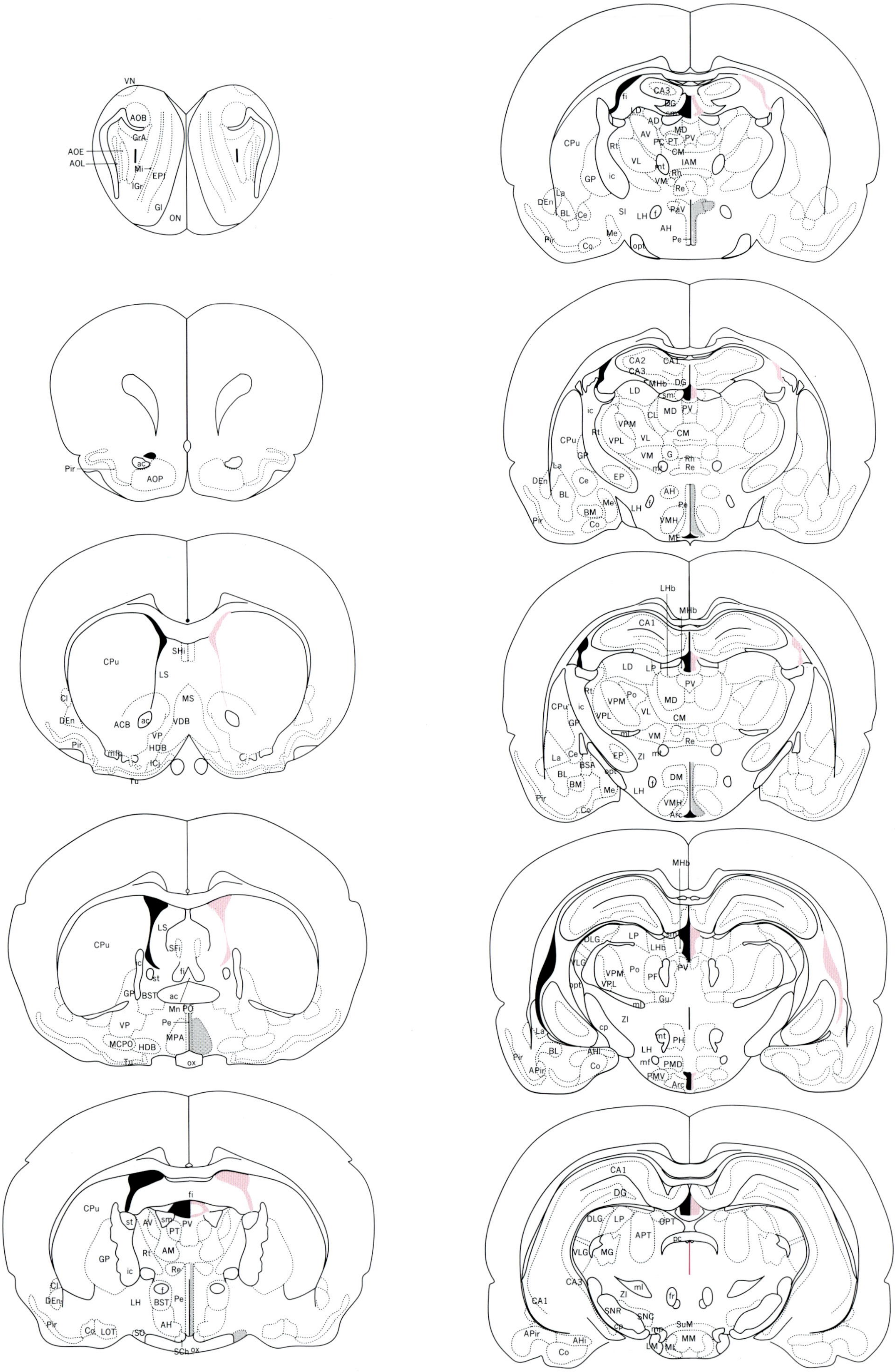

Atrial natriuretic peptide receptor (binding sites)

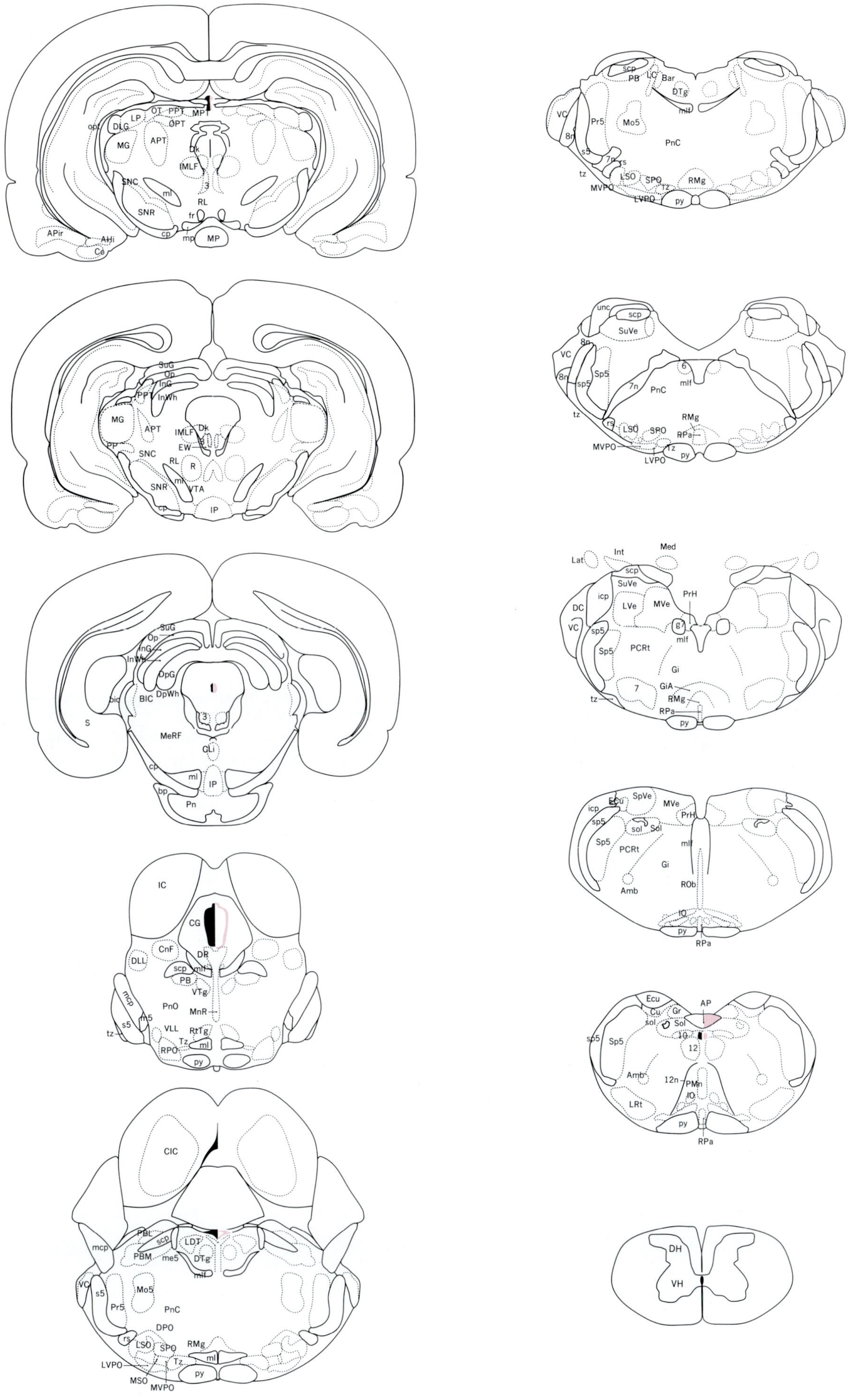

29 The vasoactive intestinal peptide system

Vasoactive intestinal peptide

Structure

Vasoactive intestinal peptide (VIP) was first isolated and purified from the porcine duodenum by Said and Mutt in 1970[9]. It was later identified in the central nervous system as well. VIP is made up of 28 amino acid residues, and the amino acid sequence shows structural homologies to that of secretin (secreted from the digestive tract) and glucagon (secreted from pancreas). The peptide histidine isoleucine (PHI), which originates from the same precursor as VIP, has histidine at the N-terminal and isoleucine at the C-terminal. These peptides show structural homology among different species (Fig. 29.1).

Distribution and physiological function

In the central nervous system, VIP immunoreactivity has been observed in the olfactory bulb, cerebral cortex, hippocampus, amygdaloid complex, hypothalamus, central grey matter, and spinal cord[1–3,5,8,10]. In the cerebral cortex, VIP-positive cells are seen concentrated in layers II–IV; morphologically, they are bipolar and spindle types. The processes from these cell bodies lengthen vertically to deep layers at one end and extend even up to the leptomeninges of the cerebral surface at the other end. Acetylcholine is co-localized in the VIP-positive cells that exist in the cerebral cortex. These VIP-positive cells, rather than being projection-type neurons, play an important role in functioning as interneurons in the cortex. Since injection of VIP into the cerebral cortex, hippocampus, and spinal cord causes excitation of the neurons, it may be inferred that VIP possibly functions as a local excitatory neurotransmitter[7]. In addition, since VIP-positive fibres are identified around the leptomeninges of the cerebral cortex, it is also believed that VIP may be involved in the local regulation of blood flow. Also, VIP accelerates glucose utilization in the cerebral cortex. Thus, VIP is widely perceived to participate in the brain energy metabolism[4,6].

In the hypothalamus, VIP-positive cells exist in the suprachiasmatic nucleus. VIP-positive fibres from this nucleus are known to project dorsally to the paraventricularis hypothalamic nucleus and periventricular nucleus. This tract is thought to participate in the circadian rhythm initiated by the suprachiasmatic nucleus. One study has demonstrated that the mRNA and peptides VIP, PHI, and gastrin-releasing peptide (GRP) coexist in a cell of the suprachiasmatic nucleus. This nucleus plays a central role in maintaining the circadian rhythm. Simultaneous injection of all three co-existing peptides (VIP, PHI, and GRP) results in phase delay of circadian rhythm. This effect is more prominent than when peptides are injected individually. The VIP and PHI contents in the suprachiasmatic nucleus show variation according to the light–dark cycle. The VIP and PHI contents, decrease during the light hours and increase during hours of darkness. The mRNA levels of VIP and PHI also show similar variation. There are two visual tracts that transfer photic information to the suprachiasmatic nucleus. Neurotransmitters present in these tracts are glutamic acid, substance P (SP), and neuropeptide Y (NPY). The fibres containing neurotransmitter project into the suprachiasmatic nucleus and synapse leading to cells containing VIP, PHI, and GRP. Injection of NPY into the suprachiasmatic nucleus also results in phase advance and phase delay of circadian rhythm. Thus, the function of VIP and PHI in the suprachiasmatic nucleus has been postulated to be correlated strongly with the oscillation of circadian rhythm. The functional relationship of VIP and PHI to other neurotransmitters needs additional study. VIP-positive fibres exist in the preoptic area and paraventricular hypothalamic nucleus in the hypothalamus. In this region, VIP is believed to regulate the hormone production of the anterior lobe of the hypophysis. In the hypophysis, VIP accelerates the secretion of prolactin, growth hormone, and luteinizing hormone, and inhibits the secretion of somatostatin.

Many VIP-positive cells exist in the amygdaloid complex also, and these form the principal amygdaloid complex–hypothalamus tract. This projection tract passes through stria terminalis and terminates at the bed nucleus of stria terminalis, medial preoptic area, and anterior nuclei of hypothalamus.

VIP-positive cells have been identified in the hippocampus. These are numerous in the CA1 region, and their numbers decrease in the following order: CA2 > CA3 > dentate gyrus. The VIP-positive cells in the hippocampus are also interneurons, similar to those of the cerebral cortex and those involved in excitability.

Apart from the above-mentioned location, VIP-positive cells and fibres are relatively less common, although they exist in the corpus striatum, thalamus, superior colliculus, and central grey matter of the mesencephalon. In the spinal cord VIP-positive cells are also seen, and they project into the upper brain. In a normal state, the number of VIP-positive cells in the dorsal root

Porcine VIP	His-Ser-Asp-Ala-Val-Phe-Thr-Asp-Asn-Tyr-Thr-Arg-Leu-Arg-Lys-Gln-Met-Ala-Val-Lys-Lys-Tyr-Leu-Asn-Ser-Ile-Leu-Asn-NH_2
Porcine PHI	His-Ala-Asp-Gly-Val-Phe-Thr-Ser-Asp-Phe-Ser-Arg-Leu-Leu-Gly-Gln-Leu-Ser-Ala-Lys-Lys-Tyr-Leu-Glu-Ser-Leu-Ile-NH_2
Human PHI	His-Ala-Asp-Gly-Val-Phe-Thr-Ser-Asp-Phe-Ser-Lys-Leu-Leu-Gly-Gln-Leu-Ser-Ala-Lys-Lys-Tyr-Leu-Glu-Ser-Leu-Met-NH_2

Fig. 29.1 Amino acid sequences of porcine vasoactive intestinal peptide (VIP) and peptide histidine isoleucine (PHI) and human PHI.

ganglion is less. However, it has been confirmed that the number of VIP-positive cells increases after excision of the peripheral branch of the ganglion, using a tagged label of protein and the mRNA. VIP is also implicated in the acceleration of glycogen degradation. Increase of VIP content in the ganglion after peripheral nerve excision also suggests an important role for VIP in the nerve regenerative process. In layers I–II of the dorsal column of the spinal cord, VIP-positive fibres exist densely, and these are supplied from the dorsal root ganglion.

In the peripheral nervous system, VIP-positive cells and fibres exist in the autonomic nervous system, Meissner's plexus, and Auerbach's plexus. VIP-positive cells and fibres which exist in the digestive tract, respirative tract, urogenital organs, and exocrine glands are involved in nerve regulation to smooth muscle, blood vessel, and secretor cells, and influence smooth muscle relaxation, vasodilation, and acceleration of secretory organs. In the postganglionic fibres that control sweat glands and salivary glands, VIP and acetylcholine coexist and these coordinate vasodilation and acceleration of secretions from these glands.

Helospectin (HS) and the two forms of pituitary adenylate cyclase-activating polypeptide, (PACAP 27, and PACAP 38) are newly discovered peptides isolated from the salivary gland venom of the lizard *Heloderma horridum* and the ovine hypothalamus, respectively. They show chemical similarities to VIP, appear to have similar functions, and are present in the brain.

References

1. Abrams, G.M., Nilaver, G., and Zimmerman, E.A. (1985). *Handbook of chemical neuroanatomy*, Vol. 4, *GABA and neuropeptides in the CNS*, Part 1, pp. 335–54. (ed. A. Björklund and T. Hökfelt).
2. Gall, C., Seroogy, K.B., and Brecha, N. (1986). *Brain Res.*, **374**, 389–94.
3. Gibson, S.I., Polak, J.M., Bloom, S.R. and Wall, P.D. (1981). *J. Comp. Neurol.*, **201**, 65–79.
4. Heistad, D.D., Marcus, M.L., Said, S.I., and Gross, P.M. (1980). *Am. J. Physiol.*, **239**, H73–H80.
5. Loren, I., Emson, P.C., Fahrenkrug, J. *et al.* (1979) *Neuroscience*, **4**, 1953–76.
6. Magistretti, P.J., Morrison, J.H., Shoemaker, W.J. *et al.* (1981) *Proc. Natl Acad. Sci. USA*, **78**, 6536–9.
7. Phillis, J., Kirkpatrick, J.R., and Said, S.I. (1978). *J. Physiol. Pharmacol.*, **56**, 337–40.
8. Roberts, G.W., Woodhams, P.L., Bryant, M.G. *et al.* (1980) *Histochemistry*, **65**, 103–19.
9. Said, S.I. and Mutt, A. (1970). *Science*, **169**, 1217–18.
10. Sims, K.B., Hoffmann, D.L., Said, S.I., and Zimmerman, E.A. (1980). *Brain Res.*, **186**, 165–83.

Vasoactive intestinal peptide receptor

Subtypes

Experiments in synaptosome preparations have shown that two binding sites exist for VIP receptors; one with a high affinity and a small dissociation constant, and the other with a low affinity and a large dissociation constant. The molecular weight of VIP binding protein has been reported, and the molecular weight of protein at the high-affinity binding site is 46–73 kDa. The molecular weight of the low-affinity binding site is 30–117 kDa[7]. G protein is present in this binding site.

Receptor and second messenger

Experiments in synaptosomal preparations and hypothalamus slices from rat brain have provided evidence that VIP release is strongly influenced by increased K^+ concentration, and this release is Ca^{2+} dependent[4,5]. VIP binds to its receptor and links with G proteins, which activate adenylic cyclase and which leads to an increase in intracellular cAMP[2].

Physiological function

VIP coexists with various neuroactive substances at the same interneurons. This is especially so in the post ganglionic neurons of the autonomic nervous system and cerebral cortex, where VIP coexists with acetylcholine and enhances its actions. In the salivary gland, VIP enhances sialorrhea of acetylcholine that is mediated by the muscarinic acetylcholine receptor[6]. In the cerebral cortex, acetylcholine enhances the inositol phosphoric acid metabolism[10]. These facts suggest that there is a post-synaptic link between the acetylcholine receptor and the VIP receptor. In the cerebral cortex, it has been shown that VIP enhances the increase of intracellular cAMP more strongly than noradrenaline[8]. This suggests that noradrenaline- and VIP-containing nerve endings terminate at same cortex neuron, and this may have postsynaptic functional implications. Thus, there are many examples where the expression and function of VIP receptors depend on connection with other neuroactive substances. An analysis of the VIP activation system should therefore take into consideration the interlinks with other neurotransmission systems, as well as the structural studies of the receptor.

Vasoactive intestinal peptide (immunocytochemistry)

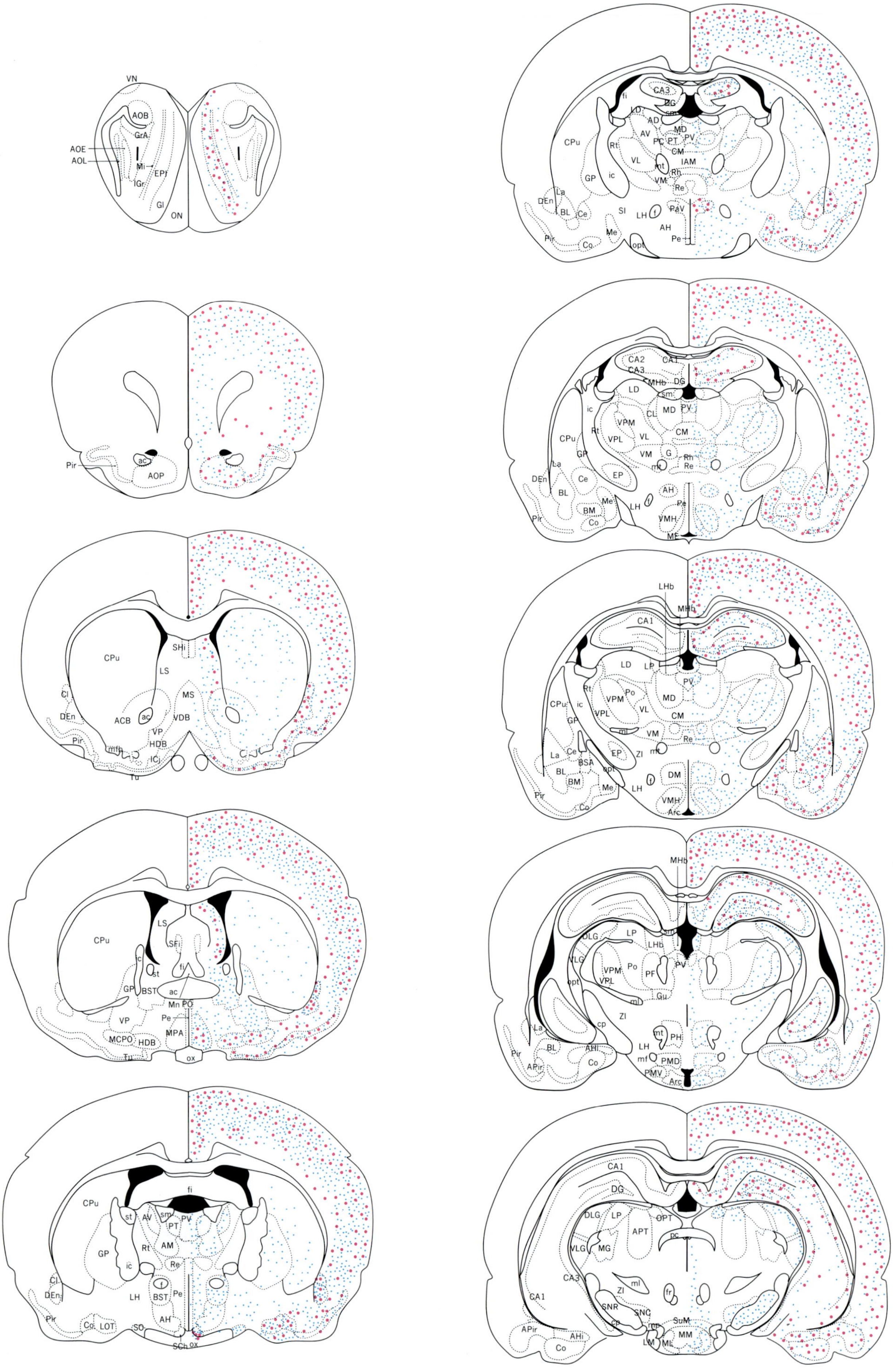

Vasoactive intestinal peptide (immunocytochemistry)

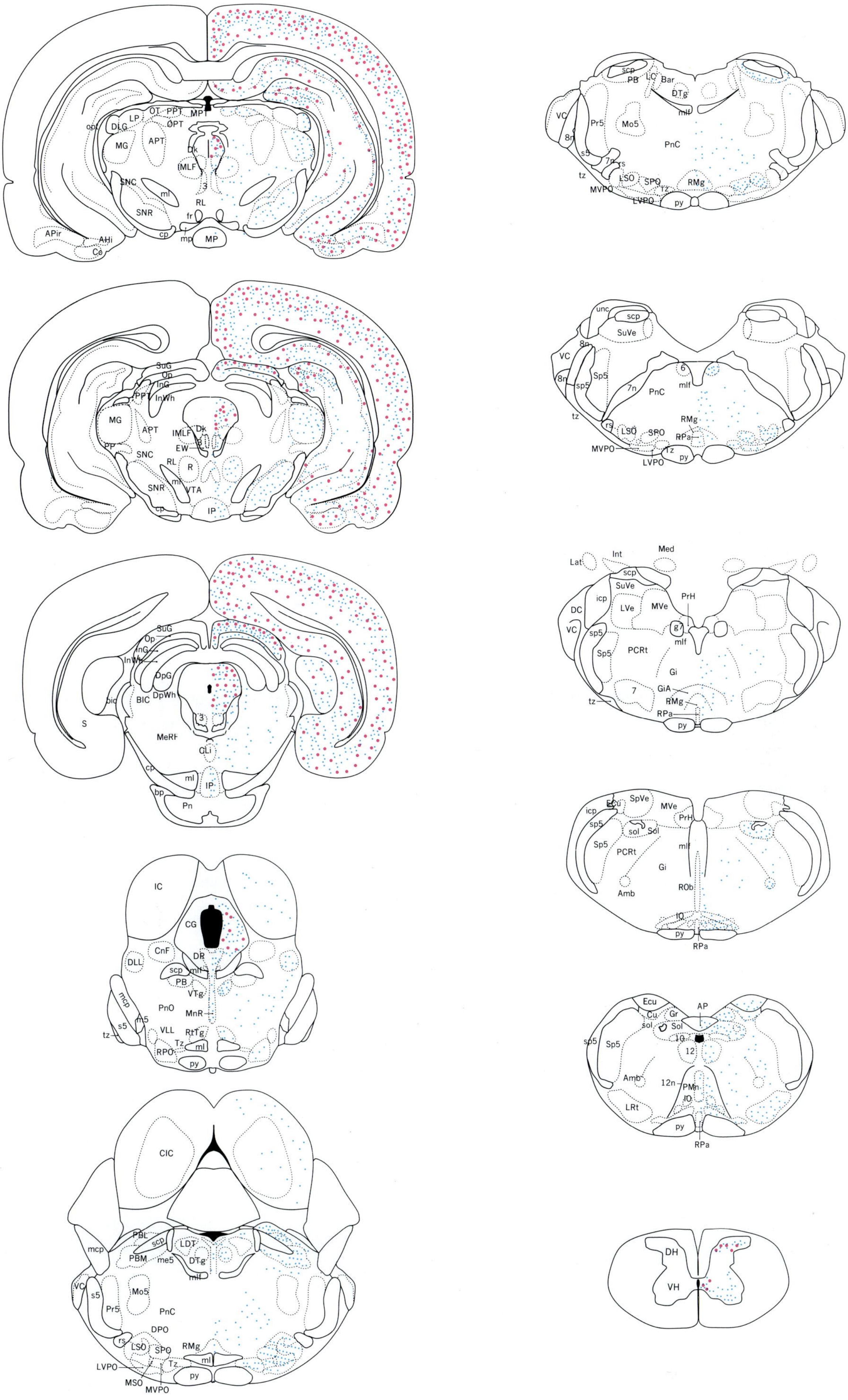

Vasoactive intestinal peptide receptor

(Continued from p. 311)

Distribution

The structure of the VIP receptor has not yet been clarified, and no specific ligand to the VIP receptor has yet been discovered. Studies on the brain distribution of VIP receptor have been carried out by labelling the binding sites with radioactive (mono [^{125}I]-iodo-Tyr^{10}, Met-O^{17}) VIP as a marker[1,3,9,11]. The distribution of binding sites in rat is not uniform. Regions with a high density of binding sites include the cerebral cortex, corpus striatum, preoptic area of hypothalamus, suprachiasmatic nucleus, supraoptic hypothalamic nucleus, hippocampus, dentate gyrus, lateral thalamus, superior colliculus, medial geniculate body, superior olivar nucleus, parapeduncular nucleus, locus ceruleus, and cochlear nuclei. Regions with moderate density binding sites are hippocampus CA1, CA2, and CA3, cingulate gyrus, and cerebellar cortex. Binding sites of low density are observed in regions of the hypothalamus, pons, and reticular formation of medulla oblongata. Among these, in the dorsomdial part of the corpus striatum, dorsal cochlear nuclei, and cerebellar cortex, VIP-positive fibres are extremely rare or do not exist, suggesting that these are sites of typical mismatch. Doubts therefore remain whether the binding sites identified by the use of radioactive ligand indicate the practical action sites or functional sites.

The VIP receptor was isolated from cDNA library of rat lung by Ishihara *et al.* (*Neuron*, **8**, 811, 1992) and has been cloned to determine its amino acid sequence. The rat VIP receptor consists of 459 amino acid residues and has a molecular wight of approximately 52 kDa (Fig. 29.2). This receptor has seven transmembrane domains and has 48 per cent homology with the secretin receptor amino acid sequence. The VIP receptor is structurally related to the receptors of secretin, calcitonin, and parathyroid hormone, and these form a new subfamily of G-protein binding protein receptors. The mRNA of VIP receptor is observed strongly in lung, liver, intestine, and brain, but is not found in the atrium, kidney, pancreas, stomach, and adrenal gland. In the brain, gene expression of the VIP receptor is strongly observed in the mitral cell layer of olfactory bulb, cerebral cortex, and hippocampus. Gene expression of the VIP receptor is weak in the granular cell layer of cerebellum, medulla oblongata, pons, superior colliculus, thalamus, hypothalamus, corpus striatum, and septal area. The localized sites of mRNA of these VIP receptors match the localization data for the VIP receptor, obtained by radioactive marker studies.

The VIP$_2$ receptor was identified by PCR of rat pituitary cDNA, and a full-length clone was isolated from a rat olfactory bulb cDNA library. The VIP$_2$ receptor mRNA was present mainly in the thalamus , hippocampus, and in the suprachiasmatic nucleau (Lutz *et al.*, *FEBS Lett.*, **334**, 8, 1993).

Two forms of pituitary adenylate cyclase-activating polypeptide, (PACAP 27 and PACAP 38) are members of the vasoactive intestinal peptide (VIP)/secretin/glucagon family of peptides. PACAP receptors that are positively coupled to adenylate cyclase and phospholipase C have been recently identified. Autoradiography revealed that [^{125}I]PACAP binding sites were dense in the piriform cortex, diagonal band, accumbens nucleus, anterior part of the striatum, hippocampal formation, habenular nucleus, lateral hypothalamic area, superior colliculus, and dorsal raphe nucleus. Moderate to high labelling was observed in the medial septal nucleus, olfactory tubercle, caudal part of the striatum, most parts of the thalamus, supraoptic and substantia nigra pars compacta, locus coeruleus, pontine reticular nucleus, and cerebellum (Masuo Y., *et al.*, *Brain Res.*, **575**, 113, 1992). The expression of PACAP receptors was found in the rat cortex, hippocampus, cerebellum, and hypothalamus during postnatal development (D'Agate V. *et al.*, *Elit. J. Neurosci.*, **8**, 310, 1996).

References

1. Besson, J., Sarrieau, A., Vial, M., Marie, J.-C. *et al.* (1986) *Brain Res.*, **398**, 329–36.
2. Deschodt-Lanckman, M., Robberecht, P., and Christophe, J. (1977). *FEBS Lett.*, **83**, 76–80.
3. Dietl, M.M., Hof, P.R., Martin, J.-L. *et al.* (1990). *Brain Res.*, **520**, 14–26.
4. Emson, P.C., Fahrenkrug, J., Schaffalitzky de Muckadel, O.B. *et al.* (1978). *Brain Res.*, **143**, 174–8.
5. Giachetti, A., Said, S.I., Reynolds, R.D., and Koniges, F.C. (1977). *Proc. Natl Acad. Sci. USA*, **74**, 3424–8.
6. Laburthe, M. and Couvineau, A. (1988). *Ann. N.Y. Acad. Sci.*, **627**, 296–313.
7. Lundberg, J.M., Hedlund, B., and Bartfai, T. (1982). *Nature*, **295**, 147–8.
8. Magistretti, P.J. and Shorderet, M. (1984). *Nature (Lond.)*, **308**, 280–2.
9. Martin, J.L., Dietl, M.M., Hof, R.P. *et al.* (1987). *Neuroscience*, **23**, 539–65.
10. Raiteri, M., Marchi, M., and Paudice, P. (1987). *Eur. J. Pharmacol.*, **133**, 127–8.
11. Staum-Olsen, P., Ottersen, B., Gammeltofts, S., and Fahrenkrug, J. (1985). *Brain Res.*, **330**, 317–21.

```
VIP receptor (1–49)                                              ASPQHECEYLQLIEIQRQQCLEE----------------AQLENET-----------TGCCKMWDNLTCWPTTPRG
Secretin receptor (1–60)                                     AHTVGVPPRLCDVRRVLLEERAHCLQQL----------SKEKKGALGPET----------ASGCEGLWDNMSCWPSSAPA
Parathyroid hormone receptor (1–95)  DADDVITKEEQIILLRNAQAQCEQRLKEVLRVPEAESAKDWMSRSAKTKKEKPAEKLYPQAEESREVSDRSRLQDGFCLPEWDNIVCWPAGVPG
Calcitonin receptor (1–60)                                 AHTPTLEPEPFLYILGKQMLEAQHRCYDRM------------QKLPPYQGE-----------GLYCNRTWDGWSCWDDTPAG

VIP receptor (50–140)                QAVVLDCPLIFQLFAPIHGYNISRSCTEEG-WSQLEPGPYHIACGLNDRASSLDEQQQTKFYNTVKTG--YTIGYSLSLASLLVAMAILSLFRK
Secretin receptor (61–147)           RTVEVQCPKFLLMLSNKNG-SLFRNCTQDG-WSETFPRP-DLACGVN-INNSFNERRHA-YLLKLKVM--YTVGYSSSLAMLLVALSILCSFRR
Parathyroid hormone receptor (96–185) KVVAVPCPDYFYDFNHK-G-RAYRRCDSNGSWEL-VPGN-NRTWANYSECVKFLTNETREREVFDRLGMIYTVGYSISLGSLTVAVLILGYFRR
Calcitonin receptor (61–147)         VLAEQYCPDYFPDFDAA-E-KVTKYCGEDGDWYR-HPES-NISWSNYTMCNAFTPDKLQNAYILYYLAI---VGHSLSILTLLISLGIFMFLRS

VIP receptor (141–217)                LHCTRNYIHMHLFMSFILRATAVFIKDMALFNSGEIDHC-----------------SEASVGCKAAVVFFQYCVMANFFWLLVEGLYLYTLLAV
Secretin receptor (148–224)           LHCTRNYIHMHLFVSFILRALSNFIKDAVLFSSDDVTYC-----------------DAHKVGCKLVMIFFQYCIMANYAWLLVEGLYLHTLLAI
Parathyroid hormone receptor (186–280) LHCTRNYIHMHLFVSFMLRAVSIFIKDAVLYSGVSTDEIERITEEELRAFTEPPPADKAGVGCRVAVTVFLYFLTTNYYWILVEGLYLHSLIFM
Calcitonin receptor (148–222)         ISCQRVTLHKNMFLTYVLNSIIIIVHLVVI---VPNGELVK----------------RDPPICKVLHFFHQYMMSCNYFWMLCEGVYLHTLIVV

VIP receptor (218–310)                SFFSERKYFWGYILIGWGVPSVFITIWTVVRIYFEDFGCWDTIINSSLWWIIKAPILLSILVNFVLFICIIRILVQKLRPPDIGKNDS-SPYSR
Secretin receptor (225–317)           SFFSERKYLQAFVLLGWGSPAIFVALWAITRHFLENTGCWDINANASVWWVIRGPVILSILINFIFFINILRILMRKLRTQETRGSET-NHYKR
Parathyroid hormone receptor (281–373) AFFSEKKYLWGFTLFGWGLPAVFVAVWVTVRATLANTECWLSSGN-KKWIIQV-PILAAIVVNFILFINIIRVLATKLRETNAGRCDTRQQYRK
Calcitonin receptor (223–312)         SVFAEGQRLWWYHVLGWGFPLIPTTAHAITRAVLFNDNCWLSVDT-NLLYIIHGPVMAALVVNFFFLLNILRVLVKKLKESQ--EAES-HMYLK

VIP receptor (311–387)                LAKSTLLLIPLFGIHYVMFAFFPDNFKAQVKMVFELVVGS-----FQGFVVAILYCFLNGEVQAELRRKWRRWHLQGVLGWS
Secretin receptor (318–393)           LAKSTLLLIPLFGIHYIVFAFSPED-AMEVQLFFELALGS-----FQGLVVAVLYCFLNGEVQLEVQKKWRQWHLQEFPLRP
Parathyroid hormone receptor (374–454) LLKSTLVLMPLFGVHYIVFMATPYTE-VSGILWQVQMHYEMLFNSFQGFFVAIIYCFCNGEVQAEIKKSWSRWTLALDFKRK
Calcitonin receptor (313–386)         AVRATLILVPLLGVQFVVLPWRPST--PLLGKIYDYVVHS--LIHFQGFFVAIIYCFCNHEVQGALKRQWNQY----QAQRW
```

Fig. 29.2 Amino acid sequences of rat VIP receptor, and secretin, parathyroid hormone, and calcitonin receptors (Ishihara *et al*., *Neuron*., **8**, 811, 1992).

Vasoactive intestinal peptide receptor (binding sites)

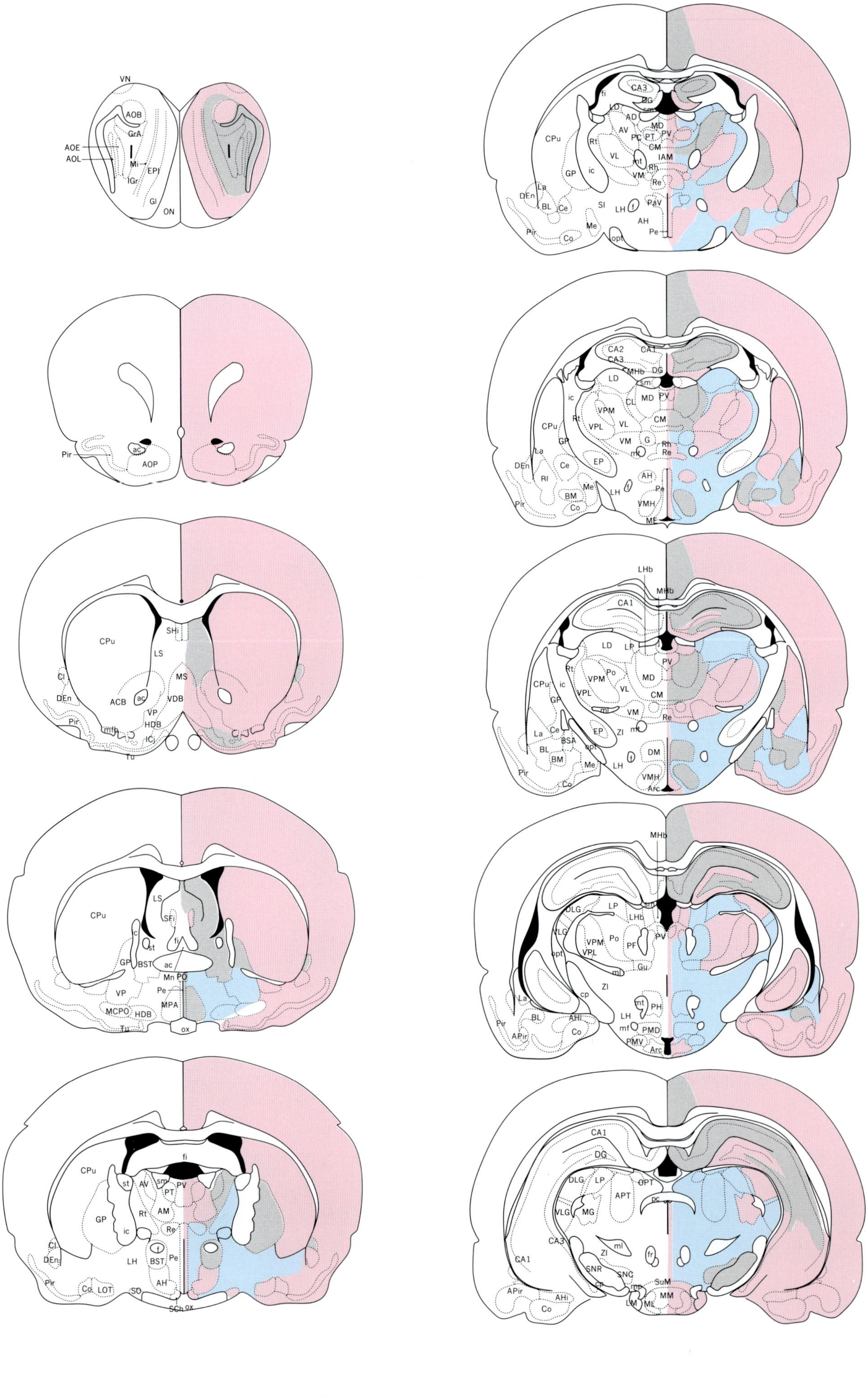

Vasoactive intestinal peptide receptor (binding sites)

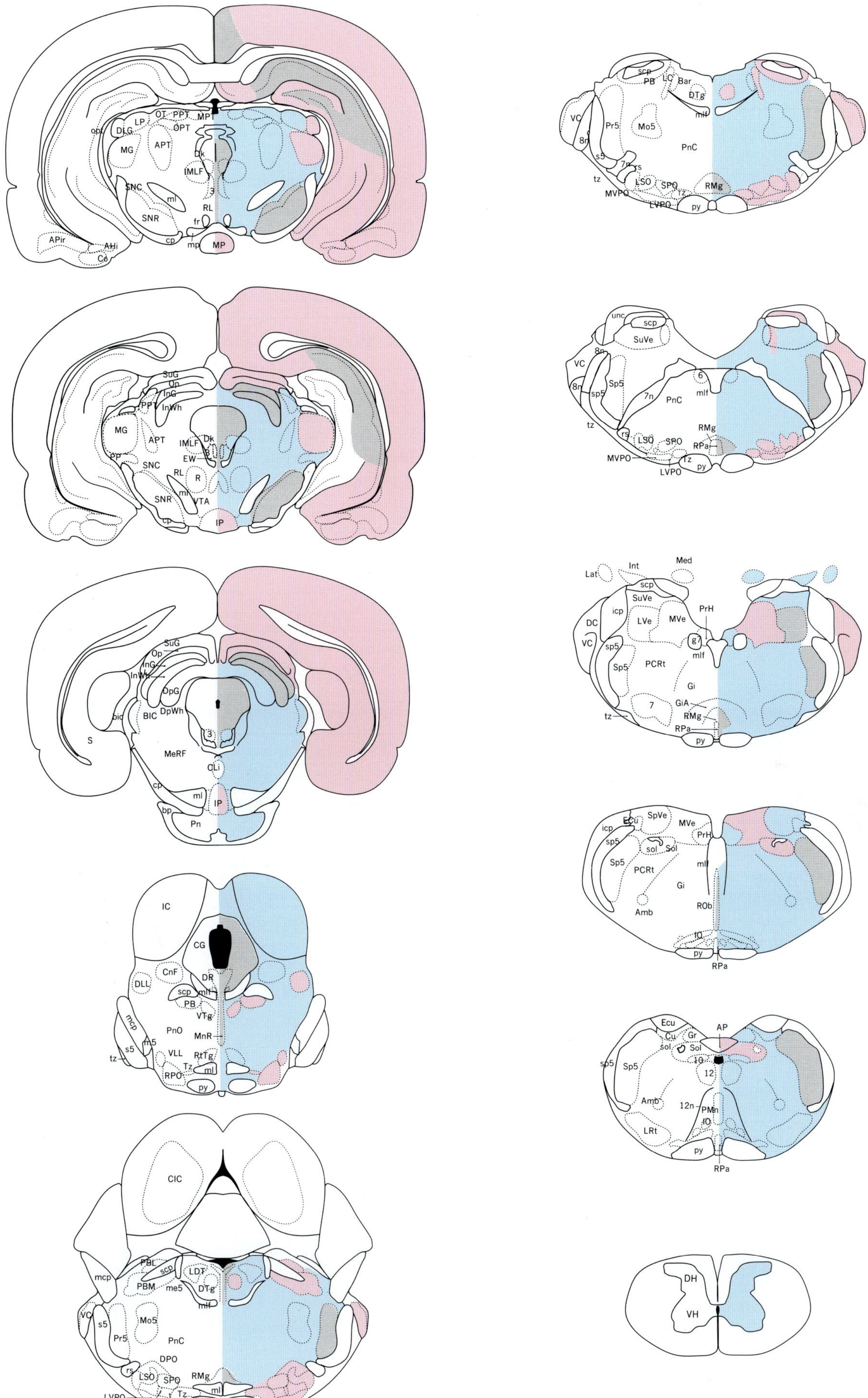

30 The argininosuccinate system

Argininosuccinate

Biosynthesis

L-Argininosuccinate was originally thought to be one of the intermediate metabolites of the arginine biosynthesis system. It is synthesized by the condensation of citrulline and aspartate. This condensation is catalysed by argininosuccinate synthetase enzyme (EC 6.3.4.5):

$$\text{citrulline} + \text{asparte} \xrightarrow{\text{argininosuccinate synthetase}} \text{argininosuccinate}$$

The structural formula of argininosuccinate is shown in Fig. 30.1.

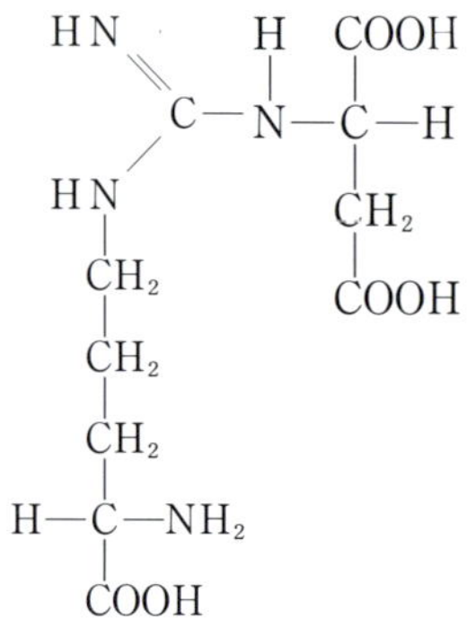

Fig. 30.1 Chemical structure of argininosuccinate.

Distribution

In considering the functions of argininosuccinate, the distribution *in vivo* could be an index, but because the quantities present are minuscule, determination by amino acid analysis is somewhat difficult (references cited in Table 30.1). For this reason, the distribution of argininosuccinate synthetase in the organs has become an index. The liver and kidney are the two organs in which argininosuccinate synthetase is present in high concentration. In the kidney, the argininosuccinate that is made from citrulline by argininosuccinate synthetase is cleared to form arginine by argininosuccinate lyase (EC 4.3.2.1):

$$\text{argininosuccinate} \xrightarrow{\text{argininosuccinate lyase}} \text{arginine} + \text{fumarate}$$

Arginine that is synthesized in the kidney is transported by the blood stream to other organs, and therefore the kidney is considered to be the major depot for arginine supply.

Meanwhile, in the liver, arginine is hydrolysed again to urea and ornithine by the action of arginase, thus participating in the formation of urea:

$$\text{arginine} \xrightarrow{\text{arginase}} \text{urea} + \text{ornithine}$$

Physiological function

As shown in Table 30.2, the enzyme activity of argininosuccinate synthetase in the brain is low: so much so that it went unnoticed until recently. However, from recent investigations using the immunohistochemical techniques, it has become clear that this enzyme exists in specific nerve cell groups in the central nervous system of rat and has been suggested to participate in the synthesis of neurotransmitters or neuromodulators[3].

By Fura-2 microfluorometric studies for measuring intracellular free calcium concentration, Nakamura *et al.*[4] demonstrated that argininosuccinate may function as a transmitter or modulator in the brain. Nakamura *et al.* found out that argininosuccinate reduced the increase in intracellular free calcium produced by glutamate in isolated cerebellar neurons of immature rats (Fig. 30.2). It was also reported that argininosuccinate reduced the response to quisqualate, but not the response to *N*-methyl-D-aspartate (NMDA) or kainate. However, L-argininosuccinate alone did not cause any remarkable physiological change. From these observations, it is sug-

Table 30.1. Rat liver levels of aspartate, argininosuccinate, and arginine.

	μmol/g wet weight
Aspartate	0.3–3.5
Argininosuccinate	0.034
Arginine	0.02–0.1

Table 30.2. Levels of argininosuccinate synthetase enzyme activity reported in some rat organs.

	enzyme activity (nmole/min. per mg protein)
Liver	82.5 ± 16.6
Kidney	29.7 ± 1.7
Brain	0.174 ± 0.015

From ref. 5.

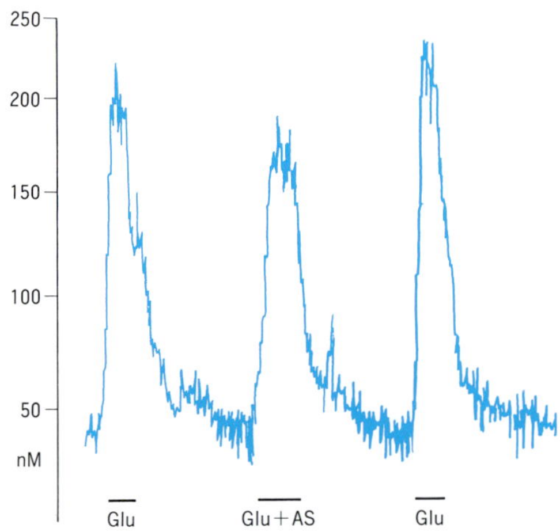

Fig. 30.2 The effects of L-argininosuccinate (AS) on the amplitude of the increase in intracellular-free calcium concentration produced by L-glutamate (Glu) using Fura-2 microfluorometry (from ref. 4).

gested that the argininosuccinate may have a physiological action of making a transmitter modification in the specific neurons of the central nervous system. Differences in this modifying action have been observed in varying sites and types of cells, leading to the postulation that a variety of actions may be plausible.

Furthermore, argininosuccinate is degraded to arginine by argininosuccinate lyase. In the central nervous system, neurons may have this enzyme at the existing sites of argininosuccinate synthetase enzyme, as reported by Nakamura *et al.* In addition, cloning of argininosuccinate lyase distributed in the central nervous system of rat has also been carried out[2]. From these observations, some kinds of nerve cells have been thought to synthesize arginine. It has been reported that when the arginine supply from the kidney is stopped and the arginine synthesis in various organs of rat is determined, there will be less in the brain; but a certain concentration can be determined, as compared to those organs which are not considered to have arginine synthesizing ability. Meanwhile, since there is an experimental report which shows arginine's modifying action in inhibiting the effect of acetylcholine in the peripheral nervous system where argininosuccinate synthetase enzyme exists, the possibility that arginine itself can act as a neuromodulator in the central nervous system cannot be denied.

In addition, it has already been demonstrated that, in certain discrete neuronal populations, arginine produces nitric oxide by the action of nitric oxide synthetase (EC1.14.23)[1]:

$$\text{arginine} + \text{oxygen} \xrightarrow{\text{nitric oxide synthetase}} \text{nitric oxide} + \text{citrulline}$$

Nitric oxide is different from other neurotransmitters in that it is used immediately after its synthesis, without being stored. Nitric oxide could diffuse to the adjoining nerve cells and activate guanyl cyclase of the cell. The distribution of nitric oxide synthetase (NOS) in the rat brain has been reported by Bredt *et al.*[1]. The sites that are observed to have the highest concentrations are the molecular and granular layers of the cerebellar cortex; Purkinje layer cells have a somewhat lesser concentration. The granular cell layer of the olfactory bulb also has a high concentration of NOS. Other than these sites, the regions that are high in NOS include the superior colliculus and inferior colliculus, dentate gyrus of hippocampus, horizontal part of Broca's diagonal band, islands of Calleja and bed nucleus of stria terminalis. The surface layer of cerebral cortex has a lower concentration of NOS.

Among these sites, those which do not provide a corresponding distribution of argininosuccinate synthetase enzyme include, the granular layer of cerebellar cortex and olfactory bulb. In both of these sites, neurons that have argininosuccinate synthetase are hardly observed, and most parts of the existing sites of neurons with argininosuccinate synthetase do not record the existence of NOS. Thus, it can be concluded that most regions in which argininosuccinate is made by argininosuccinate synthetase are not involved in the direct synthesis of nitric oxide.

Polyclonal antibody to argininosuccinate was developed, and when we used this antibody to map immunohistochemically the relevant regions of interest, satisfactory results were not obtained. However, the existence of argininosuccinate synthetase in brain is a fact and it serves as a marker for the distribution of argininosuccinate as well as arginine.

References

1. Bredt, D., Hwang, P.M., and Snyder, S.H. (1990). *Nature*, **347**, 768–70.
2. Kawamoto, S., Kaneko, T., Mizuki, N. *et al.* (1989). *Mol. Brain Res.*, **5**, 235–41.
3. Nakamura, H., Saheki, T., and Nakagawa, S. (1990). *Brain Res.*, **530**, 108–12.
4. Nakamura, H., Yada, T., Saheki, T. *et al.* (1991). *Brain Res.*, **539**, 312–15.
5. Saheki, T., Imamura, Y., and Nakagawa, S. (1994). *Dementia*, **8**, 133–8.

Argininosuccinate (immunocytochemistry)

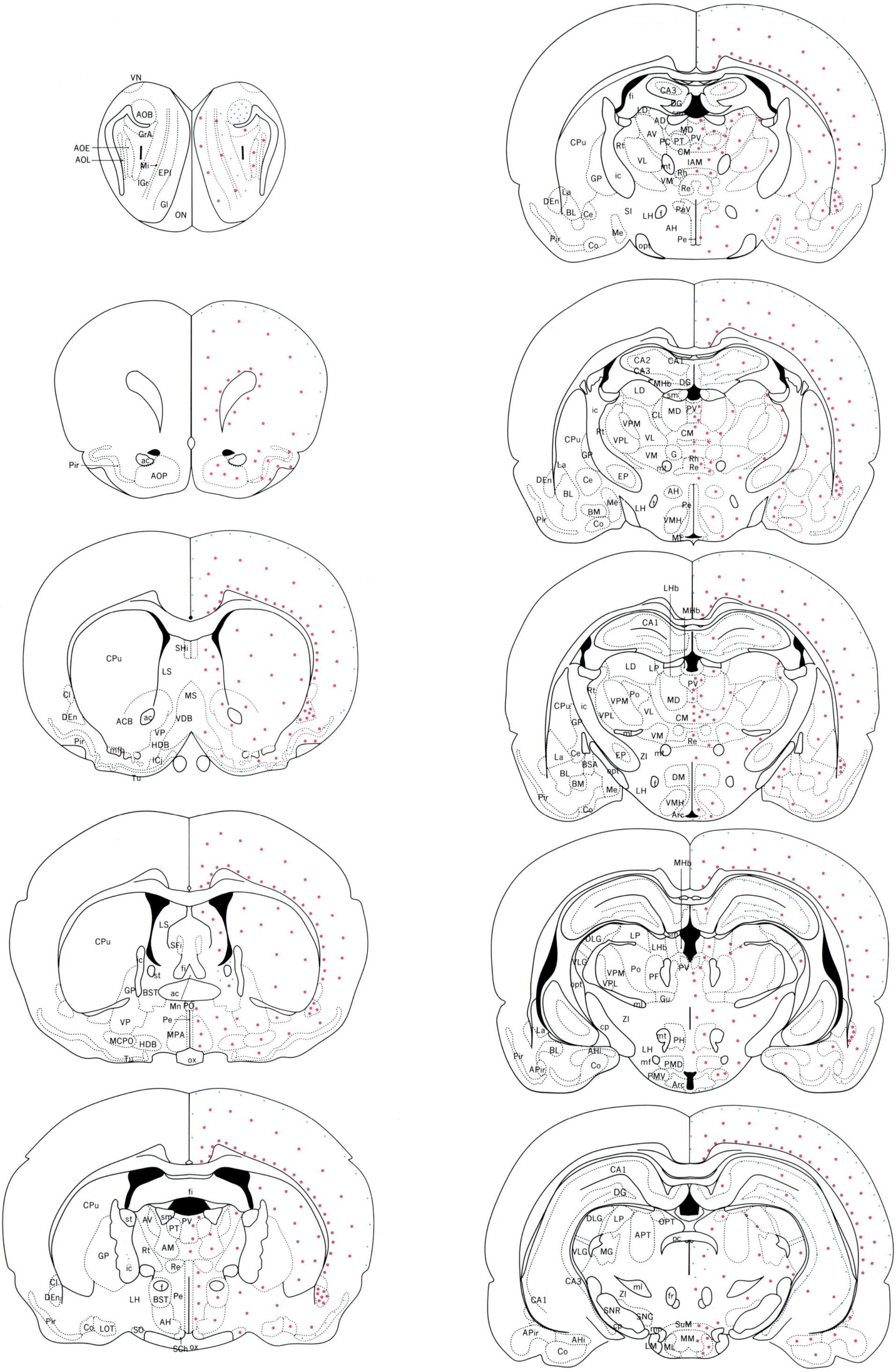

Argininosuccinate (immunocytochemistry)

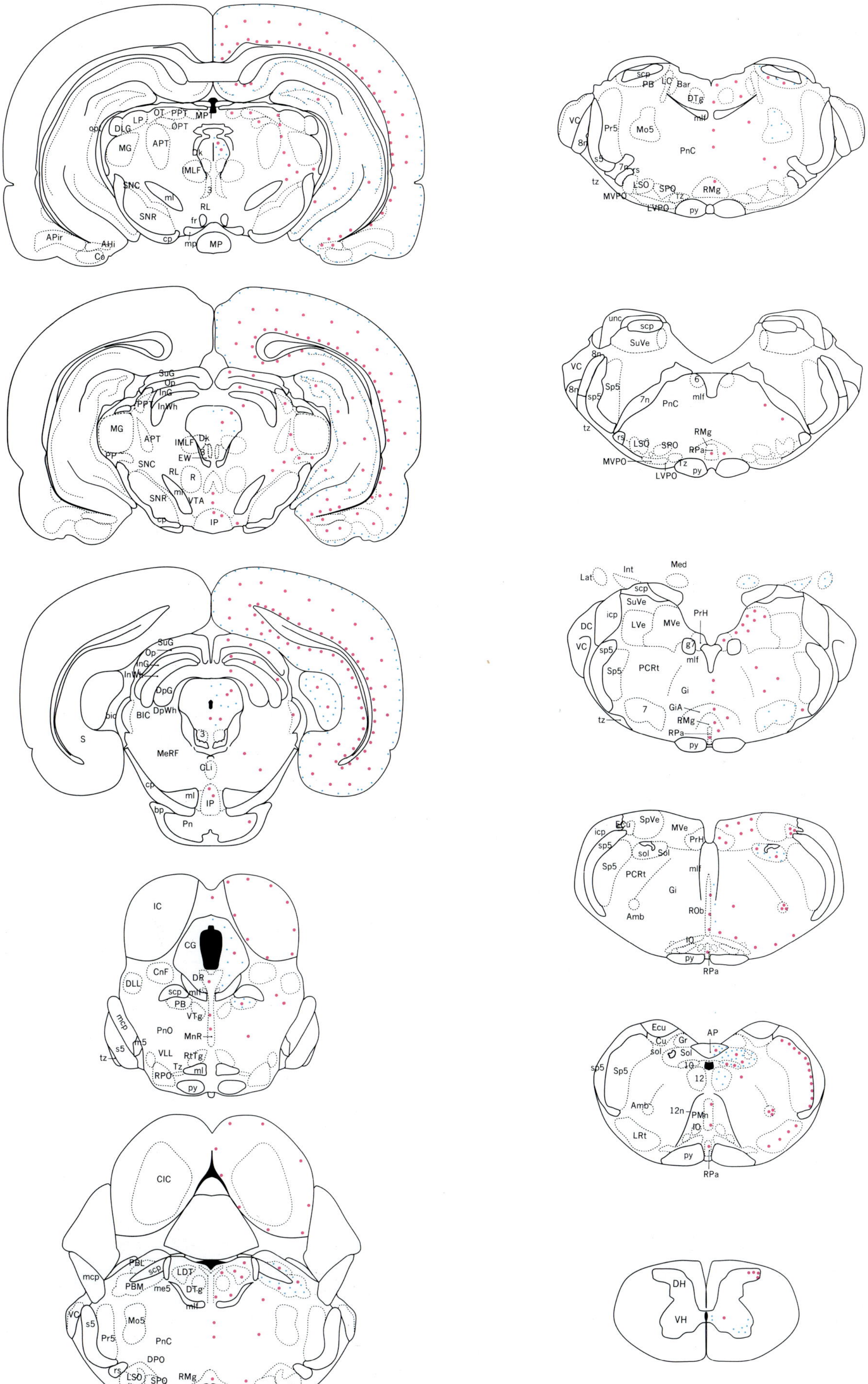

Cerebellum

1 Cerebellofugal neurotransmitters and neuromodulators

It has been known for some time that climbing fibres and mossy fibres exist as input pathways to the cerebellum. The core of the climbing fibres originates in the inferior olivary nucleus, and exists as a mixture from part of the reticular formation or the nuclei pontis. The climbing fibres pass through the inferior cerebellar peduncle to reach the cerebellum and then form a synapse with primary and secondary dendritic shaft of Purkinje cells in the molecular layer (Fig. 1). Climbing fibres cause strong excitation in Purkinje cells. The co-laterals of climbing fibres form a synapse with other cells (such as stellate cells or Golgi cells). Glutamic acid has been identified as a candidate neurotransmitter (neuromodulator) of climbing fibres. In fact, Purkinje cells, the main target cells of climbing fibres, express various glutamate receptors, as shown in Table 1. It is not clear whether all the glutamate receptors expressed in the Purkinje cells are those of glutamic acid from climbing fibres. Apart from glutamic acid, the presence of somatostatin, calcitonin-gene related peptide (CGRP), and enkephalin in the cerebellum have been reported. Simultaneous stimulation of climbing fibres and parallel fibres (axons of granule cells) depresses the synaptic transfer efficiency of parallel fibres for a long period. This phenomenon is known as long-term depression. Nitric oxide, which is produced via a glutamic acid receptor in climbing fibres, is thought to play an important role in this phenomenon.

Mossy fibres originate in the spinal cord (spinocerebellar tracts), pontine nuclei (pontocerebellar tract), vestibular nuclei (mainly inferior nucleus), vestibular nerve, lateral reticular nucleus, paramedian reticular nucleus, and sensory nuclei of trigeminal nerve. Mossy fibres form a swelling in the granule cell layer, referred to as a rosette. At the centre of this rosette, dendrites of granule cells and axons of Golgi cells form the cerebellar glomeruli. Granule cells are the target cells of mossy fibres and Golgi cells. As of now, the neurotransmitter (or neuromodulator) present in mossy fibres has not been clarified. However, the granule cells express GluR 2 and 4 of AMPA receptors, kainate type receptor, NMDA receptor, and various subunits of $GABA_A$ receptor (Table 1). This suggests the possibility that glutamic acid and GABA could be the neurotransmitters (neuromodulators) in the mossy fibres. However, GABA is an inhibitory neurotransmitter (neuromodulator) and the input from mossy fibres to granule cells is excitatory. Thus, GABA could be excluded as a neurotransmitter (neuromodulator) that is present in mossy fibres. In mouse, CGRP and enkephalins have been shown to be present in mossy fibres.

Apart from mossy fibres and climbing fibres, noradrenaline (NA) fibres of locus ceruleus origin or serotonin fibres of dorsal raphe nucleus origin have been suggested as the third input pathway in the cerebellum. The NA fibres pass through the unciform fasciculus to reach the cerebellum. In the cerebellum, these fibres synapse with dendrites of Purkinje cells and those of granular cells at sites other than cerebellar glomeruli. They terminate at deep cerebellar nuclei. The target cell of serotonin in the cerebellum is almost the same as that of noradrenaline. The locus ceruleus of mouse contains CGRP, so the CGRP fibres of locus ceruleus origin are also present in mouse cerebellum. In addition, juvenile rats have substance P-containing fibres temporarily in the cerebellum, though their origin is not clear.

2 Neurons and neurotransmitters/neuromodulators in cerebellar circuits

In the molecular layer of the cerebellar cortex, there are stellate and basket cells, which form a circuit network with dendrites of Purkinje cells and axons of granule cells (known as parallel fibres). The granular layer is densely packed with granule cells, and interstitially cerebellar glomeruli (small clear spaces) are present. Mixed within the granule cells, large cells, identified as Golgi cells, are also located. Among the Golgi cells, those which have short processes are referred to as Golgi II cells, and these have the highest frequency among Golgi cells. Those which protrude dendrites horizontally at the most superior regions of the granular layer are called horizontal cells, and are distinguished from the other Golgi cells. The existence of cells known as Bergmann glia should also be noted.

The neuronal cell layer where the Purkinje cells are aligned is referred to as the Purkinje cell layer.

The granule cells that receive input from mossy fibres lengthen the axon to the molecular layer and bifurcate to a T-shaped type at various heights. The fibres that bifurcate to T-shaped types are called parallel fibres, and these synapse with dendrites of Purkinje cells, Golgi cells, stellate cells, and basket cells. A major role of granule cells is in the transfer of excitatory input received from mossy fibres to Purkinje cells via the parallel fibres. Meanwhile, the Golgi cells that receive excitatory input from granule cells pass it to the axon end and then act inhibitorily on the granule cells. Thus, Golgi cells control granule cells by a negative feedback mechanism. Similarly, stellate and basket cells that receive excitatory input from granule cells give rise to the axons that terminate on Purkinje cells lying extrinsic to the region, where parallel fibres of granule cells make them excitatory, and thus inhibit the action of Purkinje cells. Therefore, the response pattern is designed to excite only Purkinje cells in the only region where parallel fibres exist to excite the Purkinje cells.

The neurotransmitter (neuromodulator) present in the granule cells has not yet been verified. However, Golgi cells that receive input from the parallel fibres of granule cells express Glu-R3 and Glu-R2 subunits of AMPA-type receptors. In the molecular layer where stellate and basket cells exist to receive input of granule cells, mRNA for various $GABA_A$ receptor subunits, ion channel type glutamate receptor subunits, metabotropic type glutamate receptor mGluR2 subtype, and glycine receptor β subunit are abundantly observed (Table 1). These facts suggest glutamate, GABA, and glycine as strong candidates for the neurotransmitter (neuromodulator) in this region. Also, Purkinje cells that are targets of granule cells make mRNA for various $GABA_A$ receptor subunits, NMDA type glutamic receptor subunits, metabotropic glutamate receptor (mGluR1), and glycine receptor β subunit.

Since Purkinje cells are the principal target cells of climbing fibres, as has been mentioned already, it is not easy to comprehend that the above-mentioned receptors expressed in Purkinje cells are receptors of neurotransmitter (neuromodulator) from granule cells. However, they align exceedingly well with the type of receptors expressed in Golgi cells, stellate and basket cells (with allowance for the fact that subunit composition differs widely among the cell types). This supports the possibility that glutamic acid, GABA, and glycine are involved in the functional aspects of granule cells. However, the input from granule cells to Purkinje cells, stellate and basket cells, or Golgi cells is of excitatory type. Among the candidate neurotransmitters discussed here, only glutamic acid is an excitatory neurotransmitter. GABA and glycine are inhibitory neurotransmitters. Thus, one can identify glutamic acid as the probable neurotransmitter (neuromodulator) contained in the granule cells. If this is the case, the question arises which cells supply GABA or glycine to the $GABA_A$ receptor or glycine receptor of stellate or basket cells. It can probably be interpreted that input is received from interneurons itself as an autoreceptor or from other GABA and glycine interneurons.

It has been shown that all the stellate and basket cells as well as Golgi cells of interneurons are inhibitory and contain GABA. In fact Purkinje cells and granule cells do express various $GABA_A$ receptor subunits. Simultaneously, both Purkinje cells and granular cells which are targets of the stellate, basket, and Golgi cells express glycine receptor β subunit. Considering that glycine is an inhibitory neurotransmitter (neuromodulator) and that interneurons exhibit inhibitory action, it is highly probably that Golgi cells, stellate cells, and basket cells should contain glycine as well. Both Purkinje cells and granule cells express glutamic acid receptor. However, since glutamic acid is an excitatory neurotransmitter (neuromodulator) and the interneurons belong to the inhibitory neuron type, the possibility of glutamic acid functioning as a neurotransmitter (neuromodulator) in interneurons is very remote.

3 Output system from cerebellum and neurotransmitters/neuromodulators

The output system from the cerebellum is categorized into two types. In one type, Purkinje cells project directly into the exterior of the cerebellum. This constitutes only a small proportion, and the cerebellovestibular tracts belongs to this type. In the second type, information from Purkinje cells is relayed into deep cerebellar nuclei and from here the message is projected to the outside of cerebellum. A greater proportion of the cerebellar output is relayed in this manner.

Fibres originating from the dentate, emboliform, and globosus deep cerebellar nuclei pass through the superior cerebellar peduncle and terminate contralaterally in the red nucleus, ventrolateral thalamic nucleus, and intralaminar thalamic nuclei of thalamus. Fibres from the fastigiate nucleus pass through the unciform fasciculus as well as the inferior cerebellar penduncle and terminate in the vestibular nuclei.

GABA has been established as the neurotransmitter (neuromodulator) present in the Purkinje cells. As shown in Table 1, in deep cerebellar nuclei, expression of various $GABA_A$ receptor subunits has been identified. Also reported are the expressing of glycine α_1 and β subunits in deep cerebellar nuclei. The interneurons in deep cerebellar nuclei are probably regulated by glycine. Various subunits of ion-channel type glutamate receptor are also expressed in the neurons of deep cerebellar nuclei. The possibility that Purkinje cells contain glutamic acid cannot be denied. However, it seems more appropriate to consider that input containing glutamate from the brain stem terminates in deep cerebellar nuclei.

4 Bargmann glia

It is well known that Bargmann glia have an important role in layer formation of the cerebellum. In the cerebellum of normal adult animals, it has been clarified that Bargmann glia express glycine-degrading enzymes, glycine transporter type I, Glu-R1 and Glu-R4 subunits of AMPA type glutamate receptor, and GLAST (one kind of glutamate transporter). This suggests the possibility that Bargmann glia participate in neurotransmission by regulating the amount of neurotransmitter present in the synaptic cleft.

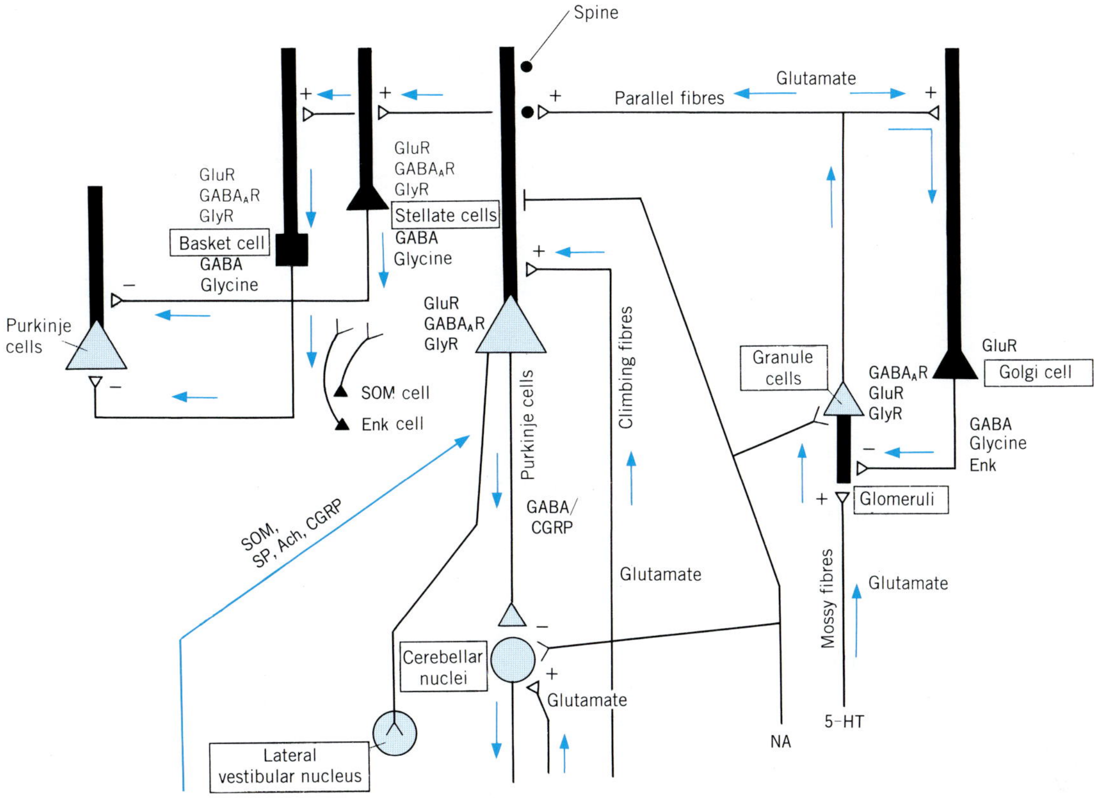

Fig. 1 Chemical neuronal network in cerebellum.

Table 1. Gene expression subunits/subtypes of amino acid receptors in cerebellum

	Subunits/subtypes	**Molecular layer**	**Purkinje cell layer**	**Granular layer**	**Cerebellar nuclei**	**Bergmann glia**
Glutamate receptors	NMDA-R 1 (NR1)	+	++++	++++	++++	–
	Glu-R2 (AMPA)	–	+	–	–	+++
	2 (AMPA)	+++	++++	+++	+++	–
	3 (AMPA)	+++	+++	–(Golgi+)	+++	–
	4 (AMPA)	+	+	++	++	++++
	5/6 (Kainate)	+	++++	+++	+	–
	mGluR1 (metabotropic)	±	++	–	–	
	2 (metabotropic)	–	–	–(Golgi++)	–	
$GABA_A$ receptors	α_1	+	++	+	++	
	α_3	++	–	–	++	
	α_4	++	–	–	+	
	β_1	+	++	+	–	
	β_2	+	++	++	++	
	β_3	++	–	++	+	
	γ_1	–	++	+	–	
	γ_2	+	++	++	++	
Glycine receptors	α_1	–	–	–	+–++	
	β	+	++	±	++	

Tyrosine hydroxylase (immunocytochemistry)

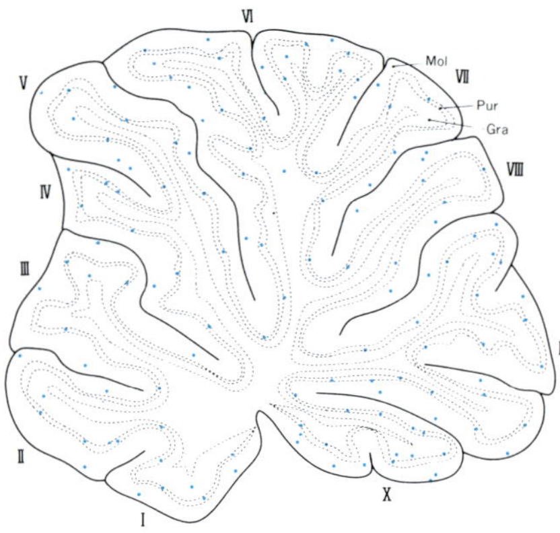

Dopamine receptor (ISH)

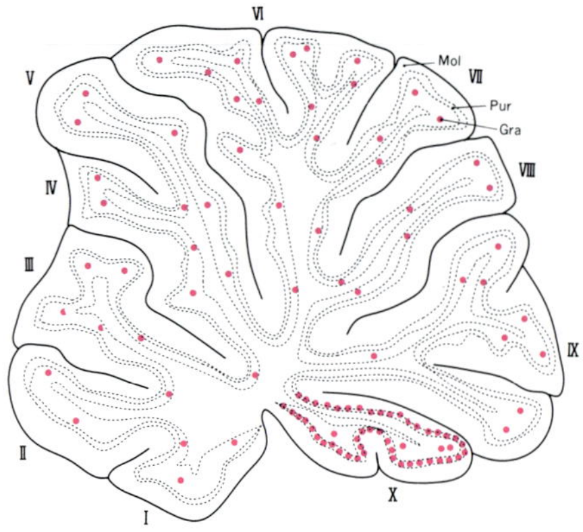

Noradrenaline (immunocytochemistry)

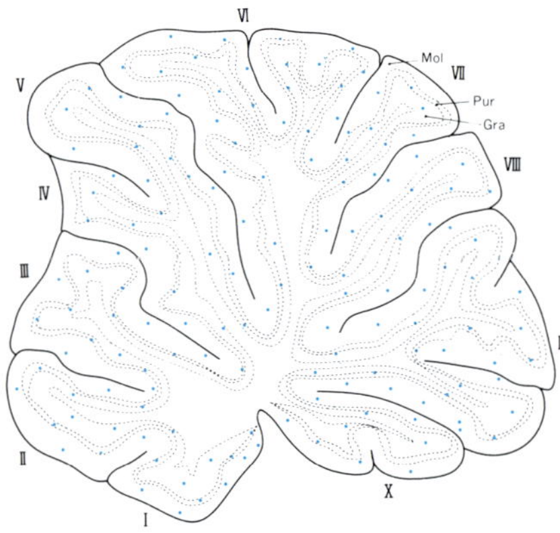

Adrenaline (immunocytochemistry)

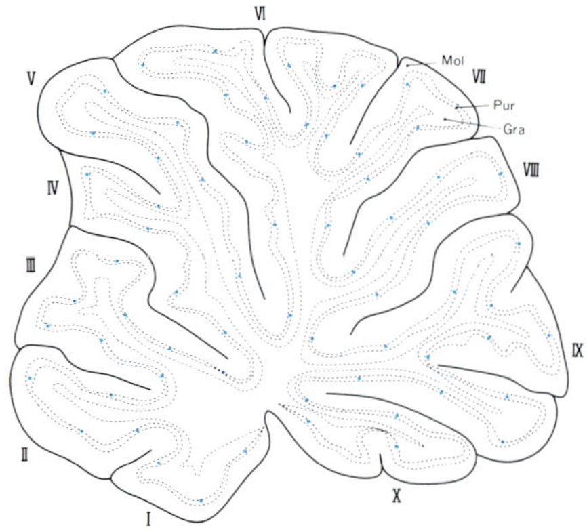

Adrenaline β_1 receptor (binding sites)

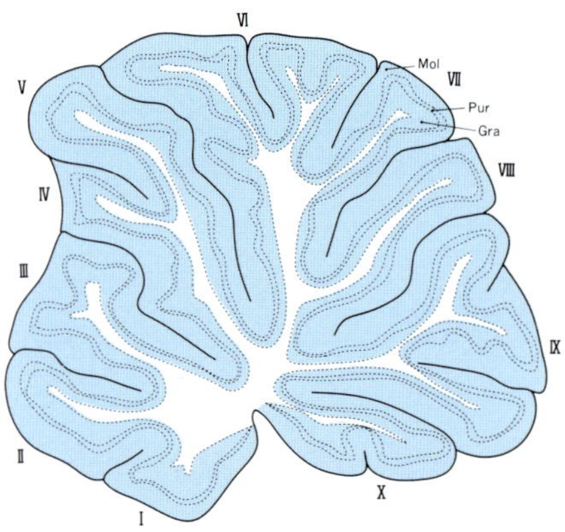

Adrenaline β_2 receptor (binding sites)

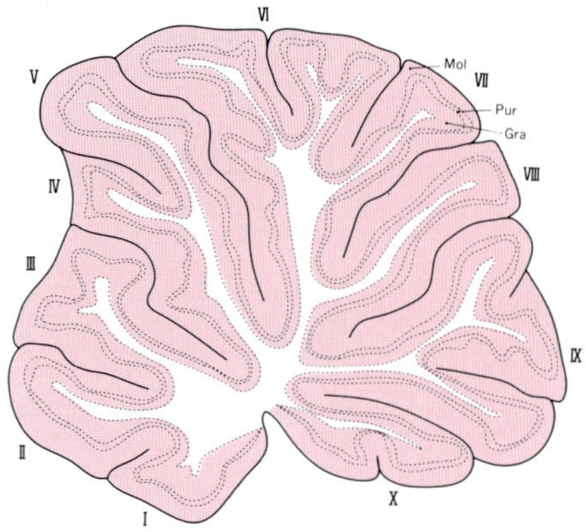

Adrenaline β receptor (immunocytochemistry)

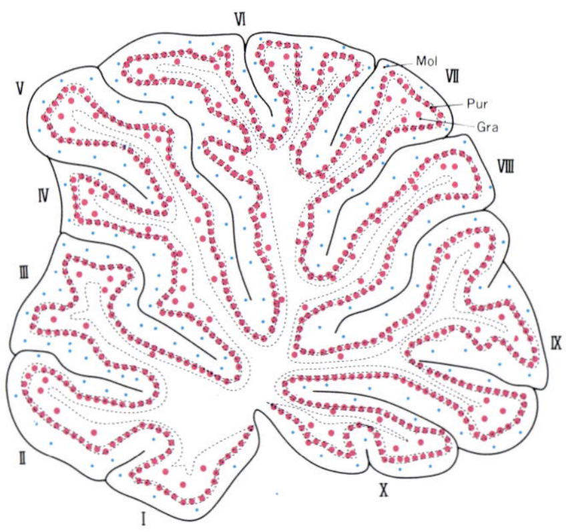

Serotonin (ISH)

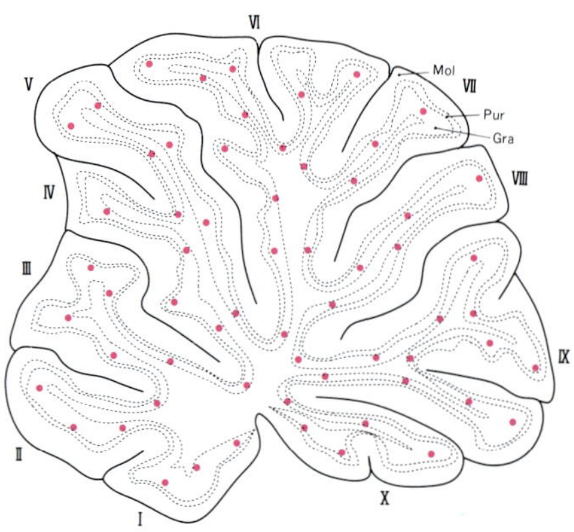

Histamine H_2 receptor (binding sites)

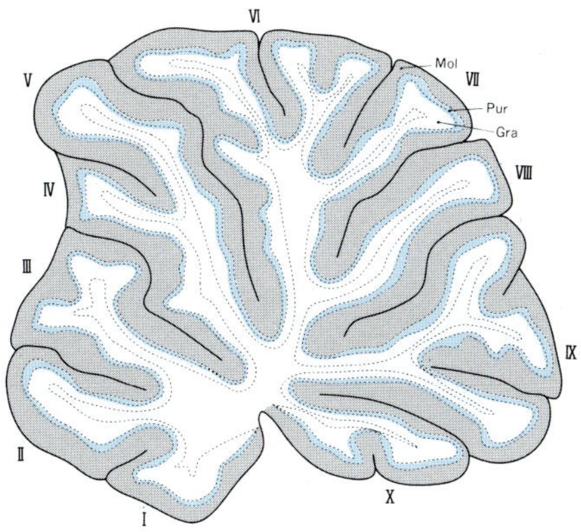

Nicotinic acetylcholine receptor (binding sites)

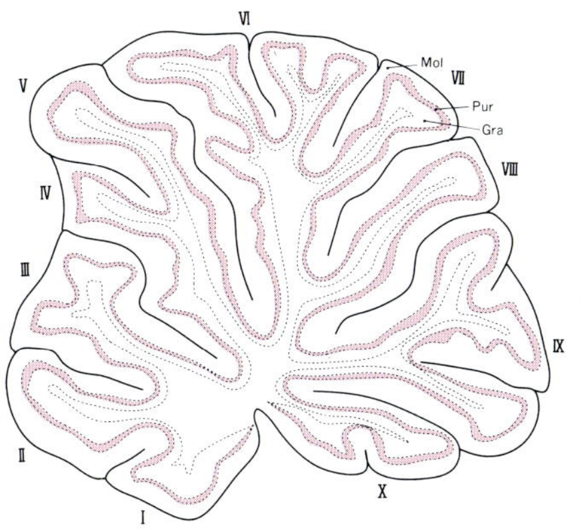

Nicotinic acetylcholine receptor (binding sites)

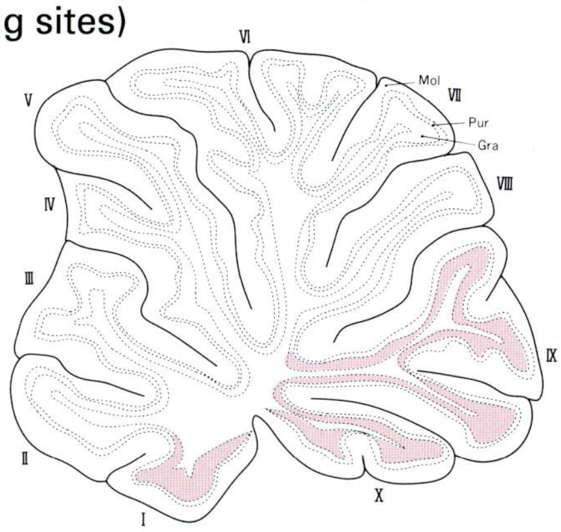

Nicotinic acetylcholine receptor (immunolabelling)

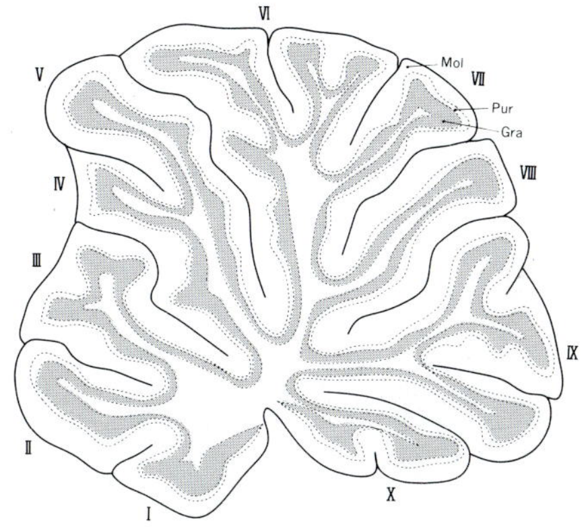

Nicotinic acetylcholine receptor (ISH) α_{4-2}

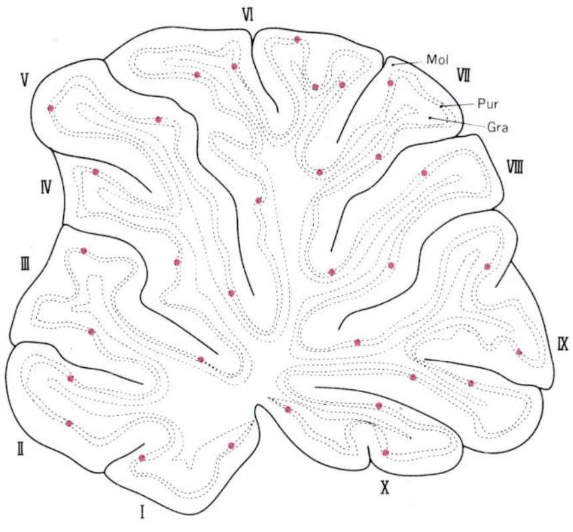

Nicotinic acetylcholine receptor (ISH) α_{4-2}

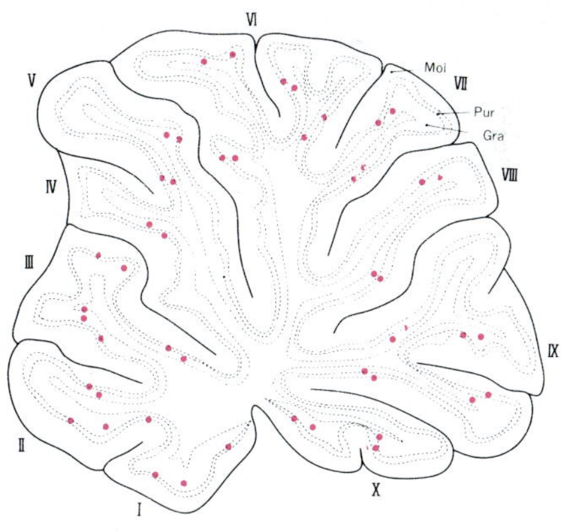

Nicotinic acetylcholine receptor (ISH) α_2

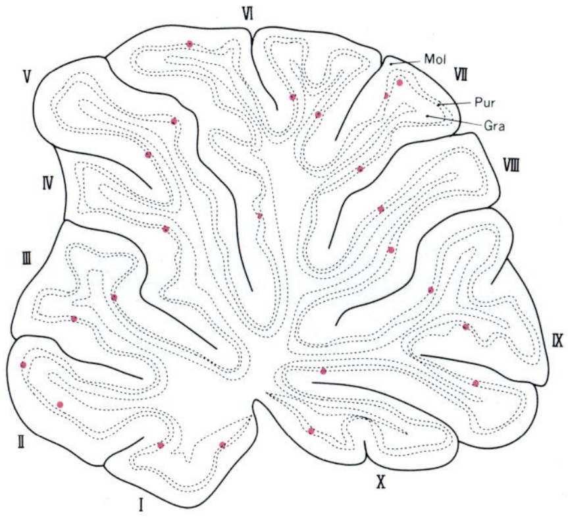

Nicotinic acetylcholine receptor (ISH) α_3

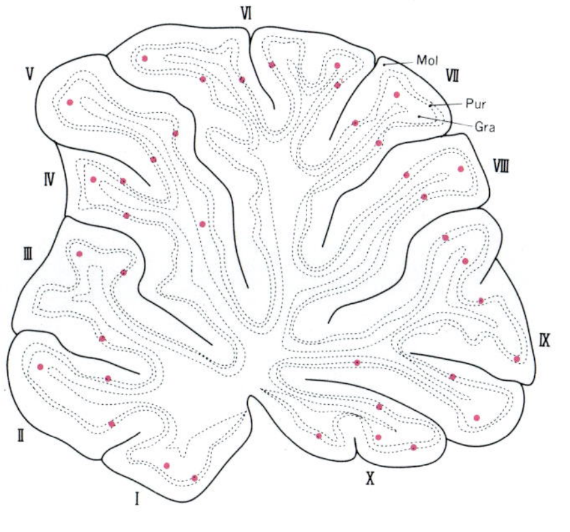

Nicotinic acetylcholine receptor (ISH) β_2

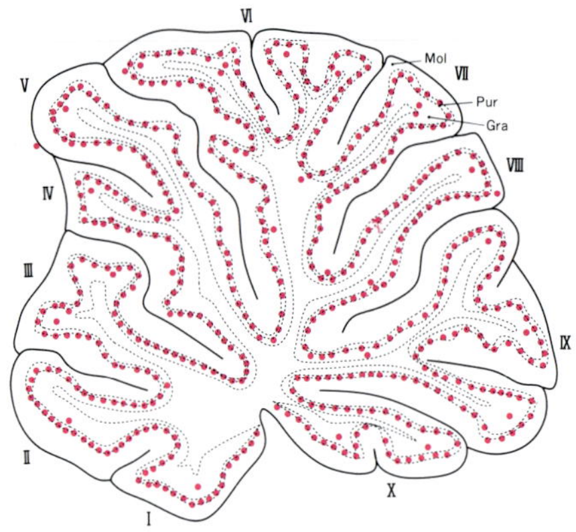

Muscarinic acetylcholine receptor (binding sites)

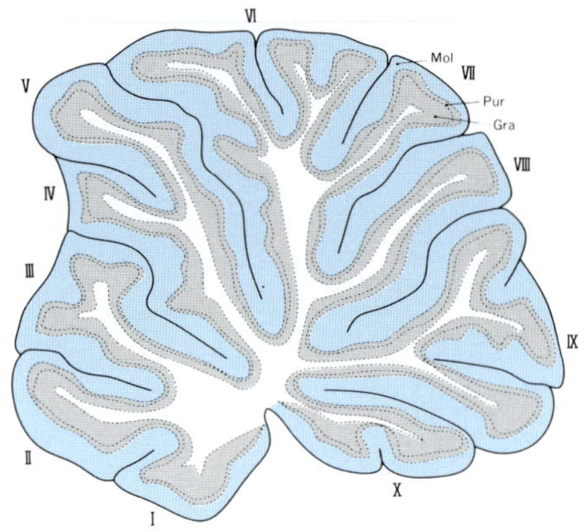

GABA (immunocytochemistry)

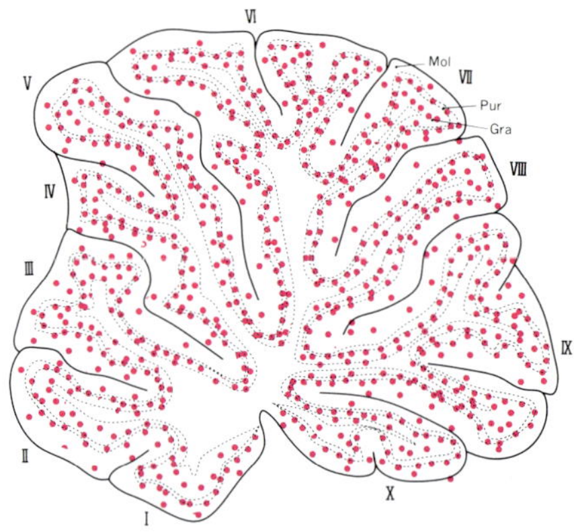

$GABA_A$ receptor (binding sites)

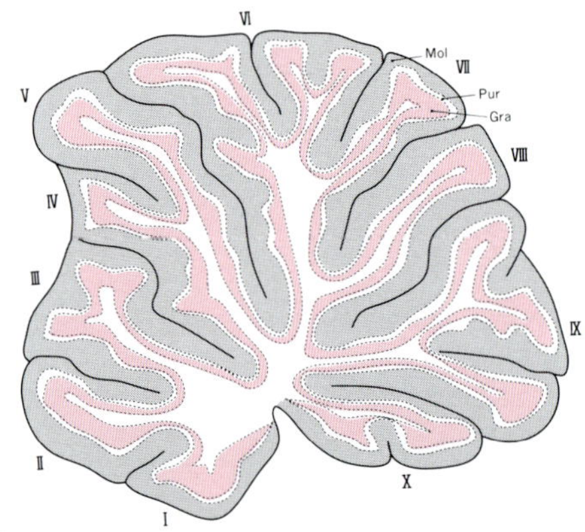

$GABA_A$ receptor (immunocytochemistry)

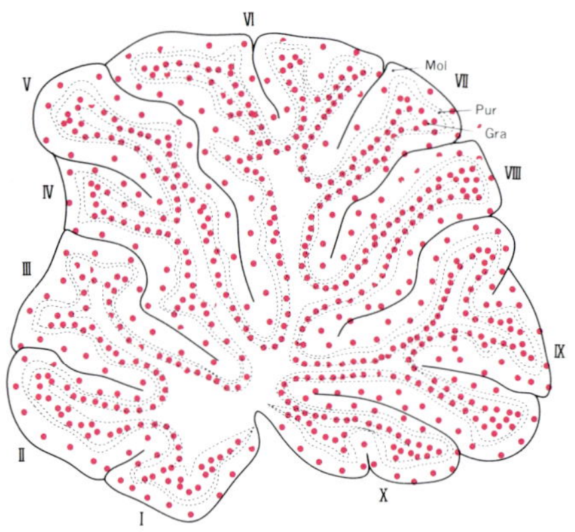

$GABA_A$ receptor (ISH) α_1

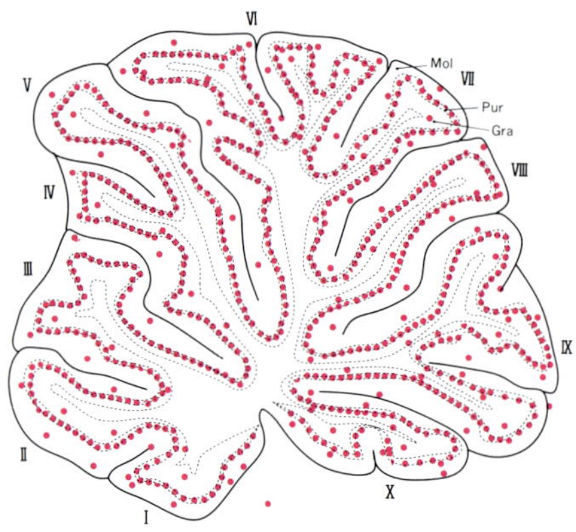

$GABA_A$ receptor (ISH) α_3

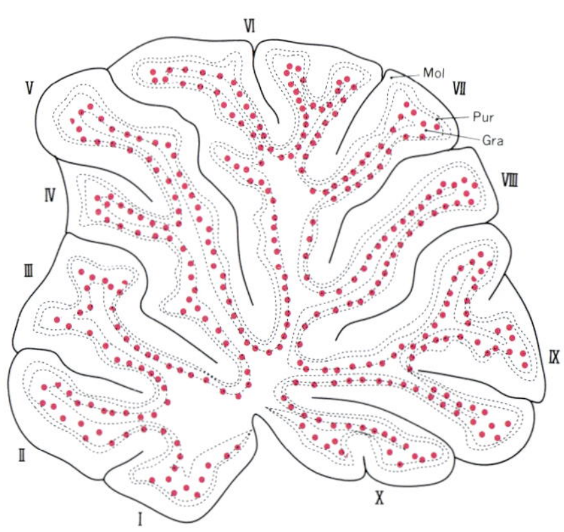

$GABA_A$ receptor (ISH) α_4

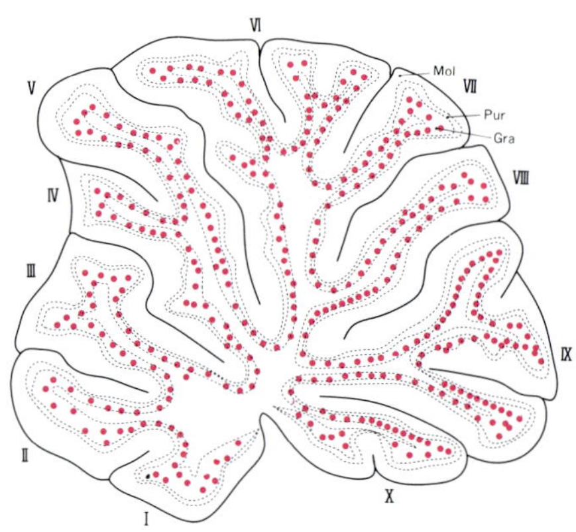

$GABA_A$ receptor (ISH) β_1

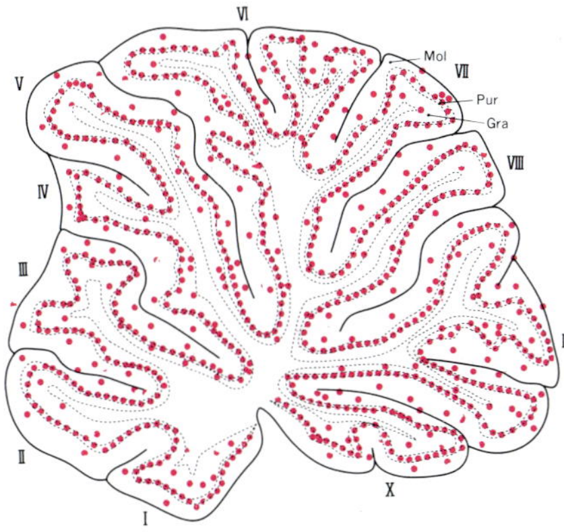

$GABA_A$ receptor (ISH) β_2

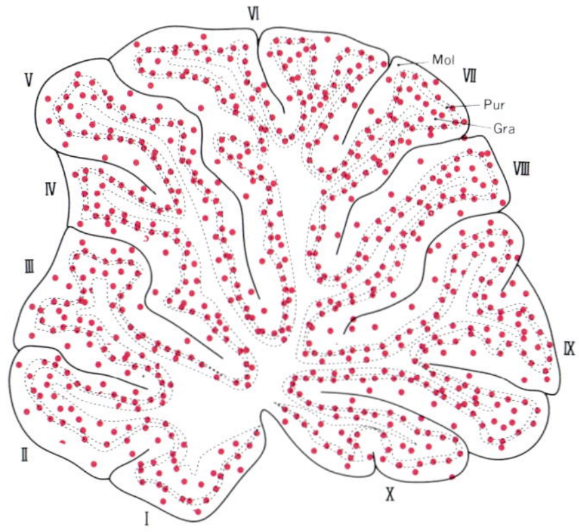

$GABA_A$ receptor (ISH) β_3

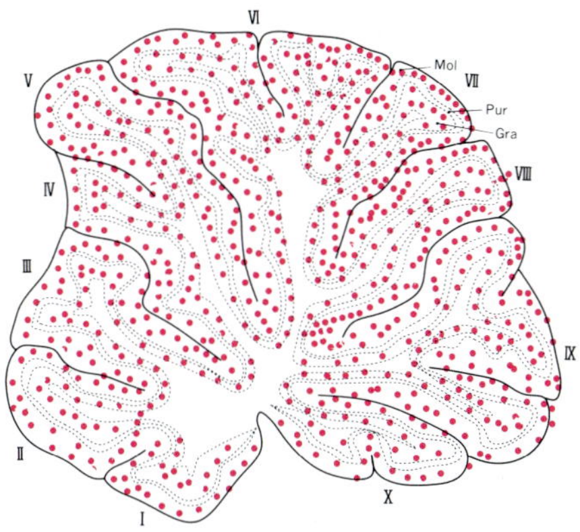

$GABA_A$ receptor (ISH) γ_1

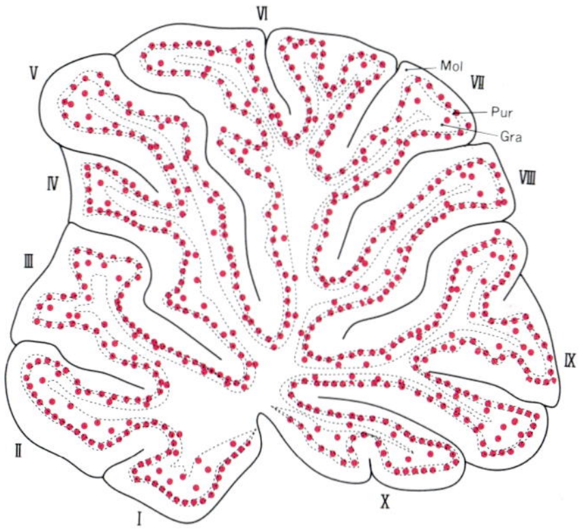

$GABA_A$ receptor (ISH) γ_2

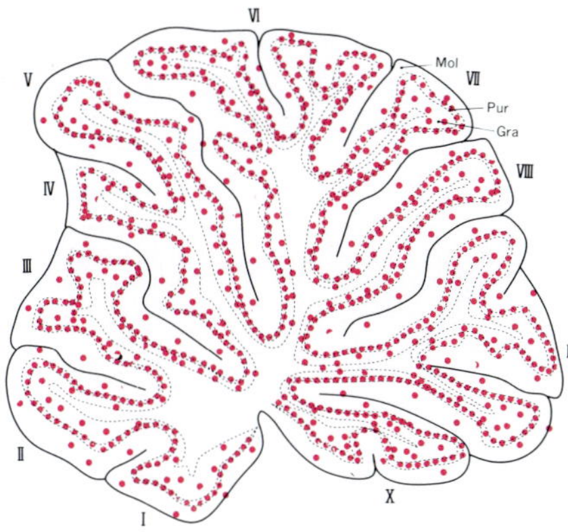

$GABA_B$ receptor (binding sites)

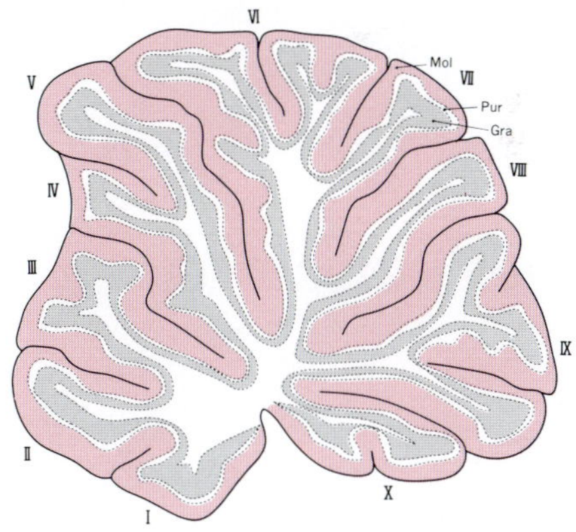

Glycine (immunocytochemistry)

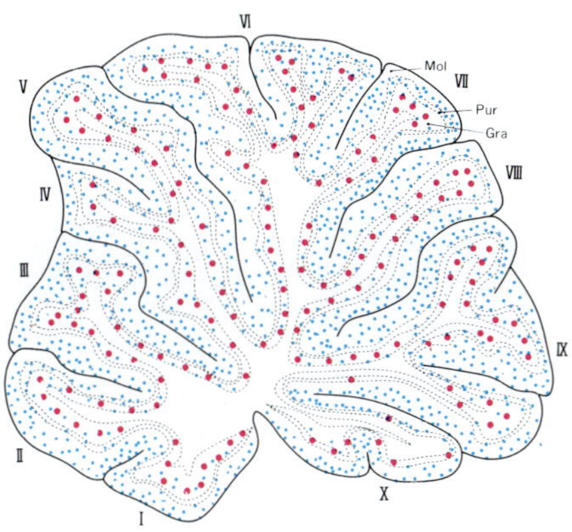

Glycine receptor (binding sites)

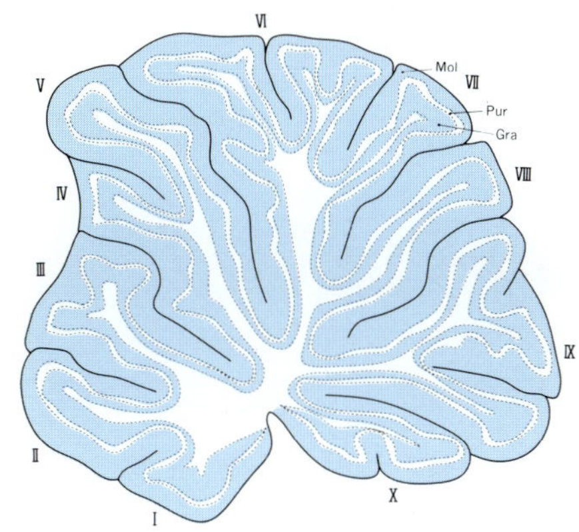

Glycine receptor (ISH)

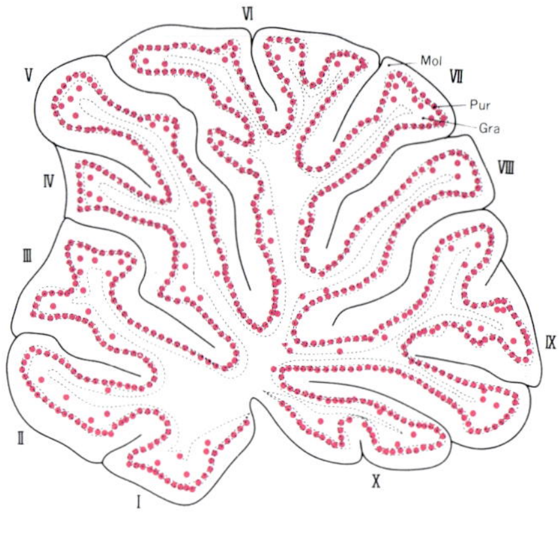

Glycine receptor (immunocytochemistry)

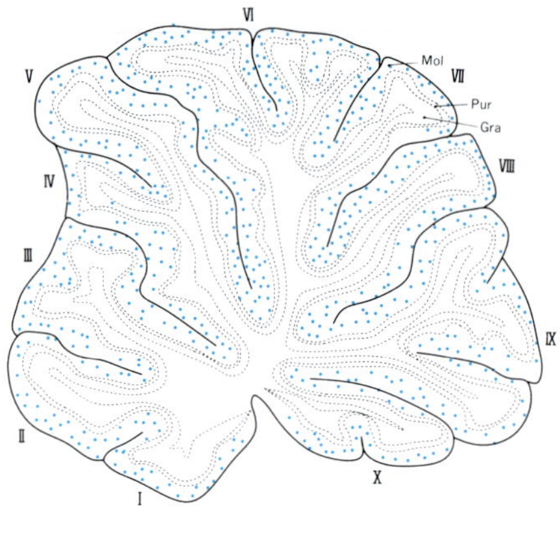

Glutamate (immunocytochemistry)

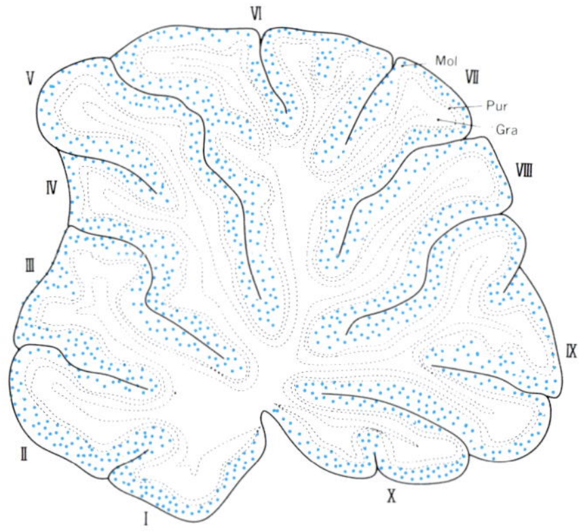

[^{3}H]-AMPA (binding sites)

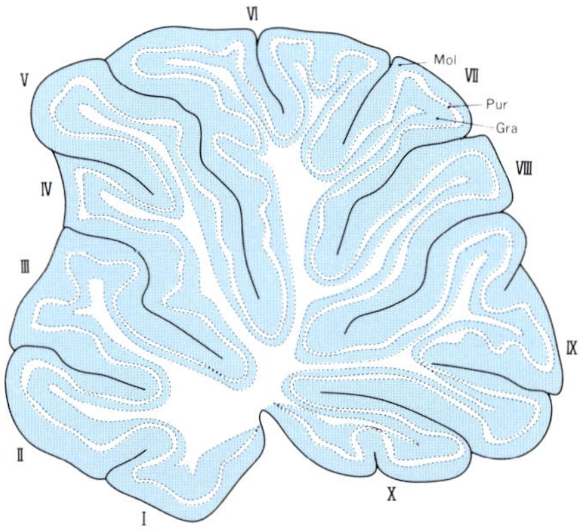

[^{3}H]-Kainate (binding sites)

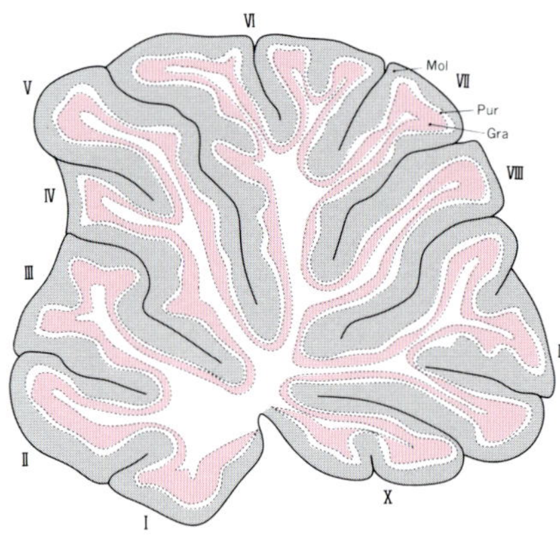

NMDA-[^{3}H] Glutamate (binding sites)

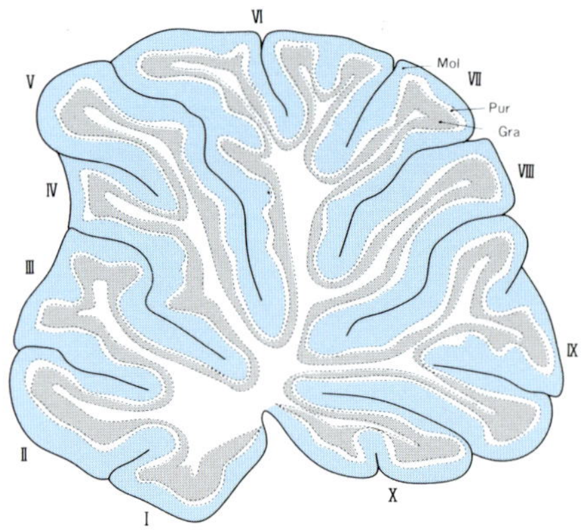

Glutamate receptor (ISH) GluR1 (AMPA)

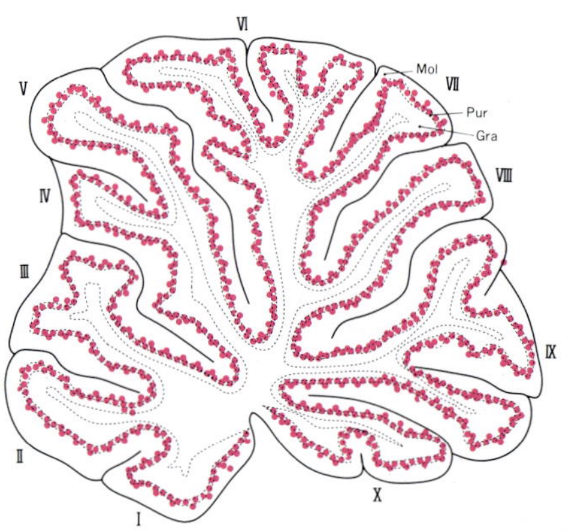

Glutamate receptor (ISH) GluR2 (AMPA)

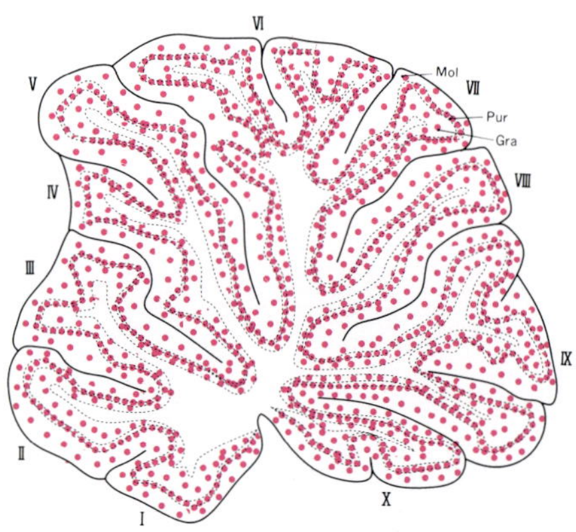

Glutamate receptor (ISH) GluR3 (AMPA)

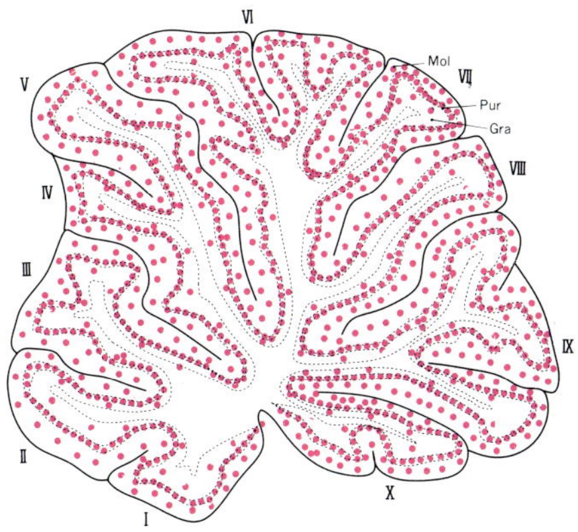

Glutamate receptor (ISH) GluR4 (AMPA)

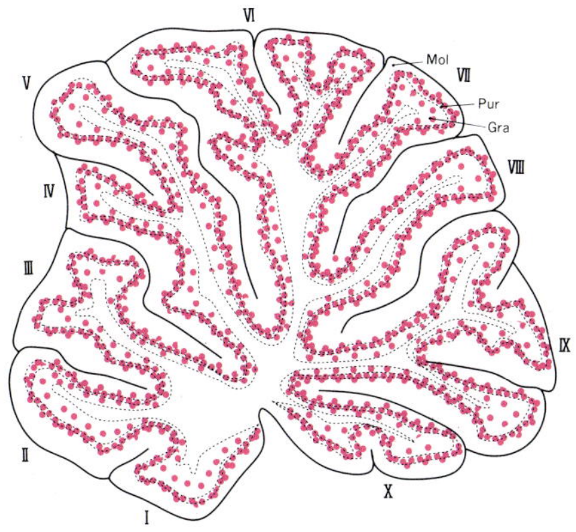

Aspartate (immunocytochemistry)

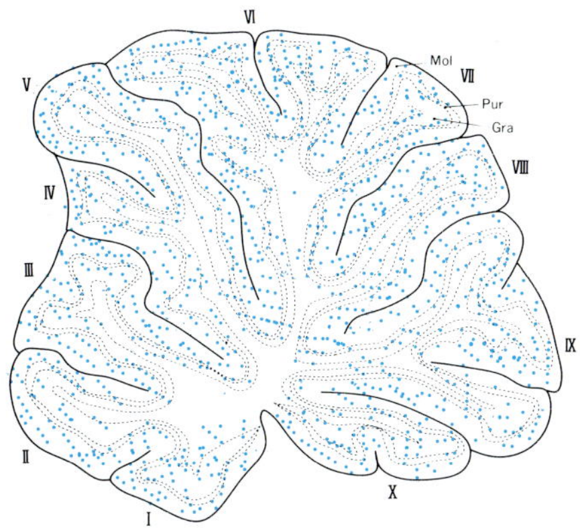

Taurine (immunocytochemistry)

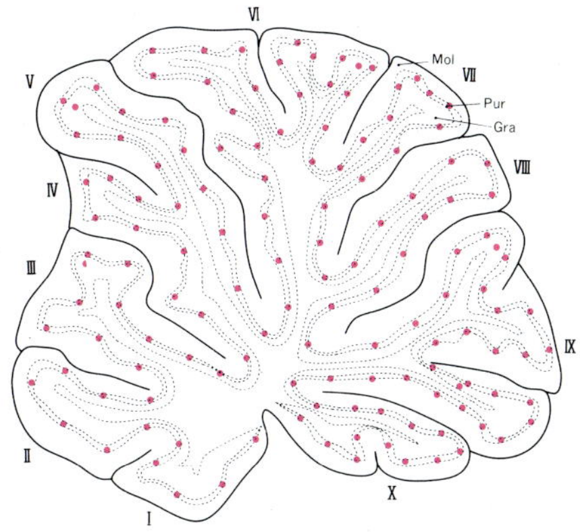

Adenosine (immunocytochemistry)

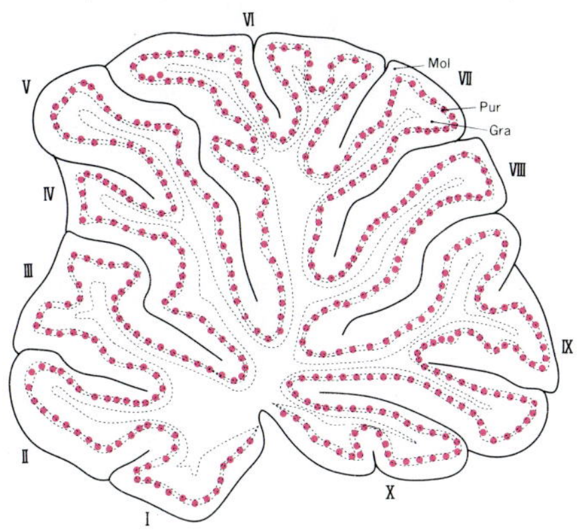

Adenosine A_1 receptor (binding sites)

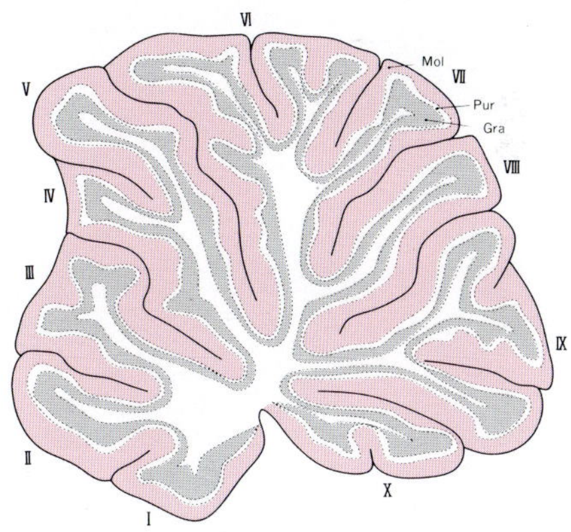

FGF (immunocytochemistry)

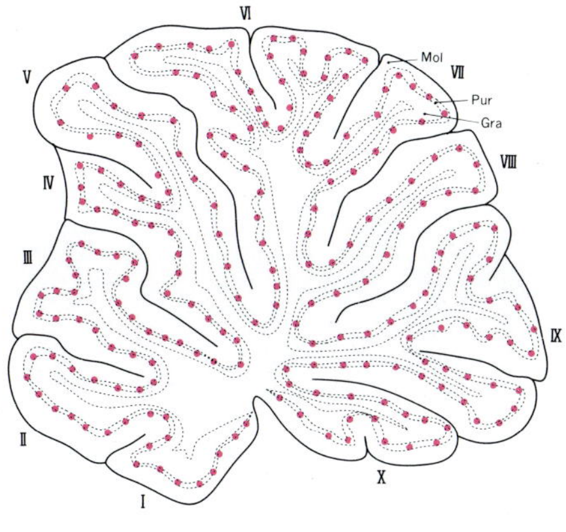

FGF receptor (ISH)

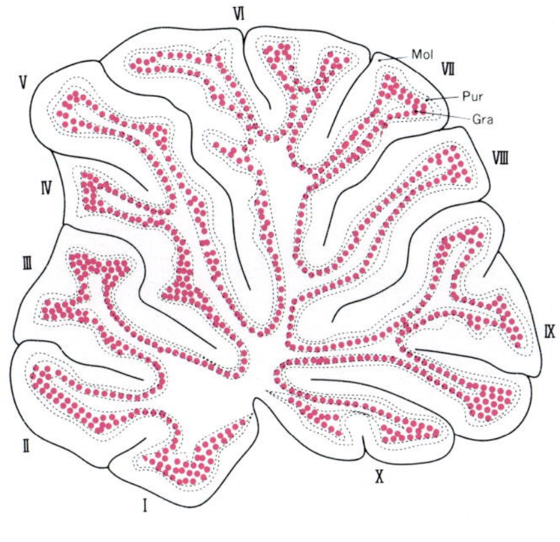

Pleiotrophin (ISH)

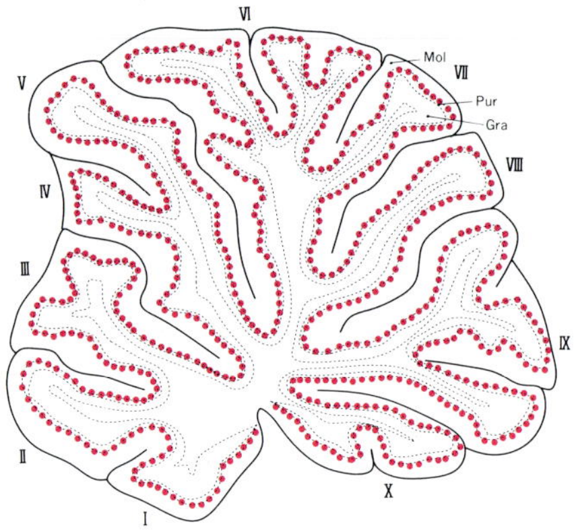

Calcitonin gene related peptide (immunocytochemistry)

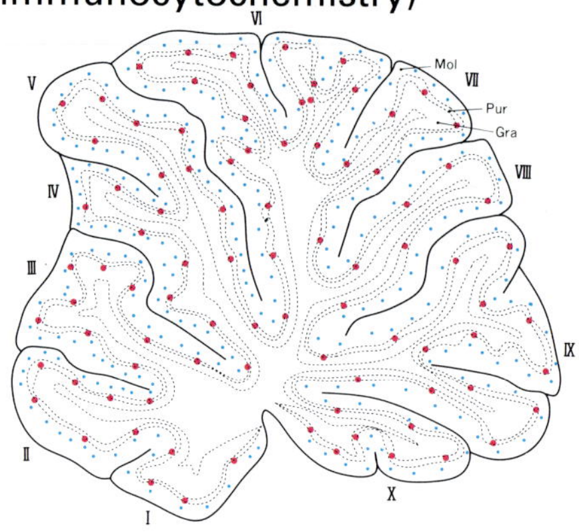

Calcitonin gene related peptide receptor (binding sites)

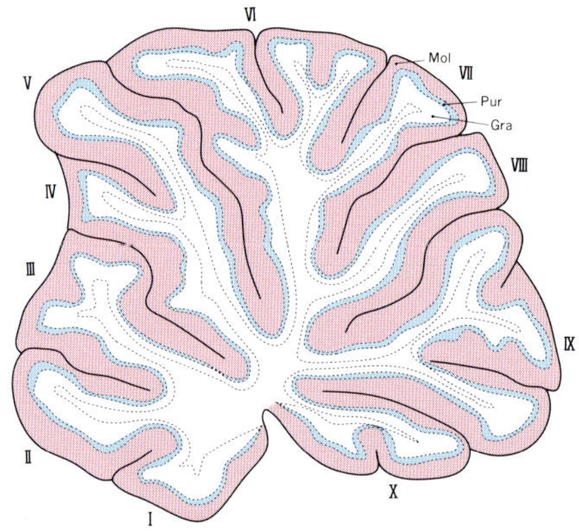

Corticotropin releasing hormone (immunocytochemistry)

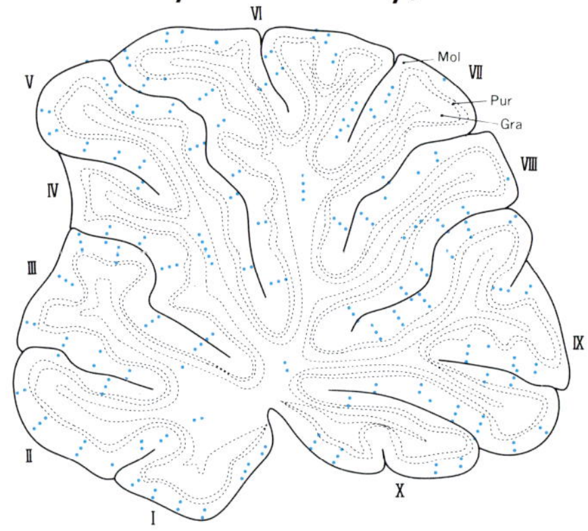

Corticotropin releasing hormone receptor (binding sites)

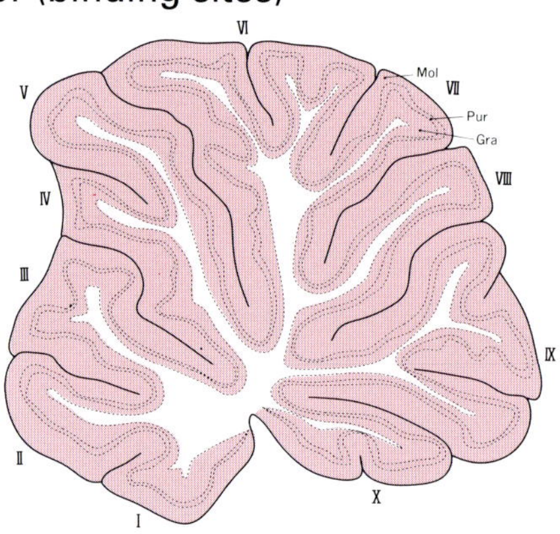

Enkephalin (immunocytochemistry)

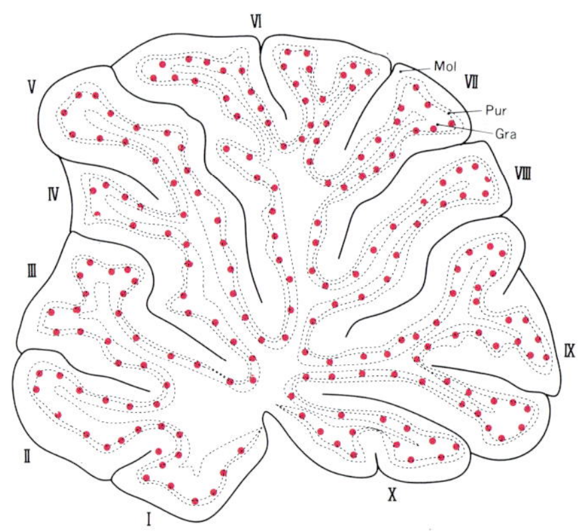

Somatostatin (immunocytochemistry)

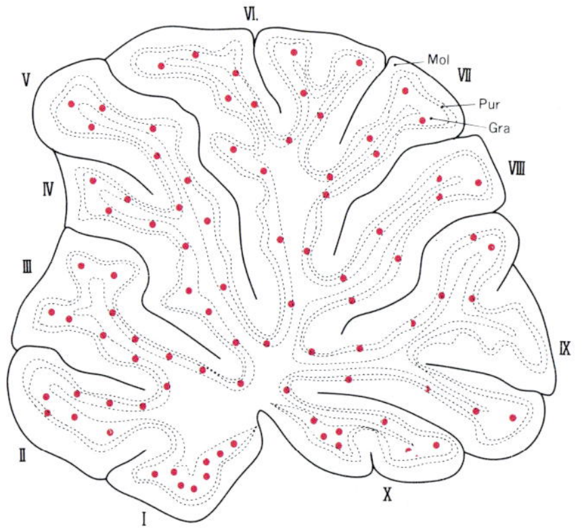

Neuropeptide Y (immunocytochemistry)

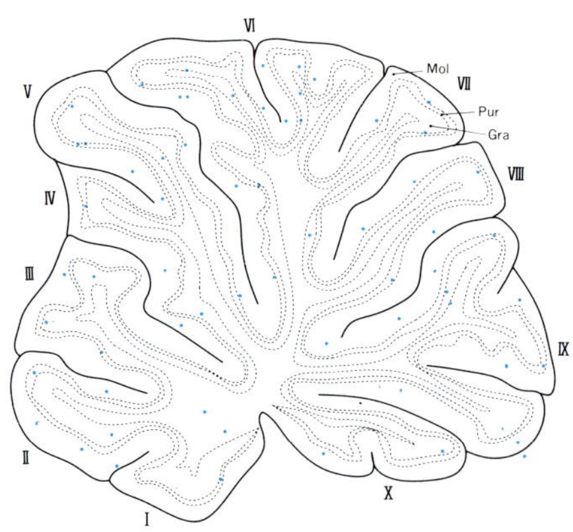

Neuropeptide Y receptor (binding sites)

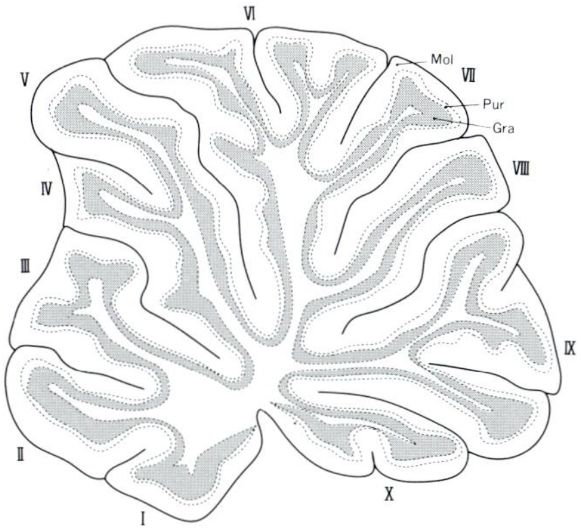

Vasoactive intestinal peptide receptor (binding sites)

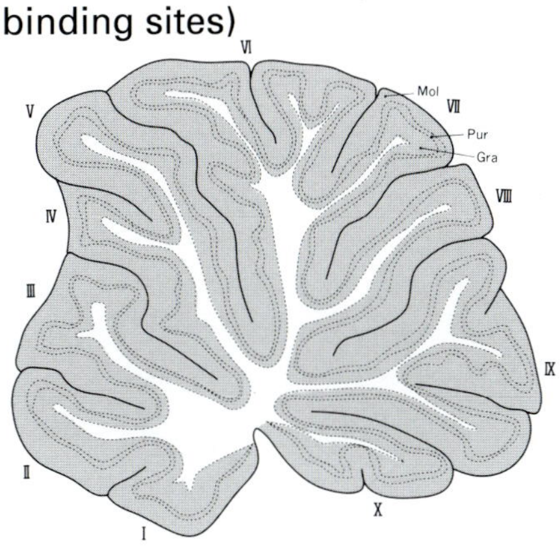

Argininosuccinate (immunocytochemistry)

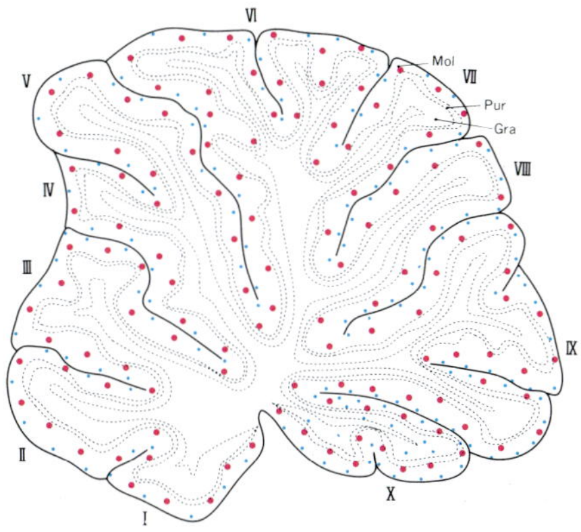

Index

A kinase 14, 58
abductor nucleus 119
acceptors 17
accumbens nucleus 275, 315
acethylcholine 66, 260, 310, 314
acethylcholine esterase 66
acetyl-cholinergic system 2
acetyl-CoA 66
acidic FGF 168
adenosine 154, 159
adenosine deaminase deficiency 154
adenosine recepors 158
adenylate cyclase 14, 36, 264, 274
adrenal cortex 218
adrenal medulla 32
adrenaline 20, 32
adrenergic receptors 36
aldosterone 218
α-ketoglutaric acid 130
Alzheimer's disease 260
Ammom's horn 281
AMP 154
AMPA/kainate (non-NMDA) receptor 130 134
AMPA receptors 10
AMPA-type receptor 13
amygdala 54, 182, 291
amygdaloid complex 32, 159, 196, 201, 208, 230, 242, 255, 260, 264, 269, 274, 275, 280, 281, 291 310
amygdaloid nucleus 158, 227
analgesic action 10
analgesic effect 10
angiotensin 218
angiotensin II 218
ANP receptor 306
anterior olfactory nucleus 275
antinociception 208
arcuate nucleus 186, 242, 254, 260
area postrema 274
Arg-vasopressin 268
arginine 319
Arginiosuccinate 318
aspartate 144
astrocytes 17, 119
athrocytosis 196
ATP 154
atrium 302
Auerbach's plexus 311
autoreceptors 16, 24
axons 2

Bargmann glia 324
basal ganglia 103, 122, 149
basal nuclei 54
basal nucleus of Meynert 67
basic FGF 168
basket cells 323, 324
bed nucleus of stria terminalis 99, 242, 254, 255, 260, 310, 319
bed nucleus stria terminals 299
β-adrenergic 158
β-adrenergic receptors 148
β-endorphin 242
bek 172
Bergmann glia 17, 323
blood pressure 182
blood–brain barrier 130
bombesin 234
bradykinin 12
brain derived neurotrophic factor 164
brain stem 98
Broca's diagonal band 208, 319

C kinase 58
Ca^{2+}-calmodulin dependent protein kinase II 14
calcitonin 314
calcitonin gene-related peptides 3
calcitonin-gene related peptide 192, 323
calcium concentration 212
cAMP 3, 14, 58
cAMP response element 209
cAMP-dependent protein kinase 102
catecholamine 20
caudatenucleus 290, 302
caudateputamen 103
central branches 10
central grey matter 10, 119
central grey matter of mesencephalon 260
central nucleus of amygdaloid complex 254
cerebellar cortex 230
cerebellum 98, 103, 119, 135, 148, 154, 158, 159, 254, 298
cerebral cortex 32, 42, 54, 66, 67, 98, 103, 122, 131, 148, 154, 158, 201, 227, 255, 260, 264, 290, 294, 295, 314, 319, 298, 310
cGMP-dependent protein kinase 14
chemical synapses 2
cholecystokinin 200
choline 66
choline acetyl transferase 66
cingulate gyrus 275
claustrum 230, 260, 264
climing fibres 323
cochlea 67
cochlear nuclei 314
cochlear nucleus 119
coexistence 3
colliculus inferior 182
colliculus superior 182
commissural nucleus 242
contraction of gallbladder 200
contraction of smooth muscles 182
contractions of the gallbladder 205
corpus striatum 154, 158, 159, 182, 186, 260, 264, 294, 298, 302, 310, 314
corticotropin releasing factor 226
corticotropin releasing factor receptor 230
cyclic AMP 158
dendrites 2
dentate gyrus 103, 291
diacylglycerol 58, 205
diacylglycerol inositol triphosphate 14
diagonal band 280, 315
diagonal band nucleus 254
dopamine 20
dopamine receptor 20, 21, 25
dorsal and ventral cochlear nuclei 260
dorsal horn 10
dorsal raphe nuclei 208
dorsal raphe nucleus 42, 254, 274, 315
dorsal root ganglia 5, 10, 260
dorsolateral tegmental nucleus 260
dorsomedial hypothalamic nucleus 254, 260
dynorphin 242

electrical synapses 2
en passent 70
endoplastic reticulum 14
enkephalin 10, 323
entorhinal cortex 281

facial nucleus 260, 274
FGF-5 168
fibroblast growth factor 168
fibroblasts 12
fig 172
follicle stimulating hormone 280
formalin 10

G-protein coupled receptors 2, 36, 58, 70
G-protein coupled type 13
GABA transporter 98
GABAA 102
GABAA receptor 13
GABAB 102
GABAC receptor 102
galanin 67, 254
γ-aminobutyric acid 2, 98
gap juntions 2
glial cells 2
globs pallidus 99, 242
glucagon 310
glucocorticoid response element 209
glucose 130
glutamate 5, 10
glutamate ion-channel type receptor 15
glutamic acid 98, 130, 323
glutamic acid decarboxylase 98, 154
glutaminase 131
glycine 2, 17, 118
glycine cleavage system 119
glycine receptor 13
Golgi cells 323, 324
gonadoliberin 254
growth factors 2
growth hormone release inhibiting factor 280

growth hormone releasing hormone 280
GRP receptor 238

H1 receptors 58
H2 receptors 58
H3 receptors 58
helospectin 311
hind limbs 10
hippocampus 32, 42, 103, 119, 122, 135, 148, 154, 158, 159, 201, 227, 254, 255, 260, 264, 274, 280, 281, 291, 294, 295, 298, 310, 314, 315, 319
histamine 12, 54
histamine H1 receptor 12
histamine *N*-methyl-transferase 54
histamine receptors 58
histidine decarboxylase 154
hst-2 168
HST1 168
5-HT receptors 46
hypoglossal nucleus 208, 274
hypophysis 254
hypothalamus 42, 54, 119, 154, 159, 182, 238, 254, 255, 260, 275, 280, 281, 286, 290, 294, 298, 302, 310
hypothalamus-mammillary body 54
hypothermia 208

immunocytochemistry 3, 5
in situ hybridization 3
inferior cerebellar peduncle 323
inferior colliculus 98, 260, 264, 274, 319
inferior olivary nucleus 323
inferior olive nucleus 274
inositol triphosphate 205
insulin-dependent diabetes 99
int-2 168
interpeduncular nucleus 103
ion-channel type 13
ion-channel type receptors 15
ionotropic 70
ionotropic receptors 2

KGF 168
kinase A 14
kinase C 14

L-arginie 17
L-histidine decarboxylase 54
lateral geniculate body 98, 294
lateral lemniscus 260
lateral nucleus of hypothalamus 269
lateral olfactory tract nucleus 264
lateral reticural nucleus 323
lateral septal nucleus 281
lenticular nucleus 302
leucine zipper structure 46
leuenkephalin 242
locus ceruleus 32, 254, 264, 269, 314, 315, 323
luteinizing hormone releasing hormone 280
luteinizing hormone 280
Lys-vasopressin 268

mACh receptor 70
macrophages 12
mammillary body 154, 201, 208, 275
mast cells 12
medial geniculate body 314
medial geniculate nucleus 103
medial habenula nucleus 103, 280
medial preoptic area 254
medial septal nucleus 254, 315
medial septum 98
median eminence 260, 280
medulla oblongata 254, 295, 302, 314
Meissner's plexus 311
mesencephalon 254, 264
met-enkephalin 242
metabolism of inositol phosphate 24
metabotropic receptor 130, 134
midbrain 10
MK 176
monoaminergic system 2
mossy fibres 323
motor neurons 3
mRNAs 3
muscarinic Ash receptor 158
muscle contraction 192
myenteric plexus 205

nACh receptor 70
natriuretic peptide 302
nerve growth factor 164
neuroendocrine 268
neurokinin A 15, 182
neurokinin α 182
neurokinin A receptor 186
neurokinin B 15, 182
neurokinin β 182
neurokinin B receptor 186
neuromedin N 212
neuromodulators 3
neurons 2
neuropeptide Y (NPY) 3, 294
neurophysin I 268
neurophysin II 268
neurotensin 208
neurotensin receptor 212
neurotransmission 2
neurotransmitters 3
neurtrophin-3 164
nicotinic(n) Ach receptors 3
nitric oxide 2, 16, 17, 319, 323
nitric oxide synthase 17
nitric oxide synthetase 319
NMB receptor 238
NMDA receptor 130, 134
noradrenaline 20, 32
noradrenergic (NA) neuron 3
noxious stimuli 10
NPY receptor 298
nucleus accumbens 102, 103, 158, 159, 186, 274
nucleus of Darkschewitsch 98
nucleus tractus solitarii 242

oculomotor nucleus 119
olfactory bulb 119, 135, 148, 154, 158, 182, 227, 230, 254, 260, 302, 310, 319
olfactory tubercle 98, 103, 158, 255, 315
opioid peptides 242, 248
organovasculosum lamina terminals 218
oxytocin 268
oxytocin receptor 274

pain transmission 182
pancreatic enzyme secretion 205
pancreatic peptide 294
parabrachial nucleus 254, 294
parallel fibres 323
parapeduncular nucleus 314
parasympathetic system 5, 67
parathyroid hormone 314
paraventricular hypothalamic nucleus 226, 227, 254, 294
paraventricular nucleus 159, 218, 268
paraventricular nucleus of the hypothalamus 209
paraventricular parvocellular nucleus 269
paraventricularis hypothalamic nucleus 242, 310
Parkinson's disease 20
peptide histidine isoleucine 310
peptide YY 294
peripheral branch 10
periventricular nucleus 260, 310
phosphatidylinositol diphosphate 14
phospholipase 274
phospholipase A2 14
phospholipase C 36, 58, 281
physalaemin 254
pituitary adenylate cyclase-activating polypeptide 315
pituitary adenylate cyclase-actuating polypeptide 311
pituitary gland 205, 260
PKC 281
pleiotrophin 176
pons 98, 254, 314
pontine laterodorsal tegmentum 67
pontine nuclei 323
positron emission tomography 21, 58
post-synaptic receptor 298
posterior horn 10
posterior hypothalamic nucleus 269
postganglionic neuron 5, 314
postganglionic sympathetic neurons 3
postgranglionic neurons 2, 67
potentiation of insulin secretion 205
preoptic area of hypothalamus 314
Preprotachykinin A 182
presynaptic membrane 2
presynaptic receptor 298
presynaptic site 16
primary sensory afferent fibres 10
prodynorphin 242
proenkephalin 242
propiomelanocortin 242, 254
protein kinase A 37
protein kinase C 14
purinergic neurons 154
Purkinje cells 103, 149, 154, 323, 324

ranatensin 234
raphe nuclei 10, 42, 182
relaxation of the sphincter of Oddi 205
renal-type receptor 274
reticular formation 119, 260, 275
reuniens nucleus 159
rhomboid nucleus 159

saliva secretion 182
salivary glands 67
schizophrenia 205
secretin 310, 314

septal 32
septal area 255, 269, 280
septum 54, 275, 290
septum pellucidum 290
serine hydroxymethyl-transferase 118
serotonin 42
skin 12
solitary tract nucleus 227, 254, 260, 274, 294
somatostatin 260, 294
somatostatin-14 260, 264
somatostatin-28 260, 264
spinal cord 10, 122, 230, 243, 254, 255, 310, 323
spinal trigeminal nucleus 260
spinothalamic tract 10
stellate cells 323
stiff-man syndrome 99
stria terminalis 269, 275
  the bed nucleus of 260
striatum 99, 275
strychnine 122
subiculum 264, 275
substance P 5, 182, 254
substance P receptor 186
substantia innominata 98
substantia nigra 155, 159, 186, 201, 242
succinic hemialdegyde dehydrogenase 98
superior colliculus 154, 158, 159, 255, 274, 310, 314, 315, 319
superior olivary nucleus 67, 119, 314
suprachiasmatic nucleus 238, 269, 274, 294, 295, 299, 310, 314, 315
supraoptic hypothalamic nucleus 314
supraoptic nucleus 238, 254, 268, 295
sympathetic ganglia 294
sympathetic nervous system 32
synapse 2
synapsin I 15
synaptic cleft 2, 16
Synaptic transmission 2

tachykinin receptor 186
tachykinins 182
taurine 148
telencephalon 254
tetrodotoxin 154
thalamus 10, 32, 54, 98, 148, 154, 158, 159, 186, 254, 255, 274, 275, 315
thyroid stimulating hormone 260, 280, 286
thyrotropin releasing hormone 280, 286
transporters 2, 16
trapezoid nucleus 119
TRH receptor 290
tricarboxylic acid cycle 98
trigeminal ganglion 260
trigeminal spinal tract nucleus 274
trk 164
trochlear nucleus 119
tryptophan 42
tuberomammillary nucleus 54
tyrosine 20
tyrosine kinase type 13

vascular-type receptor 274
vasoactive intestinal peptide 5, 66, 310
vasoactive peptide 315
vasodilation 182, 192
vasomotor centers 218
vasopressin 268
vasopressin receptors 274
ventral pallidum 255
vastibular nerve 323
vestibular nuclei 323
VIP receptor 314

zona incerta 183, 260, 274